Sexual Selection in Primates
New and Comparative Perspectives

Sexual Selection in Primates provides an up-to-date account of all aspects of sexual selection in primates, combining theoretical insights, comprehensive reviews of the primate literature and comparative perspectives from relevant work on other mammals, birds and humans. Topics include sex roles, sexual dimorphism in weapons, ornaments and armaments, sex ratios, sex differences in behaviour and development, mate choice, sexual conflict, sex-specific life-history strategies, sperm competition and infanticide. The outcome of the evolutionary struggle between the sexes, the flexibility of roles and the leverage of females are discussed and emphasised throughout. *Sexual Selection in Primates* is aimed at graduates and researchers in primatology, animal behaviour, evolutionary biology and comparative psychology.

PETER KAPPELER is Head of the Department of Behaviour and Ecology at the German Primate Centre in Göttingen, Germany.

CAREL VAN SCHAIK is a professor in the Department of Biological Anthropology and Anatomy at Duke University, Durham, NC, USA.

T0215223

Sexual Selection in Primates

New and Comparative Perspectives

Edited by

Peter M. Kappeler
German Primate Centre, Göttingen, Germany

Carel P. van Schaik
Department of Biological Anthropology and Anatomy, Duke University, USA

CAMBRIDGE
UNIVERSITY PRESS

CAMBRIDGE UNIVERSITY PRESS
Cambridge, New York, Melbourne, Madrid, Cape Town, Singapore, São Paulo

Cambridge University Press
The Edinburgh Building, Cambridge CB2 8RU, UK

Published in the United States of America by Cambridge University Press, New York

www.cambridge.org
Information on this title: www.cambridge.org/9780521537384

First published 2004

A catalogue record for this publication is available from the British Library

Library of Congress Cataloguing in Publication data
Sexual selection in primates: new and comparative perspectives / edited by Peter M. Kappeler & Carel P. van Schaik.
      p.   cm.
Includes bibliographical references.
ISBN 0 521 53738 X (paperback)
1. Primates – (Behavior).   2. Sexual selection in animals.   I. Kappeler, Peter M.   II. van Schaik, Carel P.
QL737.P9S444   2003
599.8′1562–dc22   2003055314

ISBN 978-0-521-53738-4 paperback

Transferred to digital printing 2007

Contents

Contributors

SONIA M. ALTIZER
Department of Environmental Studies
Emory University
Atlanta, GA, USA
saltize@emory.edu

TIM R. BIRKHEAD
Department of Animal and Plant Sciences
University of Sheffield
Sheffield, UK
t.r.birkhead@sheffield.ac.uk

GILLIAN R. BROWN
Department of Zoology
University of Cambridge
Cambridge, UK
grb4@psych.st-andrews.ac.uk

TIM H. CLUTTON-BROCK
Department of Zoology
University of Cambridge
Cambridge, UK
thcb@cam.ac.uk

STEVEN W. GANGESTAD
Department of Psychology
University of New Mexico
Albuquerque, NM, USA
sgangest@unm.edu

PATRICIA ADAIR GOWATY
Institute of Ecology
University of Georgia
Athens, GA, USA
gowaty@sparrow.ecology.uga.edu

MACE HACK
Department of Ecology and Evolutionary Biology
Princeton University
Princeton, NJ, USA
mhack@princeton.edu

PETER M. KAPPELER
Department of Behaviour and Ecology
German Primate Centre
Göttingen, Germany
pkappel@gwdg.de

PHYLLIS C. LEE
Department of Anthropology
University of Cambridge
Cambridge, UK
pcl1@hermes.cam.ac.uk

CHARLES L. NUNN
Department of Biology
University of California
Davis, CA, USA
cnunn@ucdavis.edu

J. MICHAEL PLAVCAN
Department of Anthropology
University of Arkansas
Fayetteville, AR, USA
mplavcan@uark.edu

GAURI R. PRADHAN
Department of Biological Anthropology and Anatomy
Duke University
Durham, NC, USA
gau_pradhan@hotmail.com

DANIEL I. RUBENSTEIN
Department of Ecology and Evolutionary Biology
Princeton University
Princeton, NJ, USA
dir@princeton.edu

JOANNA M. SETCHELL
Department of Biological Anthropology
University of Cambridge
Downing Street
Cambridge, UK
mandrills@yahoo.co.uk

JOAN B. SILK
Department of Anthropology
University of California
Los Angeles, CA, USA
jsilk@anthro.ucla.edu

CHARLES T. SNOWDON
Department of Psychology
University of Wisconsin, Madison
Madison, WI, USA
snowdon@facstaff.wisc.edu

RANDY THORNHILL
Department of Biology
University of New Mexico
Albuquerque, NM, USA
rthorn@mail.unm.edu

ROBERT L. TRIVERS
Department of Anthropology
Rutgers University
New Brunswick, NJ, USA
trivers@rci.rutgers.edu

SUCI ATMOKO UTAMI
Faculty of Biology
National University
Jakarta, Indonesia
tiua@cbn.net.id

JAN A.R.A.M. VAN HOOFF
Ethology and Socio-ecology
Utrecht University
Utrecht, The Netherlands
jaramvanhooff@planet.nl

MARIA A. VAN NOORDWIJK
Department of Biological Anthropology
and Anatomy
Duke University
Durham, NC, USA
MariavanN@aol.com

CAREL P. VAN SCHAIK
Department of Biological Anthropology
and Anatomy
Duke University
Durham, NC, USA
vschaik@acpub.duke.edu

DIETMAR P. ZINNER
Department of Behaviour
and Ecology
German Primate Centre
Göttingen, Germany
dzinner@gwdg.de

Foreword

ROBERT L. TRIVERS

The past 30 years have seen a sea change in the study of sexual selection in primates. With few exceptions–such as Goodall's work on chimpanzees and DeVore's on baboons–the earlier work usually consisted only of descriptive field studies giving such crude social and ecological parameters as group size, age and sex composition, preferred food species, and whether reproduction was annual or not. The literature was often dull and justified only by the phylogenetic closeness of the species studied to ourselves.

Now we have studies of male paternity using DNA analysis, experimental analyses of the attractiveness of female swellings, comparative work on sexual dimorphism in size across various primates characterised for intensity of sexual selection, comparative studies of the primary sex ratio in primates and studies of sperm competition as well. On the theoretical side, we have passed well beyond the initial model linking sex differences in variance in reproductive success to sex differences in parental investment. We now have theory regarding reproductive skew, sex-antagonistic genes and sex-antagonistic selection, female choice biased toward daughters, and extra-pair and sperm competition as major variables in sexual selection. This is the new world that the present volume introduces us to. What follows is one person's selective summary, meant to highlight the key findings. Much is left out, of course, and there are certainly some important results that I have overlooked.

REVIEW, THEORY, HISTORY

Reviews are always welcome, and Peter Kappeler and Carel van Schaik have given us a very valuable review of the literature on sexual selection in primates, with more than 300 references, and covering all aspects of the subject, including those not covered elsewhere in this volume. A clear view of history and of relevant conceptual distinctions is also valuable, here provided by Tim Clutton-Brock, who also shows the value of the kind of within-species and comparative work in mammals that he has helped to pioneer. He shows, for example, that in some species lifetime reproduc-

tive success (RS) typically varies less than does within-year RS, while birth weight may be positively associated with later RS in males but not in females. Across species of mammals sexual dimorphism in adult body size predicts sexual dimorphism at birth, and sexual dimorphism is also associated, as expected, with greater differential male mortality among juveniles. Theoretical advances have been many and varied, and Patricia Gowaty, who has herself contributed, provides a useful review. She emphasises that random, non-heritable factors can mimic variance in reproductive success due to sexual selection. Flexible mating strategies are implied by the fact that variety of variables, such as change in risk of predation, can be shown to switch an individual from choosy to indiscriminate. Sexually antagonistic genes are obviously important, as is sexually antagonistic selection. For primates, this often means male control pitted against female resistance.

COMMUNICATION, SEXUAL SWELLINGS, MATE CHOICE IN HUMANS

Charles Snowdon emphasises that the study of communication in primates is still rudimentary compared to similar work in birds and frogs. Sounds are mostly involved in male aggression, visual signals with female advertisement, while olfactory cues are associated with both female advertisement and female competition. The best-studied cases are the sexual swellings found, for example, in baboons, and Dietmar Zinner *et al.* conclude that the evidence on sexual swelling in primates supports the hypothesis that these do not serve to advertise female quality (e.g. genetic) but instead act as a graded signal of fertility, which serves to manipulate the male degree of paternity-certainty to female advantage. The presence of sexual swellings across species is not associated with indicators of female conflict in ways predicted by the quality indicator model. Steven Gangestad and Randy Thornhill review the important progress that has recently been achieved in understanding human mate choice. They emphasise the utility of studying female choice throughout the menstrual cycle (and in the context of short- vs. long-term relationships) for teasing out the relative importance of selection for good genes and selection for parental investment. As a woman approaches ovulation she places a greater value on bodily symmetry and on the relative masculinity of a man's face. A woman's cycle also affects her interest in extra-pair copulation and it affects a man's behavioural styles in her presence. One could only wish we had similar quality data from humans for other major systems, such as the way in which degree of relatedness (or degree of reciprocal tendencies) affects human behaviour.

INFANTICIDE, SPERM COMPETITION

Infanticide has been a uniquely important topic in primatology for some 30 years now. Here, Carel van Schaik *et al.* present a very welcome development of Sarah Hrdy's intuitive insight that male infanticide may select for polyandrous matings by females in order to confuse paternity and spread it around among a variety of males. Since a gain in one male's paternity is a loss in another's, how is there a net increase in safety for the infants? One possibility is that total paternity is 1.0 but total perception of paternity can be manipulated to be >1.0. Another is that as long as there is a non-negligible chance of paternity, the cost of a mistake may be greater than the gain from infanticide of an unrelated youngster. A male counter-strategy of harassing females may evolve to reduce the latters' tendencies toward extra mating, which may, in turn, select on females for longer oestrous cycles, a longer follicular stage and less predictable times of ovulating. Sperm competition and post-copulatory selection are also important aspects of sexual selection that have come into prominence in the last 20 years. Tim Birkhead and Peter Kappeler integrate the more meagre primate data on sperm competition with the more detailed bird literature. Sperm competition occurs when more than one male mates with a female during her cycle. It generates mate guarding, multiple copulations, larger testes in relation to body size, an increase in size of the sperm midpiece (containing the mitochondria to power the sperm) and cryptic female choice, including (probably) female orgasm.

SEXUAL SELECTION AND ITS CORRELATES

While drawing attention to the paucity of developmental data, Joanna Setchell and Phyllis Lee show that high sexual size dimorphism is associated with earlier female reproduction and shorter interbirth intervals. They also show that the risk of predation appears to affect the relationship between adult size dimorphism and size of neonate, as well as age at weaning. Using genetic data, Maria van Noordwijk and Carel van Schaik show that paternity concentration (or, roughly, variance in male RS) is lower with larger group size and in seasonal species, which, in turn, show higher extra-group paternity. Also, the lower the number of females in a male's natal group, the higher his chance of migrating out. The paternity benefits of occupying the top position predict when and how often males disperse and how they acquire top rank. The method of independent taxonomic contrasts has emerged as a key tool in comparative work to give a more fine-grained test of otherwise significant trends. Michael Plavcan employs it to show that sexual dimorphism is enhanced only in polygynous vs. monogamous/polyandrous species, with no effect in single- vs. multi-male groups. These differences are true for sexual dimorphism in body size but not in canine length. Is this partly because female size is also affected by female–female competition? Group size is inversely related to size of female canines (which suggests that female–female competition is more intense in smaller groups).

THE SEX RATIO, SEXUALLY TRANSMITTED DISEASES

The primary sex ratio is an important variable that can, in theory, be socially revealing it reflects, for example, as Joan Silk and Gillian Brown show, the degree to which, after the period of parental investment, one sex imposes a cost on parents (or gives a benefit) not provided by the opposite sex. After a useful review of the relevant theory, we are offered a cautionary tale on the possible over-interpretation of Trivers and Willard's findings: the notion that whichever sex converts parental investment better into future RS should be produced in greater numbers when more investment is available. Regarding maternal dominance and tendency to produce one sex or the other, associations are as often toward one sex as the other, and the larger the sample size, the smaller the deviations are from no effect at all. Thus, random effects may be generating what appear to be significant associations, and the dreadful possibility looms that most or all of the Trivers–Willard effects in primates may be an illusion! Charles Nunn and Sonia Altizer treat the little-studied but important subject of sexually transmitted diseases (STDs) and sexual selection. They point out that we expect higher variance in male copulatory success to be associated across species with a lower per capita infection rate, as shown with data on HIV and STLV-1. Mate choice for avoiding STDs may select for parasites with benign effects on the external phenotype. Across primate species, there is no correlation between either oral–genital grooming or urination after sex and females having multiple male partners.

A MALE DIMORPHISM, A PARALLEL UNGULATE

Sexual selection can also generate multiple morphs within a sex. While male dimorphisms are widespread in insects and fish, they are nearly unknown in mammals. Suci Utami and Jan van Hooff provide a revealing glimpse into one such case: the flanged and unflanged orangutan males. Although many unflanged males may eventually mature into flanged ones, some unflanged ones live many years, reproduce often, and die without ever becoming flanged. The relative frequency of the two forms is presumably held in some kind of frequency-dependent equilibrium. Unflanged males are silent, small, search over large areas and have copulations more often employing force than do the larger, more dominant flanged ones, who give calls and are relatively stationary. Multi-level societies, in which smaller groups are embedded within larger ones, occur in different forms in primates. As Daniel Rubenstein and Mace Hack argue, the zebra provides a useful parallel to hamadryas baboons, and here they give an overview of their long-term work on zebras. A noteworthy achievement is that by relating herd size to risk of predation, a measure of vegetation quality and the number of bachelor males nearby, they are able to produce a model of decision-making that regulates social behaviour.

While some topics are not covered in individual chapters, the review chapter by Peter Kappeler and Carel van Schaik gives a good summary and the relevant references. The present volume gives us an excellent view of where the field stands today. Let us hope that it also helps provide a springboard for further work along similar lines, more in-depth studies of individual species, more detailed comparative work, more DNA analyses of paternity, more field and laboratory experimentation, and more integration with human data. This should also give us a deeper understanding of the forces that have moulded primate evolution, including our own.

Preface

Anybody following the literature in behavioural and evolutionary biology has noticed that sexual selection studies have dominated the journals in this field more than any other single topic for the last decade or so. Why has sexual selection been such a sexy topic? First, this boom coincides with important methodological innovations. The invention of DNA-fingerprinting, for example, has led to the development of new tools for the measurement of reproductive success and the outcome of mate choice. Similarly, the development of new comparative methods has stimulated numerous tests of evolutionary hypotheses that address key predictions of sexual selection theory. Second, theoretical advances of the theory itself have been astounding. There are at least three areas where new ideas have generated disproportional interest and significant new insights. They concern sexual conflict, sperm competition and the study of various indicators of good genes in the context of mate choice.

Sexual conflict has been recognised as an arena of intense intersexual coevolution driven by the fundamental genetic conflict between males and females. Many adaptations of both sexes are now being recognised as a result of an intersexual arms race. Using many examples from the primate literature, Smuts and Smuts (1993) identified sexual coercion as a behavioural mechanism employed by males to resolve the conflict between the sexes to their advantage. Sexual coercion continues to be one area of investigation where primatologists make important general contributions, especially with respect to the multiple pervasive effects of the risk of infanticide on behaviour and physiology, but other aspects of sexual conflict have remained virtually unexplored in primates.

The importance of sperm competition as a key mechanism of sexual selection has also only been fully appreciated in the 1990s, even though it had been identified and recognised at least another decade earlier. Examination of numerous behavioural, physiological and anatomical traits from the perspective that they may reflect adaptations to sperm competition have provided some of the most impressive examples for the evolutionary arms race between the sexes, and, at least to us, of the fascination of organismal biology. A paper by Roger Short (1979) on genital selection in the great apes convinced primatologists as early as the 1970s that sperm competition is an important evolutionary force in our closest biological relatives, but disappointingly little research on sperm competition by primatologists has been stimulated by him or the many recent beautiful studies on the topic in other taxa, even though most primates are notoriously promiscuous.

Finally, most recent theoretical and empirical studies of sexual selection have been concerned with indirect, mostly genetic, benefits of mate choice. The question as to the criteria on which females base their choice in species where they receive no resources from mates, has naturally focused on genetic benefits, either in the form of improved genetic compatibility, increased heterozygosity or high-quality genes. Because genetic quality cannot be assessed directly, much research has subsequently focused on phenotypic indicators thereof, such as fluctuating asymmetry and correlates of immunocompetence. This approach has also been adopted by many students of mate choice in humans, but, interestingly, it has not figured prominently in studies of non-human primates.

Thus, sexual selection is an exciting and active field of modern research driven by methodological innovations and conceptual progress. Nevertheless, primatologists appear to be largely excluded from this particular mainstream of evolutionary biology. The reasons for this have to do with biological differences between radiations, along with the usual problems of small sample sizes and difficulty with experiments on primates. Thus, the study of indirect benefits of female choice is, by and large, not applicable to primates, among whom there is little-to-no evidence of male ornaments for which females could express preferences. Instead, sexually dimorphic traits expressed in males tend to be armaments only, a clear trend among mammals more generally. In addition, compared to students of sexual selection in other lineages, primatologists have focused more on sexual coercion (but see Clutton-Brock & Parker, 1995). This difference provides another example of the importance of appreciating an organism's natural and life history for identifying meaningful questions. Because big differences in natural histories are typically related to qualitative differences in key life history traits, asking the same questions about all organisms clearly does not make sense. Finally, promiscuity and sperm competition are ubiquitous among primates, despite the fact that primates have relatively slow life histories and each female may only produce a few single young during her reproductive career, so that they should select their mates carefully. There is some indication that female promiscuity in primates serves both to confuse and to concentrate paternity because they experience an unusual risk of infanticide – another example of a major difference between primates and birds, where female promiscuity has been explained by factors other than infanticide risk.

Nonetheless, many important details concerning the causes and mechanisms of promiscuity and sperm competition remain unstudied. We still lack precise ideas about possible cryptic mechanisms female primates may employ to bias paternity, perhaps because required experimental approaches are rarely, if ever, possible with primates. Finally, genetic paternity studies, which are required to identify potential skew in male reproductive success, are only beginning to be published for primates, while ornithologists are already conducting meta-analysis of the results of many different studies. As a result, we do not know whether females in different taxa value male phenotypic and genotypic traits differently.

These open questions on aspects of sexual selection in primates and the apparent under-representation of primate sexual selection studies in mainstream behavioural and evolutionary biology has provided us with the motivation to take stock and to begin filling some of these gaps. To this end, we organised a conference (the 3. Göttinger Freilandtage) at the German Primate Centre in December 2001 to discuss these issues among more than 200 participants. Various aspects of sexual selection in birds and mammals as well as human and non-human primates were presented in 55 oral and poster papers, including 16 talks by invited speakers. Following the conference, 15 contributions were solicited in written form, and each one was subjected to rigorous peer review. They constitute a representative sample of the contributions to the conference, encompassing specific case studies, comprehensive reviews and theoretical analyses, as well as studies of non-primates that provide important comparative perspectives on general principles related to the issues raised above. We think that together they provide an up-to-date account of research on sexual selection in primates as well as numerous stimulating suggestions for future research on these topics.

The conference, as well as the resulting volume, would not have been possible without the support of many people and organisations. The 3. Göttinger Freilandtage conference was made possible by generous grants from the Deutsche Forschungsgemeinschaft (DFG), the Niedersächsisches Ministerium für Wissenschaft und Kultur, the German Primate Centre (DPZ) and the Sparkasse Göttingen. Michael Lankeit not only came up with the original idea of organising Freilandtage-meetings, but he also supported every instalment from the first moment on in crucial ways. Christoph Knogge did an amazing job in organising every logistical detail before and during the meeting to everyone's satisfaction. The members of the Abteilung Verhaltensforschung/Ökologie (Department of Behaviour and Ecology) at the DPZ, in particular Manfred Eberle, Eckhard Heymann, Roland Hilgartner, Julia Ostner, Oliver Schülke, Ulrike Walbaum and Dietmar Zinner, helped beyond the call of duty with the preparation of this conference.

The quality of the present volume is to a large extent due to the constructive comments of all contributors, who served as internal referees, as well as Susan Alberts, Roberto Delgado, Franz-Josef Kaup, Rebecca Lewis, Wayne Linklater, Patrik Lindenfors, Ian Penton-Voak, Michael Pereira, Elisabeth Sterck and Serge Wich, who provided additional comments on individual chapters. Ulrike Walbaum had the incredible patience to format every chapter and figure and double-check every reference. As always, Tracey Sanderson at Cambridge University Press provided crucial advice and support and did everything to keep this volume on its ambitious schedule. We thank all of them wholeheartedly. Finally, it is our pleasure to dedicate this volume to Claudia and Maria, Theresa and Anna, and Jakob and Jaap, for their understanding, support and inspiration during the preparation of this volume.

Part I
Introduction

1 • Sexual selection in primates: review and selective preview

PETER M. KAPPELER
Department of Behaviour and Ecology
German Primate Centre
Göttingen, Germany

CAREL P. VAN SCHAIK
Department of Biological Anthropology and Anatomy
Duke University
Durham, NC, USA

INTRODUCTION

Sexual selection theory has provided a successful framework for studying sex differences in behaviour, morphology, development and reproductive strategies, as well as the resulting mating systems (Campbell, 1972; Emlen & Oring, 1977; Bradbury & Andersson, 1987; Andersson, 1994). Humans and other primates have played an important stimulating role in the original development of the second pillar of modern evolutionary theory (Darwin, 1871; Clutton-Brock, this volume), and primates were the subjects of the first scientific paper devoted to sexual selection (Darwin, 1876). After languishing for nearly a century, sexual selection theory was re-discovered by evolutionary biologists and students of animal behaviour, who made it one of the most active fields of organismal biology by refining the theoretical framework and testing it with many new empirical data, mainly from insects and birds (Andersson, 1994). Studies of sexual selection in primates, in contrast, were not resumed with the same general enthusiasm and vigour, but there were notable exceptions (e.g. Crook, 1972; Clutton-Brock *et al.*, 1977; Harvey *et al.*, 1978; Hrdy, 1979; Smuts & Smuts, 1993; van Schaik *et al.*, 1999; van Schaik, 2000a). The first aim of this chapter is to provide an introductory summary of the main concepts and mechanisms of sexual selection theory in a way that highlights these contributions and questions of general interest from primatology, and to introduce the subsequent chapters of this volume.

A second aim of this chapter, and this volume in general, is to encourage integration of studies of sexual selection in primates with related work on other taxa and in other disciplines, including evolutionary theory (see also Maestripieri & Kappeler, 2002). Students of primates have addressed numerous specific questions about aspects of sexual selection, but the salient results and general conclusions emerging from these studies have not been systematically synthesised for more than 15 years (see Hrdy & Whitten, 1987; Smuts, 1987a, b). Moreover, recent studies of human sexual behaviour have addressed new, evolutionary questions (Buss, 1994; Thornhill & Gangestad, 1996), but the resulting new discipline of evolutionary psychology has remained largely isolated from primatology, even though comparative evidence from non-human primates, in particular, could advance the study and interpretation of human sexual strategies. Finally, observational and experimental studies of aspects of sexual selection in other mammals continue to provide an important comparative basis for characterising taxon-specific reproductive strategies and constraints that need to be more fully integrated into future research on primates and humans. We therefore hope that the chapters in this book stimulate exchange of new developments in theoretical, human, primate and mammalian sexual selection studies and will foster discussion among these largely separated sub-disciplines.

SELECTION IN RELATION TO SEX

Having developed the theory of evolution by natural selection (Darwin, 1859), Charles Darwin realised that a number of traits, mostly sexually dimorphic ornaments and armaments, did not seem to promote survival or were not primarily involved in the production of offspring. He suggested that these secondary sexual traits confer an advantage with respect to acquiring mates (Darwin, 1871), by intimidating rivals and/or by attracting mates, and he developed a theory of selection in relation to sex that was linked to natural selection. Primates show abundant evidence of such dimorphism (recently summarised by Dixson, 1998) in maternal investment; growth and developmental patterns; body and canine size; scent glands and scent-marking behaviour; vocalisations and various visual ornaments such as manes, flanges,

Sexual Selection in Primates: New and Comparative Perspectives, ed. Peter M. Kappeler and Carel P. van Schaik. Published by Cambridge University Press. © Cambridge University Press 2004.

coloured skin and fur; as well as interspecific variation in traits such as relative testes size and penile morphology, and the presence of sexual swellings. All these sex differences suggest that the traits involved evolved under the influence of sexual selection. The same is true for a host of sex differences in behavioural, physiological and even ecological and life-history characteristics (Dixson, 1998; van Schaik et al., 1999; Kappeler et al., 2003), including adaptive adjustment of birth sex ratios (Clutton-Brock & Iason, 1986; Dittus, 1998; but see Silk & Brown, this volume), a body of theory to which primate studies have contributed importantly (Clark, 1978; Silk, 1984; van Schaik & Hrdy, 1991; Brown, 2001).

Whereas Darwin's conceptual distinction between natural and sexual selection continues to be appreciated, more recent analyses and arguments summarised by Clutton-Brock (this volume) support the consideration of sexual selection as a form of natural selection that acts differently on the two sexes. This perspective leads to an emphasis on studying causes and consequences of sex-specific effects of selection. Why selection operates differentially on males and females has traditionally been related to fundamental aspects of their reproductive physiology and biology. As reviewed in detail by Gowaty (this volume), traditional sex roles, characterised by discriminating females and competitive males, were theoretically based on an elaboration of the consequences of anisogamy (Williams, 1966; Parker, 1979; Bulmer & Parker, 2002) and gained initial empirical support through a simple experiment with fruit flies (Bateman, 1948). Robert Trivers (1972) subsequently decoupled sex roles from gamete size and linked them instead to relative parental investment. In the end, variation in sex roles was convincingly linked to sex differences in potential reproductive rates (Clutton-Brock & Parker, 1992).

Sexual selection can be dissected into distinct components. Right from the start, Darwin identified mate choice (intersexual selection) and competition for mating partners (intrasexual selection) as driving forces behind adaptations in response to sexual selection (Bradbury & Andersson, 1987; Kirkpatrick, 1987; Andersson, 1994; Andersson & Iwasa, 1996; Cunningham & Birkhead, 1998; Kokko et al., 2002; see Clutton-Brock, this volume, for a historical perspective). However, it took until the 1970s for a third major component to be recognised: mating or intersexual conflict (Parker, 1979). This conflict varies from disagreement over the identity of mates or the frequency of mating to the chemical composition of seminal fluids. The outcome of this conflict may vary widely, producing stable equilibria in which one

sex has the advantage, or complex arms races between the sexes (e.g. Hammerstein & Parker, 1987; Johnstone & Keller, 2000). We will return to these mechanisms with an emphasis on primate examples below.

Despite these theoretical clarifications, it is difficult in practice to demonstrate that sexual selection was responsible for the presence of a trait. For that, precise quantification and documentation of inter-individual variation, their effects on conspecifics, as well as their relation to variation in reproductive success is still required (see Snowdon, this volume).

COMPETITION FOR MATES

Fitness of male primates is generally limited by access to receptive females, and males compete among themselves either to exclude rivals from access to females altogether, or to mate more often and/or at the right time (Kappeler, 1999). As a result, several genetic studies have shown that male primates generally exhibit greater variance in reproductive success than females (e.g. de Ruiter et al., 1994; Gerloff et al., 1999; Launhardt et al., 2001; Soltis et al., 2001; Vigilant et al., 2001), even though some of the variance among males is reduced when the entire lifespan is considered (Altmann et al., 1996; van Noordwijk & van Schaik, this volume). Determinants and consequences of variation in male monopolisation potential and variation in competitive mechanisms employed under different circumstances are therefore central questions for a better understanding of male reproductive strategies.

The evolutionary impact of mating competition among primate males on their morphology, life history, physiology and behaviour has been recognised for a long time (Schultz, 1938), although it took until the 1970s for comparative evidence to emerge (Crook & Gartlan, 1966; Crook, 1972; Clutton-Brock et al., 1977; Harvey et al., 1978; Alexander et al., 1979; Short, 1979; Harcourt et al., 1981; Harvey & Harcourt, 1984; Clutton-Brock, 1985). Several of the examples of systematic and predicted relationships between mating systems and morphological and behavioural correlates were first documented in primates and have become classic textbook examples (e.g. Krebs & Davies, 1992).

Sexual dimorphism in body and canine size, in particular, have fascinated primatologists for a long time because detailed data for a large number of species showed a rich variation, including some of the most spectacular cases found among mammals, such as mandrills, orangutans and gorillas (Short, 1979; Rodman & Mitani, 1987; Weckerly, 1998;

Setchell *et al.*, 2001). Morphological differences between human males and females, as well as their development over evolutionary times, have likewise occupied generations of anthropologists (e.g. Ghesquiere *et al.*, 1985; Plavcan & van Schaik, 1997). It has long been evident that body and canine size are good indicators of the intensity of male–male competition, but only recently are we actually beginning to explain most of the variance, especially among polygynous species (Plavcan, 2001, this volume; Lindenfors, 2002). Studies of primate sexual dimorphism have also stimulated similar work in other mammals (Alexander *et al.*, 1979; Gittleman & van Valkenburgh, 1997; Weckerly, 1998; Lindenfors *et al.*, 2002). A second recent development in studies of primate sexual dimorphism has been a focus on developmental questions (Leigh, 1995; Leigh & Shea, 1995; Leigh & Terranova, 1998; Smith & Leigh, 1998; Pereira & Leigh, 2003) that has been extended to other aspects of sexual selection by Setchell and Lee (this volume; see also Badyaev, 2002).

Partly as a result of this additional focus on ontogeny, there has been increasing recognition of the fact that sexual selection does not only target adult males in the context of mating competition. Rather, the entire life history of males, including their morphological and physiological development, as well as their timing of transfers into other groups, has apparently been shaped by sexual selection (Alberts & Altmann, 1995a, b). Setchell and Lee (this volume) summarise and evaluate the effects of intrasexual selection during various phases of the ontogeny of male primates (see also Pereira & Leigh, 2003), whereas Utami and van Hooff (this volume) discuss the enigmatic special case of bimaturism in orangutans. Along the same lines, van Noordwijk and van Schaik (this volume) focus on behavioural decisions by male primates related to the way they can achieve high reproductive success, showing that multiple aspects of their career choices, from the risk taken in rank acquisition to the nature and timing of transfer decisions, have been shaped by intrasexual selection. Convergent processes and outcomes in an independent taxon are described by Rubenstein and Hack (this volume) in their analysis of zebra stallions' behavioural and evolutionary responses to trade-offs created by natural and sexual selection.

THE MALE PERSPECTIVE OF COMPETITIVE SCENARIOS

The nature of competition among primate males is determined by the monopolisability of females, which in turn is determined by various female features: their spatial distribution, the size of their groups and the degree of synchrony of their receptive periods (Mitani *et al.*, 1996a; Nunn, 1999; Kappeler, 2000; see also Emlen & Oring, 1977; Jarman, 1983; Ims, 1988, 1990; Clutton-Brock, 1989; Carranza *et al.*, 1995; Creel & Macdonald, 1995; Michener & McLean, 1996). How many members of the same and opposite sex live together is both a determinant and outcome of sexual strategies, in particular those of males (Kappeler & van Schaik, 2002), but also, indirectly, those of females, whose behaviours affect the parameters of male–male competition (Wiley & Poston, 1996; Gowaty, 1997).

From the males' perspective, the most basic question is whether females are dispersed in space, or not. Whenever reproductively active females are not associated with each other, males must make a strategic decision about dispersal and ranging behaviour (Dunbar, 2000). Depending on the males' decision, two fundamentally different types of social organisations can be distinguished: males either also range independently, often trying to encompass ranges of several females within their home range, or they associate permanently with either a single female or a group of spatially clumped females (van Schaik & Kappeler, 1997, 2003).

Given a particular spatial distribution of females and the males' decision to associate with them or not, three hierarchical levels of intermale competition can be distinguished (Kappeler, 1999; Setchell & Kappeler, 2004). First and foremost, males ought to be concerned with gaining access to as many receptive females as possible, while at the same time preventing rivals from doing so. Complete monopolisation of several females should always be the most successful male reproductive strategy and, thus, the top male priority. Second, if complete monopolisation is not possible, a male should try to maximise his number of copulations, while keeping the number of copulations by rivals at a minimum. Finally, if males cannot skew the number of copulations in their favour, they may rely on various mechanisms of post-copulatory selection to maximise their chances of fertilisation with just one or a few copulations. At each level of competition, mechanisms of both scramble and contest competition can be employed and combined.

MECHANISMS OF MATING COMPETITION

Primate males employ a number of mechanisms before, during and even after mating to out-compete their rivals in the race for fertilisations (summarised in Dixson, 1998; Setchell & Kappeler, 2004). Apart from the timing in relation to mating (before, during, after), the distinction between contest (or interference) and scramble (or exploitation) mechanisms is useful to characterise competitive mechanisms. Along this spectrum, the use of physical force, aggression and threat gives way to indirect mechanisms not requiring physical contact, such as effective mate searching and sperm competition. Below, we briefly characterise the best-known mechanisms in primates (see also Setchell & Kappeler, 2004 for a more detailed discussion).

Receiving and sending signals

First, in species in which males are not permanently associated with females, males need to find receptive females. Males with increased endurance and improved abilities to detect female acoustic and olfactory signals from long range should have an advantage over others. However, the sensory sensitivities and capabilities of solitary primates, for which this ability is most important, remain virtually unstudied (cf. Schwagmeyer & Woonter, 1986; Ims, 1990; Schwagmeyer et al., 1998).

Second, because of the inherent risks of overt fighting, there is selection for displays and signals during male encounters. Visual signals emanating from body and canine size may play a role in this context, as well as the coloured skins, manes and capes of hair, cheek flanges and other facial adornments found in males of many primate species (Dixson, 1998; Snowdon, this volume). However, in contrast to birds (Andersson, 1986; Zuk et al., 1990; Zuk, 1991; Møller & Pomiankowski, 1993), there is very little evidence demonstrating that the comparatively rare sexually-dimorphic ornaments of primates are actually condition-dependent (but see Setchell & Dixson, 2001b; see also West & Packer, 2002) or that male rivals even attend to these signals (but see Gerald, 2001). Moreover, no studies to date have systematically investigated female preference for male ornaments in non-human primates, which is a potential additional or alternative function of these visual signals.

Sexually dimorphic acoustic signals, such as loud calls, could also function in repelling and deterring non-resident rivals or in aiding species recognition and influencing female mating decisions, but there has been very little support for these hypotheses in recent reviews and comparative tests (Wich & Nunn, 2002; Snowdon, this volume). Furthermore, there is some evidence to suggest that the transfer of olfactory signals among primate males has been shaped by sexual selection. Males often have more and bigger scent glands than females and they also mark and investigate scents more

often (Dixson, 1998; Heymann, 1998). In some primates, the frequency, quality and effect on the receiver of male signals have been shown to be status-dependent (Kappeler, 1990c, 1998; Fornasieri & Roeder, 1992; Perret, 1992; Kraus et al., 1999; Maggioncalda et al., 1999), but much more comparative and experimental work is required to illuminate the function of pheromones in intrasexual selection further.

Physical prowess and dominance

Whenever males fight for access to receptive females, sexual selection theory predicts that characters contributing to physical superiority, such as large body size and/or weaponry will be favoured (Darwin, 1871). Most polygynous primates are indeed sexually dimorphic in body size and weaponry because selection for physical superiority is more intense among males, presumably because they compete for non-shareable benefits directly linked to fitness (van Schaik, 1996), even though selection for such traits is not limited to males (Plavcan et al., 1995). The resulting sexual dimorphism in body and canine size has been analysed mostly among species (but see Bercovitch, 1989), using mating system classification (Clutton-Brock et al., 1977; Kappeler, 1990a, 1996; Lindenfors & Tullberg, 1998), operational sex ratio (Mitani et al., 1996b) or behavioural competition levels (Kay et al., 1988; Plavcan & van Schaik, 1992; Plavcan, 2001) to determine the relative contribution of intrasexual selection vis-à-vis other ecological factors and evolutionary by-products and developmental mechanisms (reviewed in Plavcan, 2001, this volume; Pereira & Leigh, 2003; Setchell & Lee, this volume).

Dominance, which is typically based on physical superiority, is an important behavioural mechanism used by male primates to obtain access to receptive females (Packer, 1979; Samuels et al., 1984; Shively & Smith, 1985; Bercovitch, 1988). The priority-of-access model, which postulates that the dominance hierarchy functions as a queue and that the number of simultaneously fertile females determines male access, provides a theoretical framework for the analysis of dominance effects (Altmann, 1962). This mechanism of reproductive competition is often manifested as mate guarding or consortships. In species where consortships occur, most copulations take place during this time, thus increasing the consorting male's probability of mating, in addition to providing an opportunity for preventing rivals from doing so. High-ranking males are often more successful in forming consortships (Bercovitch, 1991; Cowlishaw & Dunbar, 1991), but some females consort with several males in succession (Hrdy & Whitten, 1987; Hrdy, 2000), indicating that females can affect the competitive regime for males both directly and indirectly.

There has been much debate over the relationships among dominance, mating and reproductive success in male primates (Cowlishaw & Dunbar, 1991; de Ruiter & van Hooff, 1993). The genetic measures of reproductive skew among males needed to examine the crucial relationship between mating success and reproductive success are gradually becoming available (e.g. Pope, 1990; de Ruiter et al., 1992; Altmann et al., 1996; Launhardt et al., 2001). The priority-of-access model, which incorporates the effects of indirect female choice (Wiley & Poston, 1996), provides the basic expectation for reproductive skew among males. Deviations from it are due to the success of alternative mating tactics, sperm competition and female mating preferences (Møller & Ninni, 1998; Petrie & Kempenaers, 1998; Johnstone et al., 1999; Engh et al., 2002). Future studies should strive to determine the relative contributions of these processes in primates.

Inhibition and alternative tactics

In several primates, subordinate males show signs of partial physiological suppression of sexual function, characterised by reduced body mass and condition; reduced testis size; smaller, less active scent glands; reduced development of secondary sexual traits; decreased levels of circulating testosterone, growth hormone and luteinising hormone; lower frequencies of sexual and olfactory behaviours, or any combination thereof (Schilling et al., 1984; Perret, 1992; Kraus et al., 1999; Maggioncalda et al., 1999, 2000; Setchell & Dixson, 2001a; Utami & van Hooff, this volume). This phenomenon can be interpreted as an adaptation of the subordinate (see Gross, 1996): inhibition allows him to remain in the presence of dominant males, thereby avoiding aggression and the costs of high testosterone for a period (Utami et al., 2002; Setchell, 2003). Similarly, juvenile males may prolong growth and delay maturation to achieve larger size and competitive ability upon entering the mating arena (Alberts & Altmann, 1995b). Because testes are typically already spermatogenic in maturing males, they may obtain a few low-risk sneaky copulations while in this phase (e.g. Berard et al., 1994). On the other hand, because inhibited males may suffer a disadvantage in sperm competition and because they are less attractive to females (e.g. van Hooff & Utami, this volume), the inhibition may reflect pressures exerted by the dominant male(s). More information on primates in nature, as well as other mammals (e.g. Arnold & Dittami, 1997), is needed to determine the relative importance of self-imposed inhibition and dominant-imposed suppression. One possibility is that where stress is a major mechanism, we are dealing with true suppression imposed by the dominant (see Sapolsky, 1985; von Holst, 1998), whereas in the absence of stress the inhibition may reflect an adaptation on the part of the subordinate.

At least four behavioural mechanisms of alternative mating strategies have been recognised. First, males in poor physical condition or social position can form coalitions to force a superior male to relinquish a receptive female (Bercovitch, 1988; Noë & Sluijter, 1990), whereas dominant males may need to form coalitions to improve their effectiveness at mate guarding in the presence of too many rivals (Watts, 1998). Second, adult males living in all-male bands throughout the year can raid groups containing multiple females and a single resident male during the brief mating season. By doing so, they may avoid the costs of being around dominants for most of the year (Borries, 2000; Cords, 2000). The presence of such all-male groups may have profound effects on social organisation (see Rubenstein & Hack, this volume). Third, young, less powerful males may associate with a fully developed male and the group of females associated with him, thereby obtaining occasional mating opportunities and prospects of succession, at reduced costs of female defence to the dominant (Pope, 1990; Robbins, 1999; Watts, 2000). Finally, by forming friendships with particular females, some males obtain access to at least one female at little risk of aggression from dominant males (Smuts, 1985). Despite these many diverse examples, it is likely that not all alternative mating tactics of male primates have been discovered yet (cf. Koprowski, 1993; Stockley et al., 1994; Gemmell et al., 2001), and we clearly need detailed long-term studies to understand their stability and genetic payoffs.

Copulatory and post-copulatory mechanisms

Once a male has successfully gained access to a receptive female, variable aspects of copulation may influence his reproductive success. First, copulation frequency is much higher in promiscuous than in monandrous species, both reflecting and creating different intensities of sperm competition (Dixson, 1998). Second, copulatory patterns – including variation in intromission and ejaculatory patterns, number of thrusts before ejaculation, length of intromission, duration of copulation, and the need for single or multiple intromissions prior to ejaculation – exhibit similar variation among species that appear to correlate with broad mating system categories (Dixson, 1998). Because most primates are notoriously promiscuous, presumably as a result of several female benefits and trade-offs (Gangestad & Thornhill, this volume; van Schaik et al., this volume), sperm competition among primates is widespread and intense (Birkhead & Kappeler, this volume), constituting a powerful and ubiquitous mechanism of mating competition. With the exception of testes size (Harcourt & Gardiner, 1994; Harcourt et al., 1995; Harcourt, 1996; Kappeler, 1997a), however, physiological and morphological determinants and correlates of success in sperm competition in primates are still relatively poorly known (Dixson, 1998; Birkhead & Kappeler, this volume). To what extent individual variation in primate testes size is positively correlated with competitive potential (rank, body size) on the one hand, and mating and reproductive success on the other hand, independent of potentially confounding co-variables, also remains largely unresolved by the existing interspecific studies (but see Bercovitch, 1989). As with studies of sexual dimorphism, studies of variation in testes size among primates have inspired and influenced similar investigations in other mammals (Kenagy & Trombulak, 1986; Dewsbury & Pierce, 1989; Møller, 1989; Møller & Birkhead, 1989; Heske & Ostfeld, 1990; Rose et al., 1997; Gomendio et al., 1998; Hosken, 1998).

Even after fertilisation has been achieved, male reproductive competition can continue with two additional mechanisms at the intersection of male–male competition and intersexual conflict. First, females may terminate investment in the developing foetus in the presence of a new dominant male. This 'Bruce effect', which is well known in rodents (Schwagmeyer, 1979; Mahady & Wolff, 2002), is advantageous for the male inducing it, because it will create a mating opportunity in the near future, so that it was originally considered a product of male–male competition (Trivers, 1972). However, whenever the risk of infanticide or the loss of paternal care following birth are high (Labov, 1981), resorption or abortion is primarily adaptive from the female perspective so that it may ultimately represent more of a female reproductive strategy. Despite the difficulties of detecting the early termination of reproduction, there is some evidence that it may occur among primates under the right circumstances (Pereira, 1983; see also Forbes, 1997).

Second, males may interfere with the reproductive success of rivals, while at the same time improving their own, by committing sexually selected infanticide. The evidence in support of this hypothesis has recently been summarised (van Schaik & Janson, 2000) and is overwhelming. Briefly, males in dozens of primate species have now been observed to kill unrelated dependent infants, leading to a faster resumption of the affected mother's reproductive activity, thereby creating additional mating opportunities for the male (Hrdy, 1979; van Schaik, 2000a). Due to their particular life history characteristics, primates may be especially vulnerable to infanticide (van Schaik & Kappeler, 1997), but the same principles have been demonstrated in other mammalian lineages (van Noordwijk & van Schaik, 2000; van Schaik, 2000b). This strong selective force on male and female reproductive success has shaped many other aspects of primate reproductive physiology and social behaviour (van Schaik & Janson, 2000; van Schaik et al., this volume).

MATE CHOICE

Mate choice by females has been the aspect of Darwin's original theory initially meeting the most scepticism. Since the 1970s, however, female choice has been one of the most intensively and productively studied topics in evolutionary biology (e.g. Searcy, 1979; O'Donald, 1980; Andersson, 1982; Kirkpatrick, 1982; summarised in Andersson, 1994). Whereas the sophistication of theoretical models is as yet not matched by equally sophisticated empirical evidence, we now have a good idea about the underlying behavioural and genetic mechanisms, as well as the benefits females may derive from careful choice of their mates (e.g. Hamilton & Zuk, 1982; Zuk, 1991; Johnstone, 1995; Gibson & Langen, 1996; Møller, 1997, 2000).

In some cases, females obtain direct, material benefits from choosing a particular male that have positive effects on their fecundity (Thornhill & Alcock, 1983). More commonly, however, females obtain indirect benefits in the form of advantages accruing to their offspring, because they choose males of the right species (Panhuis et al., 2001), reduce the effects of inbreeding by discriminating against closely related males (Perrin & Mazalov, 2000), or prefer mates with genotypes that maximise offspring heterozygosity (McCracken & Bradbury, 1977). Phenotypic indicators of male quality that are preferred by females can be colourful patches, length of tails, symmetry of structures, aspects of male calls, and age or the morphological correlates thereof (Andersson, 1986; McComb, 1991; Møller, 1992; Swaddle & Cuthill, 1994; Gangestad & Thornhill, 1998; this volume; Widemo & Saether, 1999; Waynforth, 2001). Nonetheless, some doubt persists as to whether the currently most popular models provide a complete explanation, for instance because the presence of some pre-existing female preferences may reflect sensory exploitation by males (Ryan & Keddy-Hector, 1992; Endler & Basolo, 1998; Ryan, 1998), thus prompting additional modelling (see e.g. Kokko et al., 2002).

FEMALE CHOICE IN PRIMATES

As in other mammals, fitness of female primates is limited by access to resources and the quality of parental care (Trivers, 1972; Emlen & Oring, 1977). Because primates have relatively slow life histories and each female may only produce a few single young during her reproductive career (Lee, 1996; Ross, 1998), female primates are expected to select their mates carefully. The potential importance of female choice for female fitness is further accentuated by the fact that males in most primate species provide little or no direct infant care. However, apart from the active avoidance of inbreeding (Grob et al., 1998), there is surprisingly little evidence for female choice in primates, either in terms of the exclusive selection of particular mates or the consequences of such persistent choices on male phenotypes (Small, 1989; Keddy-Hector, 1992). Here we examine some of the possible reasons for this, from the demographic context, to mating behaviour, to post-mating discrimination.

Living in permanent social units with long-term membership – a near-universal feature of primates – imposes potential demographic constraints on choice. Thus, species living in pairs are predisposed towards monandry, although recent behavioural and genetic studies revealed that extra-pair copulations are possible, sometimes common, and often initiated by female primates (Palombit, 1994; Fietz et al., 2000; Sommer & Reichard, 2000). Hence, females may in fact often select or reject mates despite demographic or social constraints. Females may also exercise choice by affecting the composition of their social units. Transfer between social units is present in some species (e.g. Sterck & Korstjens, 2000), but is likely to be an expensive alternative option relative to the fitness benefits obtained by unconstrained female choice of mates. Females of some species are known to exert influence on the success of male immigration attempts (e.g. Smuts, 1987a) or may not accept a male who has taken over the social unit they live in (e.g. Dunbar, 1984), but it is not clear how widespread these choices are. However, social units in which females have mating access to multiple males are remarkably common among primates (van Schaik, 2000b).

Within the demographic constraints imposed upon them by the nature of their social organisation, primate females show a striking tendency toward active promiscuity (Hrdy & Whitten, 1987; Hrdy, 2000; Paul, 2002) and employ still poorly studied mechanisms and criteria to choose mates, if they exercise any direct choice at all. In species with substantial direct male care, through female mate choice – or more precisely in this case, choice of polyandry – females

may obtain important direct benefits (Goldizen & Terborgh, 1986; Goldizen, 1987). It is possible that the fitness benefits in terms of infanticide avoidance and protection by likely fathers (see van Schaik et al., this volume) outweigh any genetic benefits gained from choosing particularly attractive males in other social settings. Promiscuity is found in spite of the serious risks of disease transmission (Nunn & Altizer, this volume) and the tendency among mammalian males, prominently displayed by male primates, of trying to curtail the behavioural freedom of females to exert mate choice (see Gowaty, this volume) – an expression of mating conflict (see below).

Their reproductive physiology provides female primates with several mechanisms and levels of control to affect the number and identity of available mates (Wiley & Poston, 1996). First, seasonal reproduction, usually dictated by environmental factors, provides a necessary, albeit insufficient, condition for the synchronisation of cycles among several associated females. The more females are receptive at or during the same time, the smaller is the ability of a single dominant male to monopolise access to all of them (Emlen & Oring, 1977; Dunbar, 2000; Eberle & Kappeler, 2002). As a result, females may increase their opportunities to choose additional mates. Second, females can apparently vary the length of their follicular phase within certain limits, which has the same consequences for male monopolisation potential and female choice (van Schaik et al., 1999). Third, females in different species either advertise or conceal ovulation (Burley, 1979; Andelman, 1987; Sillén-Tullberg & Møller, 1993; Converse et al., 1995; van Schaik et al., 1999; Heistermann et al., 2001), in effect manipulating males. By clearly advertising receptive periods, females can incite male–male competition without bearing the costs of prolonged and/or repeated matings (e.g. Hoelzel et al., 1999). By concealing ovulation, females reduce male monopolisation potential and increase the pool of potential mates because no male can restrict his mating effort to a particular time. The sexual swellings of some Old World monkeys and apes may fulfil both functions simultaneously. Increasing sexual swellings advertise the approach of a receptive period, but they do not allow pinpointing of the exact date of ovulation (see Nunn et al., 2001; Zinner et al., this volume). There is some evidence that these signals can also be used strategically (Zinner & Deschner, 2000).

An important behavioural mechanism of female choice is mating cooperation. With the possible exception of orangutans (Galdikas, 1985a, b; Schuermann & van Hooff, 1986; Fox, 2002; Utami et al., 2002), mating in primates requires active female cooperation. By simply walking away or sitting down, females can prevent matings at a particular time or by a particular male, albeit at some potential cost. Lunging or cuffing in response to the advances of males is an effective mechanism in many species without pronounced sexual dimorphism. Active solicitations and presentations towards particular males are common forms of positive female choice in other species (Janson, 1984, 1986). Lemurs constitute a particularly interesting case because lemur females have more freedom to choose than other primate females as they match males in size and weaponry (Kappeler, 1990a, 1991, 1996; Richard, 1992; Plavcan et al., 1995) and dominate them socially in all contexts (Kappeler, 1990b, 1993; Pereira et al., 1990; Kubzdela et al., 1992). Interestingly, polyandrous matings are nevertheless the rule among non-pair-living lemurs (Jolly, 1967; Richard, 1974; Sauther, 1991; Overdorff, 1998; Ostner & Kappeler, 1999; Radespiel et al., 2001), even though females do have and exhibit certain preferences (Pereira & Weiss, 1991). Bonobos present a similar, puzzling case (Wrangham, 1993; Parish, 1994). Future research will have to determine whether factors other than infanticide avoidance (Hrdy, 2000; Jolly et al., 2000), perhaps also proximate ones (de Waal, 1987; Hrdy, 1995), can explain the sexual behaviour of these species.

Active promiscuity will move the arena of mate choice to the post-mating period. Cryptic female choice (Eberhard, 1996) will tend to produce a discrepancy between mating and siring success. It is an adaptation to situations where females cannot keep males from mating or have other reasons to mate with multiple males (Gowaty, 1997; Tregenza & Wedell, 2000; Zeh & Zeh, 2001). Mechanisms of cryptic choice in primates remain largely unknown, but selective orgasm is one promising possibility deserving more investigation (see Birkhead & Kappeler, this volume).

VARIANT SEX ROLES AND SEXUAL CONFLICT

Whenever one sex has become the limiting factor for the reproductive success of the other sex, the latter will compete for access, and the former will exert choice. As discussed above, in many animals, and especially among mammals, females are the limiting factor for males, who therefore compete for access, allowing the females to exert their choice. However, this neat dichotomy is sometimes upset by cases of females competing for access to mates, or males exerting mate choice, as well as by a ubiquitous conflict of interest between the sexes (Gowaty, 1997, and this volume). We now address these complications.

COMPETITIVE FEMALES

Female–female mating competition for sexual access to mates has so far mostly been observed in artificial or rare situations (summarised in Nunn et al., 2001; but see Gowaty, this volume). It is not to be confused with various other phenomena that may superficially resemble it. Thus, the active role taken by females in soliciting matings (Hrdy, 1981; Hrdy & Whitten, 1987; van Schaik et al., 1999) does not necessarily reflect competition for access to mates (see below). Likewise, examples of inhibition of female ovarian activity in the presence of a dominant female in various callitrichids (Abbott, 1989) is generally thought to reflect female competition for access to helpers, rather than mates. Finally, females may compete over social access to dominant males, as in gorillas (Watts, 1992, 2000), but again there is no evidence for females competing for sexual access as well (Robbins, 1999).

If females actually compete for access to males, we expect them also to show the features that go with it: enlarged armaments or special ornaments where this competition is intense. Relatively increased female canine size is indeed associated with increased intensity and frequency of competition among females (Plavcan et al., 1995), which can have lethal effects (McGraw et al., 2002), but this interspecific effect is due to competition for resources and not for mates. Moreover, reversed sexual dimorphism, with females being larger or heavier than males, has also been documented in a few primate species, notably among lemurs (Kappeler, 1991; Wright, 1995; Schmid & Kappeler, 1998; Richard et al., 2002) and some New World monkeys (Ford, 1994), but again there is no evidence for escalated female competition in the context of mating in these taxa (Richard, 1987; Kappeler, 1993; Jolly, 1998).

Special female ornaments are expected where they enhance their owners being chosen by males. Pagel (1994) offered such an indicator explanation for female sexual swellings, and a recent field study of baboons presented the first support for this hypothesis (Domb & Pagel, 2001). However, Zinner et al. (2002, and this volume) conclude that the evidence for female competition for access to mates is weak. Moreover, their re-analysis and interpretation of the baboon data indicate that they actually offer more support for the idea that swellings indicate proximity to ovulation. Scent gland size and/or use in callitrichids may represent another potential example (Heymann, 1998), but these taxa typically contain only a single reproductive female. Thus, as theoretically expected, given their much higher parental investment and particularly slow reproductive rates, there is very little evidence for female–female competition in primates as of yet. However, incidental observations of females preferentially attacking oestrous females (Wasser, 1983) or harassing matings of other females (e.g. Linn et al., 1995) are nonetheless puzzling and require an explanation.

CHOOSY MALES

Male mate choice in polygynous mating systems is expected whenever there are direct or opportunity costs to mating (Berger, 1989; Altmann et al., 1996; Gowaty, this volume). Male mating costs are either direct or indirect and may surface before or during mating. First, males in the majority of primate species with male transfer experience substantial costs in the form of increased risks of starvation, predation and injury during transfer and immigration (Alberts & Altmann, 1995a, b). The analysis of the available data on male transfer decisions clearly indicates that they are driven by improved access to mates (Jones, 1983; van Noordwijk & van Schaik, 1985, 1988, 2001, and this volume; Sussman, 1992; Borries, 2000; Olupot & Waser, 2001), so that they should be considered as part of the indirect mating costs. Similar, though not yet quantified, costs may be incurred by males in solitary species searching for dispersed mates (e.g. Kappeler, 1997b). Establishment of dominance relations prior to mating can be interpreted as a reproductive investment, especially whenever dominance and reproductive success are positively related (see above), and may therefore constitute part of the direct costs. The size and shape of the maxillary canines of most male primates illustrate these potential costs vividly (see Plavcan, this volume).

Second, once receptive females are located or identified, costs of mate guarding may accrue. These costs may again be indirect, as through reduced opportunities for foraging (e.g. Alberts et al., 1996) and a concomitant deterioration of condition and health (cf. Clutton-Brock et al., 1982), or direct in the form of increased aggression from potential mates and/or rivals (Enomoto, 1981; Colquhoun, 1987; Pereira & Weiss, 1991; Manson, 1994; Drews, 1996; Fawcett & Muhumuza, 2000). Moreover, physiological costs of sperm production, which appear much more significant than previously thought (Dewsbury, 1982; Preston et al., 2001; Wedell et al., 2002), and sperm delivery, which can be compromised by consecutive ejaculations (Dixson, 1995, 1998), constitute constraints that should also favour mate selectivity. Such strategic mate choice has been directly demonstrated in captive rhesus monkeys (Wallen, 2001): when kept as male–female pairs, some mating occurred on every day of the female's cycle (Goy, 1979); however, when multiple females were kept with a single male, mating was limited to the fertile period of each

female's cycle (Wallen et al., 1984). More such experiments with a diversity of species are clearly desirable.

Finally, variation in female fecundity or quality should also contribute to differential allocation of male mating effort because competition for adolescent, nulliparous or low-ranking females may yield relatively smaller reproductive benefits (see Altmann, 1980; van Noordwijk & van Schaik, 1999) at the same high costs. Male primates may also exhibit particular mating (and social) preferences independent of rank or fecundity (e.g. Smuts, 1985; Palombit et al., 1997; Pereira & McGlynn, 1997; van Schaik & Aureli, 2000) that may also reflect some form of male choice based on yet unknown criteria.

In species where males invest heavily in reproduction, for instance in pair-living species with extensive male care for infants, females are expected to compete for access to males just as males compete over females, and may even signal variation in their quality to potential mates through visual (Domb & Pagel, 2001) or olfactory (Heymann, 1998) indicators. Such male choice appears unusually prominent in humans, where phenotypic signals of female health and fecundity are considered universally attractive (Buss & Schmitt, 1993; Thornhill & Gangestad, 1996; Manning et al., 1999).

SEXUAL CONFLICT

Sexual conflict occurs whenever the genetic interests of females and males diverge (Chapman et al., 2003). Perhaps because the frequency of monogamy was initially overestimated, reproduction was traditionally viewed as a cooperative endeavour between the sexes. However, more recent studies building on seminal contributions by Trivers (1972) and Parker (1979) confirmed these authors' divergent assessment that the reproductive interests of the sexes are in most cases asymmetrical. Because the reproductive rates and optima differ between the sexes (Clutton-Brock & Parker, 1992; Bulmer & Parker, 2002), a dynamic conflict arises in which competitive males are selected to override the mate preferences by choosy females, leading to an evolutionary arms race between the sexes (Holland & Rice, 1998; Johnstone & Keller, 2000). Fuelled most importantly by elegant experiments with Drosophila (Rice, 1992, 1996; Holland & Rice, 1999), sexual conflict has since been demonstrated to have a genetic basis with a host of behavioural, morphological and physiological manifestations and measurable consequences for offspring fitness (Royle et al., 2002) and even speciation (Parker & Partridge, 1998; Gavrilets, 2000; but see Gage et al., 2002). Sexually antagonistic allelic evolution may even affect the expression of sexual dimorphism (Rice & Chippindale, 2001).

Among the many mechanisms of sexual conflict, sexual coercion by males in the form of physical attack, intimidation, harassment, and interruption of copulation and forced matings was first recognised and described as a behavioural expression of intersexual mating conflict in primates (Smuts & Smuts, 1993). These expressions of sexual coercion represent part of a general tendency for most male primates and mammals to dominate females in all contexts (Fedigan, 1982; Wright, 1993; Moore et al., 2001) and thus to control their reproduction. Similar mechanisms have since been described in other taxa (Clutton-Brock & Parker, 1995), but our knowledge of other male coercive strategies in primates has not greatly increased since then (but see Soltis et al., 1997, 2001; Fox, 2002). Variation in sexual dimorphism and intersexual relationships among primates therefore provides a rich and still under-used basis for comparative studies of the causes, contexts and mechanisms of sexual coercion, as well as the corresponding female counter-strategies.

The evolutionary dynamics of the sexual arms race have been illuminated by recent analyses of infanticide and its consequences. Infanticide of dependent young by strange males can be interpreted as an extreme form of male sexual coercion that occurs post-mating (Hrdy, 1979). Van Schaik et al. (this volume) suggest that where the male strategy of infanticide poses a threat to the survival of offspring, females have responded with various behavioural and physiological counter-strategies to confuse paternity (van Schaik et al., 1999; van Schaik, 2000a; van Schaik & Janson, 2000). Pre-copulatory male sexual coercion may have evolved in response to these counter-strategies to restrict female reproductive behaviour. However, there is evidence to suggest that effective sexual harassment is limited to sexually-dimorphic Old World primates, where the intersexual arms race may have led to further counter-adaptations, including longer mating periods, longer follicular phases with more unpredictable ovulation and exaggerated swelling (van Schaik et al., this volume).

CONCLUSIONS AND PROSPECTS

Sexual selection is a rich and complex field of research. The three main mechanisms – mating competition, mate choice and intersexual conflict – are clearly separate conceptually, but, as with natural and sexual selection, their relative contributions may be hard to disentangle. Thus, indirect female choice affects the way males compete for sexual access to

females, and hence the nature of male traits (Wiley & Poston, 1996). Likewise, traits favoured by intrasexual competition may also serve males to overcome female mating preferences, serving them in intersexual conflict (Smuts & Smuts, 1993; Clutton-Brock & Parker, 1995). Finally, female mate choice may be for male traits that serve in male–male competition, but males with these preferred traits may also reduce harassment by being effective in mate guarding (Wrangham, 1979; Mesnick, 1997).

Over the years, attempts to explain sexual dimorphism and social organisation in primates have shifted from an initial near-exclusive emphasis on male–male competition to an approach that also includes female choice (especially the indirect component: Wiley & Poston, 1996), and in particular mating conflict (Smuts & Smuts, 1993; Clutton-Brock & Parker, 1995; Gowaty, 1997, and this volume). Sexual conflict presumably pervades male and female adaptations in many traits and operates on several levels that remain to be elucidated by the next wave of studies of selection in relation to sex in primates. We believe primate studies can contribute to the development of sexual selection theory by studying real-life complexity, both through naturalistic observation and experimental manipulation of group composition. The strength of primate studies is their attention to behavioural detail, allowing them to unravel the underlying strategies and mechanisms.

The behavioural expression of sexual conflict is probably stronger in primates, and mammals more generally, than in birds, fish and many insects. We noted earlier that differences in life history and life style may be responsible for the different research foci in the different lineages (Kappeler & van Schaik, 2002). Nonetheless, this broad difference may explain why, for instance, handicapping ornaments are common among male birds but virtually absent among male mammals. This is not to say that primate females do not express mating preferences, but rather that their preferences concern the direct benefits of protection against harassment and infanticide, rather than the indirect ones provided by selection on male handicaps.

Mate choice by female primates remains woefully understudied, and such study may well produce surprising results. For instance, there are a few primate species, such as mandrills, in which males do show bright coloration (the general absence of colour vision in mammals makes primates a good taxon to contrast with birds). How are we to understand the phylogenetic distribution of these species (cf. Paul, 2002)? So far, there has been no convincing answer, and many factors are potentially involved, including the small number of males that females can sample among primates, and the long-term familiarity with most males, which might obviate the need for viability indicators. However, one interesting feature of mandrill society that has recently come to light is that groups are enormous, and that males tend to range separately from females much of the time (Abernethy *et al.*, 2002). Hence, infanticide is unlikely to be a problem, and females are freed from the need to select mates on the basis of potential to protect them and their dependent infants, allowing them to select for intrinsic viability indicators, i.e., ornaments and armaments. Infanticide is similarly absent among another taxon of diurnal primates: the small platyrrhines with extensive male care. Many of these have strikingly bright colours compared to other platyrrhines; they are not clearly sexually dimorphic because both males and females heavily invest in offspring and thus should carefully select their mates. Systematic study of male ornaments in primates is therefore badly needed. This study should also include consideration of sexually dichromatic species, whose phylogenetic distribution remains poorly understood.

Primatologists continue their detailed naturalistic observations of primate socioecology, and we suspect that increased attention to sexual selection may reveal further impacts on socioecology. Consider, for instance, sexual dimorphism in feeding time. Competition among males should lead to the minimisation of feeding time, so as to allow more time for monitoring the movements of females and rivals. Indeed, males in polygynous groups spend far less time feeding than females in the same groups, whereas males in pair-living species, who are under less intense pressure to minimise their feeding time, actually spend as much or even more time feeding than the females. Sexual selection may affect sex differences in diet as well. These are usually ascribed to natural selection in reproductive investment, favouring females selecting foods with higher nutrient densities and lower toxicity. However, sex differences in diet among chimpanzees also reflect male reproductive strategies, with males focusing on foods occurring in large patches, thus allowing them to remain gregarious (Pandolfi *et al.*, 2003). We therefore also expect more work in the future on the influence of male mating competition on details of their socioecology (see also Clutton-Brock, this volume).

In conclusion, our closest biological relatives continue to provide challenging and rewarding subjects for the study of various aspects of sexual selection. Such studies provide a natural basis against which studies of human sexual strategies can be interpreted. Similarly, primates constitute an interesting group for broader comparisons with other mammals because of their enormous diversity of social and mating systems. We suggest that the apparent under-representation of primate studies in key areas of contemporary evolutionary biology has interesting biological reasons that remain to be more fully explored. We hope that the present volume will serve as a stimulating and inspiring point of departure.

ACKNOWLEDGEMENTS

We thank Claudia Fichtel, Maria van Noordwijk, Dietmar Zinner, Eckhard Heymann, Manfred Eberle, Oliver Schülke, Julia Ostner and Roland Hilgartner for many stimulating discussions on sexual selection and/or constructive comments on this manuscript.

REFERENCES

Abbott, D. H. 1989. Social suppression of reproduction in primates. In *Comparative Socioecology*, ed. V. Standen & R. A. Foley. Oxford: Blackwell, pp. 285–304.

Abernethy, K. A., White, L. J. T. & Wickings, E. J. 2002. Hordes of mandrills (*Mandrillus sphinx*): extreme group size and seasonal male presence. *Journal of Zoology, London*, **258**, 131–7.

Alberts, S. C. & Altmann, J. 1995a. Balancing costs and opportunities: dispersal in male baboons. *American Naturalist*, **145**, 279–306.

1995b. Preparation and activation: determinants of age at reproductive maturity in male baboons. *Behavioral Ecology and Sociobiology*, **36**, 397–406.

Alberts, S. C., Altmann, J. & Wilson, M. L. 1996. Mate guarding constrains foraging activity of male baboons. *Animal Behaviour*, **51**, 1269–77.

Alexander, R. D., Hoogland, J. L., Howard, R. D., Noonan, K. M. & Sherman, P. W. 1979. Sexual dimorphism and breeding systems in pinnipeds, ungulates, primates, and humans. In *Evolutionary Biology and Human Social Behavior*, ed. N. A. Chagnon & W. Irons. North Scituate, MA: Duxbury, pp. 402–35.

Altmann, J. 1980. *Baboon Mothers and Infants*. Cambridge, MA: Harvard University Press.

Altmann, J., Alberts, S. C., Haines, S. A. *et al.* 1996. Behavior predicts genetic structure in a wild primate group. *Proceedings of the National Academy of Sciences, USA*, **93**, 5797–801.

Altmann, S. A. 1962. A field study of the sociobiology of the rhesus monkey, *Macaca mulatta. Annals of the New York Academy of Sciences*, **102**, 338–435.

Andelman, S. 1987. Evolution of concealed ovulation in vervet monkeys (*Cercopithecus aethiops*). *American Naturalist*, **129**, 785–99.

Andersson, M. 1982. Female choice selects for extreme tail length in a widowbird. *Nature*, **299**, 818–20.

1986. Evolution of condition-dependent sex ornaments and mating preferences: sexual selection based on viability differences. *Evolution*, **40**, 804–16.

1994. *Sexual Selection*. Princeton: Princeton University Press.

Andersson, M. & Iwasa, Y. 1996. Sexual selection. *Trends in Ecology and Evolution*, **11**, 53–8.

Arnold, W. & Dittami, J. 1997. Reproductive suppression in male alpine marmots. *Animal Behaviour*, **53**, 53–66.

Badyaev, A. V. 2002. Growing apart: an ontogenetic perspective on the evolution of sexual size dimorphism. *Trends in Ecology and Evolution*, **17**, 369–78.

Bateman, A. J. 1948. Intrasexual selection in *Drosophila*. *Heredity*, **2**, 349–68.

Berard, J. D., Nürnberg, P., Epplen, J. T. & Schmidtke, J. 1994. Alternative reproductive tactics and reproductive success in male rhesus macaques. *Behaviour*, **129**, 177–201.

Bercovitch, F. B. 1988. Coalitions, cooperation and reproductive tactics among adult male baboons. *Animal Behaviour*, **36**, 1198–209.

1989. Body size, sperm competition, and determinants of reproductive success in male savanna baboons. *Evolution*, **43**, 1507–21.

1991. Social stratification, social strategies, and reproductive success in primates. *Ethology and Sociobiology*, **12**, 315–33.

Berger, J. 1989. Female reproductive potential and its apparent evaluation by male mammals. *Journal of Mammalogy*, **70**, 347–58.

Borries, C. 2000. Male dispersal and mating season influxes in Hanuman langurs living in multi-male groups. In *Primate Males: Causes and Consequences of Variation in Group Composition*, ed. P. M. Kappeler. Cambridge: Cambridge University Press, pp. 146–58.

Bradbury, J. W. & Andersson, M. B. (eds.) 1987. *Sexual Selection: Testing the Alternatives*. New York, NY: Springer.

Brown, G. R. 2001. Sex-biased investment in nonhuman primates: can Trivers and Willard's theory be tested? *Animal Behaviour*, **61**, 683–94.

Bulmer, M. G. & Parker, G. A. 2002. The evolution of anisogamy: a game-theoretic approach. *Proceedings of the Royal Society of London, Series B*, **269**, 2381–8.

Burley, N. 1979. The evolution of concealed ovulation. *American Naturalist*, **114**, 835–58.

Buss, D. M. 1994. *The Evolution of Desire: Strategies of Human Mating*. New York, NY: Basic Books.

Buss, D. M. & Schmitt, D. P. 1993. Sexual strategies theory: an evolutionary perspective on human mating. *Psychological Reviews*, **100**, 204–32.

Campbell, B. (ed.) 1972. *Sexual Selection and the Descent of Man, 1871–1971*. Chicago, IL: Aldine.

Carranza, J., Garcia-Munoz, A. J. & De Dios Vargas, J. D. 1995. Experimental shifting from harem defence to territoriality in rutting red deer. *Animal Behaviour*, **49**, 551–4.

Chapman, T., Arnqvist, G., Bangham, J. & Rowe, L. 2003. Sexual conflict. *Trends in Ecology and Evolution*, **18**, 41–7.

Clark, A. B. 1978. Sex ratio and local resource competition in a prosimian primate. *Science*, **201**, 163–5.

Clutton-Brock, T. H. 1985. Size, sexual dimorphism, and polygyny in primates. In *Size and Scaling in Primate Biology*, ed. W. L. Jungers. New York: Plenum Press, pp. 51–60.

1989. Mammalian mating systems. *Proceedings of the Royal Society of London, Series B*, **236**, 339–72.

Clutton-Brock, T. H. & Iason, G. 1986. Sex ratio variation in mammals. *Quarterly Review of Biology*, **61**, 339–74.

Clutton-Brock, T. H. & Parker, G. A. 1992. Potential reproductive rates and the operation of sexual selection. *Quarterly Review of Biology*, **67**, 437–56.

1995. Sexual coercion in animal societies. *Animal Behaviour*, **49**, 1345–65.

Clutton-Brock, T. H., Harvey, P. H. & Rudder, B. 1977. Sexual dimorphism, socionomic sex ratio and body weight in primates. *Nature*, **269**, 797–800.

Clutton-Brock, T. H., Guinness, F. E. & Albon, S. D. 1982. *Red Deer: Behavior and Ecology of Two Sexes*. Chicago, IL: University of Chicago Press.

Colquhoun, I. C. 1987. Dominance and 'fall fever': the reproductive behavior of male brown lemurs (*Lemur fulvus*). *Canadian Reviews of Physical Anthropology*, **6**, 10–19.

Converse, L. J., Carlson, A. A., Ziegler, T. E. & Snowdon, C. T. 1995. Communication of ovulatory state to mates by female pygmy marmosets, *Cebuella pygmaea*. *Animal Behaviour*, **49**, 615–21.

Cords, M. 2000. The number of males in guenon groups. In *Primate Males: Causes and Consequences of Variation in Group Composition*, ed. P. M. Kappeler. Cambridge: Cambridge University Press, pp. 84–96.

Cowlishaw, G. & Dunbar, R. I. M. 1991. Dominance rank and mating success in male primates. *Animal Behaviour*, **41**, 1045–56.

Creel, S. R. & Macdonald, D. W. 1995. Sociality, group size, and reproductive suppression among carnivores. *Advances in the Study of Behavior*, **24**, 203–57.

Crook, J. H. 1972. Sexual selection, dimorphism, and social organization in the primates. In *Sexual Selection and the Descent of Man*, ed. B. G. Campbell. Chicago, IL: Aldine, pp. 180–230.

Crook, J. H. & Gartlan, J. C. 1966. Evolution of primate societies. *Nature*, **210**, 1200–3.

Cunningham, E. J. A. & Birkhead, T. R. 1998. Sex roles and sexual selection. *Animal Behaviour*, **56**, 1311–22.

Darwin, C. 1859. *On the Origin of Species*. London: Murray.

1871. *The Descent of Man and Selection in Relation to Sex*. London: Murray.

1876. Sexual selection in relation to monkeys. *Nature*, **15**, 18–19.

de Ruiter, J. R. & van Hooff, J. A. R. A. M. 1993. Male dominance rank and reproductive success in primate groups. *Primates*, **34**, 513–23.

de Ruiter, J. R., Scheffrahn, W., Trommelen, G. J. J. M. *et al.* 1992. Male social rank and reproductive success in wild long-tailed macaques: paternity exclusions by blood protein analysis and DNA fingerprinting. In *Paternity in Primates: Genetic Tests and Theory*, ed. R. D. Martin, A. F. Dixson & E. J. Wickings. Basel: Karger, pp. 175–90.

de Ruiter, J. R., van Hooff, J. A. R. A. M. & Scheffrahn, W. 1994. Social and genetic aspects of paternity in wild long-tailed macaques (*Macaca fascicularis*). *Behaviour*, **129**, 203–24.

de Waal, F. B. M. 1987. Tension regulation and nonreproductive functions of sex in captive bonobos (*Pan paniscus*). *National Geographic Research*, **3**, 318–35.

Dewsbury, D. A. 1982. Ejaculate cost and male choice. *American Naturalist*, **119**, 601–10.

Dewsbury, D. A. & Pierce, J. P., Jr. 1989. Copulatory patterns of primates as viewed in broad mammalian perspective. *American Journal of Primatology*, **17**, 51–72.

Dittus, W. P. J. 1998. Birth sex ratios in toque macaques and other mammals: integrating the effects of maternal condition and competition. *Behavioral Ecology and Sociobiology*, **44**, 149–60.

Dixson, A. F. 1995. Sexual selection and ejaculate frequencies in primates. *Folia Primatologica*, **64**, 146–52.

1998. *Primate Sexuality*. Oxford: Oxford University Press.

Domb, L. G. & Pagel, M. 2001. Sexual swellings advertise female quality in wild baboons. *Nature*, **410**, 204–6.

Drews, C. 1996. Contexts and patterns of injuries in free-ranging male baboons (*Papio cyncocephalus*). *Behaviour*, **133**, 443–74.

Drickamer, L. C., Gowaty, P. A. & Holmes, C. M. 2000. Free female mate choice in house mice affects reproductive success and offspring viability and performance. *Animal Behaviour*, **59**, 371–8.

Dunbar, R. I. M. 1984. *Reproductive Decisions: An Economic Analysis of Gelada Baboon Social Strategies*. Princeton, NJ: Princeton University Press.

2000. Male mating strategies: a modeling approach. In *Primate Males: Causes and Consequences of Variation in Group Composition*, ed. P. M. Kappeler. Cambridge: Cambridge University Press, pp. 259–68.

Eberhard, W. G. 1996. *Female Control: Sexual Selection by Cryptic Female Choice*. Princeton, NJ: Princeton University Press.

Eberle, M. & Kappeler, P. M. 2002. Mouse lemurs in space and time: a test of the socioecological model. *Behavioral Ecology and Sociobiology*, **51**, 131–9.

Emlen, S. T. & Oring, L. W. 1977. Ecology, sexual selection, and the evolution of mating systems. *Science*, **197**, 215–23.

Endler, J. A. & Basolo, A. L. 1998. Sensory ecology, receiver biases and sexual selection. *Trends in Ecology and Evolution*, **13**, 415–20.

Engh, A. L., Funk, S. M., van Horn, R. C. *et al.* 2002. Reproductive skew among males in a female-dominated mammalian society. *Behavioral Ecology*, **13**, 193–200.

Enomoto, T. 1981. Male aggression and the sexual behavior of Japanese monkeys. *Primates*, **22**, 15–23.

Fawcett, K. & Muhumuza, G. 2000. Death of a wild chimpanzee community member: possible outcome of intense sexual competition. *American Journal of Primatology*, **51**, 243–7.

Fedigan, L. 1982. *Primate Paradigms: Sex Roles and Social Bonds*. Montreal: Eden Press.

Fietz, J., Zischler, H., Schwiegk, C. *et al.* 2000. High rates of extra-pair young in the pair-living fat-tailed dwarf lemur, *Cheirogaleus medius*. *Behavioral Ecology and Sociobiology*, **49**, 8–17.

Forbes, L. 1997. The evolutionary biology of spontaneous abortion in humans. *Trends in Ecology and Evolution*, **12**, 446–50.

Ford, S. M. 1994. Evolution of sexual dimorphism in body weight in Platyrrhines. *American Journal of Primatology*, **34**, 221–44.

Fornasieri, I. & Roeder, J. J. 1992. Marking behaviour in two lemur species (*L. fulvus* and *L. macaco*): relation to social status, reproduction, aggression and social change. *Folia Primatologica*, **59**, 137–48.

Fox, E. A. 2002. Female tactics to reduce sexual harassment in the Sumatran orangutan (*Pongo pygmaeus abelii*). *Behavioral Ecology and Sociobiology*, **52**, 93–101.

Gage, M. J. G., Parker, G. A., Nylin, S. & Wiklund, C. 2002. Sexual selection and speciation in mammals, butterflies and spiders. *Proceedings of the Royal Society of London, Series B*, **269**, 2309–16.

Galdikas, B. M. F. 1985a. Adult male sociality and reproductive tactics among orangutans at Tanjung Puting. *Folia Primatologica*, **45**, 9–24.

1985b. Subadult male orangutan sociality and reproductive behavior at Tanjung Puting. *American Journal of Primatology*, **8**, 87–99.

Gangestad, S. W. & Thornhill, R. 1998. Menstrual cycle variation in women's preferences for the scent of symmetrical men. *Proceedings of the Royal Society of London, Series B*, **265**, 927–33.

Gavrilets, S. 2000. Rapid evolution of reproductive barriers driven by sexual conflict. *Nature*, **403**, 886–9.

Gemmell, N. J., Burg, T. M., Boyd, I. L. & Amos, W. 2001. Low reproductive success in territorial male Antarctic fur seals (*Arctocephalus gazelle*) suggests the existence of alternative mating strategies. *Molecular Ecology*, **10**, 451–60.

Gerald, M. 2001. Primate colour predicts social status and aggressive outcome. *Animal Behaviour*, **61**, 559–66.

Gerloff, U., Hartung, B., Fruth, B., Hohmann, G. & Tautz, D. 1999. Intracommunity relationships, dispersal pattern and paternity success in a wild living community of bonobos (*Pan paniscus*) determined from DNA analysis of faecal samples. *Proceedings of the Royal Society of London, Series B*, **266**, 1189–95.

Ghesquiere, J., Martin, R. D. & Newcombe, F. 1985. *Human Sexual Dimorphism*. London: Taylor & Francis.

Gibson, R. & Langen, T. 1996. How do animals choose their mates? *Trends in Ecology and Evolution*, **11**, 468–70.

Gittleman, J. L. & van Valkenburgh, B. 1997. Sexual dimorphism in the canines and skulls of carnivores: effects of size, phylogeny, and behavioral ecology. *Journal of Zoology, London*, **242**, 97–117.

Goldizen, A. W. 1987. Facultative polyandry and the role of infant-carrying in wild saddle-back tamarins (*Saguinus fuscicollis*). *Behavioral Ecology and Sociobiology*, **20**, 99–109.

Goldizen, A. W. & Terborgh, J. 1986. *Cooperative Polyandry and Helping Behavior in Saddle-backed Tamarins (Saguinus fuscicollis)*. Cambridge: Cambridge University Press.

Gomendio, M., Harcourt, A. H. & Roldan, E. R. S. 1998. Sperm competition in mammals. In *Sperm Competition and Sexual Selection*, ed. T. R. Birkhead & A. P. Møller. London: Academic Press, pp. 667–751.

Gowaty, P. A. 1997. Sexual dialectics, sexual selection, and variation in reproductive behavior. In *Feminism and Evolutionary Biology: Boundaries, Intersections, and Frontiers*, ed. P. A. Gowaty. New York, NY: Chapman & Hall, pp. 351–84.

Goy, R. 1979. Sexual compatibility in rhesus monkeys: predicting sexual performance of oppositely sexed pairs of adults. *Ciba Foundation Symposium*, **62**, 227–55.

Grob, B., Knapp, L. A., Martin, R. D. & Anzenberger, G. 1998. The major histocompatibility complex and mate choice: inbreeding avoidance and selection of good genes. *Experimental and Clinical Immunogenetics*, **15**, 119–29.

Gross, M. R. 1996. Alternative reproductive strategies and tactics: diversity within sexes. *Trends in Ecology and Evolution*, **11**, 92–7.

Hamilton, W. D. & Zuk, M. 1982. Heritable true fitness and bright birds: a role for parasites? *Science*, **218**, 384–7.

Hammerstein, P. & Parker, G. A. 1987. Sexual selection: games between the sexes. In *Sexual Selection: Testing the Alternatives*, ed. J. W. Bradbury & M. B. Andersson. New York, NY: John Wiley, pp. 119–42.

Harcourt, A. H. 1996. Sexual selection and sperm competition in primates: what are male genitalia good for? *Evolutionary Anthropology*, **5**, 121–9.

Harcourt, A. H. & Gardiner, J. 1994. Sexual selection and genital anatomy of male primates. *Proceedings of the Royal Society of London, Series B*, **255**, 47–53.

Harcourt, A. H., Harvey, P. H., Larson, S. G. & Short, R. V. 1981. Testis weight, body weight, and breeding system in primates. *Nature*, **293**, 55–7.

Harcourt, A. H., Purvis, A. & Liles, L. 1995. Sperm competition: mating system, not breeding season, affects testes size of primates. *Functional Ecology*, **9**, 468–76.

Harvey, P. H. & Harcourt, A. H. 1984. Sperm competition, testes size, and breeding system in primates. In *Sperm Competition and the Evolution of Animal Mating Systems*, ed. R. L. Smith. New York, NY: Academic Press, pp. 589–600.

Harvey, P. H., Kavanaugh, M. & Clutton-Brock, T. H. 1978. Sexual dimorphism in primate teeth. *Journal of Zoology, London*, **186**, 475–85.

Heistermann, M., Ziegler, T. Z., van Schaik, C. P. *et al.* 2001. Loss of oestrus, concealed ovulation and paternity confusion in free-ranging Hanuman langurs. *Proceedings of the Royal Society of London, Series B*, **268**, 2445–51.

Heske, E. J. & Ostfeld, R. S. 1990. Sexual dimorphism in size, relative size of testes, and mating system in North American voles. *Journal of Mammalogy*, **71**, 510–9.

Heymann, E. W. 1998. Sex differences in olfactory communication in a primate, the moustached tamarin, *Saguinus mystax* (Callitrichinae). *Behavioral Ecology and Sociobiology*, **40**, 37–45.

Hoelzel, A. R., Le Boeuf, B. J., Reiter, J. & Campagna, C. 1999. Alpha-male paternity in elephant seals. *Behavioral Ecology and Sociobiology*, **46**, 298–306.

Holland, B. & Rice, W. R. 1998. Perspective: chase-away sexual selection – antagonistic seduction versus resistance. *Evolution*, **52**, 1–7.

  1999. Experimental removal of sexual selection reverses intersexual antagonistic coevolution and removes a reproductive load. *Proceedings of the National Academy of Sciences, USA*, **96**, 5083–8.

Hosken, D. J. 1998. Testes mass in megachiropteran bats varies in accordance with sperm competition theory. *Behavioral Ecology and Sociobiology*, **44**, 169–78.

Hrdy, S. B. 1979. Infanticide among animals: a review, classification, and examination of the implications for the reproductive strategies of females. *Ethology and Sociobiology*, **1**, 13–40.

  1981. *The Woman that Never Evolved*. Cambridge, MA: Harvard University Press.

  1995. The primate origins of female sexuality, and their implications for the role of nonconceptive sex in the reproductive strategies of woman. *Human Evolution*, **10**, 131–44.

  2000. The optimal number of fathers. Evolution, demography, and history in the shaping of female mate preferences. In *Evolutionary Perspectives on Human Reproductive Behavior*, ed. D. LeCroy & P. Moller. New York, NY: The New York Academy of Sciences, pp. 75–96.

Hrdy, S. B. & Whitten, P. L. 1987. Patterning of sexual activity. In *Primate Societies*, ed. B. B. Smuts, D. L. Cheney, R. M. Seyfarth, R. W. Wrangham & T. T. Struhsaker. Chicago, IL: University of Chicago Press, pp. 370–84.

Ims, R. A. 1988. The potential for sexual selection in males: effect of sex ratio and spatiotemporal distribution of receptive females. *Evolutionary Ecology*, **2**, 338–52.

  1990. Mate detection success of male *Clethrionomys rufocanus* in relation to the spatial distribution of sexually receptive females. *Evolutionary Ecology*, **4**, 57–61.

Janson, C. H. 1984. Female choice and mating system of the brown capuchin monkey *Cebus apella* (Primates: Cebidae). *Zeitschrift für Tierpsychologie*, **65**, 177–200.

  1986. The mating system as a determinant of social evolution in capuchin monkeys (*Cebus*). In *Proceedings of the Xth International Congress of Primatology*, ed. J. Else & P. C. Lee. Cambridge: Cambridge University Press, pp. 169–79.

Jarman, P. J. 1983. Mating system and sexual dimorphism in large, terrestrial, mammalian herbivores. *Biological Reviews*, **58**, 485–520.

Johnstone, R. A. 1995. Sexual selection, honest advertisement and the handicap principle: reviewing the evidence. *Biological Reviews*, **70**, 1–65.

Johnstone, R. A. & Keller, L. 2000. How males can gain by harming their mates: sexual conflict, seminal toxins, and the cost of mating. *American Naturalist*, **156**, 368–77.

Johnstone, R. A., Woodroffe, R., Cant, M. A. & Wright, J. 1999. Reproductive skew in multimember groups. *American Naturalist*, **153**, 315–31.

Jolly, A. 1967. Breeding synchrony in wild *Lemur catta*. In *Social Communication among Primates*, ed. S. A. Altman. Chicago, IL: University of Chicago Press, pp. 3–14.

  1998. Pair-bonding, female aggression and the evolution of lemur societies. *Folia Primatologica*, **69**, suppl. 1, 1–13.

Jolly, A., Caless, S., Cavigelli, S. *et al.* 2000. Infant killing, wounding and predation in *Eulemur* and *Lemur*. *International Journal of Primatology*, **21**, 21–40.

Jones, K. C. 1983. Inter-troop transfer of *Lemur catta* males at Berenty, Madagascar. *Folia Primatologica*, **40**, 145–60.

Kappeler, P. M. 1990a. The evolution of sexual size dimorphism in prosimian primates. *American Journal of Primatology*, **21**, 201–14.

  1990b. Female dominance in *Lemur catta*: more than just female feeding priority? *Folia Primatologica*, **55**, 92–5.

  1990c. Social status and scent marking behaviour in *Lemur catta*. *Animal Behaviour*, **40**, 774–6.

  1991. Patterns of sexual dimorphism in body weight among prosimian primates. *Folia Primatologica*, **57**, 132–46.

  1993. Female dominance in primates and other mammals. In *Perspectives in Ethology*. Vol. 10: *Behaviour and Evolution*, ed. P. P. G. Bateson, P. H. Klopfer & N. S. Thompson. New York, NY: Plenum Press, pp. 143–58.

  1996. Intrasexual selection and phylogenetic constraints in the evolution of sexual canine dimorphism in strepsirrhine primates. *Journal of Evolutionary Biology*, 9, 43–65.

1997a. Intrasexual selection and testis size in strepsirrhine primates. *Behavioral Ecology*, 8, 10–19.

1997b. Intrasexual selection in *Mirza coquereli*: evidence for scramble competition polygyny in a solitary primate. *Behavioral Ecology and Sociobiology*, 41, 115–28.

1998. To whom it may concern: transmission and function of chemical signals in *Lemur catta*. *Behavioral Ecology and Sociobiology*, 42, 411–21.

1999. Primate socioecology: new insights from males. *Naturwissenschaften*, 86, 18–29.

2000. Primate males: history and theory. In *Primate Males: Causes and Consequences of Variation in Group Composition*, ed. P. M. Kappeler. Cambridge: Cambridge University Press, pp. 3–7.

Kappeler, P. M. & van Schaik, C. P. 2002. Evolution of primate social systems. *International Journal of Primatology*, 23, 707–40.

Kappeler, P. M., Pereira, M. E. & van Schaik, C. P. 2003. Primate socioecology and life history. In *Primate Socioecology and Life History*, ed. P. M. Kappeler & M. E. Pereira, in press. Chicago, IL: University of Chicago Press, pp. 1–23.

Kay, R. F., Plavcan, J. M., Glander, K. E. & Wright, P. C. 1988. Sexual selection and canine dimorphism in New World monkeys. *American Journal of Physical Anthropology*, 77, 385–97.

Keddy-Hector, A. C. 1992. Mate choice in non-human primates. *American Zoologist*, 32, 62–70.

Kenagy, G. J. & Trombulak, S. C. 1986. Size and function of mammalian testes in relation to body size. *Journal of Mammalogy*, 67, 1–22.

Kirkpatrick, M. 1982. Sexual selection and the evolution of female choice. *Evolution*, 36, 1–12.

1987. Sexual selection by female choice in polygynous animals. *Annual Reviews in Ecology and Systematics*, 18, 43–70.

Kokko, H., Brooks, R., McNamara, J. M. & Houston, A. L. 2002. The sexual selection continuum. *Proceedings of the Royal Society of London, Series B*, 269, 1331–40.

Koprowski, J. L. 1993. Alternative reproductive tactics in male eastern grey squirrels: 'making the best of a bad job'. *Behavioral Ecology and Sociobiology*, 4, 165–71.

Kraus, C., Heistermann, M. & Kappeler, P. M. 1999. Physiological suppression of sexual function of subordinate males: a subtle form of intrasexual competition among male sifakas (*Propithecus verreauxi*)? *Physiology and Behavior*, 66, 855–61.

Krebs, J. R. & Davies, N. B. 1992. *Behavioural Ecology*. Sunderland: Sinauer.

Kubzdela, K., Richard, A. F. & Pereira, M. E. 1992. Social relations in semi-free-ranging sifakas (*Propithecus verreauxi coquereli*) and the question of female dominance. *American Journal of Primatology*, 28, 139–45.

Labov, J. 1981. Pregnancy blocking in rodents: adaptive advantages for females. *American Naturalist*, 118, 361–71.

Launhardt, K., Borries, C., Hardt, C., Epplen, J. T. & Winkler, P. 2001. Paternity analysis of alternative reproductive routes among the langurs (*Semnopithecus entellus*) of Ramnagar. *Animal Behaviour*, 61, 53–64.

Lee, P. C. 1996. The meaning of weaning: growth, lactation, and life history. *Evolutionary Anthropology*, 5, 87–96.

Leigh, S. R. 1995. Socioecology and the ontogeny of sexual size dimorphism in anthropoid primates. *American Journal of Physical Anthropology*, 97, 339–56.

Leigh, S. R. & Shea, B. T. 1995. Ontogeny and the evolution of adult body size dimorphism in apes. *American Journal of Primatology*, 36, 37–60.

Leigh, S. R. & Terranova, C. 1998. Comparative perspectives on bimaturism, ontogeny, and dimorphism in lemurid primates. *International Journal of Primatology*, 19, 723–49.

Lindenfors, P. 2002. Sexually antagonistic selection on primate size. *Journal of Evolutionary Biology*, 15, 595–607.

Lindenfors, P. & Tullberg, B. S. 1998. Phylogenetic analyses of primate size evolution: the consequences of sexual selection. *Biological Journal of the Linnean Society*, 64, 413–47.

Lindenfors, P., Tullberg, B. S. & Biuw, M. 2002. Phylogenetic analyses of sexual selection and sexual size dimorphism in pinnipeds. *Behavioral Ecology and Sociobiology*, 52, 188–93.

Linn, G. S., Mase, D., Lafrançois, D., O'Keeffe, R. T. & Lifshitz, K. 1995. Social and menstrual cycle phase influences on the behavior of group-housed *Cebus apella*. *American Journal of Primatology*, 35, 41–57.

Maestripieri, D. & Kappeler, P. M. 2002. Evolutionary theory and primate behavior. *International Journal of Primatology*, 23, 703–5.

Maggioncalda, A. N., Sapolsky, R. M. & Czekala, N. M. 1999. Reproductive hormone profiles in captive male orangutans: implications for understanding developmental arrest. *American Journal of Physical Anthropology*, 109, 19–32.

Maggioncalda, A. N., Czekala, N. M. & Sapolsky, R. M. 2000. Growth hormone and thyroid stimulating hormone concentrations in captive male orangutans: implications for understanding developmental arrest. *American Journal of Primatology*, **50**, 67–76.

Mahady, S. J. & Wolff, J. O. 2002. A field test of the Bruce effect in the monogamous prairie vole (*Microtus ochrogaster*). *Behavioral Ecology and Sociobiology*, **52**, 31–7.

Manning, J. T., Trivers, R. L., Singh, D. & Thornhill, R. 1999. The mystery of female beauty. *Nature*, **399**, 214–15.

Manson, J. H. 1994. Male aggression: a cost of female mate choice in Cayo Santiago rhesus macaques. *Animal Behaviour*, **48**, 473–5.

McComb, K. E. 1991. Female choice for high roaring rates in red deer, *Cervus elaphus*. *Animal Behaviour*, **41**, 79–88.

McCracken, G. F. & Bradbury, J. W. 1977. Paternity and genetic heterogenety in the polygynous bat, *Phyllostomus hastatus*. *Science*, **198**, 303–6.

McGraw, W. S., Plavcan, J. M. & Adachi-Kanazawa, K. 2002. Adult female *Cercopithecus diana* employ canine teeth to kill another adult female *C. diana*. *International Journal of Primatology*, **23**, 1301–8.

Mesnick, S. L. 1997. Sexual alliances: evidence and evolutionary implications. In *Feminism and Evolutionary Biology: Boundaries, Intersections, and Frontiers*, ed. P. A. Gowaty. New York, NY: Chapman & Hall, pp. 207–60.

Michener, G. R. & McLean, I. G. 1996. Reproductive behaviour and operational sex ratio in Richardson's ground squirrels. *Animal Behaviour*, **52**, 743–58.

Mitani, J. C., Gros-Louis, J. & Manson, J. H. 1996a. Number of males in primate groups: comparative tests of competing hypotheses. *American Journal of Primatology*, **38**, 315–32.

Mitani, J. C., Gros-Louis, J. & Richards, A. F. 1996b. Sexual dimorphism, the operational sex ratio, and the intensity of male competition in polygynous primates. *American Naturalist*, **147**, 966–80.

Møller, A. P. 1989. Ejaculate quality, testes size and sperm production in mammals. *Functional Ecology*, **3**, 91–6.

1992. Patterns of fluctuating asymmetry in weapons: evidence for reliable signalling of quality in beetle horns and bird spurs. *Proceedings of the Royal Society of London, Series B*, **248**, 199–206.

1997. Immune defence, extra-pair paternity, and sexual selection in birds. *Proceedings of the Royal Society of London, Series B*, **264**, 561–6.

2000. Male parental care, female reproductive success, and extra-pair paternity. *Behavioral Ecology*, **11**, 161–8.

Møller, A. P. & Birkhead, T. R. 1989. Copulation behaviour in mammals: evidence that sperm competition is widespread. *Biological Journal of the Linnean Society*, **38**, 119–31.

Møller, A. P. & Ninni, P. 1998. Sperm competition and sexual selection: a meta-analysis of paternity studies of birds. *Behavioral Ecology and Sociobiology*, **43**, 345–58.

Møller, A. P. & Pomiankowski, A. 1993. Why have birds got multiple sexual ornaments? *Behavioral Ecology and Sociobiology*, **32**, 167–76.

Moore, A. J., Gowaty, P. A., Wallin, W. G. & Moore, P. J. 2001. Sexual conflict and the evolution of female mate choice and male social dominance. *Proceedings of the Royal Society of London, Series B*, **268**, 517–23.

Noë, R. & Sluijter, A. A. 1990. Reproductive tactics of male savanna baboons. *Behaviour*, **113**, 117–70.

Nunn, C. L. 1999. The number of males in primate social groups: a comparative test of the socioecological model. *Behavioral Ecology and Sociobiology*, **46**, 1–13.

Nunn, C. L., van Schaik, C. P. & Zinner, D. 2001. Do exaggerated sexual swellings function in female mating competition in primates? A comparative test of the reliable indicator hypothesis. *Behavioral Ecology*, **12**, 646–54.

O'Donald, P. 1980. Sexual selection by female choice in a monogamous bird: Darwin's theory corroborated. *Heredity*, **45**, 201–17.

Olupot, W. & Waser, P. M. 2001. Correlates of intergroup transfer in male grey-cheeked mangabeys. *International Journal of Primatology*, **22**, 169–87.

Ostner, J. & Kappeler, P. M. 1999. Central males instead of multiple pairs in redfronted lemurs, *Eulemur fulvus rufus* (Primates, Lemuridae)? *Animal Behaviour*, **58**, 1069–78.

Overdorff, D. J. 1998. Are *Eulemur* species pair-bonded? Social organization and mating strategies in *Eulemur fulvus rufus* from 1988–1995 in southeast Madagascar. *American Journal of Physical Anthropology*, **105**, 153–66.

Packer, C. 1979. Male dominance and reproductive activity in *Papio anubis*. *Animal Behaviour*, **27**, 37–45.

Pagel, M. 1994. The evolution of conspicuous oestrus advertisement in Old World monkeys. *Animal Behaviour*, **47**, 1333–41.

Palombit, R. A. 1994. Extra-pair copulations in a monogamous ape. *Animal Behaviour*, **47**, 721–3.

Palombit, R. A., Seyfarth, R. M. & Cheney, D. L. 1997. The adaptive value of 'friendships' to female baboons: experimental and observational evidence. *Animal Behaviour*, **54**, 599–614.

Pandolfi, S. S., van Schaik, C. P. & Pusey, A. E. 2003. Sex differences in termite fishing among gombe chimpanzees are due to socioecological factors. In *Animal Social Complexity: Intelligence, Culture, and Individualized Societies*, ed. F. B. M. de Waal & P. L. Tyack. Cambridge, MA: Harvard University Press, pp. 414–18.

Panhuis, T. M., Butlin, R. K., Zuk, M. & Tregenza, T. 2001. Sexual selection and speciation. *Trends in Ecology and Evolution*, **16**, 364–71.

Parish, A. R. 1994. Sex and food control in the 'uncommon chimpanzee': how bonobo females overcome a phylogenetic legacy of male dominance. *Ethology and Sociobiology*, **15**, 157–79.

Parker, G. A. 1979. Sexual selection and sexual conflict. In *Sexual Selection and Reproductive Conflict in Insects*, ed. M. S. Blum & N. A. Blum. New York, NY: Academic Press, pp. 123–66.

Parker, G. A. & Partridge, L. 1998. Sexual conflict and speciation. *Philosophical Transactions of the Royal Society of London, Series B*, **353**, 261–74.

Paul, A. 2002. Sexual selection and mate choice. *International Journal of Primatology*, **23**, 877–904.

Pereira, M. E. 1983. Abortion following the immigration of an adult male baboon (*Papio cynocephalus*). *American Journal of Primatology*, **4**, 93–8.

Pereira, M. E. & Leigh, S. R. 2003. Modes of primate development. In *Primate Life History and Socioecology*, ed. P. M. Kappeler & M. E. Pereira, in press. Chicago, IL: University of Chicago Press, pp. 146–76.

Pereira, M. E. & McGlynn, C. A. 1997. Special relationships instead of female dominance for redfronted lemurs, *Eulemur fulvus rufus*. *American Journal of Primatology*, **43**, 239–58.

Pereira, M. E. & Weiss, M. L. 1991. Female mate choice, male migration, and the threat of infanticide in ringtailed lemurs. *Behavioral Ecology and Sociobiology*, **28**, 141–52.

Pereira, M. E., Kaufman, R., Kappeler, P. M. & Overdorff, D. J. 1990. Female dominance does not characterize all of the Lemuridae. *Folia Primatologica*, **55**, 96–103.

Perret, M. 1992. Environmental and social determinants of sexual function in the male lesser mouse lemur (*Microcebus murinus*). *Folia Primatologica*, **59**, 1–25.

Perrin, N. & Mazalov, V. 2000. Local competition, inbreeding, and the evolution of sex-biased dispersal. *American Naturalist*, **155**, 116–27.

Petrie, M. & Kempenaers, B. 1998. Extra-pair paternity in birds: explaining variation between species and populations. *Trends in Ecology and Evolution*, **13**, 52–8.

Plavcan, J. M. 2001. Sexual dimorphism in primate evolution. *Yearbook of Physical Anthropology*, **44**, 25–53.

Plavcan, J. M. & van Schaik, C. P. 1992. Intrasexual competition and canine dimorphism in primates. *American Journal of Physical Anthropology*, **87**, 461–77.

1997. Interpreting hominid behavior on the basis of sexual dimorphism. *Journal of Human Evolution*, **32**, 345–74.

Plavcan, J. M., van Schaik, C. P. & Kappeler, P. M. 1995. Competition, coalitions and canine size in female primates. *Journal of Human Evolution*, **28**, 245–76.

Pope, T. R. 1990. The reproductive consequences of male cooperation in the red howler monkey: paternity exclusion in multi-male and single-male troops using genetic markers. *Behavioral Ecology and Sociobiology*, **27**, 439–46.

Preston, B. T., Stevenson, I. R., Pemberton, J. M. & Wilson, K. 2001. Dominant rams lose out by sperm depletion. A waning success in siring counters a ram's high score in competition for ewes. *Nature*, **409**, 681–2.

Radespiel, U., Ehresmann, P. & Zimmermann, E. 2001. Contest versus scramble competition for mates: the composition and spatial structure of a population of gray mouse lemurs (*Microcebus murinus*) in North-west Madagascar. *Primates*, **42**, 207–20.

Rice, W. R. 1992. Sexually antagonistic genes – experimental evidence. *Science*, **256**, 1436–9.

1996. Sexually antagonistic male adaptation triggered by experimental arrest of female evolution. *Nature*, **381**, 232–4.

Rice, W. R. & Chippindale, A. K. 2001. Intersexual ontogenetic conflict. *Journal of Evolutionary Biology*, **14**, 685–93.

Richard, A. F. 1974. Patterns of mating in *Propithecus verreauxi verreauxi*. In *Prosimian Biology*, ed. R. D. Martin, G. A. Doyle & A. C. Walker. London: Duckworth, pp. 49–74.

1987. Malagasy prosimians: female dominance. In *Primate Societies*, ed. B. B. Smuts, D. L. Cheney, R. M. Seyfarth, R. W. Wrangham & T. T. Struhsaker. Chicago, IL: University of Chicago Press, pp. 25–33.

1992. Aggressive competition between males, female-controlled polygyny and sexual monomorphism in a Malagasy primate, *Propithecus verreauxi*. *Journal of Human Evolution*, **22**, 395–406.

Richard, A. F., Dewar, R. E., Schwartz, M. & Ratsirarson, J. 2002. Life in the slow lane? Demography and life histories of male and female sifaka (*Propithecus verreauxi verreauxi*). *Journal of Zoology, London*, **256**, 421–36.

Robbins, M. M. 1999. Male mating patterns in wild multimale mountain gorilla groups. *Animal Behaviour*, **57**, 1013–20.

Rodman, P. S. & Mitani, J. C. 1987. Orangutans: sexual dimorphism in a solitary species. In *Primate Societies*, ed. B. B. Smuts, D. L. Cheney, R. M. Seyfarth, R. W.

Wrangham & T. T. Struhsaker. Chicago, IL: University of Chicago Press, pp. 146–54.

Rose, R. W., Nevison, C. M. & Dixson, A. F. 1997. Testes weight, body weight and mating systems in marsupials and monotremes. *Journal of Zoology, London*, **243**, 523–31.

Ross, C. 1998. Primate life histories. *Evolutionary Anthropology*, **6**, 54–63.

Royle, N. J., Hartley, I. R. & Parker, G. A. 2002. Sexual conflict reduces offspring fitness in zebra finches. *Nature*, **416**, 733–6.

Ryan, M. J. 1998. Sexual selection, receiver biases, and the evolution of sex differences. *Science*, **281**, 1999–2003.

Ryan, M. J. & Keddy-Hector, A. 1992. Directional patterns of female mate choice and the role of sensory biases. *American Naturalist*, **139**, S4–35.

Samuels, A., Silk, J. B. & Rodman, P. S. 1984. Changes in the dominance rank and reproductive behaviour of male bonnet macaques (*Macaca radiata*). *Animal Behaviour*, **32**, 994–1003.

Sapolsky, R. 1985. Stress-induced suppression of testicular function in the wild baboon: role of glucocorticoids. *Endocrinology*, **116**, 2273–8.

Sauther, M. L. 1991. Reproductive behavior of free-ranging *Lemur catta* at Beza Mahafaly Special Reserve, Madagascar. *American Journal of Physical Anthropology*, **84**, 463–77.

Schilling, A., Perret, M. & Predine, J. 1984. Sexual inhibition in a prosimian primate: a pheromone-like effect. *Journal of Endocrinology*, **102**, 143–51.

Schmid, J. & Kappeler, P. M. 1998. Fluctuating sexual dimorphism and differential hibernation by sex in a primate, the gray mouse lemur (*Microcebus murinus*). *Behavioral Ecology and Sociobiology*, **43**, 125–32.

Schuermann, C. L. & van Hooff, J. A. R. A. M. 1986. Reproductive strategies of the orang-utan: new data and a reconsideration of existing sociosexual models. *International Journal of Primatology*, **7**, 265–87.

Schultz, A. H. 1938. The relative weight of the testes in primates. *Anatomical Record*, **72**, 387–94.

Schwagmeyer, P. L. 1979. The Bruce effect: an evaluation of male/female advantages. *American Naturalist*, **114**, 932–9.

Schwagmeyer, P. L. & Woonter, S. J. 1986. Scramble competition polygyny in thirteen-lined ground squirrels: the relative contributions of overt conflict and competitive mate searching. *Behavioral Ecology and Sociobiology*, **19**, 359–64.

Schwagmeyer, P. L., Parker, G. A. & Mock, D. W. 1998. Information asymmetries among males: implications for fertilization success in the thirteen-lined ground squirrel. *Proceedings of the Royal Society of London, Series B*, **265**, 1861–5.

Searcy, W. A. 1979. Female choice of males: a general model for birds and its application to red-winged blackbirds (*Agelaius phoeniceus*). *American Naturalist*, **114**, 77–100.

Setchell, J. M. 2003. The evolution of alternative reproductive morphs in male primates. In *Sexual Selection and Reproductive Competition in Primates: New Perspectives and Directions*, ed. C. Jones, in press. American Society of Primatologists Special Topics in Primatology.

Setchell, J. M. & Dixson, A. F. 2001a. Arrested development of secondary sexual adornments in subordinate adult male mandrills (*Mandrillus sphinx*). *American Journal of Physical Anthropology*, **115**, 245–52.

2001b. Changes in the secondary sexual adornments of male mandrills (*Mandrillus sphinx*) are associated with gain and loss of alpha status. *Hormones and Behavior*, **39**, 177–84.

Setchell, J. M. & Kappeler, P. M. 2004. Selection in relation to sex in primates. *Advances in the Study of Behavior*, in press.

Setchell, J. M., Lee, P. C., Wickings, E. J. & Dixson, A. F. 2001. Growth and ontogeny of sexual size dimorphism in the mandrill (*Mandrillus sphinx*). *American Journal of Physical Anthropology*, **115**, 349–60.

Shively, C. & Smith, D. G. 1985. Social status and reproductive success of male *Macaca fascicularis*. *American Journal of Primatology*, **9**, 129–35.

Short, R. V. 1979. Sexual selection and its component parts, somatic and genital selection, as illustrated by man and the great apes. *Advances in the Study of Behavior*, **9**, 131–58.

Silk, J. 1984. Local resource competition and the evolution of male-biased sex ratios. *Journal of Theoretical Biology*, **108**, 203–13.

Sillen-Tullberg, B. & Møller, A. P. 1993. The relationship between concealed ovulation and mating systems in anthropoid primates: a phylogenetic analysis. *American Naturalist*, **141**, 1–25.

Small, M. F. 1989. Female choice in nonhuman primates. *Yearbook of Physical Anthropology*, **32**, 103–27.

Smith, R. & Leigh, S. R. 1998. Sexual dimorphism in primate neonatal body mass. *Journal of Human Evolution*, **34**, 173–201.

Smuts, B. B. 1985. *Sex and Friendship in Baboons*. Hawthorne, CA: Aldine.

1987a. Gender, aggression, and influence. In *Primate Societies*, ed. B. B. Smuts, D. L. Cheney, R. M. Seyfarth, R. W. Wrangham & T. T. Struhsaker. Chicago, IL: University of Chicago Press, pp. 400–12.

1987b. Sexual competition and mate choice. In *Primate Societies*, ed. B. B. Smuts, D. L. Cheney, R. M. Seyfarth, R. W. Wrangham & T. T. Struhsaker. Chicago, IL: University of Chicago Press, pp. 385–99.

Smuts, B. B. & Smuts, R. W. 1993. Male aggression and sexual coercion of females in nonhuman primates and other mammals: evidence and theoretical implications. *Advances in the Study of Behavior*, **22**, 1–63.

Soltis, J., Mitsunaga, F., Shimizu, K., Yanagihara, Y. & Nozaki, M. 1997. Sexual selection in Japanese macaques. I. Female mate choice or male sexual coercion? *Animal Behaviour*, **54**, 725–36.

Soltis, J., Thomsen, R. & Takenaka, O. 2001. The interaction of male and female reproductive strategies and paternity in wild Japanese macaques, *Macaca fuscata*. *Animal Behaviour*, **62**, 485–94.

Sommer, V. & Reichard, U. 2000. Rethinking monogamy: the gibbon case. In *Primate Males: Causes and Consequences of Variation in Group Composition*, ed. P. M. Kappeler. Cambridge: Cambridge University Press, pp. 159–68.

Sterck, E. H. M. & Korstjens, A. 2000. Female dispersal and infanticide avoidance in primates. In *Infanticide by Males and Its Implications*, ed. C. P. van Schaik & C. H. Janson. Cambridge: Cambridge University Press, pp. 293–321.

Stockley, P., Searle, J. B., Macdonald, D. W. & Jones, C. S. 1994. Alternative reproductive tactics in male common shrews: relationships between mate-searching behaviour, sperm production, and reproductive success as revealed by DNA fingerprinting. *Behavioral Ecology and Sociobiology*, **34**, 71–8.

Sussman, R. W. 1992. Male life history and intergroup mobility among ringtailed lemurs (*Lemur catta*). *International Journal of Primatology*, **13**, 395–414.

Swaddle, J. P. & Cuthill, I. C. 1994. Preference for symmetric males by female zebra finches. *Nature*, **367**, 165–6.

Thornhill, R. & Alcock, J. 1983. *The Evolution of Insect Mating Systems*. Cambridge, MA: Harvard University Press.

Thornhill, R. & Gangestad, S. 1996. The evolution of human sexuality. *Trends in Ecology and Evolution*, **11**, 98–102.

Tregenza, T. & Wedell, N. 2000. Genetic compatibility, mate choice and patterns of parentage: invited review. *Molecular Ecology*, **9**, 1013–27.

Trivers, R. L. 1972. Parental investment and sexual selection. In *Sexual Selection and the Descent of Man*, ed. B. Campbell. Chicago, IL: Aldine, pp. 136–79.

Utami, S. A., Goossens, B., Bruford, M. W., de Ruiter, J. & van Hooff, J. A. R. A. M. 2002. Male bimaturism and reproductive success in Sumatran orang-utans. *Behavioral Ecology*, **13**, 643–52.

van Noordwijk, M. A. & van Schaik, C. P. 1985. Male migration and rank acquisition in wild long-tailed macaques (*Macaca fascicularis*). *Animal Behaviour*, **33**, 849–61.

1988. Male careers in Sumatran long-tailed macaques (*Macaca fascicularis*). *Behaviour*, **107**, 24–43.

1999. The effects of dominance rank and group size on female lifetime reproductive success in wild long-tailed macaques, *Macaca fascicularis*. *Primates*, **40**, 105–30.

2000. Reproductive patterns in Eutherian mammals: adaptations against infanticide? In *Infanticide by Males and Its Implications*, ed. C. H. van Schaik & C. H. Janson. Cambridge: Cambridge University Press, pp. 322–60.

2001. Career moves: transfer and rank challenge decisions by male long-tailed macaques. *Behaviour*, **138**, 359–95.

van Schaik, C. P. 1996. Social evolution in primates: the role of ecological factors and male behaviour. *Proceedings of the British Academy*, **88**, 9–31.

2000a. Infanticide by male primates: the sexual selection hypothesis revisited. In *Infanticide by Males and Its Implications*, ed. C. P. van Schaik & C. H. Janson. Cambridge: Cambridge University Press, pp. 27–60.

2000b. Social counterstrategies against male infanticide in primates and other mammals. In *Primate Males: Causes and Consequences of Variation in Group Composition*, ed. P. M. Kappeler. Cambridge: Cambridge University Press, pp. 34–52.

van Schaik, C. P. & Aureli, F. 2000. The natural history of valuable relationships in primates. In *Natural Conflict Resolution*, ed. F. Aureli & F. B. M. de Waal. Berkeley, Los Angeles, London: California Press, pp. 307–33.

van Schaik, C. P. & Hrdy, S. B. 1991. Intensity of local resource competition shapes the relationship between maternal rank and sex ratios at birth in cercopithecine primates. *American Naturalist*, **138**, 1555–62.

van Schaik, C. P. & Janson, C. H. 2000. *Infanticide by Males and Its Implications*. Cambridge: Cambridge University Press.

van Schaik, C. P. & Kappeler, P. M. 1997. Infanticide risk and the evolution of male–female association in primates. *Proceedings of the Royal Society of London, Series B*, **264**, 1687–94.

2003. The evolution of pair-living in primates. In *Monogamy: Partnerships in Birds, Humans and Other Mammals*, ed. U. Reichard & C. Boesch. Cambridge: Cambridge University Press, pp. 59–80.

van Schaik, C. P., van Noordwijk, M. A. & Nunn, C. L. 1999. Sex and social evolution in primates. In *Comparative Primate Socioecology*, ed. P. C. Lee. Cambridge: Cambridge University Press, pp. 204–40.

Vigilant, L., Hofreiter, M., Siedel, H. & Boesch, C. 2001. Paternity and relatedness in wild chimpanzee communities. *Proceedings of the National Academy of Sciences, USA*, **98**, 12890–5.

von Holst, D. 1998. The concept of stress and its relevance for animal behavior. *Advances in the Study of Behavior*, **27**, 1–131.

Wallen, K. 2001. Sex and contest: hormones and primate sexual motivation. *Hormones and Behavior*, **40**, 339–57.

Wallen, K., Winston, L. A., Gaventa, S., Davis-DaSilva, M. & Collins, D. C. 1984. Periovulatory changes in female sexual behavior and patterns of steroid secretion in group-living rhesus monkeys. *Hormones and Behavior*, **18**, 431–50.

Wasser, S. K. 1983. Reproductive competition and cooperation among female yellow baboons. In *Social Behavior of Female Vertebrates*, ed. S. K. Wasser. New York, NY: Academic Press, pp. 349–90.

Watts, D. P. 1992. Social relationships of immigrant and resident female mountain gorillas. I. Male–female relationships. *American Journal of Primatology*, **28**, 159–82.

  1998. Coalitionary mate guarding by male chimpanzees at Ngogo, Kibale National Park, Uganda. *Behavioral Ecology and Sociobiology*, **44**, 43–56.

  2000. Causes and consequences of variation in male mountain gorilla life histories and and group membership. In *Primate Males: Causes and Consequences of Variation in Group Composition*, ed. P. M. Kappeler. Cambridge: Cambridge University Press, pp. 169–79.

Waynforth, D. 2001. Mate choice tradeoffs and women's preference for physically attractive men. *Human Nature*, **12**, 207–19.

Weckerly, F. W. 1998. Sexual-size dimorphism: influence of mass and mating systems in the most dimorphic mammals. *Journal of Mammalogy*, **79**, 33–52.

Wedell, N., Gage, M. J. G. & Parker, G. A. 2002. Sperm competition, male prudence and sperm-limited females. *Trends in Ecology and Evolution*, **17**, 313–20.

West, P. M. & Packer, C. 2002. Sexual selection, temperature, and the lion's mane. *Science*, **297**, 1339–43.

Wich, S. A. & Nunn, C. L. 2002. Do male 'long-distance calls' function in mate defense? A comparative study of long-distance calls in primates. *Behavioral Ecology and Sociobiology*, **52**, 474–84.

Widemo, F. & Saether, S. A. 1999. Beauty is in the eye of the beholder: causes and consequences of variation in mating preferences. *Trends in Ecology and Evolution*, **14**, 26–31.

Wiley, R. H. & Poston, J. 1996. Indirect mate choice, competition for mates, and coevolution of the sexes. *Evolution*, **50**, 1371–81.

Williams, G. C. 1966. *Adaptation and Natural Selection*. Princeton, NJ: Princeton University Press.

Wrangham, R. W. 1979. On the evolution of ape social systems. *Social Science Information*, **18**, 334–68.

  1993. The evolution of sexuality in chimpanzees and bonobos. *Human Nature*, **4**, 447–80.

Wright, P. C. 1993. Variations in male-female dominance and offspring care in non-human primates. In *Sex and Gender Hierarchies*, ed. B. D. Miller. Cambridge: Cambridge University Press, pp. 127–45.

  1995. Demography and life history of free-ranging *Propithecus diadema edwardsi* in Ranomafana National Park, Madagascar. *International Journal of Primatology*, **16**, 835–54.

Zeh, J. A. & Zeh, D. W. 2001. Reproductive mode and the genetic benefits of polyandry. *Animal Behaviour*, **61**, 1051–63.

Zinner, D. & Deschner, T. 2000. Sexual swellings in female Hamadryas baboons after male take-overs: 'deceptive' swellings as a possible female counter-strategy against infanticide. *American Journal of Primatology*, **52**, 157–68.

Zinner, D., Alberts, S. C., Nunn, C. L. & Altmann, J. 2002. Significance of primate sexual swellings. *Nature*, **420**, 142–3.

Zuk, M. 1991. Sexual ornaments as animal signals. *Trends in Ecology and Evolution*, **6**, 228–31.

Zuk, M., Thornhill, R., Ligon, J. D. *et al.* 1990. The role of male ornaments and courtship behavior in female mate choice of red jungle fowl. *American Naturalist*, **136**, 459–73.

2 • What is sexual selection?

TIM H. CLUTTON-BROCK
Department of Zoology
University of Cambridge
Cambridge, UK

In the discussion on Sexual Selection in my Descent of Man, no case interested and perplexed me so much as the brightly-coloured hinder ends and adjoining parts of certain monkeys. As these parts are more brightly coloured in one sex than the other, and as they become more brilliant during the season of love, I concluded that the colours had been gained as a sexual attraction. I was well aware that I thus laid myself open to ridicule; though in fact it is not more surprising that a monkey should display his bright-red hinder end than that a peacock should display his magnificent tail.

C. Darwin, *Nature*, 2 November 1876, p. 18

INTRODUCTION

As this is a book about sexual selection, it is worth starting by considering what it means and how it differs from natural selection. The first section of this chapter briefly reviews the early history of ideas about the evolution of sex differences, while the second examines current definitions of sexual selection and the distinction between natural and sexual selection. The third section synthesises some of the developments in our understanding of the evolution of sex differences since Darwin's day. Finally, the fourth section provides a rough guide to some problems and pitfalls that scientists investigating sexual selection have encountered that are relevant to research on sexual selection in primates.

A BRIEF HISTORY OF SEXUAL SELECTION

In the *Origin of Species* (1859), Darwin provided a framework for explaining the evolution of adaptive differences between species, many of which he attributed to selection for traits that increased the survival of the individuals that carried them. However, he was aware of the need to extend his theory to provide an explanation of the evolution of the striking sex differences in body size, morphology or coloration that are a conspicuous feature of many animal species which cannot easily be attributed to selection operating through survival. Over the previous 30 years, these differences between the sexes had been the focus of papers by John Hunter, the eminent surgeon, anatomist and classifier of monsters (Hunter, 1837, 1861). Hunter proposed that differences between the sexes were of two kinds: those involving the sexual organs themselves, which were evident from birth and did not change during an individual's lifetime; and those that did not develop until the animal approached breeding age (such as differences in body size, plumage and the tendency to be fat), which he termed 'secondary' marks or characters of sex because they appeared after the primary sex differences (Hunter, 1861).

Hunter appreciated both that 'secondary' sexual characters were functionally related to fighting or display and that their extent varied with ecology.

> The males of almost every class of animals are probably disposed to fight, being, as I have observed, stronger than the females; and in many of these are parts destined solely for that purpose, as the spurs of the cock, and the horns of the bull . . . One of the most general marks (of sex) is the superior strength of make in the male; and another circumstance, perhaps equally so, is this strength being directed to one part more than another, which parts [sic] is that most immediately employed in fighting. This difference in external form is more particularly remarkable in the animals whose females are of a peaceable nature, as are the greatest number of those which feed on vegetables, and the marks to discriminate the sexes are in them very numerous.
>
> (Hunter, 1837)

Sexual Selection in Primates: New and Comparative Perspectives, ed. Peter M. Kappeler and Carel P. van Schaik. Published by Cambridge University Press. © Cambridge University Press 2004.

Darwin adopted Hunter's distinction between primary and secondary sexual differences with an important difference. Instead of using secondary to refer to sexually dimorphic traits that develop some time after hatching or birth, Darwin drew a functional distinction. His primary sexual characters were those connected with the act of reproduction itself, while secondary sexual characters were used in acquiring mating partners. In addition, Darwin recognised a third category of differences between the sexes – those that were '*related to different habits of life, and not at all, or only indirectly, to the reproductive functions*', among which he included structures associated with sex differences in feeding behaviour (Darwin, 1871). He regarded secondary sexual characters as the product of sexual selection; and primary sexual characters, as well as sex differences in the 'habits of life', as the result of natural selection:

> . . . when the females and males of any animal have the same general habits of life, but differ in structure, colour, or ornament, such differences have been mainly caused by sexual selection: that is by individual males having had, in successive generations, some slight advantage over other males, in their weapons, means of defence, or charms which they have transmitted to their male offspring alone.
>
> Darwin, *The Descent of Man* (1871)

Darwin distinguished sexual selection from natural selection on two main grounds: first, that it was a consequence of competition among members of the same sex rather than among members of different sexes or species; and, second, that it depended on variation in reproductive success rather than survival: '. . . *this form of selection depends, not on a struggle for existence in relation to other organic beings or to external conditions, but on a struggle between the individuals of one sex, generally the males, for the possession of the other sex. The result is not death to the unsuccessful competitor but few or no offspring.*' He also realised that it involved two different processes, which are now commonly referred to as intrasexual and intersexual selection:

> The sexual struggle is of two kinds; in the one it is between the individuals of the same sex, generally the males, in order to drive away or kill their rivals, the females remaining passive; whilst in the other, the struggle is likewise between the individuals of the same sex, in order to excite or charm those of the opposite sex, generally the females, which no longer remain passive, but select the more agreeable partners.
>
> Darwin, Chapter XXI (1871)

Like Hunter, Darwin (1871) appreciated that secondary sexual characters tended to be more highly developed in polygynous species: '*That some relation exists between polygamy and the development of secondary sexual characters, appears nearly certain; and this supports the view that a numerical preponderance of males would be eminently favourable to the action of sexual selection*', Chapter VIII. However, he did not spell out precisely why males compete more strongly for access to females than vice versa and it was left to biologists of this century to provide a detailed answer (see below). In addition, his treatment of intersexual selection in *The Descent of Man* focuses on the consequences of female choice for the evolution of male characteristics rather than on the reasons why selection might favour and maintain choice in females. Where he refers to this issue, he most commonly suggests that males with highly developed weapons or adornments are likely to be more vigorous than other males and that females are likely to gain various direct benefits by mating with vigorous males and to raise more offspring.

Darwin's theory of sexual selection was less readily accepted by scientists than the theory of natural selection. Wallace (1889) agreed that combat between males was an important source of selection pressures leading to sexual dimorphism, but regarded this as a form of *natural* selection on the grounds that it increased '*the vigour and fighting power of the male animals, since, in every case, the weaker are either killed, wounded or driven away*'. He regarded Darwin's second mode of sexual selection – female choice of particular males – as unimportant on the grounds that any consequences that female choice might have would be annulled by natural selection – unless females selected the fittest males, in which case the results of sexual and natural selection would be inseparable. He also pointed to the lack of evidence of consistent female choice for mates carrying particular characteristics. Some 50 years later, much the same points were reaffirmed in two influential papers by Huxley (1938a, b).

Wallace's objection that sexual selection is a form of natural selection is semantically correct – after all, Darwin originally coined the term 'natural selection' in order to mark its relation to *man*'s power of selection, and the opposite of natural selection is not sexual but artificial or human selection (see Brown, 1975; Halliday, 1978; Fig. 2.1). However, his insistence that the process of sexual selection described by Darwin could only increase the survival of males or their average reproductive success is clearly wrong, for female choice for heritable male characteristics (such as tail size) can cause them to develop to a point at which they reduce the average fitness of males (see below).

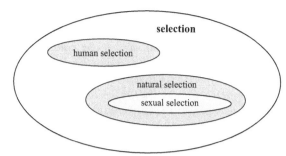

Fig. 2.1   A simple clarification of selection. Both natural and sexual selection can be further divided either on the basis of the components of fitness affected (e.g. fecundity selection versus survival selection) or on the basis of the mechanisms involved (e.g. intrasexual versus intersexual selection).

THERE IS ONLY ONE SELECTION

One of the problems in writing about sexual selection today is that the term is used in so many different ways. Since the 1970s, Darwin's second mode of sexual selection – intersexual selection – has been the focus of a large body of research, partly because it had previously been discounted (see Huxley, 1938a, b) and partly because its operation and consequences are more complex than those of intrasexual selection, and it has consequently attracted a larger share of the attention of theorists (see Andersson, 1994). So intense has the interest in intersexual selection been that many biologists have come to use sexual selection to refer to intersexual selection alone (see Bradbury & Andersson, 1987). Even where writers use 'sexual selection' to include both intrasexual and intersexual selection, the term is still employed in a wide variety of different meanings. In some cases, it is used to refer to the behavioural processes responsible for variation in mating success (direct competition for males or variation in their capacity to attract mating partners); in others, it is used to refer to the relationships between phenotypic variation and mating success that the same processes generate. In other circumstances, sexual selection is used to refer to the evolution of sex differences in behaviour or morphology (see also Snowdon, this volume). The situation is further complicated by the fact that some evolutionary biologists distinguish between selection and the response to selection (which requires that the trait should be heritable), while others only consider selection to operate on traits that have a heritable basis (see Endler, 1986) – a difference which regularly complicates discussions between behavioural ecologists (who typically adopt the first position) and many population geneticists (who adopt the second).

Table 2.1   *Which of the following do you mean by sexual selection?*

1. Mating competition between males.
2. Variation in male mating success.
3. Sex differences in variation in breeding success or in the opportunity for sexual selection.
4. Consistent patterns of mate choice by members of either sex.
5. Selection on particular traits arising from competition with members of the same sex to acquire mates.
6. Selection on particular traits arising from reproductive competition with members of the same sex.
7. Sex differences in traits that confer advantages in competition for mates.
8. Sex differences in the intensity of selection on particular traits.
9. Sex differences in the intensity of selection on particular traits caused by contrasting effects of the trait on mating success in the two sexes.

The range of meanings commonly attached to sexual selection is illustrated in the multiple-choice question in Table 2.1. All of the nine phenomena listed are commonly referred to as sexual selection, and a student might reasonably have some difficulty in answering. However, only two (number 5: 'Selection on particular traits arising from competition with members of the same sex to acquire mates', and number 6: 'Selection on particular traits arising from reproductive competition with members of the same sex') closely reflects Darwin's description of the process. Of the others, 1 to 4 are processes that can contribute to sexual selection, while 7 to 9 are common (but not inevitable) consequences of sexual selection. In particular, it is important to appreciate that sex differences in variation in reproductive success (which is sometimes referred to as the *opportunity for sexual selection*) will not necessarily reflect the relative intensity of selection of particular traits nor will they necessarily predict the degree of sexual dimorphism that develops.

In some cases, this ambiguity in the use of sexual selection does not matter too much but there are others where the meaning attached to sexual selection affects the logic of evolutionary arguments or the interpretation of data. First, the definition of sexual selection can be important where an attempt is being made to distinguish between the process or consequences of selection operating through competition for reproductive opportunities and selection operating through survival. There are good reasons for wishing to make comparisons of this kind, for both the process of evolution and its consequences may be affected by whether selection for survival or for breeding success is dominant (see Lande, 1980, 1982; Endler, 1986). To take one example, some evidence suggests that sex differences in body size may lead to sex differences in survival where they have evolved through sexual selection, but do not necessarily do so where they have evolved through natural selection for ecological separation between the sexes (Newton & Marquiss, 1979; Clutton-Brock et al., 1985; Clutton-Brock, 1991a, b, 1994).

Second, precisely what is meant by sexual selection is important where a distinction is being drawn between selection operating through intrasexual competition and selection operating through intersexual selection or mate choice. Variable or shifting definitions of sexual selection can easily blur this distinction and can lead to evidence that a trait has evolved through sexual selection in the broad sense being used to support arguments that it has evolved through mate choice.

Third, the meaning that we attach to sexual selection is important when we wish to predict or compare the opportunity for sexual selection in different species or populations. For example, if our aim is to estimate the relative contribution of differences in mating success to variation in fitness, we may wish to use some estimate of variance in mating success relative to variance in other fitness components in the same sex. Conversely, if we wish to estimate the extent to which selection pressures can diverge between the two sexes, we may use some index of the relative variance in breeding success in males and females. In the past, not only have attempts to estimate the opportunity for sexual selection used a wide variety of indices incorporating different components of fitness (see Clutton-Brock, 1988), but some have used indices of variance in mating success relative to variance in other components of fitness in the same sex (see Howard, 1979; Wade, 1979; Wade & Arnold, 1980), while others have used indices of relative variance in mating or breeding success in the two sexes (see Ralls, 1977; Payne, 1979). In both cases, it is important that the aim of the exercise is clear and that an appropriate index is used.

Some theoretical and practical problems in identifying the boundaries of sexual selection still persist. In particular, should we confine sexual selection to cases where variation in fitness arises through differences in mating success, and follow definition 5 in Table 2.1 ('Selection on particular traits arising from competition with members of the same sex to acquire mates')? Or should we broaden it to include cases where variation in fitness arises through other components of reproductive success and adopt definition 6 ('Selection on particular traits arising from reproductive competition between members of the same sex')? This distinction may look like hair splitting, but it has practical implications, particularly for our view of the selection processes operating in females, where competition for breeding positions and for the capacity to rear offspring is common. If we follow the first of the two options described above, we should not regard cases where selection operates through competition between members of the same sex to enter breeding condition, or to suppress reproduction by other members of the same sex, or to ensure the survival of their own progeny as examples of sexual selection (see Faulkes & Abbot, 1997; French, 1997; Clutton-Brock et al., 1998, 2001). Reproductive competition of this kind shares many similarities with competition for mating partners, including potentially high costs to survival, and a strong case can be made that it should be regarded as a form of sexual selection. Moreover, it is often difficult to distinguish clearly between selection operating through variation in the capacity to acquire mates and selection operating through variation in other components of reproductive success or survival – for example, where members of one sex compete for territories the quality of which affects both their survival as well as their ability to acquire mates (see Newton, 1998) or where females prefer older males (e.g. Borgerhoff Mulder, 1988; Le Boeuf & Reiter, 1988). However, following the second option (which involves classifying all forms of selection arising through reproductive competition between members of the same sex as components of sexual selection) has the potential disadvantage that it may blur the distinction between natural and sexual selection. For example, definition 6 would include selection arising from competition for resources that allow individuals to reproduce, as well as selection operating through variation in fecundity that was related to social rank.

There are arguments for and against both definitions but my own preference is for the second, broader option, since it is clear that selection on males and females operates in diverse ways through a wide range of different components of fitness, so that there are benefits in avoiding the restrictions imposed by a simple dichotomy between sexual and natural selection. Indeed, a case can be made that there would be advantages in abandoning the distinction between natural selection and sexual selection altogether and focusing instead on comparing the selection pressures operating on males and females and their consequences in different species. Natural selection is, after all, a single process that can operate on variation in fitness generated for different reasons at many stages of the life history (see Fig. 2.1). I like to think that it may have been for this reason that, in the title of his book, Darwin replaced the term 'sexual selection' with the more precise 'selection in relation to sex'.

TWENTIETH-CENTURY SEXUAL SELECTION

Since the debates of the nineteenth century, there have been important developments in our understanding of both Darwin's forms of sexual selection (Andersson, 1994). The reasons why males generally compete more strongly for females than vice versa and the distribution of sex differences in competition for males have been shown to be related to sex differences in the costs of producing gametes and rearing progeny (Fisher, 1930; Bateman, 1948; Parker *et al.*, 1972; Trivers, 1972; Emlen & Oring, 1977; Sutherland, 1985a, b; Clutton-Brock & Vincent, 1991; Clutton-Brock & Parker, 1992; Kokko & Monaghan, 2001; Gowaty, this volume). Consistent relationships have been demonstrated between the development of secondary sexual characters and the extent of polygamy (Clutton-Brock *et al.*, 1977; Alexander *et al.*, 1979). A wide range of empirical studies has shown that the fitness costs of male combat are commonly high and that the development of many secondary sexual characters, such as increased male size and the development of weaponry and adornments, can also have substantial costs (Clutton-Brock *et al.*, 1985; Clutton-Brock, 1991b, 1994). Contrary to Darwin's original predictions, these costs have proved to extend beyond the period of active reproduction to affect the relative survival of males and females during their early development (Clutton-Brock *et al.*, 1985; Clutton-Brock, 1991b). In addition, theoretical models of sexual selection have confirmed that intrasexual selection (as well as intersexual selection) can lead to the development of traits that increase mating success at a cost to survival, and may lower the mean fitness of the population (Fisher, 1930; O'Donald, 1980; Lande, 1981; Kirkpatrick, 1982; Andersson, 1994).

The evolutionary causes and consequences of female mate choice have been extensively explored, both in theory and practice. There is now substantial empirical evidence that, as Darwin suggested, females can increase their own survival as well as the number of offspring they produce by mating selectively with 'vigorous' males that can guard mates, ensure their access to resources or help to provision or guard their offspring (Andersson, 1994). It is also clear that the development of male secondary sexual characters often affects the male's attractiveness to females and

male mating success (Andersson, 1982, 1994). Where male characteristics are heritable, mating with selected partners can increase the fitness of a female's offspring and this process may favour and maintain the evolution of consistent mating preferences in females, leading to the evolution and maintenance of 'exaggerated' characters in males (Fisher, 1930; O'Donald, 1980; Lande, 1981; Bradbury & Andersson, 1987; Andersson, 1994).

Several new themes unknown to Darwin have emerged. First, it is now appreciated that females often mate with more than one partner and that competition between sperm from more than one male in the female tract can be intense (Birkhead & Møller, 1992), favouring high mating rates, large testes and ejaculates, and large sperm size (Harcourt *et al.*, 1981; Gomendio & Roldan, 1993). Second, it is clear that, in some species, males compete by killing the offspring of competitors, especially where this reduces latency to subsequent conception in females (Hrdy, 1972; Hausfater & Hrdy, 1984; van Schaik & Janson, 2000; van Schaik *et al.*, this volume). Infanticidal male strategies have, in turn, led to the evolution of physiological and behavioural adaptations in females that limit the ability of males to kill their infants or reduce costs to their own fitness (Hrdy, 1972; van Schaik & Janson, 2000; van Schaik *et al.*, this volume). Third, it is clear that sexual coercion is widespread and that, in some species, males commonly attempt to force females to mate with them through forced copulation, repeated harassment, punishment and intimidation, with costs to female fitness (Smuts & Smuts, 1993; Clutton-Brock & Parker, 1995a, b). Here, too, females have developed a range of behavioural and morphological counter-strategies that enable them to avoid male control or to reduce the costs of male harassment (Smuts & Smuts, 1993; Clutton-Brock & Parker, 1995a, b; Eberhard, 1996; van Schaik *et al.*, this volume).

Another development in our understanding of sexual selection has been in our knowledge of its operation among females. While Darwin (1871) appreciated that secondary sexual characters were occasionally more highly developed in females than males, it is not clear that he understood the reasons for this. Studies of polyandrous wading birds have shown that, in these species, intrasexual competition among females is unusually intense and that females are commonly larger, more aggressive and more highly ornamented than males (Oring, 1982; Erckmann, 1983). Females of these species may also show behavioural traits, such as infanticidal tendencies following the take-over of males (see Emlen *et al.*, 1989), that are more usually characteristic of males. There may also be other cases where high variance in female breeding success and intense reproductive competition between females have led to the greater development of traits that enhance competitive ability in females than males. For example, in obligately cooperative mammals, such as naked mole-rats and meerkats, reproductive skew and variance in female fecundity are unusually high because the young are fed by non-breeding helpers, so that the fecundity of dominant females may be less closely constrained by their own capacity to rear offspring. In several of these species, females are larger and more aggressive than males, over whom they are often dominant (Alexander *et al.*, 1979; Clutton-Brock *et al.*, 2001).

In some polygynous species, too, selection for competitive success or for the ability to acquire mates may favour the evolution of secondary sexual characters in females. In spotted hyaenas, female breeding success and dominance rank are closely related and subordinate females often fail to breed (Frank, 1986; Frank *et al.*, 1995). Females are larger and more aggressive than males. In addition, their genitalia have become modified to resemble those of males, possibly because this deflects aggression from dominant females. In other polygynous species where females compete for access to preferred mating partners, females have developed exaggerated adornments that appear to attract males. For example, pronounced sexual swellings around the time of ovulation occur in the females of a number of cercopithecine primates that live in multi-male–multi-female groups, where females have breeding access to several unrelated males (Clutton-Brock & Harvey, 1976; Nunn, 1999; van Schaik *et al.*, 1999; Zinner *et al.*, this volume), as well as in some birds with similar breeding systems (Davies *et al.*, 1996). Recent studies of olive baboons suggest that males compete more intensely for females with larger swellings because such females are more likely to conceive and their offspring are more likely to survive (Domb & Pagel, 2001; though see also Zinner *et al.*, 2002). In these species, females may gain direct benefits from signalling their reproductive status or quality, either because this raises the chance that they will be consorted by dominant males or because this helps to maximise the number of different males that mate with them, generating benefits from increased paternal care or reduced risk of infanticide (Clutton-Brock & Harvey, 1976; Nunn, 1999; van Schaik *et al.*, 1999). Alternatively, they may gain genetic benefits for their offspring as a result of mating competition or sperm competition between males (see Andersson, 1994). Exaggerated female adornments that are absent in males appear to be rare among mammals, suggesting that benefits to females must be unusually strong among cerco-

pithecine primates (though another possibility is that female mammals more commonly signal their reproductive status or quality to males through olfaction and that exaggerated traits of this kind are commoner than we appreciate). In many of these species, variance in breeding success may still be greater among males than females, emphasising the point that the evolution of sex differences depends on the relative intensity of selection on particular traits in the two sexes rather than on the relative opportunity for sexual selection (see above).

Finally, we now know more about the kind of sex differences in feeding behaviour and habitat use that Darwin identified as his third category of sex differences (see above). While, in some cases, these may have evolved as a result of natural selection favouring resource partitioning between the sexes (Selander, 1972), in others it is clear that sexual selection has played a dominant role in their evolution (Clutton-Brock, 2001). For example, among wading birds, the wintering grounds of males are often closer to the breeding grounds than those of females in polygynous and monogamous species, while this situation is reversed in polyandrous ones, presumably because travel distance to the breeding grounds affects arrival dates and there is stronger selection for arrival dates in whichever sex competes most intensely for mates (Erckmann, 1981, 1983; Clutton-Brock, 1983). In other cases, sex differences in feeding behaviour appear to be by-products of sexual selection, favouring enlarged body size in one sex. In dimorphic arboreal primates, for example, females may spend more time feeding in the terminal twigs of trees than males because they are lighter (Clutton-Brock, 1973). Similarly, sex differences in habitat use among ungulates may often be a consequence of sex differences in body size that have evolved as a result of strong selection for large body size in males (Clutton-Brock & Harvey, 1983; Ruckstuhl & Neuhaus, 2000; see also Rubenstein & Hack, this volume).

A ROUGH GUIDE TO SEXUAL SELECTION

However we define sexual selection, there are problems in investigating the evolution of sex differences and the operation of selection in the two sexes, which are common to many different animals, as follows:

(1) *Where mating success varies widely with age, estimates of variation in mating success calculated across individuals of different ages within single breeding seasons may substantially overestimate the opportunity for sexual selection.*

Long-term studies of many mammals now provide extensive evidence of the extent to which mating success varies with age (see, for example, Le Boeuf & Reiter, 1988; Clutton-Brock *et al.*, 1988; Fig. 2.2). Where this is the case, estimates of variation in mating success calculated across samples of animals of different ages may substantially overestimate variation in lifetime mating success and the opportunity for sexual selection. In these circumstances, attempts to compare the opportunity for sexual selection need to be based on comparisons between animals of the same age or to use estimates of variation in mating success over the lifespan (Altmann *et al.*, 1988; Clutton-Brock, 1988; Altmann *et al.*, 1997).

(2) *Estimates of the relative opportunity for sexual selection will not necessarily predict the degree of sexual dimorphism in particular traits.*

Measures of relative variance in mating success or proxies (such as variance in harem size) are sometimes used to predict the intensity of sexual selection on traits associated with mating success. Since stochastic factors play a large role in generating variance in mating success (Sutherland, 1985b), estimates of the opportunity for sexual selection will not necessarily reflect the extent of sex differences in the intensity of selection on particular traits. Moreover, different traits affect competitive ability in different species, generating interspecific contrasts in selection pressures on particular traits. For example, while male mating success is related to male size and dominance in cervids, which fight by pushing, it is unrelated to male size or dominance in equids, which fight by biting (Feh, 1990). As would be expected, sexual dimorphism for body size has evolved in cervids but not in equids. Many similar examples occur in birds, where males of some species use bright plumage to attract females and often show limited vocal repertoires while, in others, males use elaborate songs and possess drab plumage (Catchpole, 1987). As a result, consistent correlations between measures of the opportunity for sexual selection and the development of particular sexual differences should not necessarily be expected, and their absence does not indicate that breeding systems do not affect the opportunity for sexual selection or that sexual selection is weak. This has implications for research strategies. For example, if we wish to investigate why sexual dimorphism in body size is highly developed in cercopithecine primates but not in lemurs (Kappeler, 1990) we may need to understand the factors affecting fighting success in both groups, as well as the comparative costs of large body size or rapid growth in each sex.

(3) *The costs and benefits of sexually dimorphic traits commonly appear at different stages of the lifespan, so that attempts to*

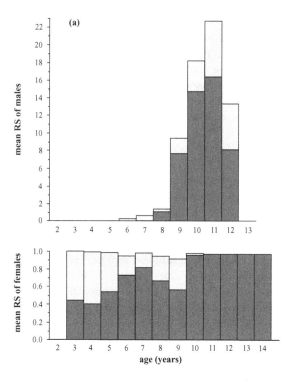

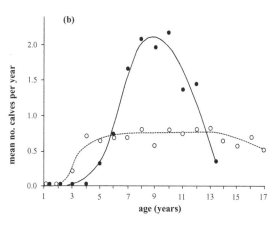

Fig. 2.2   Age and reproductive success (RS) in males and females in polygynous mammals: (a) elephant seals (from Le Boeuf & Reiter, 1988); (b) red deer (from Clutton-Brock *et al.*, 1988).

measure costs or benefits that are restricted to a single stage of the lifecycle may miss effects that occur at other stages and may produce misleading results.

Where a trait (such as body size) confers substantial advantages on mating competition in one sex, sexual selection is likely to favour its development until a point is reached where further increases are balanced by costs to some other

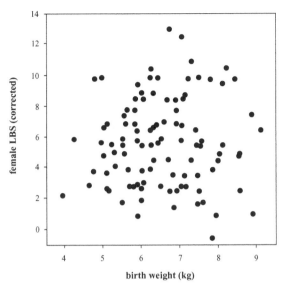 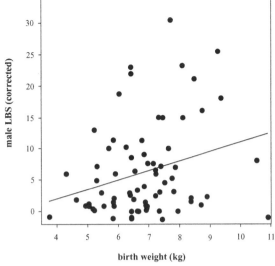

Fig. 2.3   Lifetime breeding success (LBS) in relation to birth weight for female and male red deer (from Kruuk *et al.*, 1999b).

component of fitness. Where benefits are large, substantial costs are consequently likely to be present – but costs and benefits may occur at different stages of the lifespan (see also Setchell & Lee, this volume). In particular, the development of larger body size in adult males than females is commonly associated with stronger selection for fast pre- and post-natal growth rates in males (Clutton-Brock, 1991a, b). For example, in red deer, where adult size has a stronger effect on fitness in males than females, lifetime breeding success is more closely related to birth weight in males than females (Kruuk *et al.*, 1999b: Fig. 2.3). In many sexually dimorphic species this has apparently led to the evolution of faster growth rates in juvenile males (Fig. 2.4), which are commonly associated with higher mortality rates in males compared to females (see Fig. 2.5). In contrast, there is limited evidence that large body size has costs to the survival of adult males and it seems likely that the evolution of large male body size and sexual size dimorphism may be constrained largely by its costs during development (Promislow, 1992; Promislow *et al.*, 1992).

(4) *Sex differences in development may affect the relative costs to parents of raising sons and daughters generating conflicts of interest between offspring and parents as well as between sibs, which may constrain the evolution of sex differences.*

Sex differences in early development also affect the costs to parents of raising sons and daughters (Trivers & Willard, 1973; Trivers, 1980). Both in birds and mammals, there is

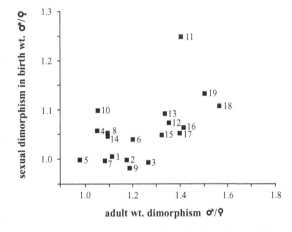

Fig. 2.4   Sexual dimorphism in birth weight in relation to adult weight across 19 species (from Clutton-Brock, 1991a): (1) and (2) *Onchomys*; (3) *Proechimys*; (4) coypu; (5) ringtail possum; (6) pig; (7) horse; (8) roe deer; (9) moose; (10) Chinese water deer; (11) mouflon; (12) Soay sheep; (13) pigtail macaque; (14) rhesus macaque; (15) fallow deer; (16) red deer; (17) wapiti; (18) reindeer; (19) white-tailed deer.

evidence that sexual dimorphism in early growth rates is associated with differences in the energetic costs to parents of rearing sons and daughters (Clutton-Brock, 1991a), as Trivers and Willard (1973) predicted. In most cases, these differences appear to be caused by sex differences in the rate

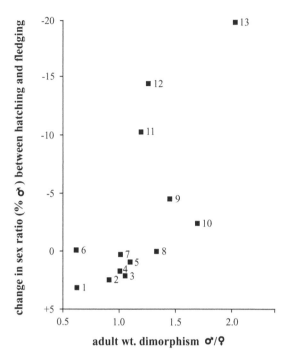

Fig. 2.5   Sex differences in juvenile mortality between hatching and fledging in bird species showing different degrees of weight dimorphism: (1) peregrine falcon; (2) American kestrel; (3) eastern bluebird; (4) starling; (5) snow goose; (6) European sparrow hawk; (7) red-cockaded woodpecker; (8) blue grouse; (9) red-winged blackbird; (10) yellow-headed blackbird; (11) rook; (12) common grackle; (13) capercaillie (from Clutton-Brock *et al.*, 1985). These differences are probably more marked in birds than mammals because weight dimorphism among fledgelings is typically as large as in adults (from Clutton-Brock, 1991a).

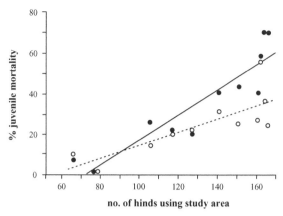

Fig. 2.6   Mortality during the first two years of life among male (●) and female (○) red deer in relation to population density (from Clutton-Brock *et al.*, 1985).

of parent–offspring conflict may vary with the sex of the offspring, though this will not necessarily be translated into variation in the intensity of behavioural conflict (Clutton-Brock, 1991a; Mock & Parker, 1997; Lessells, 2002).

Differences in the costs of raising sons and daughters could also affect the optimal sex ratio that parents should produce (Trivers & Willard, 1973), especially where the relative costs of raising sons and daughters vary with parental phenotype (Gomendio *et al.*, 1990). Recent studies provide evidence of adaptive variation in the sex ratio in a number of vertebrates (Clutton-Brock *et al.*, 1984; Komdeur *et al.*, 1997; Sheldon *et al.*, 1999) though not all sex ratio trends are likely to be adaptive (see below; and Silk & Brown, this volume).

(5) *The costs and benefits of particular traits may only be evident under particular ecological circumstances or in some categories of animals.*

In many cases, the costs of sexually selected traits are only evident when resources are in short supply. In polygynous mammals, increased mortality in juvenile males before and after birth is only apparent where resources are scarce or external conditions are harsh (Clutton-Brock, 1991a). For example, in red deer, sex differences in mortality during gestation as well as in the first winter of life only appear when conditions are harsh and population density is high (Clutton-Brock *et al.*, 1985; Kruuk *et al.*, 1999a; Fig. 2.6). In other cases, the costs of sexually selected traits to juveniles or their parents are only apparent among the offspring of inferior parents. In red deer, sex differences in juvenile mortality are only present among the offspring of subordinate mothers (Clutton-Brock *et al.*, 1985). Similarly, subordinate

of demand and there is limited evidence that parents treat their sons and daughters differently (Clutton-Brock, 1991a). Differences in the energetic costs of rearing sons and daughters can translate into differential costs to parental survival or subsequent fitness (Clutton-Brock *et al.*, 1985), though they do not always do so (Clutton-Brock, 1991a). In some cases, this may be because parents can accommodate the increased energetic costs of raising the more expensive sex. For example, in red deer, where male calves suckle more frequently than females, raising sons depresses the survival and breeding success of subordinate mothers more than raising daughters while, in dominant females, the costs of raising sons and daughters do not differ (Gomendio *et al.*, 1990). Where the benefits of increased investment are paid largely by the parent, while direct benefits accrue to offspring, the intensity

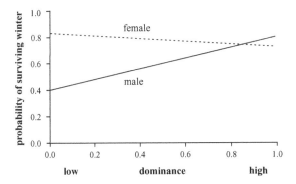

Fig. 2.7 Maternal dominance and the costs of raising sons and daughters in red deer (from Gomendio *et al.*, 1990). Survival amongst subordinates but not dominants is depressed after rearing males.

mothers are more likely to die after rearing sons than after rearing daughters, while no differences are present among dominants (see Fig. 2.7). One implication of these results is that negative evidence should always be treated with caution. Another is that the costs and benefits of particular traits to laboratory animals maintained on easily available unlimited food supplies may not reflect their magnitude in natural populations.

(6) *Not all sex differences are adaptive.*

Since strong sexual selection commonly generates costs to other components of fitness, it is often associated with sex differences that are by-products of sexually selected adaptations but are not themselves adaptive. For example, the scarcity of evidence that mothers respond differently to suckling attempts by their sons and daughters suggests that sex differences in rearing costs may be a by-product of sex differences in growth rather than of adaptive discrimination by parents. Similarly, though sex differences in juvenile mortality are sometimes interpreted as a consequence of maternal discrimination against sons, several studies have now shown that these differences persist when juveniles are reared alone (McClure, 1981). This suggests that they are probably a by-product of sexual selection for fast growth rates in juvenile males rather than of parental discrimination against males (Trivers, 1972; Clutton-Brock, 1991a). Finally, many consistent trends in birth sex ratios may occur because resource shortage generates higher rates of mortality among male than female embryos and may not be a consequence of adaptive manipulation of the sex ratio by parents (Clutton-Brock, 1991a; Kruuk *et al.*, 1999a). In order to avoid accusations of adaptive story-telling, evolutionary biologists need to be alert to the consequences of ecological differences between the sexes and to avoid assuming that sex differences are necessarily adaptive.

REFERENCES

Alexander, R. D., Hoogland, J. L., Howard, R. D., Noonan, K. M. & Sherman, P. W. 1979. Sexual dimorphisms and breeding systems in pinnipeds, ungulates, primates and humans. In *Evolutionary Biology and Human Social Behaviour*, ed. N. A. Chagnon & W. Irons. Belmont, CA: Wadsworth, pp. 402–35.

Altmann, J., Hausfater, G. & Altmann, S. A. 1988. Determinants of reproductive success in savannah baboons. In *Reproductive Success*, ed. T. H. Clutton-Brock. Chicago, IL: Chicago University Press, pp. 403–18.

Altmann, J., Alberts, S. C. & Haines, S. A. 1997. Behaviour predicts genetic structure in a wild primate group. *Proceedings of the National Academy of Sciences, USA*, **93**, 5797–801.

Andersson, M. 1982. Female choice selects for extreme tail length in a widowbird. *Nature*, **299**, 818–20.

1994. *Sexual Selection*. Princeton, NJ: Princeton University Press.

Bateman, A. J. 1948. Intrasexual selection. In *Drosophila*. *Heredity*, **2**, 349–68.

Birkhead, T. R. & Møller, A. P. 1992. *Sperm Competition in Birds*. London: Academic Press.

Borgerhoff Mulder, M. 1988. Reproductive success in three Kipsigi cohorts. In *Reproductive Success*, ed. T. H. Clutton-Brock. Chicago, IL: Chicago University Press, pp. 419–38.

Bradbury, J. W. & Andersson, M. B. (eds.) 1987. *Sexual Selection: Testing the Alternatives*. London: John Wiley and Sons.

Brown, J. L. 1975. *The Evolution of Behavior*. New York, NY: Norton.

Catchpole, C. K. 1987. Bird song, sexual selection and female choice. *Trends in Ecology and Evolution*, **2**, 94–7.

Clutton-Brock, T. H. 1973. Feeding levels and feeding sites of red colobus (*Colobus badius tephrosceles*) in the Gombe National Park. *Folia Primatologica*, **19**, 368–79.

1983. Selection in relation to sex. In *Evolution from Molecules to Men*, ed. B. J. Bendall. Cambridge: Cambridge University Press, pp. 457–81.

1988. Reproductive success. In *Reproductive Success*, ed. T. H. Clutton-Brock. Chicago, IL: Chicago University Press, pp. 472–86.

1991a. *The Evolution of Parental Care*. Princeton: Princeton University Press.

1991b. The evolution of sex differences and the consequences of polygyny in mammals. In *The Development and Integration of Behaviour: Essays in Honour of Robert Hinde*, ed. P. Bateson. Cambridge: Cambridge University Press, pp. 229–53.

1994. The costs of sex. In *The Differences between the Sexes*, ed. R. V. Short & E. Balaban. Cambridge: Cambridge University Press, pp. 347–62.

2001. Sociality and population dynamics. In *Ecology: Achievement and Challenge*, ed. M. C. Press, N. J. Huntly & S. Levin. Oxford: Blackwell Science, pp. 47–66.

Clutton-Brock, T. H. & Harvey, P. H. 1976. Evolutionary rules and primate societies. In *Growing Points in Ethology*, ed. P. P. G. Bateson & R. A. Hinde. Cambridge: Cambridge University Press, pp. 195–237.

1977. Primate ecology and social organisation. *Journal of Zoology, London*, **183**, 1–39.

1983. The functional significance of variation in body size among mammals. In *Advances in the Study of Mammalian Behavior*, ed. J. F. Eisenberg & D. G. Kleiman. Shippenburg, PA: The American Society of Mammalogists, pp. 632–63.

Clutton-Brock, T. H. & Parker, G. A. 1992. Potential reproductive rates and the operation of sexual selection. *Quarterly Review of Biology*, **67**, 437–56.

1995a. Punishment in animal societies. *Nature*, **373**, 209–16.

1995b. Sexual coercion, harassment and intimidation in animal societies. *Animal Behaviour*, **49**, 1345–65.

Clutton-Brock, T. H. & Vincent, A. C. J. 1991. Sexual selection and the potential reproductive rates of males and females. *Nature*, **351**, 58–60.

Clutton-Brock, T. H., Harvey, P. H. & Rudder, B. 1977. Sexual dimorphism, socionomic sex ratio and body weight in primates. *Nature*, **269**, 797–800.

Clutton-Brock, T. H., Albon, S. D. & Guinness, F. E. 1984. Maternal dominance, breeding success and birth sex ratios in red deer. *Nature*, **308**, 358–60.

1985. Parental investment and sex differences in juvenile mortality in birds and mammals. *Nature*, **313**, 131–3.

1988. Reproductive success in male and female red deer. In *Reproductive Success*, ed. T. H. Clutton-Brock. Chicago, IL: Chicago University Press, pp. 325–43.

Clutton-Brock, T. H., Brotherton, P. N. M., Smith, R. *et al.* 1998. Infanticide and expulsion of females in a cooperative mammal. *Proceedings of the Royal Society, Series B*, **265**, 2291–5.

Clutton-Brock, T. H., Brotherton, P. N. M., Russell, A. F. *et al.* 2001. Cooperation, control and concession in meerkat groups. *Science*, **291**, 478–81.

Darwin, C. 1859. *On the Origin of Species*. London: John Murray.

1871. *The Descent of Man and Selection in Relation to Sex*. London: John Murray.

Davies, N. B., Harley, I. R., Hatchwell, B. J. *et al.* 1996. Female control of ovulations to maximise male help: a comparison of polygynandrous alpine accentors *Prunella collaris* and dunnocks *Prunella modularis*. *Animal Behaviour*, **51**, 27–47.

Domb, L. G. & Pagel, M. 2001. Sexual swellings advertise female quality in wild baboons. *Nature*, **410**, 204–6.

Eberhard, W. G. 1996. *Female Control: Sexual Selection by Cryptic Female Choice*. Princeton, NJ: Princeton University Press.

Emlen, S. T. & Oring, L. W. 1977. Ecology, sexual selection, and the evolution of mating systems. *Science*, **197**, 215–23.

Emlen, S. T., Demong, N. J. & Emlen, D. J. 1989. Experimental induction of infanticide in female wattled Jacanas. *Auk*, **106**, 1–7.

Endler, J. A. 1986. *Natural Selection in the Wild*. Princeton, NJ: Princeton University Press.

Erckmann, W. J. 1981. The evolution of sex role reversal and monogamy in shore birds. Ph.D. thesis, University of Washington, Seattle.

1983. The evolution of polyandry in shore birds: an evaluation of hypotheses. In *Social Behavior of Female Vertebrates*, ed. S. K. Wasser. London: Academic Press, pp. 114–68.

Faulkes, C. G. & Abbott, D. H. 1997. The physiology of a reproductive dictatorship: regulation of male and female reproduction by a single breeding female in colonies of naked mole-rats. In *Cooperative Breeding in Mammals*, ed. N. G. Solomon & J. A. French. Cambridge: Cambridge University Press, pp. 268–301.

Feh, C. 1990. Long-term paternity data in relation to different aspects of rank for Camargue stallions, *Equus caballus*. *Animal Behaviour*, **40**, 995–6.

Fisher, R. A. 1930. *The Genetical Theory of Natural Selection*. Oxford: Clarendon Press.

Frank, L. G. 1986. Social organisation of the spotted hyaena *Crocuta crocuta*. II. Dominance and reproduction. *Animal Behaviour*, **35**, 1510–27.

Frank, L. G., Holekamp, K. E. & Smale, L. 1995. Dominance, demography and reproductive success of spotted hyaenas.

In *Serengeti II: Dynamics, Management and Conservation of an Ecosystem*, ed. A. R. E. Sinclair & P. Arcese. Chicago, IL: Chicago University Press, pp. 364–84.

French, J. A. 1997. Proximate regulation of singular breeding in Callitrichid primates. In *Cooperative Breeding in Mammals*, ed. N. G. Solomon & J. A. French. Cambridge: Cambridge University Press, pp. 34–75.

Gomendio, M. & Roldan, E. R. S. 1993. Mechanisms of sperm competition: linking physiology and behavioural ecology. *Trends in Ecology and Evolution*, **8**, 95–100.

Gomendio, M., Clutton-Brock, T. H., Albon, S. D., Guinness, F. E. & Simpson, M. J. 1990. Mammalian sex ratios and variation in costs of rearing sons and daughters. *Nature*, **343**, 261–3.

Halliday, T. R. 1978. Sexual selection and mate choice. In *Behavioural Ecology: An Evolutionary Approach*, ed. J. R. Krebs & N. B. Davies. Oxford: Blackwell Scientific Publishers, pp. 180–213.

Harcourt, A. H., Harvey, P. H., Larson, S. G. & Short, R. V. 1981. Testis weight, body weight and breeding system in primates. *Nature*, **293**, 55–7.

Hausfater, G. & Hrdy, S. B. (eds.) 1984. *Infanticide: Comparative and Evolutionary Perspectives*. New York, NY: Aldine.

Howard, R. D. 1979. Estimating reproductive success in natural populations. *American Naturalist*, **114**, 221–31.

Hrdy, S. B. 1972. *The Langurs of Abu: Female and Male Strategies of Reproduction*. Cambridge, MA: Harvard University Press.

Hunter, J. 1837. An account of an extraordinary pheasant. In *Observations on Certain Parts of the Animal Oeconomy* with notes by Richard Owen. London: Longman, Orme, Brown, Green and Longmans.

1861. Observations on generation. In *Essays on Observations on Natural History, Anatomy, Physiology and Geology*, Vol. I. London: John van Voorst.

Huxley, J. S. 1938a. The present standing of the theory of sexual selection. In *Evolution: Essays on Aspects of Evolutionary Biology*, ed. G. R. de Beer. Oxford: Clarendon Press, pp. 11–42.

1938b. Darwin's theory of sexual selection and the data subsumed by it, in the light of recent research. *American Naturalist*, **72**, 416–33.

Kappeler, P. M. 1990. The evolution of sexual size dimorphism in prosimian primates. *American Journal of Primatology*, **21**, 201–14.

Kirkpatrick, M. 1982. Sexual selection and the evolution of female choice. *Evolution*, **36**, 1–12.

Kokko, H. & Monaghan, P. 2001. Predicting the direction of sexual selection. *Ecology Letters*, **4**, 159–65.

Komdeur, J., Daan, S., Tinbergen, J. & Mateman, C. 1997. Extreme adaptive modification in the sex ratio of the Seychelles warbler's eggs. *Nature*, **385**, 522–5.

Kruuk, L. E. B., Clutton-Brock, T. H., Albon, S. D., Pemberton, J. M. & Guinness, F. E. 1999a. Population density affects sex ratio variation in red deer. *Nature*, **399**, 459–61.

Kruuk, L. E. B., Clutton-Brock, T. H., Rose, K. E. & Guinness, F. E. 1999b. Early determinants of lifetime reproductive success differ between the sexes in red deer. *Proceedings of the Royal Society, Series B*, **266**, 1655–61.

Lande, R. 1980. Sexual dimorphism, sexual selection and adaptation in polygenic characters. *Evolution*, **34**, 294–305.

1981. Models of operation of sexual selection on polygenic traits. *Proceedings of the National Academy of Sciences, USA*, **78**, 3721–5.

1982. A quantitative genetic theory of life-history evolution. *Ecology*, **63**, 607–15.

Le Boeuf, B. J. & Reiter, J. 1988. Lifetime reproductive success in northern elephant seals. In *Reproductive Success*, ed. T. H. Clutton-Brock. Chicago, IL: Chicago University Press, pp. 344–62.

Lessells, C. M. 2002. Parental investment in relation to offspring sex. In *The Evolution of Begging*, ed. J. Wright & M. L. Leonard. The Netherlands: Klumer, pp. 65–85.

McClure, P. A. 1981. Sex-biased litter reduction in food-restricted wood rats (*Neotoma floridana*). *Science*, **211**, 1058–60.

Mock, D. W. & Parker, G. A. 1997. *The Evolution of Sibling Rivalry*. Oxford: Oxford University Press.

Newton, I. 1998. *Population Limitation in Birds*. London: Academic Press.

Newton, I. & Marquiss, M. 1979. Sex ratio among nestlings of the European sparrowhawk. *American Naturalist*, **113**, 309–15.

Nunn, C. 1999. The evolution of exaggerated sexual swellings in primates and the graded signal hypothesis. *Animal Behaviour*, **58**, 229–46.

O'Donald, P. 1980. *Genetic Models of Sexual Selection*. Cambridge: Cambridge University Press.

Oring, L. W. 1982. Avian mating systems. In *Avian Biology*, vol. VI., ed. D. S. Farner, J. R. King & K. C. Parkes. New York, NY: Academic Press, pp. 1–92.

Parker, G. A., Baker, R. R. & Smith, V. G. F. 1972. The origin and evolution of gamete dimorphism and the male–female phenomenon. *Journal of Theoretical Biology*, **36**, 529–53.

Payne, R. B. 1979. Sexual selection and intersexual differences in variance of breeding success. *American Naturalist*, **114**, 447–52.

Promislow, D. E. L. 1992. Costs of sexual selection in natural populations of mammals. *Proceedings of the Royal Society*, *Series B*, **247**, 203–10.

Promislow, D. E. L., Montgomerie, R. & Martin, T. E. 1992. Mortality costs of sexual dimorphism in birds. *Proceedings of the Royal Society*, *Series B*, **250**, 143–50.

Ralls, K. 1977. Sexual dimorphism in mammals: avian models and unanswered questions. *American Naturalist*, **111**, 917–38.

Ruckstuhl, K. E. & Neuhaus, P. 2000. Sexual segregation in ungulates: a new approach. *Behaviour*, **137**, 361–77.

Selander, R. K. 1972. Sexual selection and dimorphism in birds. In *Sexual Selection and the Descent of Man, 1871–1971*, ed. B. C. Campbell. Chicago, IL: Aldine-Altherton, pp. 180–220.

Sheldon, B. C., Andersson, S., Griffith, S. C., Ornborg, J. & Sendecka, J. 1999. Ultraviolet colour variation influences blue tit sex ratios. *Nature*, **402**, 874–7.

Smuts, B. B. & Smuts, R. W. 1993. Male aggression and sexual coercion of females in non-human primates and other mammals: evidence and theoretical implications. *Advances in the Study of Animal Behaviour*, **22**, 1–63.

Sutherland, W. J. 1985a. Measures of sexual selection. *Oxford Surveys in Evolutionary Biology*, **1**, 90–101.

1985b. Chance can produce a sex difference in variation in mating success and explain Bateman's data. *Animal Behaviour*, **33**, 1348–52.

Trivers, R. L. 1972. Parental investment and sexual selection. In *Sexual Selection and the Descent of Man, 1871–1971*, ed. B. C. Campbell. Chicago, IL: Aldine-Atherton, pp. 136–79.

1980. *Social Evolution*. New York, NY: Benjamin Cummings.

Trivers, R. L. & Willard, D. E. 1973. Natural selection of parental ability to vary the sex ratio of offspring. *Science*, **179**, 90–2.

van Schaik, C. P. & Janson, C. H. 2000. *Infanticide by Males and Its Implications*. Cambridge: Cambridge University Press.

van Schaik, C. P., van Noordwijk, M. A. & Nunn, C. L. 1999. Sex and social evolution in primates. In *Comparative Primate Sociobiology*, ed. P. C. Lee. Cambridge: Cambridge University Press, pp. 204–31.

Wade, M. J. 1979. Sexual selection and variance in reproductive success. *American Naturalist*, **114**, 742–6.

Wade, M. J. & Arnold, S. J. 1980. The intensity of sexual selection in relation to male behaviour, female choice and sperm precedence. *Animal Behaviour*, **28**, 446–61.

Wallace, A. R. 1889. *Darwinism, an Exposition of the Theory of Natural Selection*. London: Macmillan.

Zinner, D., Alberts, S. C., Nunn, C. L. & Altmann, J. 2002. Significance of primate sexual swellings. *Nature*, **420**, 142–3.

3 • Sex roles, contests for the control of reproduction, and sexual selection

PATRICIA ADAIR GOWATY

Institute of Ecology
University of Georgia
Athens, GA, USA

INTRODUCTION

Sexual selection is a vast topic – the most researched set of ideas in the biology of social behaviour since the 1970s. A thorough, modern, single-source review of sexual selection is Andersson (1994). In the ten years since Andersson, much has been learned about sexual selection. Papers bearing on sexual selection in the primary literature (see, for example, *Animal Behaviour, Behavioral Ecology, Behavioral Ecology and Sociobiology, Proceedings of the Royal Society*) are overwhelmingly common. Their breadth and depth will not disappoint interested neophytes. Because the range of scholarship in sexual selection is so large, it is probably impossible for a single author in a single paper to present more than an idiosyncratic review of the study of sexual selection. My long-standing sensitivity to the sometimes left-out roles of females in sexual selection has affected my view of studies on the subject, and focused my attention particularly on sex roles, which is what I discuss in this chapter. I make no claims that the view here is more than my own; though, of course, I hope it is useful to others. I urge readers – newcomers, especially – to sample widely in the old and new titles of sexual selection, to think critically about all the assumptions of sexual selection theories, and to devise experiments or controlled observational tests capable of rejecting our dearest assumptions, if they are in fact false. Sexual selection theory is so rich that, with such tests, understanding of adaptive social behaviour will continue to advance, as it has so remarkably since the 1970s.

DISCRIMINATING FEMALES AND COMPETITIVE MALES MAY RESULT IN MALE SEXUAL SELECTION

In *The Descent of Man and Selection in Relation to Sex*, Darwin (1871) defined sexual selection as a type of natural selection concerned only with reproductive competition. This defi-nition contrasted with 'survival selection'. Much has been made of the differences between natural and sexual selection; however, in practice and in theory it is often difficult to distinguish sexual selection from natural selection (Andersson, 1994; Clutton-Brock, this volume).

After publication of *The Origin of Species* (1859), in which he described natural selection, Darwin was challenged to explain how selection could account for traits that lowered, rather than enhanced, their bearers' survival probabilities. His contemporaries asked how natural selection could explain the obvious, bizarre and elaborate traits of many males. Darwin's answer was that selection favoured some traits despite their costs to survival, because they enhanced their bearers' probabilities of successful reproduction. His answer was an implicit cost–benefit analysis. What was costly to survival could nevertheless evolve if it enhanced individual reproduction (keep in mind, however, that some traits may evolve due to sexual selection and not lower survival). Darwin (1871) defined sexual selection as selection that 'depends on the advantage which certain individuals have over others of the same sex and species solely in respect of reproduction' (p. 209).

Selection occurs whenever survival and/or reproductive success differ(s) among individuals. Sexual selection is due to variance in reproductive success among members of the same sex and species (Fig. 3.1), while natural selection is due to variance in survival among individuals of the same species independent of their sex. Variance in reproduction due to differential access to resources should be thought of as sexually selected whenever such differential access is due to reproductive competition. For example, variation among females in their abilities to avoid manipulative males could affect variation among them in their abilities to accrue essential nutrients or other resources for reproduction, or their abilities to express their mate preferences, and such mechanisms should be considered under female–female sexual

Sexual Selection in Primates: New and Comparative Perspectives, ed. Peter M. Kappeler and Carel P. van Schaik. Published by Cambridge University Press. © Cambridge University Press 2004.

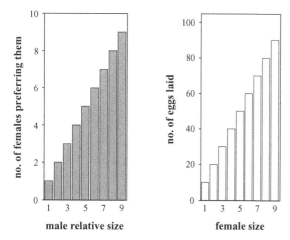

Fig. 3.1 Within-sex trait variation associated with variation in reproductive success (sexual selection). Whether one attributes size variation to survival selection or to sexual selection is often not obvious. If female size facilitates competition among females for preferred mates, it may be a sexually selected trait.

selection. Thus, it is not always clear whether female–female competition over resources is necessarily 'simply natural selection'.

Darwin's lengthy discussion of sexual selection focused on two categories of social interactions: within-sex and between-sex. He discussed two mechanisms; one (female mate choice) fits in the category of between-sex interactions and the other (male–male combat) fits in the category of within-sex interactions. Both were mechanistic explanations for the evolution in males of traits favoured because they enhanced reproduction of males relative to their rivals. Darwin's treatment of female sexual selection was partial, at best, and understandable, given that the challenge he faced in the 1871 volume was to explain bizarre and elaborate traits that one typically observes in males only.

Virtually all modern discussions of sex roles begin with Darwin in tribute to his insights explaining how 'typical' sexual behaviour – sexually discreet and discriminating females and indiscriminate, competitive males – could lead to the evolution of bizarre and elaborate traits in males. The vast majority of modern studies of sexual selection focus on the many unsolved problems of the evolution of traits in males. Among the things that I stress in this paper are the extraordinary number of unsolved problems associated with sexual selection among females, including the fact that there may be benefits of female discrimination beyond those suggested by variation in male traits.

A BRIEF INTELLECTUAL HISTORY OF SEX ROLES

Darwin's interest in sexual behaviour was broad and *The Descent of Man and Selection in Relation to Sex* contains many examples, even in species with higher female than male parental investment, inconsistent with passive, universally discriminating females. Nevertheless, discriminating females and randy, competitive males continue to be in some sub-fields the stereotypical sex-differentiated ideals. For example, as far as I have been able to see, the dramatic observations demonstrating that female mammals, particularly primates, in some situations seemed anything but discriminating and males in some situations anything but competitive had little effect on the questions ornithologists and others asked until the mid 1990s. Despite evidence that Darwin held a broader view than many of his contemporaries – perhaps broader than was convenient for his theory of sexual selection via female mate choice – many modern Darwinists since the 1960s have consistently promulgated a narrow, stereotypical view of females' behavioural variation.

Almost all modern discussions of sex roles also include citations to Bateman's (1948) experiment with *Drosophila melanogaster*, which reinforced the Darwinian view of discriminating females and indiscriminate, competitive males. Bateman realised that if these typical sex roles are responsible for selection on male traits, male mating success variance arising from competitive males and choosy females would lead to larger mating success variance among males than among females. Bateman reported for the first time that variances in mating success among males and among females were what one would expect if females were discriminating and males competitive and profligate. Using flies with male-specific phenotypic markers, he tested this prediction by estimating the mating success of males and females in vials by counting the number of offspring sired by particular males in each female's brood. In sexual species the average reproductive success of males and females must be the same, but the variances in mating and reproductive success may differ, as they did in Bateman's experiment (Fig. 3.2). Most of the females had one or two mates, so that their mating success was more like each other's than the males. Males as often had no mates as one or two, and some males had four. Thus, the variance in reproductive success of the males was greater than the females. Quoting Darwin, Bateman attributed the variance differences to the 'undiscriminating eagerness in males and discriminating passivity in females' (Bateman, 1948, p. 367), although it was unclear whether he actually watched

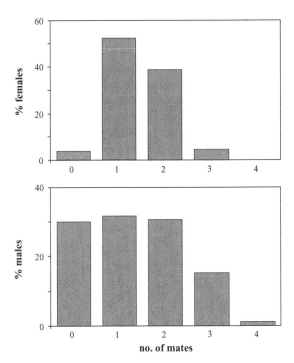

Fig. 3.2   In Bateman's (1948) experiment with *Drosophila melanogaster*, mating success varied less among females than among males.

the behaviour of the flies in the mating vials. Additionally, it is curious that although Bateman based his estimate of number of mates on evaluation of genetic paternity, he did not report reproductive success in terms of numbers of offspring sired per male, but only in terms of number of mates.

Neither Darwin nor Bateman adequately explained the selection pressures working on females to be choosy, though implicit in Darwin's discussions about the advantages of polygyny for males were some hints to the idea that sexual differences might owe their evolution to the selective forces of relative parental investment. That task remained until Williams (1966), Parker *et al*. (1972) and Trivers (1972) each discussed the selective pressures favouring typical sex roles.

ANISOGAMY AND PARENTAL INVESTMENT FAVOURED CHOOSY FEMALES AND INDISCRIMINATE MALES

George Williams (1966), a naturalist and an evolutionary theorist, generalised Bateman's theories and provided a powerful, intuitive description of the selective pressures that may

have favoured choosy females and competitive, indiscriminate males. He is often quoted:

> It is commonly observed that males have a greater readiness for reproduction than females. This is understandable as a consequence of the greater physiological sacrifice made by females for the production of each surviving offspring. A male mammal's essential role may end with copulation, which involves a negligible expenditure of energy and materials on his part, and only a momentary lapse of attention from matters of direct concern to his safety and well-being. The situation is markedly different for the female, for which copulation may mean a commitment to a prolonged burden, in both the mechanical and physiological sense, and its many attendant stresses and dangers. Consequently, the male, having little to lose in his primary reproductive role, shows an aggressive and immediate willingness to mate with as many females as may be available. If he undertakes his reproductive role and fails, he has lost very little. If he succeeds, he can be just as successful for a minor effort as a female could be only after a major somatic sacrifice. Failure for a female mammal may mean weeks or months of wasted time. The mechanical and nutritional burden of pregnancy may mean increased vulnerability to predators, decreased disease resistance, and other dangers for a long time. Even if she successfully endures these stresses and hazards she can still fail completely if her litter is lost before weaning. Once she starts on her reproductive role she commits herself to a certain high minimum of reproductive effort. Natural selection should regulate her reproductive behavior in such a way that she will assume the burdens of reproduction only when the probability of success is at some peak value that is not likely to be exceeded.
>
> The traditional coyness of the female is thus easily attributed to adaptive mechanisms by which she can discriminate the ideal moment and circumstances for assuming the burdens of motherhood . . .
>
> The greater promiscuity of the male and greater caution and discrimination of the female is found in animals generally.
>
> (Williams, 1966, pp. 183–5)

Following Williams (1966), Geoff Parker (1979) and Bob Trivers (1972) galvanised modern efforts to understand selection and sex roles with their hypotheses explaining why females were choosy and males indiscriminate. The ideas of Parker *et al.* (1972) focused on the selective pressures giving rise to anisogamy – the differences in the sizes of the gametes. He argued that 'male and female are what they are because of the type of gamete they carry' (Parker, 1979). Theoretically, ancient selection pressures favouring gametes that accrued resources and those that competed over the resource-rich gametes led to disruptive selection and two types of gametes: big and small, fixing – forever after – sex differences in mating behaviour.

Trivers (1972) further generalised Bateman, Williams and Parker, noting that parental investment patterns were the selective source of typical differences in choosy and indiscriminate behaviour of the sexes. He defined parental investment as 'any investment by the parent in an individual offspring that increases the offspring's chance of surviving (and hence reproductive success) at the cost of the parent's ability to invest in other offspring' (p. 139). He further argued that the sex with the greater parental investment (all else being equal) is a limiting resource for the opposite sex. If this is so, 'what governs the operation of sexual selection is the relative parental investment of the sexes in their offspring' (p. 141). In other words, males compete over access to females because access to females limits male reproductive success. When male investment approaches female investment, Trivers predicted females should be no more discriminating than males, but when male investment exceeds female investment, females should compete among themselves for mates and males should be choosy. What Trivers did was recognise the force of selective episodes beyond those that favoured anisogamy in the first place. He moved the discussion along toward the idea that choosy-versus-indiscriminate mating strategies might be flexible.

Thus, by 1972 there were at least two adaptive hypotheses for sex-role variation: (1) Parker's anisogamy idea said that ancient selection pressures fixed the sex-typical size differences of gametes and, along with gamete size differences, selection fixed female choice and male–male competitive behaviour; (2) Trivers' idea said that variation in parental investment, independent of the relative size of gametes, which are usually larger in females, determines which sex will be choosy and which competitive. These two papers, joined with Darwin and Bateman's hypotheses, were the primary sources of the notion that when individuals of one sex were 'competitive' they would not simultaneously be 'choosy'.

Objections to the anisogamy argument include the fact that sperm are usually ejaculated in multiples (Dewsbury, 1982), so that some have argued that it is the size of the packet,

but not individual sperm size, that determines sex roles. If this argument is correct, when the bolus of sperm is equal or near equal to the size of an egg, choosy and indiscriminate behaviour should not differ between the sexes.

Trivers (1972) is a remarkable, important classic that instigated two crucial research traditions in sexual selection: one is about female choice; the other about sex roles and social organisation. Not just intuitive, Trivers' parental investment hypothesis made crucial testable predictions about sex-role variation in species with male-biased parental investment. If Trivers was right, then in these species females would be competitive and males choosy, while in species with bi-parental care males would be as choosy as females in mating systems characterised by mutual mate choice.

SEXUALLY ARDENT FEMALES CHALLENGED THE MYTH OF THE COY FEMALE

Tests of the parental investment idea in species with male-biased parental investment and with cooperative, bi-parental care began to accumulate in the 1980s. As early as Goodall (1971), primatologists had described the seemingly indiscriminate sexuality of female chimpanzees, *Pan troglodytes*. And the first formal challenges to the 'coy' female idea came from primatologist Sarah Hrdy (1974, 1977, 1979, 1981). Like other naturalists before her, Hrdy had made long-term observations of animals in the wild. Her reports of sexually ardent, promiscuous, free-ranging Hanuman langur females (Hrdy, 1977) were a direct challenge to the hypothesis that selection had favoured universally discriminating females in species with female-biased parental investment. Hrdy argued that selection favoured female promiscuity as a counter-adaptation to sexually selected infanticide by male langurs (and in other species) as a tactic to confuse paternity. This interesting idea said that female promiscuity was an effort to decrease the likelihood of infant killing by adult males recently joining a troop of females. Her idea made novel predictions. As others have recently reported (Nunn & van Schaik, 2000; van Schaik *et al.*, this volume), selection favoured a very long female receptive period, itself a successful resistance tactic: their mating with multiple males, which would confuse paternity and, assuming males were sensitive to their own mating status with particular females, also thereby inhibit infanticide. In her now classically important, but still relevant and readable 1981 book, *The Woman that Never Evolved*, Hrdy focused much-needed attention on variation among females in their abilities to compete among themselves and to avoid or inhibit infanticide; she emphasised female–female competition as a source of sexual selection on females. Her book extended this discussion to other mechanisms of male manipulation or coercion of females' reproductive decisions besides infanticide. Thus, Hrdy's contributions have gone way beyond her initial rebuke of the idea of the universally discriminating female.

Besides challenging the universality of discriminating passivity of females, the exceptions that Hrdy noted stimulated further work on sexual conflict, some of which challenged stereotypic sex roles about females directly. For example, my reading of Hrdy suggested that intersexual conflict was a primary structural factor in social organisations. In some sexual conflict, females and males directly compete with each other over access to resources that females need for reproduction and over control of female reproductive decisions. The existence of such sexual conflict implies active females, because those that remain passive in contests with rivals over resources and with mates over reproductive control are most likely selected against. Despite Hrdy, in the wake of Williams and Trivers, Darwin's world-view of choosy, passive females and indiscriminately competitive males persisted. In addition, alternative hypotheses for sex-typical mating behaviour were hard to come by. The first alternative hypothesis that was taken seriously was postulated by Sutherland (1985a).

RANDOM FORCES CREATE VARIANCE DIFFERENCES IN REPRODUCTIVE SUCCESS WITHIN AND BETWEEN THE SEXES

Until 1985, no one made the critical observation that real-time variance differences in reproductive success of individuals within each sex or between the sexes could not be – by themselves – evidence for sexual selection. Until Bill Sutherland (1985a), no one argued that Bateman's variance differences were insufficient to conclude that females were choosy and passive and that males were profligate and competitive. Sutherland's argument was that random mating combined with the differences in 'handling time' subsequent to any mating could account for the differences in variances between the sexes, so that the typical variance differences that Bateman reported might have resulted even if females were not discriminating and even if males were not ardent. 'Handling time' was a variable regarding the time it took to deal with the results of a copulation, analogous to the concept, from the optimal foraging literature, of the time it took to

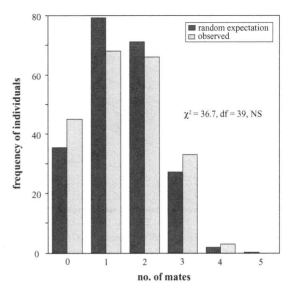

Fig. 3.3   Comparison of Bateman's (1948) results on mating success in experiments with *Drosophila melanogaster* with the random prediction of Sutherland (1985a), which was based on individual variation in latency to receptivity to re-mating. There is no significant difference between the distributions. This means that another explanation for Bateman's results besides choosy females and indiscriminate males is also viable and may even be better than sex-role variation.

process food for consumption. In Sutherland's paper, it was meant to include the physiological and behavioural burdens associated with copulation. So, for example, for males handling time might include the post-copulatory time to acquire resources that rebuild ejaculate volume, and for females the time to gain resources for provisioning of eggs or foetuses during pregnancy. Figure 3.3 shows that there is no significant difference between the variance differences reported by Bateman and those that would arise because of handling-time variation among individuals. Operationally, in Sutherland's study, 'handling time' was a component of latency from the beginning of one copulation to the next, and it might vary among individuals because of random, environmental variation and because of the time different individuals must use to 'handle' a copulation and its products. The greater variance in male than female mating success might have been entirely accounted for by males' shorter latencies than females' to receptivity rather than to ardent males and discriminating females. As far as I have been able to tell, since Sutherland's important demonstration that random forces, rather than sexual selection, could account for Bateman's results, no one has attempted a repetition of Bateman's study and no one

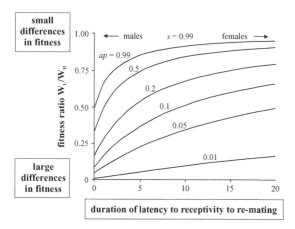

Fig. 3.4   Redrawn from Hubbell and Johnson (1987). The curves on the graph are the isoclines of equal fitness for strategists pursuing choosy (above each isocline) or indiscriminate (below each isocline) mating strategies, given a fixed survival rate, $s = 0.99$, but different values of latencies to receptivity to re-mating; fitness ratios of low- and high-quality potential mates; and strategists' encounter frequencies, ap, with high-quality potentially mating opposite-sex individuals.

has partitioned variances in mating success of any sex of any species into that attributable to mechanisms of sexual selection and that attributable to 'handling time'.

Hubbell and Johnson's (1987) work was in the new tradition of Sutherland. They argued that other factors besides handling time of a copulation combined with random mating could result in greater lifetime mating success variance in one sex than the other. But, they also showed that variation in these real-time variables, rather than any sexually selected trait, could as readily favour an evolutionarily stable strategy (ESS) for choosy or indiscriminate behaviour, independent of an individual's sex, and account for the typical variances in mating success of individuals.

Hubbell and Johnson considered individuals to be in different states: receptive and searching for a mate, encountering a potential high-quality mate, actually mating, or in a reproductive 'time out' during which a strategist is processing the current reproductive bout and is unreceptive to re-mating. Individuals in their model passed through these states until the absorbing state of death. In their model each strategist encountered a random stream of potential mates of two qualities, measured in terms of the fitness conferred on the strategist by mating with them. Strategists that mated as they encountered potential mates were 'indiscriminate'; those that mated only with individuals conferring high relative reproductive success were 'choosy'. The variables that

contributed to lifetime mating success (LMS) and lifetime reproductive success (LRS) in the model were the strategist's encounter rates, a, with potentially mating opposite-sex individuals; the probabilities of mates conferring high quality, p; the strategist's survival probability, s; and the strategist's latency from the beginning of one copulation to receptivity to the next, n (Fig. 3.4). They then derived analytical expressions for the expected lifetime mating success and lifetime reproductive success for individual females and males given time-varying environments and variation in individual life history. Next, they studied how stochastic variation in the five model parameters could contribute to an ESS for choosy versus indiscriminate strategists in either sex. They discovered that the interaction of an individual's survival probability, s, as well as its latency from the beginning of one copulation to the onset of receptivity, n, and its encounter frequencies with potential mates could affect LMS and LRS so that in many mammals, for instance, the variance in mating success or reproductive success could be larger in females than in males. Hubbell and Johnson's important bottom line echoed Sutherland (1985a, b) and emphasised a critical point for studies of sexual selection: 'Because the variation in male LMS due to chance can be large and because this is not heritable variation, we suggest that the opportunity for sexual selection on males be redefined as the residual variance in male LMS that cannot be ascribed to known and quantifiable non-genetic life-history variation' (p. 111).

Hubbell and Johnson's model also showed that the distribution of fitness differences of encountered potentially mating individuals was critical to the fitness of individuals and would contribute to an ESS for choosy and indiscriminate behaviour, typically observed as indiscriminate males and choosy females. Despite this model, I am aware of no studies that partition LMS or LRS variances into those owing to sexual selection and those owing to random, non-heritable forces.

The insights of Sutherland, and Hubbell and Johnson, inform the structure of tests of the hypothesised results of sexual selection. They stressed that to demonstrate sexual selection it is not enough to show that there are variance differences between or among the sexes, because some variance differences necessarily arise due to non-heritable environmental forces. If one finds that variance observations do not exceed those predicted by the null models, one should conclude that the variances are consistent with the force of the parameters of the null model, and that the null model is sufficient to explain the variance differences. If one observes that the variances do not match those predicted by the null model, the variances produced by the parameters of the null

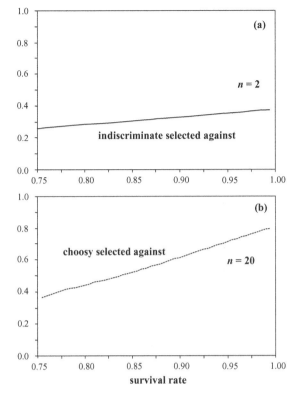

Fig. 3.5   Selection sometimes favours choosy mating strategists in the sex (usually males) with the shorter latencies to receptivity to re-mating, and sometimes favours indiscriminate mating strategists in the sex (usually females) with the longer latencies to receptivity to re-mating. The lines in these graphs represent the isoclines of equal fitness for choosy (overall fitness space above the line) and indiscriminate (overall fitness space below the line) strategists for different values of n, the latency from the beginning of one copulation until receptivity. In these two examples the encounter rate, ap, is set at 0.2. (a) For a latency to receptivity to re-mating of 2 units, $n = 2$, which may be typical of males in many species, selection would be against *indiscriminate* mating behaviour over an enormous range of survival rates and fitness ratios, of low- and high-quality potential mates, from 0 to about 0.4. (b) For a latency to receptivity to re-mating of 20 units, $n = 20$, which would be typical of females in many species, selection would be against *choosy* mating behaviour over a huge range of survival rates and fitness ratios from about 0.38 to 1.

model should be subtracted to find the variance components that may be owing to selection.

Clutton-Brock (1988) argued that the way to proceed was to investigate whether differences in breeding success are related to variation in phenotype. He then concluded that data had suggested that 'chance seldom accounted for

variation in breeding success of either sex' (p. 479). While I frequently admit to being a pan-selectionist, I believe that Clutton-Brock's conclusion was premature in 1988, just as it is today, simply because the structure of studies has so far been inadequate to test the relative contribution to variance differences of the non-heritable forces identified by Sutherland ('handling time'), and Hubbell and Johnson (a, p, s, and n), and the heritable forces of sexual selection. That is, no one, as far as I have been able to tell, has calculated the mean effects of observed s, n, a, and p on estimated variances in LMS or LRS of males and females to compare with observed variances. Until this is accomplished, it is difficult to know how much variance in LMS or LRS might be attributable to heritable trait variation.

THE HYPOTHESIS OF FLEXIBLE MATING STRATEGIES AND SWITCHABLE SEX ROLES

Hubbell and Johnson presented their model as a cautionary tale: sources of variance in mating success of males and females must be partitioned into that owing to real-time variation in environmental and social variables and that owing to trait variation in males or females that deterministically result in variance in reproductive success (sexual selection). Nevertheless, their model says much more than they claimed (Gowaty & Hubbell, unpublished).

(1) The model implicitly hypothesised that selection has acted so that individuals are exquisitely sensitive to variation in environmental parameters that would favour choosy or indiscriminate mating. It furthermore says that selection has acted so that individuals can act flexibly, and that mating behaviour is induced – at least in part – by variation in environmental and life-history circumstances. If Gowaty and Hubbell are right about this new interpretation of Hubbell and Johnson, variation in environmental and social factors *induce* flexible mating strategies.

(2) Because the model was sex-neutral – not depending on intrinsic sex-specific differences between the sexes – it predicts flexibly choosy or indiscriminate behaviour independent of an individual's sex.

(3) It showed when selection would be against indiscriminate mating in the sex with a very short latency (Fig. 3.5a).

(4) It showed when selection would act against choosy mating in the sex with a very long latency (Fig. 3.5b).

(5) It provided a theoretical, quantitative construct for efficient tests of hypotheses about fixed versus flexible sex roles. And, because virgins of both sexes in any sexual species have an $n = 0$, it suggested strongly inferential alternative tests to the predictions of parental investment theory.

(6) It suggested testable alternatives to fixed choosy females and competitive males including that sex-role switching is inducible by rivals, parasites, predators, mates and potential mates (Gowaty & Hubbell, unpublished).

Despite the fact that many overlooked Hubbell & Johnson's study (1987), which so far has been cited only 37 times, it was another interesting alternative hypothesis explaining sex roles. In 1987 their hypothesis that selection favoured an ESS for indiscriminate and choosy mating behaviour was theoretically well developed, providing quantitative alternative predictions to those positing selection that favoured universally discriminating females and indiscriminate males. In addition, as far as I have been able to tell, few have incorporated the concerns of Sutherland or the related concerns of Hubbell and Johnson into evaluations of between-sex variance in reproductive success, and no one has explicitly tested either the qualitative or quantitative predictions of the flexible sex-role hypothesis. Nevertheless, in the years since Hubbell and Johnson, observations matching their predictions, but inconsistent with hypotheses about fixed sex roles, have been accumulating. A review is in Gowaty and Hubbell (unpublished ms); there we show that as individuals' survival probability, s, declines, they become increasingly indiscriminate as predicted by the flexible mating behaviour hypothesis. For example, females change from choosy to indiscriminate under increasing predation risk in crickets (Hedrick & Dill, 1993); water striders, *Aquarius remigis* (Sih & Krupa, 1992); gobies, *Pomatoschistus minutus* (Forsgren, 1992); pipefish, *Syngnathus typhle* (Fuller & Berglund, 1996); guppies, *Poecilia reticulata* (Godin & Briggs, 1996; Gong, 1997; Dill *et al.*, 1999); and tungara frogs, *Physalaemus pustulosus* (Rand *et al.*, 1997). In upland bullies, *Gobiomorphus breviceps*, females change from choosy to indiscriminate as their own parasite load increases (Poulin, 1994), and with increasing age in house crickets, *Acheta domestics* (Gray, 1999), and in Tanzanian cockroaches *Nauphoeta cinerea* (Moore & Moore, 2001). As predicted by the hypothesis of flexible mating behaviour, individuals change from indiscriminate to choosy when their encounter frequencies with potentially mating opposite-sex individuals, ap, increases in *Drosophila melanogaster* (Chapman &

Partridge, 1996); guppies, *Poecilia reticulata* (Jirotkul, 1999); pipefish, *Syngnathus typhle* (Berglund, 1995); and male katydids (Shelly & Bailey, 1992). And, as predicted by the flexible mating behaviour hypothesis, individuals change from choosy to indiscriminate as their encounter frequencies with conspecifics decline in *Drosophila paulistorum* (Kim & Ehrman, 1995, 1997); and in mosquitofish, *Gambusia holbrooki* (Bisazza *et al.*, 2001). Sex roles even switch in African butterflies, *Acraea encedon* (Jiggins *et al.*, 2000); and in pill bugs, *Armadillidium vulgare* (Moreau *et al.*, 2001). When bacterial endosymbionts kill or alter males, so that female encounter rates with males are massively reduced, the remaining males change from indiscriminate to choosy and the females from choosy to indiscriminate. All of these results are consistent with the flexible mating behaviour hypothesis, although most authors interpreted their results in terms of trade-offs in costs and benefits of choice in particular circumstances. Our interpretation of Hubbell and Johnson generalises these sorts of explanations in terms of s, n, ap, and the distribution of fitness differences among encountered potential mates. The trade-off hypotheses are very interesting; however, they are substantially different from the flexible mating hypothesis. We have hypothesised that all individuals in sexually reproducing organisms are sensitive to any variables affecting s, n, ap, and fitness differences between encountered potentially mating individuals, and that this sensitivity combined with the ability to behave flexibly enhances instantaneous contributions to LRS. Thus, what may appear to be a specific trade-off between predation risk and the expression of choosy behaviour may in fact represent a more general ability to respond to any source of variation in s. What is nice about the claim of generality, of course, is that it is testable.

Besides emphasising that mating strategies may be flexible, induced responses to environments that individuals experience, Gowaty and Hubbell (unpublished) hypothesised that rivals, mates, potential mates and parasites may manipulate strategists into modifying their mating strategies or even into switching sex roles. This suggests that the 'typical' sex roles observed in many populations may not be a result entirely of past selection associated with anisogamy or even typical parental investment patterns, but instead owing to selection favouring adaptive responses to variable environments created by rivals, parasites, mates and potential mates, and perhaps chronically different for females and males. In some cases, rivals, parasites, mates and potential mates may simply broker information available to individuals to induce behaviour of individuals favourable to them at a fitness cost to other individuals. For example, chemical 'chastity belts' such as the secondary seminal proteins in the ejaculates of *Drosophila melanogaster* that lengthen female latencies to receptivity to re-mating, decrease the probability of sperm competition. They might simultaneously work for the male's advantage in another way as well: by lengthening the female's latencies to re-mating, the potential benefits of choosy behaviour for her may increase, giving an advantage to the last mating male whenever the probability of re-mating with that male is small or no greater than random. If this is so, it would challenge stereotypical interpretations of the determinants of sex roles and their adaptive significance. The 'discriminatingly passive' females in Bateman's (1948) experiment might not have been expressing fixed female-specific characteristics, but nevertheless have been behaving adaptively. Similar reasoning challenges interpretations of male-specific, fixed, indiscriminate mating behaviour: Bateman's experimental populations were small. He used five males and five females as mating populations. Thus, each male experienced an increasingly lower encounter frequency with potentially mating females. Low encounter frequencies with potentially mating females may have induced indiscriminate male behaviour.

EMPIRICAL FAILURES OF PARENTAL INVESTMENT THEORY

In the meantime, after parental investment theory failed to predict observed variation in pipefish and seahorses (Clutton-Brock & Vincent, 1991), another research trajectory took off from Sutherland. Clutton-Brock & Parker (1992) revised parental investment theory to argue that handling time determines the potential reproductive rate of males and females, and this, in turn, determines the operational sex ratio (ratio of receptive adults at any time in a population), which theoretically finally determines which sex is competitive and which choosy. Theirs is a supply-and-demand argument: males compete because they are the excess sex. In general, because it rested on fixed differences in potential reproductive rates of the two sexes, this idea moved attention away from other environmental, social and life-history variables that also might affect mating strategies and sex roles.

Use of relative reproductive rate to predict 'the' competitive sex is seemingly straightforward and has the appeal of being easily measured and evaluated. However, we have been unable to verify their model under the important constraint in ours: when the average reproductive success of males and females is constrained to be the same, the operational sex ratio does not predict choosy (or indiscriminate) behaviour. We have not yet studied the effect of our model's parameters on competitive behaviour. Intuitively, it remains unclear to me why choosy and competitive are alternatives in the Clutton-Brock and Parker model as noted, for example, by Owens and Thompson (1994). For one thing, Hrdy (1981, 1986, 1997) had shown that choosy females might also be competitive. And are there not indiscriminate males that do not behaviourally compete? It might be possible and profitable to predict more precisely just who within populations is likely to compete. If we are right about the causes of choosy versus indiscriminate mating behaviour, it is also likely that whether individuals of either sex fight over mating access to the opposite sex will depend upon whether any particular fight is over access to an individual with whom the strategist *would* mate, given the chance (Gowaty & Hubbell, unpublished).

COMPETITIVE FEMALES IN CONTESTS WITH MALES OVER THE CONTROL OF REPRODUCTION

Sexual conflict occurs when the fitness interests of the sexes are non-congruent. In some, but not all, cases, sexual conflict implies competitive females. Here I emphasise that competitive females also violate notions of fixed sex-typical behaviour.

Parker (1979) defined sexual conflict in terms of the evolutionary interests of the two sexes, while emphasising the effect of ancient selection pressures of anisogamy resulting in choosy, passive females and active, profligate males. He began his paper with this question: 'What happens when a characteristic of sexual selection advantage (benefit) to the male conveys a selective disadvantage (cost) to genes in the female?' Parker's paper developed explicitly genetical ideas about sexually antagonistic allelic coevolution and 'chase away sexual conflict' almost 20 years before the stunningly creative, clever experimental and conceptual work of Rice and Holland (1997; Holland & Rice, 1998), to which some modern readers attribute the 'chase away sexual conflict' idea. The 'chase away' language refers to the run-away selection that can occur when alleles advantageous in males are costly in females. An example of sexual antagonism at the level of alleles might be selection for long tails in peacocks. While peacocks with long tails may win because more females prefer them, peahens with long tails may suffer a survival disadvantage because their tails increase their predation risk. Parker concluded that males usually win in sexual conflict over alleles. His models were no doubt correct; whether the assumptions of the models match real-world selection pressures remains an empirical challenge. Parker's genetic models estimated the thresholds for conflict between the sexes over mating. For example,

> . . . consider a case in which a characteristic yielding a mating advantage to males causes some disadvantage (cost) to the females with which they mate . . . The female will always benefit from a mating with a male possessing the characteristic, provided that the cost is infinitesimal, if this means that some of her sons inherit the advantage. Similarly for the male, if the costs are felt by his own progeny via damage to his mate, then his mating advantage must be correspondingly greater than for zero costs. Hence as the cost increases there will be two thresholds of cost, one for the male and one for the female, beyond which the male characteristic (or a mating with a male possessing the characteristic) becomes disadvantageous.
>
> (Parker, 1979, pp. 124–5)

Among the assumptions of Parker's model that have yet to be adequately tested, as far as I can see, is whether selection intensity is stronger on males than on females. If this assumed asymmetry does not exist, or if it is reversed, so that females have a more effective 'quantum of adaptation against the opponent's current level of adaptation' (p. 149), females may win sexual conflicts as often as, or even more often than, males. To find out, we must understand the full suite of selection pressures that act on the sexes. While we have discovered some selection invariants acting on the sexes, I think we have as yet no adequate evaluation of the assumptions of this very interesting, pioneering and still important, model of sexual conflict.

Borgia (1979) introduced the idea, in the same volume as the Parker (1979) paper, that males could control and manipulate females' reproductive options, perhaps to the detriment of female fitness. But Borgia's idea did not depend on allelic antagonism between the sexes. It was broader and depended only on the differences between males and females in the selective pressures favouring their (sometimes) quite different reproductive decisions. Borgia's paper was an obvious forerunner to my own ideas about mating systems and constraints on females' options (Gowaty, 1996a, b, 1997c, 1999, 2002; Gowaty & Buschhaus, 1998), and my recent reading of Borgia showed me that he noted the importance of constraints acting on females' options almost 15 years before I

first started to emphasise constraints on females (Gowaty, 1992). Borgia's (1979) paper is also the first of which I am aware that discussed the different kinds of benefits (direct and indirect) that females can gain from mating decisions, and his paper explicitly considered the compromises and trade-offs among gains that females might have to make. Like Parker, Borgia thought males might often win these contests, and females might often be making the best of a bad job. The assumption that fuelled this conclusion was also like Parker's assumption: that there is little or no variation in the quality of females. Borgia (1979) is surprisingly contemporary and reading his work suggests that our current understanding of what female choice is about has not changed much over the intervening years.

In contrast to Parker (1979) and Borgia (1979), Bill Eberhard (1996) proclaimed females the winners in his important book, *Female Control*. His tour de force examined evidence from insect genitalia and internal structures of females for storing and managing sperm. He marshalled favour for his hypothesis that 'cryptic female choice' is the selective mechanism favouring the elaborate evolution of genitalia and sperm management organs. 'Cryptic' in this context means cryptic to the eye of human observers; not necessarily behavioural, it is more likely to occur as physiological parsing by females of sperm, once it is in a female's reproductive tract, well out of the sight of observers and perhaps mates. Eberhard cast females' behaviour as discriminatingly passive, despite his emphasis on cryptic mechanisms of female choice, which he said would be favoured whenever multiple males inseminated the same female. Eberhard's book is one of the most important ever written about sexual selection – it emphasises variation among females and the significance of that variation for affecting traits in males. All neophytes to the study of sexual selection should read it (Gowaty, 1997b).

Smuts and Smuts (1993) argued that male behavioural coercion of females was a mechanism of between-sex selection because it depended on variation among males in their abilities to coerce females. Thus, sexual coercion is like mate choice. Whether it is successful depends on behavioural or physiological interactions between females and males; but the outcome can result in variance among males in mating success and, when it does, sexual selection among males occurs.

It must also be so that whenever male manipulation or control is deleterious to the fitness of females and their offspring, any variation among females in their abilities to remain in control of their own reproductive decisions would lead to reproductive success variance among *females* (Gowaty, 1992, 1996a, b, 1997a, c; Gowaty & Buschhaus, 1998). Thus, sexual selection among females should be as important as male sexual selection to dynamical interactions between the sexes. Sexual selection among females will favour resistance to male attempts to manipulate and control them. And, like cryptic female choice, female resistance can be cryptic too. Imagine, for example, that males attempt to manipulate females by use of seminal peptides (Chapman *et al.*, 2000). Whenever these peptides are deleterious to female fitness, if there is variation in traits, selection is likely to favour female resistance. In such cases, resistance might be behavioural or physiological. Thus, one can easily hypothesise the evolution of peptides in females that neutralise the effects of male seminal proteins.

Ideas about male control and female resistance put within-population variation among females in the centre of the picture. Rice's (1996) experiment on *Drosophila melanogaster* stopped female evolution for about 50 generations. He did this by selective breeding. He allowed only the males with no maternal alleles to breed only with females from the source, non-evolving population. He was able to recognise and remove all males with any genes from their mothers. Rice succeeded in being able to study what would happen if an entire genome was like a non-recombining sex chromosome similar to the Y in humans that is passed more-or-less unaltered from father to son. In males there is no crossing over among the sex chromosomes, and there is another, fourth, very small chromosome in flies, so that it contributes very little. In addition he exploited a chromosomal inversion that inhibits crossing over in two of the largest of the four chromosomes that these flies have. Because he used a strain of fly in which the inversion was indicated by a mutant phenotype, he was able to remove, before breeding, all the males that did not have the inversion and therefore that had genes from their mothers. Thus, Rice removed female influence for 50 generations. When he did, males evolved into 'hyper-males' almost always able to beat females in contests over the control of reproduction. There are many conclusions to draw from his experiment. The most important one from my perspective was that it showed how powerful females are, which it did by removing female influence and comparing what happens to males and females when females do and do not have such influence. Rice interpreted his results in terms of sexually antagonistic *allelic* coevolution.

There is another, more general way to consider his results – not in terms of sexually antagonistic alleles, but in terms of sexually antagonistic *selection pressures* favouring

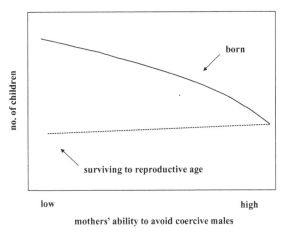

Fig. 3.6   When mothers vary in their abilities to avoid coercion of their reproductive decisions by males, their fitness and the fitness of their offspring vary. Operational variables representing variation among females in their abilities to avoid coercion could be behavioural, for example, hiding ability, running ability, or ability at coalition-building with other females or with males for protection. Operational variables might be cryptic as well, such as physiological ability to kill the sperm from coerced copulations (Gowaty & Buschhaus, 1998). Note that offspring viability is the percentage of offspring born that survive to reproductive age, so that the critical prediction from this hypothesis is that offspring-viability selection due to variation in mothers' abilities to avoid reproductive coercion drives sexual selection among females.

attempts by males to control females and females' resistance to males' control attempts (Gowaty, 2002). Sexually antagonistic selection pressures are those that favour behaviour, physiology, or morphology, enhancing individual ability to win in contests over the control of reproduction. In order for control and resistance to operate dynamically in real time, allelic antagonism (*sensu* Parker & Rice) is unnecessary. A ready example is female resistance to infanticide by incoming males, which as far as I can see is not dependent on alleles advantageous in males that when expressed in females are deleterious. Rather, sexual selection in the form of mating success variation favours infanticide by males, while sexual selection among females favours resistance in the forms of extended receptivity to copulation and promiscuous mating. I have hypothesised that the component of fitness that drives female sexual selection is offspring viability (Gowaty, 2002). I take up this topic again below.

Hrdy called the results of these contests between males and females for the control of reproduction 'strategy and counter-strategy'; I called them 'control and resistance'.

Scientists commonly use 'strategy' and 'counter-strategy' in discussing within-sex fitness contests. 'Control' and 'resistance' suggest that the dynamics of sexual conflict are different from within-sex contests, and 'female resistance' specifically emphasises active, competitive females.

Between-sex contests are special for several reasons.

(1) Not everyone readily agrees that females might successfully win behavioural contests with males. The 'male control–female resistance' language may bring detailed attention to these contests, their dynamics and their outcomes, something that depends on understanding how within-population variation among females (Fig. 3.6) affects their outcomes (Gowaty, 1996a, b, 1997b, c, 1999, 2002).

(2) While the behavioural or physiological contests are between opposite sexes, their outcomes work to affect fitness variation within each sex (i.e., sexual selection). Between-sex contests lead to within-sex variance in reproductive success (Fig. 3.7a, b), making them a bit harder for some to think about than within-sex behavioural contests over fitness.

(3) Different conclusions about the dynamics of sexual conflict may arise depending upon the theoretical origins of the between-sex conflict. Parker (1979) said males win sexually antagonistic contests over alleles. Eberhard (1996) said cryptic female choice guarantees females are the winners. But, whenever the sexually antagonistic contests are dynamic and ongoing – because selection favours males that attempt to control females' reproductive decisions and also females that remain in control of their own reproduction – it is more likely that neither sex 'wins' for long. Therefore, it is perhaps better to speak of oscillatory dynamics of behavioural and physiological contests between the sexes: sometimes male and some-times female advantage rather than 'winners'.

(4) Overall, in sexual selection winners and losers will always be the same sex, even in between-sex behavioural contests over the control of reproduction (Fig. 3.8a, b, c).

VARIATION IN OFFSPRING VIABILITY FUELS SEXUAL SELECTION FAVOURING FEMALE CONTROL OF REPRODUCTION

As Borgia (1979) originally showed, it is easy to imagine that variation among males in mating success drives sexual selection on males to control females' reproductive decisions. It is harder to show that females will experience selection to

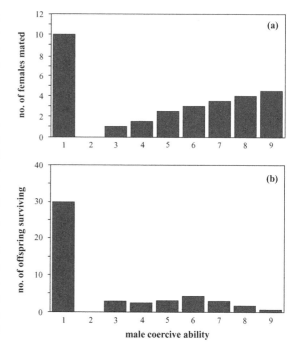

Fig. 3.7   Male–female contests over control of females' reproductive decisions affect sexual selection among males. The graphs are cartoons of potential relationships between male abilities to coerce females and (a) male mating success, and (b) male breeder productivity. Breeder productivity when offspring are at the age of first reproduction (a later measure of fitness) is the product of the number of offspring born and offspring viability (%). Complex relationships between male mating success and earlier and later fitness measures depend on females' abilities to resist male manipulation of their reproductive decisions. See Fig. 3.8 for examples of how females' abilities to resist male coercion may affect their fitness components.

resist such control. This is because females are expected to have little trouble in obtaining mates, so that variation in mating success is unlikely to drive female resistance to males. However, variation in offspring viability (quality relative to pathogen environments) is a force that is likely to exert strong selection pressure on females to remain in control of their reproduction (Gowaty, 1992, 1997c, 2002; Gowaty & Hubbell, unpublished). Rapidly evolving pathogens and parasites probably shape females' mating preferences through offspring viability. If so, whenever males attempt to reproduce with a female that does not prefer them, the opportunity for conflict over the control of reproduction arises. Winning males are those that increase their mating success relative to other males by manipulating or coercing reproduction with females that otherwise would not mate with them. Winning females are those that have higher offspring viability relative to other females because they successfully resisted reproduction with males they did not prefer. (This example should make clear that sexual selection is not just about mating success but any reproductive success that arises due to reproductive competition between members of one sex of the same species.)

The possibility that variation in offspring viability drives mate preferences has been around since Partridge (1980). She reported an offspring viability advantage for females in experimental populations with male-biased operational sex ratios compared to those reproducing in monogamous pairs. Some (Kingett *et al.*, 1981) criticised Partridge for concluding that female preferences resulted in higher offspring viability because she did not eliminate or control other mechanisms of sexual selection besides female choice. In addition, the results have been difficult to replicate. So, until recently, there were only very few demonstrations that mate preferences affected offspring viability. New experiments, specifically designed as comparisons to each other, experimentally manipulated reproduction of females so that some reproduced with males they preferred and some with males they did not prefer. We have demonstrated offspring viability deficit for females in enforced reproduction with males they did not prefer in *Drosophila pseudoobscura* (Anderson *et al.*, 2003); wild house mice, *Mus musculus* (Drickamer *et al.*, 2000); and mallards, *Anas platyrhynchos* (Bluhm & Gowaty, submitted). It is important that our preference tests were 'free'. As in other studies, we eliminated female–female and male–male competition as well as the opportunity for direct intersexual coercion. We also controlled the possibility that exaggerated traits in males manipulate or dazzle females by exploitation of their sensory biases. To control for this aspect of male manipulation of females, we drew individuals to be discriminated from virgin same-sex cultures *randomly with respect to their phenotypic variation*. This aspect of our design may be the reason we have consistently been able to observe offspring viability benefits of mate preferences: we did not assume what the cues are that females use, which is usually the methodology of mate-choice studies, particularly those designed to understand the evolution of traits in males. Taken together, the results of our experiments are strong support for the hypothesis we set out to test: females that remain in control of their own reproduction have offspring with significantly higher viability than females manipulated or coerced into reproduction with males they do not prefer. We concluded (1) that female sexual selection pivots on individuals' abilities to remain in control of their own reproduction; and (2) that rapidly evolving pathogens and

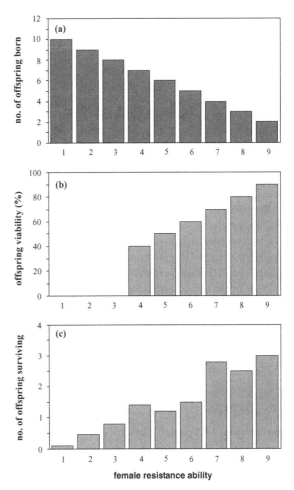

Fig. 3.8   Male–female contests over control of females' reproductive decisions affect sexual selection among females. Females' abilities to resist male coercion of their reproductive decisions may affect more than one component of fitness, if females are able to 'make the best of a bad job' by using fecundity compensation to make up for offspring viability deficits (Gowaty & Hubbell, unpublished). The graphs are cartoons of potential relationships between females' abilities to resist male coercion and (a) an early fitness measure, the number of offspring born; (b) offspring viability; and (c) breeder productivity when offspring are at the age of first breeding (a later measure of fitness). Breeder productivity is the product of the number of offspring born and offspring viability (%).

parasites are the likely drivers of female mating preferences, as Partridge (1980, 1981) originally suggested.

We reasoned that if rapidly evolving pathogens and parasites shape female mate preferences, they are likely to shape male mate preferences as well. Our parallel series of experiments on male mate preferences in mice, *Mus musculus*

(Gowaty *et al.*, 2003) and *Drosophila pseudoobscura* (Anderson *et al.*, 2002) showed similar effects to our experiments on females. Parental investment theory did not predict our results. Each of these species has typical gametic asymmetries with massively larger female gametes, and in mice parental care is by females only. Yet, when we experimentally manipulated males to reproduce with females they did not prefer, they had offspring of significantly lower viability than males that reproduced with females they did prefer. Our results are inconsistent with selection favouring stereotypically indiscriminate mating by males. Thus, we concluded that: (1) the selective powers of rapidly evolving pathogens and parasites have shaped male mate preferences, just as they shaped female mating preferences; (2) another source of sexual selection on males is from between-male variation in offspring viability that must result whenever rivals affect the ability of individuals to reproduce with females they prefer. Our results also suggested that male mate preferences may be much more widespread than usually supposed and that the results of constraints on male mating preferences may have many more selective consequences than usually thought.

THINKING ABOUT SEXUAL SELECTION

In this review I have emphasised five points that modern students of sexual selection ought to keep in mind.

First, the list of mechanisms of sexual selection is longer than just the two most famous examples of male–male combat and female choice. Male mate choice and female–female competition are two frequently noted possibilities. Other between-sex social interactions that can result in sexual selection include male coercion of females (Smuts & Smuts, 1993) and female resistance to male coercion or manipulation (Gowaty, 1997c).

Second, even when a mechanism of intersexual selection depends on interactions between members of opposite sexes, the important thing for selection is the variance in reproductive success among members of one sex. Think about female mate choice for a moment. Whenever choosers discriminate, mate choice may cause variation among the chosen in mating and reproductive success (Fig. 3.7a, b). Thus, mate choice is a mechanism of sexual selection because it theoretically results in variance among individuals of the chosen sex in mating success and perhaps other components of fitness. The between-sex social or physiological interaction of male coercion of females likewise may result in variance among

males in mating success (Fig. 3.7a, b). Female resistance to male coercion may result in variance among females in offspring viability and breeder productivity (Fig. 3.8a, b, c) and thus, it too is sexually selected.

Third, sexual selection can result in individual trade-offs among the components of fitness (Fig. 3.8a, c). The possibility that there is no correlation or even a negative correlation between components of fitness such as number of mates and the number of offspring surviving to reproductive age emphasises that sexual selection is not limited to variance in mating success among males. For example, the number of offspring surviving to the age of first reproduction is a more reliable measure of reproductive success than mating success, or even success at siring the number of eggs laid or offspring born (Gowaty, 2002).

Fourth, for a trait to be under selection, there must be variation in the trait. For sexual selection to operate the trait variation must be among individuals of the same sex. The variation must be heritable, and the trait variants must be associated with differential reproductive success among individuals of the same sex (see also Snowdon, this volume). To argue that an opportunity for sexual selection exists, variance among same-sex individuals in reproductive success must exist.

Fifth, between-sex variances in reproductive success alone are, however, an insufficient basis for the conclusion that sexual selection operates (Fig. 3.3), as within-sex variances may arise because of random, non-heritable factors (Figs. 3.4 and 3.5a, b) (Sutherland 1985a, b; Hubbell & Johnson, 1987) or natural selection in either sex. Thus, it is important to partition within-sex variance (e.g. Fig. 3.3) into that owing to known, non-heritable, stochastic forces (Hubbell & Johnson, 1987), and that owing to differential reproduction associated with within-sex trait variation.

SUMMARY AND CONCLUSIONS

Here I have provided a brief, admittedly idiosyncratic, review of the intellectual history of hypotheses about sex-differentiated social behaviour and sex roles. Ever since Darwin, sex-stereotypical notions about discriminating females and competitive males have dominated discussions about the evolution of behaviour. Darwin had argued that discriminating, passive females and indiscriminate, competitive males resulted in sexual selection among males that could account for the evolution of traits in males costly to their survival. Bateman, Williams, Parker, and Trivers each contributed to the discussions of how variation in parental investment favoured choosy females and profligate males.

Counter-observations of promiscuous females began with Hrdy's field studies of langurs and her hypothesis that indiscriminate copulation by females was a counter-adaptation to the risk of infanticide by incoming males. Observations of sexually assertive female primates (and other creatures) have continued to accumulate, challenging not only the idea that females are invariably discriminating, but also that they are passive.

Sutherland showed that mating success variance differences between males and females result from variation in 'time-out' between reproductive bouts, challenging Bateman's attribution of his results on *Drosophila melanogaster* to choosy females and competitive males. Hubbell and Johnson argued that chronic sex-differentiated behaviour could be due to current environmental conditions, rather than to sex roles fixed by anisogamy or patterns of parental investment. The Hubbell and Johnson model showed when variation in environmental and life-history parameters will select against choosy mating by females and indiscriminate mating by males, and they demonstrated that an ESS for choosy versus indiscriminate mating strategists exists, independent of their sex.

Clutton-Brock and Parker argued that which sex competes is a problem of supply and demand: when operational sex ratios favour males, they said it pays for males to compete and females to choose. I argued that females were sometimes competitive with males over access to resources essential for their reproduction and over control of females' reproductive decisions. Competitive females can compete with males for access to essential resources, but simultaneously remain choosy, preferring to reproduce with some males rather than others because of offspring viability benefits, which may be the fundamental selection pressure favouring active female resistance to manipulation and coercion of their reproductive decisions. Thus, competitive and choosy need not be alternatives.

Sex differences in behaviour may be due to flexible responses of individuals to chronic variation in environmental and life-history variables, including the rate of evolution of pathogens and parasites. I emphasise that there exist only a few descriptions of sex-typical behaviour based on systematic observations across a variety of controlled ecological contexts. Thus, the increasingly common evidence for (1) flexible sex roles along with (2) the rediscovery of selection theory predicting flexible sex roles, and (3) theories positing selection on both sexes favouring mate preferences provide promising directions for future research in a wide variety of taxa.

In summary, sex roles fixed by past selection from anisogamy or from parental investment patterns so that females are choosy and males indiscriminate are currently questionable for many species. The factors that determine whether individuals are choosy or indiscriminate seem relatively under-investigated. For that matter, the number of studies that have systematically compared male and female sexual behaviour using standardised methods to control for environmental and life-history variation that may determine facultative sexual behaviour are still rare. A reasonable approach to future investigations might begin with descriptions of female and male behaviour under similar conditions. Newer questions that now seem more important than ever include: How flexible are mating strategies? How often do chronically choosy females switch to indiscriminate and how often do chronically indiscriminate males switch to choosy? Are the chronic differences between the sexes in most species determined by strategists' encounters with potential mates, their own survival rates, their latencies to receptivity to re-mating, and the distributions of fitness differences within populations, or some combination of these factors? What are the relationships between these time-varying factors and typical patterns of sexual selection among males and *females*?

ACKNOWLEDGEMENTS

I thank Peter Kappeler for inviting me to participate in the Göttinger conference. I suspect that without his invitation, I might have remained ignorant for at least several more years of Hubbell and Johnson's (1987) paper. I thank Steve Hubbell for his suggestion, while I was preparing my talk on which this chapter is based, that I might find Hubbell and Johnson (1987) interesting in relation to sexual selection. I thank Peter Kappeler, Rebecca Lewis, Carel van Schaik, Chuck Snowden, Beth Tyler and two anonymous reviewers for comments. Grants from the US National Institute of Health (NIH) supported most of my work on sexual conflict and sexual selection among females.

REFERENCES

Anderson, W. W., Gowaty, P. A. & Kim, Y. K. 2003. Free mate choice in *Drosophila pseudoobscura* increases breeder productivity and offspring viability. *Proceedings of the National Academy of Sciences, USA*, in press.

Andersson, M. 1994. *Sexual Selection*. Princeton, NJ: Princeton University Press.

Bateman, A. J. 1948. Intrasexual selection in *Drosophila*. *Heredity*, **2**, 349–68.

Berglund, A. 1995. Many mates make male pipefish choosy. *Behaviour*, **132**, 213–8.

Bisazza, A., Vaccari, G. & Pilastro, A. 2001. Female mate choice in a mating system dominated by male sexual coercion. *Behavioral Ecology*, **12**, 59–64.

Borgia, G. 1979. Sexual selection and the evolution of mating systems. In *Sexual Selection and Reproductive Competition in Insects*, ed. M. S. Blum & N. A. Blum. New York, NY: Academic Press, pp. 19–80.

Chapman, T. & Partridge, L. 1996. Female fitness in *Drosophila melanogaster*: an interaction between the effect of nutrition and of encounter rate with males. *Proceedings of the Royal Society of London, Series B*, **263**, 755–9.

Chapman, T., Neubaum, D. M., Wolfner, M. F. & Partridge, L. 2000. The role of male accessory gland protein Acp36DE in sperm competition in *Drosophila melanogaster*. *Proceedings of the Royal Society of London, Series B*, **267**, 1097–105.

Clutton-Brock, T. H. 1988. *Reproductive Success: Studies of Individual Variation in Contrasting Breeding Systems*. Chicago, IL: University of Chicago Press.

Clutton-Brock, T. H. & Parker, G. A. 1992. Potential reproductive rates and the operation of sexual selection. *Quarterly Review of Biology*, **67**, 437–56.

Clutton-Brock, T. H. & Vincent, A. C. J. 1991. Sexual selection and the potential reproductive rates of males and females. *Nature*, **351**, 58–60.

Darwin, C. 1871. *The Descent of Man and Selection in Relation to Sex*. London: John Murray.

Dewsbury, D. A. 1982. Ejaculate cost and mate choice. *American Naturalist*, **119**, 601–10.

Dill, L. M., Hedrick, A. V. & Fraser, A. 1999. Male mating strategies under predation risk: do females call the shots? *Behavioral Ecology*, **10**, 452–61.

Drickamer, L. C., Gowaty, P. A. & Holmes, C. M. 2000. Free female mate choice in house mice affects reproductive success, offspring viability and performance. *Animal Behaviour*, **59**, 371–8.

Eberhard, W. G. 1996. *Female Control: Sexual Selection by Cryptic Female Choice*. Princeton, NJ: Princeton University Press.

Forsgren, E. 1992. Predation risk affects mate choice in a gobiid fish. *American Naturalist*, **140**, 1041–9.

Fuller, R. & Berglund, A. 1996. Behavioral responses of a sex-role reversed pipefish to a gradient of perceived predation risk. *Behavioral Ecology*, **7**, 69–75.

Godin, J. G. J. & Briggs, S. E. 1996. Female mate choice under predation risk in the guppy. *Animal Behaviour*, **51**, 117–30.

Gong, A. 1997. The effects of predator exposure on the female choice of guppies (*Poecilla reticulata*) from a high-predation population. *Behaviour*, **134**, 373–89.

Goodall, J. 1971. *In the Shadow of Man*. Boston, MA: Houghton Mifflin.

Gowaty, P. A. 1992. Evolutionary biology and feminism. *Human Nature*, **3**, 217–49

1996a. Battles of the sexes and origins of monogamy. In *Partnerships in Birds*, ed. J. L. Black. Oxford: Oxford University Press, pp. 21–52.

1996b. Field studies of parental care in birds: new data focus questions on variation in females. In *Advances in the Study of Behaviour*, ed. C. T. Snowdon & J. S. Rosenblatt. New York, NY: Academic Press, pp. 476–531.

1997a. Females' perspectives in avian behavioral ecology. *Journal of Field Ornithology*, **28**, 1–8.

1997b. Good old wine in a new bottle: a review of W. Eberhard's *Female Control: Sexual Selection by Cryptic Female Choice*. *American Zoologist*, **37**, 429–31.

1997c. Sexual dialectics, sexual selection, and variation in mating behavior. In *Feminism and Evolutionary Biology: Boundaries, Intersections, and Frontiers*, ed. P. A. Gowaty. New York, NY: Chapman & Hall, pp. 351–84.

1999. Manipulation and resistance: differential fitness among males via male exploitation of variation among females. In *Proceedings of the 22nd International Ornithological Congress, Durban*, ed. N. Adams & R. Slotow. Natal: University of Natal, pp. 2639–56.

2002. Power asymmetries between the sexes, mate preferences, and components of fitness. In *Evolution, Gender, and Rape*, ed. C. Travis. Cambridge, MA: MIT Press, pp. 61–86.

Gowaty, P. A. & Buschhaus, N. 1998. Ultimate causation of aggressive and forced copulation in birds: female resistance, the CODE hypothesis and social monogamy. *American Zoologist*, **38**, 207–25.

Gowaty, P. A., Drickamer, L. C. & Schmid-Holmes, S. 2003. Male house mice produce fewer offspring with lower viability and poorer performance when mated with females they do not prefer. *Animal Behaviour*, **65**, 95–103.

Gray, D. A. 1999. Intrinsic factors affecting female choice in house crickets: time cost, female age, nutritional condition, body size, and size-relative reproductive investment. *Journal of Insect Behavior*, 12, 691–700.

Hedrick, A. V. & Dill, L. M. 1993. Mate choice by female crickets is influenced by predation risk. *Animal Behaviour*, 46, 193–6.

Holland, B. & Rice, W. R. 1998. Perspective: chase-away sexual selection – antagonistic seduction versus resistance. *Evolution*, 52, 1–7.

Hrdy, S. B. 1974. Male–male competition and infanticide among the langurs (*Presbytis entellus*) of Abu, Rajasthan. *Folia Primatologica*, 22, 19–58.

  1977. *The Langurs of Abu*. Cambridge, MA: Harvard University Press.

  1979. Infanticide among animals: a review, classification, and examination of the implications for the reproductive strategies of females. *Ethology and Sociobiology*, 1, 13–40.

  1981. *The Woman that Never Evolved*. Cambridge, MA: Harvard University Press.

  1986. Empathy, polyandry and the myth of the 'coy' female. In *Feminist Approaches to Science*, ed. R. Bleier. New York, NY: Pergamon, pp. 119–46.

  1997. Raising Darwin's consciousness: female sexuality and the prehominid origins of patriarchy. *Human Nature*, 8, 1–49.

Hubbell, S. P. & Johnson, L. K. 1987. Environmental variance in lifetime mating success, mate choice, and sexual selection. *American Naturalist*, 130, 91–112. ·

Jiggins, F. M., Hurst, G. D. D. & Majerus, M. E. N. 2000. Sex-ratio-distorting *Wolbachia* causes sex-role reversal in its butterfly host. *Proceedings of the Royal Society of London, Series B*, 267, 63–8.

Jirotkul, M. 1999. Population density influences male–male competition in guppies. *Animal Behaviour*, 58, 1169–75.

Kim, Y.-K. & Ehrman, L. 1995. Influence of developmental isolation on courtship behavior of *Drosophila paulistorum*. *Drosophila Information Service*, 76, 126–7.

  1997. Developmental isolation and subsequent adult behavior of *Drosophila paulistorum*. IV. Courtship. *Behavior Genetics*, 28, 57–65.

Kingett, P. D., Lambert, D. M. & Telford, S. R. 1981. Does mate choice occur in *Drosophila melanogaster*? *Nature*, 293, 492.

Moore, P. J. & Moore, A. J. 2001. Reproductive aging and mating: the ticking of the biological clock in female cockroaches. *Proceedings of the National Academy of Sciences, USA*, 98, 9171–6.

Moreau, J., Bertin, A. Caubet, Y. & Rigaud, T. 2001. Sexual selection in an isopod with *Wolbachia* induced sex reversal: males prefer real females. *Journal of Evolutionary Biology*, 14, 388–94.

Nunn, C. & van Schaik, C. P. 2000. Social evolution in primates: the relative roles of ecology and intersexual conflict. In *Infanticide by Males and Its Implications*, ed. C. P. van Schaik & C. H. Janson. Cambridge: Cambridge University Press, pp. 388–419.

Owens, I. P. F. & Thompson, D. B. A. 1994. Sex differences, sex ratios and sex roles. *Proceedings of the Royal Society of London, Series B*, 258, 93–9.

Parker, G. A. 1979. Sexual selection and sexual conflict. In *Sexual Selection and Reproductive Competition in Insects*, ed. M. S. Blum & N. A. Blum. New York, NY: Academic Press, pp. 123–66.

Parker, G. A., Baker, R. R. & Smith, V. G. F. 1972. The origin and evolution of gamete dimorphism and the male–female phenomenon. *Journal of Theoretical Biology*, 36, 529–53.

Partridge, L. 1980. Mate choice increases a component of offspring fitness in fruit flies. *Nature*, 283, 290–1.

  1981. Does mate choice occur in *Drosophila melanogaster*? – reply. *Nature*, 293, 492.

Poulin, R. 1994. Mate choice decisions by parasitized female upland bullies, *Gobiomorphus breviceps*. *Proceedings of the Royal Society of London, Series B*, 256, 183–7.

Rand, A. S., Bridarolli, M. E., Dries, L. & Ryan, M. J. 1997. Light levels influence female choice in Tungara frogs: predation risk assessment? *Copeia*, 2, 447–50.

Rice, W. R. 1996. Sexually antagonistic male adaptation triggered by experimental arrest of female evolution. *Nature*, 381, 232–4.

Rice, W. R. & Holland, B. 1997. The enemies within: intergenomic conflict, interlocus contest evolution (ICE), and the intraspecific Red Queen. *Behavioral Ecology and Sociobiology*, 41, 1–10.

Shelly, T. E. & Bailey, W. J. 1992. Experimental manipulation of mate choice by male Katydids – the effect of female encounter rate. *Behavioral Ecology and Sociobiology*, 30, 277–82.

Sih, A. & Krupa, J. J. 1992. Predation risk, food deprivation and nonrandom mating by size in the stream water strider, *Aquarius remigis*. *Behavioral Ecology and Sociobiology*, 31, 51–6.

Smuts, B. & Smuts, R. W. 1993. Male aggression and sexual coercion of females in nonhuman primates and other mammals: evidence and theoretical implications. *Advances in the Study of Behavior*, **22**, 1–63.

Sutherland, W. J. 1985a. Chance can produce a sex difference in variance in mating success and account for Bateman's data. *Animal Behaviour*, **33**, 1349–52.

1985b. Measures of sexual selection. In *Surveys in Evolutionary Biology*, ed. R. Dawkins & M. Ridley. Oxford: Oxford University Press, pp. 90–101.

Trivers, R. L. 1972. Parental investment and sexual selection. In *Sexual Selection and the Descent of Man*, ed. B. Campbell. Chicago, IL: Aldine, pp. 136–79.

Williams, G. C. 1966. *Adaptation and Natural Selection*. Princeton, NJ: Princeton University Press.

Part II
Sexual signals: substrates and function

4 • Sexual selection and communication

CHARLES T. SNOWDON

Department of Psychology
University of Wisconsin, Madison
Madison, WI, USA

INTRODUCTION

Darwin first wrote about sexual selection in 1871, but only since the 1970s has it become a topic of intense scientific study. The Göttinger Freilandtage (conference) on sexual selection in primates which resulted in this chapter, was held exactly 125 years after Darwin first published his paper on sexual selection on visual signals in primates (1876). How far have we come in understanding sexual selection of communication signals since Darwin's first introduction of the topic? We have had considerable advances in studying sexually selected vocalisations in anurans (e.g. Ryan & Rand, 1990) and birds (e.g. Searcy & Andersson, 1986); sexually selected visual traits in birds (e.g. Borgia, 1985; Petrie, 1994) and invertebrates (e.g. Wilkinson & Reillo, 1994); as well as chemical signals in invertebrates (e.g. Eisner & Meinwald, 1995). The major focus of most of these studies has been on variation in male traits that provide a basis for female choice.

As I have surveyed the primate literature relating communication signals to sexual selection, my assessment is that we have very little real understanding of the role of sexual selection on primate communication that is comparable to our knowledge of other taxa. I think there are several reasons for this lack of progress since Darwin. First, primatologists have lost track of the steps needed for a truly scientific validation of sexual selection, which often requires controlled experimentation. A clear understanding of sexual selection requires not only attempting to understand ultimate causes, but also close attention to proximate mechanisms. Second, primatologists have often assumed that if phenomena superficially meet some aspects of sexual selection, then they need not entertain nor test alternative or complementary hypotheses. As I shall show, there are several explicit steps that are needed to demonstrate that a trait is sexually selected. Our null hypothesis should always be that a trait is not sexually selected.

It is important to distinguish conceptually between intrasexual selection leading to exaggerated traits within a sex that promote the reproductive success of one individual over another of the same sex, and intersexual selection leading to traits that make an individual more attractive as a mating partner for individuals of the opposite sex. Often intrasexually selected communication traits have the added function of minimising fighting or injurious aggression by signalling that one individual is likely to be more successful than the other, or by promoting avoidance of others of the same sex. Intersexual selection of communication manipulates or informs the opposite sex of the value of a potential mate, leading to differential mate choice decisions. Although it is important to maintain a conceptual differentiation between intrasexual and intersexual selection, in fact the same signal may function in both contexts. A complex or frequent song from a male bird, for example, might influence both mate choice decisions by females and avoidance by other males.

In this chapter, I focus on non-human primates, since much has been written about other taxa. I first outline what I think are the necessary criteria for saying that a signal is the result of sexual selection, both intrasexual and intersexual (see Table 4.1). Following this, I review communication in each of the major modalities used (auditory, visual and chemosensory) to evaluate the type of evidence available for either form of sexual selection.

CRITERIA FOR SEXUAL SELECTION

Sexual dimorphism

Sexual dimorphism (a difference between males and females in the expression of a trait) is usually a necessary condition for a signal to be considered the result of sexual selection. It is logically possible, but probably unlikely, that a sexually selected trait could be non-dimorphic so that selection

Sexual Selection in Primates: New and Comparative Perspectives, ed. Peter M. Kappeler and Carel P. van Schaik. Published by Cambridge University Press. © Cambridge University Press 2004.

Table 4.1 *Criteria for demonstrating sexual selection in communication.*

1. Sexual dimorphism in signals.
2. Variation between same sex individuals in the signal.
3. Conspecific discrimination and preference
   (or avoidance) for some variants over others:
   (a) by same sex individuals for intrasexual signals
   (b) by opposite sex individuals for intersexual signals.
4. Expression of preference (or avoidance) in context of
   reproduction or mating.
5. Outcomes of differential preferences (avoidances)
   based on signals must relate to reproductive success.

on one sex resulted in evolution in both sexes (e.g. Lande, 1980). It is also possible, but unlikely, that both sexes develop similar traits with the trait being sexually selected in one sex and naturally selected in the other. Hence we expect sexually selected traits to show sexual dimorphism. However, dimorphism is far from a sufficient condition for a trait to be sexually selected. A trait may be sexually dimorphic, but unrelated to direct effects of sexual selection. Thus, in those primate species with a high degree of body size dimorphism, vocalisations may also be dimorphic, but owing to the coupling of fundamental frequency with larger body size this may not necessarily be caused by selection on the signal itself. Sexual dimorphism alone cannot be used to argue for sexual selection.

Variation of dimorphic traits within a population

The next step is to demonstrate variability in some aspect of the signal within same-sex members of a population. As we all know from Darwin, selection of whatever type can only act on traits that vary within a population. One of the major obstacles that I have encountered in reviewing primate vocal signals, for example, has been the lack of quantification of variation in signals (but see Deputte & Goustard, 1980; Semple & McComb, 2000; Fischer *et al.*, 2002; Semple *et al.*, 2002 for some exceptions). Our colleagues who study bird song have developed several potential metrics of variation: song complexity as measured by the number of different note types or song types produced, song rate and song duration. Call rate and intensity are important in anuran mating calls. Yet surprisingly we have few such quantitative measures applied to primate signals. Some dimorphic vocalisations in poylgynous primates appear to be highly stereotyped (Type I and Type II loud calls; Gautier & Gautier, 1977).

Other vocalisations with the potential for great variability, such as the songs of gibbons, have received documentation of their complexity (e.g. Deputte, 1982), but rarely has this led to documentation of variability or been extended to a breeding population where effects of sexual selection might be measured. Some dimorphic vocalisations of primates – such as the songs of gibbons, the Type I and Type II loud calls of cercopithecine primates, and the roars of howler monkeys – appear to require high energetic costs. However, in comparison with our colleagues who study bird song we are mere infants in our understanding of how to measure variation in signal structure.

Conspecific discrimination of differences in distribution

It is not sufficient to demonstrate variation in the production of some signal; we must also demonstrate that these differences are perceived by conspecifics. For intrasexual selection it must be shown that members of the same sex can discriminate between different structures and rates or intensities of signals, and either act to avoid the signaller or challenge. For intersexual selection to be demonstrated there must be a correlated preference by the opposite sex for some part of the distribution of signal variation compared to other parts of the distribution. Merely demonstrating variation is of no use unless we can also demonstrate that there are consistent preferences for part of the distribution.

A preference test tells us minimally that the variation in the signal or trait is actually perceived, but useful preference tests must go well beyond this. Is the preference displayed simply because the stimuli represent something novel? In a study of putative vaginal pheromones in rhesus macaques (*Macaca mulatta*), Goldfoot *et al.* (1976) found males did show interest in vaginal secretions collected from females at the time of ovulation, but they showed equal or greater interest in odours of green peppers, and in vaginal secretions mixed with the semen of another male. Were the males attracted to a sexually selected odour? To novelty? To the fact that a potential rival had already mated with the female? Only with extremely well-controlled behavioural tests that do not assume *a priori* that a dimorphic trait is sexually selected can we really understand the basis for preferences. It is essential that the perceptual preference be based on those cues hypothesised to have been sexually selected and not on merely novel or arbitrary cues.

The choice of time frame for preference tests is also critical and may depend on the ecology of the species studied. In long-lasting, stable groups, such as those of most primate species, mate choice may be cumulative and based on integration of information from signals over a long period of time. On the other hand, the documentation of male influxes during the mating season in some guenons (e.g. Cords, 1987; Carlson & Isbell, 2001), or brief extra-group copulations in common marmosets (e.g. Lazaro-Perea, 2001) suggest that mate-choice decisions might be made very quickly (or that at times there is little real choice of mates). Similarly, whether one challenges or avoids a member of one's own sex may be based on accumulation of information through repeated encounters or over the course of development. The choice of an appropriate time frame for behavioural evaluation of differences in signals is not a trivial matter for studies on primates.

Expression of preference must be related to sexual motivation and possibility of conception

Individuals might express a variety of preferences in a variety of contexts: a particularly nurturant male might be great to have as a friend for a female with a newborn infant, but unless the female actually mates with this more nurturant male at a time when she can conceive, the preference may have little to do with sexual selection. Thus, it becomes essential to evaluate preferences for one signal or another in a reproductive context. For both intrasexual and intersexual selection the signal itself should bear some relationship to reproduction, either by repelling others of the same sex, attracting members of the opposite sex or both. Although the effects of signals may be cumulative over experience, there must be some evidence that the signal elicits behavioural choices or physiological responses that are consistent with reproduction.

A study of California mice (*Peromyscus californicus*) by Gubernick and Addington (1994) illustrates this point clearly. Female mice were placed in a choice chamber with two tethered virgin males, and the females were tested twice, first while not in oestrus and later in oestrus. Observations over 24 hours showed that females not in oestrus developed a clear associative preference with one of the two males. When subsequently tested in oestrus the females maintained the same associative preference, but fewer than 50 per cent of the females mated and of those only about half mated with their 'preferred' social partner. Thus, although the usual preference data would have indicated a preference for one male, by recording behaviour over 24 hours and testing females both out of and in oestrus, Gubernick and Addington (1994) showed a dissociation between social preferences and mating preferences. Obviously, in sexual selection, mating prefer-

ences count and mere social preferences don't count. Ideal preference tests must demonstrate avoidance or challenge of same-sex animals, as well as sexual arousal, sexual solicitation or other behaviour related to actual mating. Without these behaviours, we should be sceptical of claims for sexual selection.

An ideal signal for sexual selection to operate upon should, therefore, communicate something about reproductive condition, and be effective at a time when successful reproduction is possible. Thus, tests of immature females with male signals are useful only if it can be shown later that the same signals have a similar effect on the female when reproduction is possible. Surveys of 18-year-old college students that look at which cues from the opposite sex might lead to courtship and mating are useful only if it can be shown that the same cues are involved when adults make decisions about conception. A wide variety of primate species (from marmosets to humans) go through a phase of practice relationships where there is much non-conceptive sex (e.g. Lazaro-Perea, 2001). Signals produced and responses to them during these 'practice sessions' may or may not be informative about sexually selected signals when conception is possible.

Outcomes of differential mating and reproductive success

It should be obvious that the outcomes of differential making must be shown to result in differential reproductive success, but I found only a single study in primates that showed a correlation between a communication signal and differential reproductive success (Domb & Pagel, 2001), and serious concerns have been expressed about this study (Zinner *et al.*, this volume). In the absence of measures that differences in a signal actually result in differential reproductive success, all claims about sexual selection are simply hand-waving speculations. As biological scientists, if we wish to be taken as seriously, as are our colleagues in neuroscience, genetics and molecular biology, then we must strive for a higher level of scientific verification than we currently exhibit.

VOCAL SIGNALS: EVIDENCE OF INTRASEXUAL AND INTERSEXUAL SELECTION

Just as there is considerable size dimorphism among some primate species, so there are also sexually dimorphic calls such as the loud roars of male howler monkeys, the long calls of male orangutans, the more complex male songs of gibbons and the male loud calls of many species of cerco-pithecine monkeys. All of these calls, by virtue of being loud, frequent and broadcast over a great range, are suggestive of a sexually selected trait. However, Green (1981) cautioned against assuming that sex differences in calls would be sexually selected, noting alternatives such as differing sounds resulting from different body sizes, different social roles played by and contexts experienced by one sex versus the other, different developmental trajectories, and possible social suppression of calls.

Several studies have used playbacks of male primate long or loud vocalisations, and the nearly universal finding has been that the calls function primarily to keep groups separated from each other. Thus, in agile gibbons (*Hylobates agilis*); Mueller's gibbons (*Hylobates muelleri*) (Mitani, 1985a, b; 1988); orangutans (*Pongo pygmaeus*) (Mitani, 1985c); gray-cheeked mangabeys (*Cercocebus albigena*) (Waser, 1977; 1982); and howler monkeys (*Alouatta palliata*) (Whitehead, 1987), playback results have been most consistent with the idea that these calls serve an intergroup spacing function and thus indicate intrasexual selection. Van Schaik *et al.* (1992) have argued that across several species of langurs, the loud calls function to deter strange males and thereby defend mates. However, none of these studies documented variation among calls from different males and there was little evidence that these calls are directly involved in mate attraction. (Although they might be attractive to females, we have little experimental evidence of females displaying preferences for males based on their calls alone.) These results are especially puzzling when contrasted with parallel work on bird song and frog vocalisations where call complexity or call rate appear to be important in mate attraction (see Ryan & Rand, 1990). Why is there so little evidence of intersexual selection in primate vocal signals?

I think there are several possible explanations. First, it is commonly thought that primate vocalisations, especially loud calls of Old World primates, are stereotyped and not modifiable or variable (Gautier & Gautier, 1977). A recent review of primate vocal development (Seyfarth & Cheney, 1997) concludes that there is little evidence of plasticity in vocal production, although more recent studies using modern acoustic analysis methods have found some evidence of variation. Thus, if male primate vocalisations truly are highly stereotyped, there may be little variation upon which sexual selection can operate. Second, as noted above, we have not yet developed a metric to measure 'quality' in male vocalisations equivalent to the rate of vocalising in frogs, or both rate and song complexity in songbirds.

A third explanation is that we may rarely observe the process of mate choice by primates in nature. Frogs form ephemeral breeding groups and a female may be gravid on only a single night, so mate-choice decisions are frequent and not long-lasting. Even most of the songbirds that are studied are migratory and form new breeding pairs each spring when song rates are highest. Most primate groups, in contrast, are stable and long-lasting with relatively low rates of dispersal, with members of both sexes able to evaluate each other over a long time. In this context the spectacular vocal displays so critical to frogs and songbirds may simply be irrelevant for primates. Finally, many primates simply do not have many mate-choice decisions. A female in the polygynous species, where male long or loud calls are common, may simply have no other options. The highly specialised male vocalisations may therefore be more important for intrasexual selection than intersexual selection.

Some studies have begun to look closely at the structure of vocal signals with respect to sexual selection. Fischer *et al.* (2002) measured characteristics of the loud calls of male baboons, and found lower fundamental frequency and longer durations of calls in adult males. The males in the upper half of the dominance hierarchy had longer calls than other males (J. Fischer, personal communication), suggesting that low frequency and long duration of long calls might be a sexually selected trait. A next step would be to use playbacks to measure female responses to the loud calls of dominant versus subordinate males.

Semple and his colleagues have shown that female Barbary macaques attracted males and received more matings from higher-ranked males when they gave copulation calls (Semple, 1998). Males discriminated between copulation calls given by females at different stages of their ovulatory cycle by approaching more during playbacks of calls from females in late oestrus. The calls of females have more call units, are longer and more highly pitched in late oestrus compared to calls from the same females in early oestrus (Semple & McComb, 2000).

In subsequent research on yellow baboons (*Papio cynocephalus*) Semple *et al.* (2002) found changes in the structure of female copulation calls that correlated with the stage in the ovulatory cycle and with the competitive strength of the mating partner. Thus, female Barbary macaques and yellow baboons can vary their individual copulation calls both to attract and to regulate the quality of mates. Although we have no convincing evidence of female attraction to variations in male vocal cues, we do have evidence of variation in female vocal signals to influence mating.

SEXUAL SWELLINGS IN FEMALES: A MODEL OF INTERSEXUAL SELECTION

Darwin (1876) wrote, '*No case interested me and perplexed me as much as the brightly-coloured hinder ends and adjoining parts of certain monkeys. As these parts are more brightly coloured in one sex than the other, and as they become more brilliant during the season of love, I concluded that the colours had been gained as a sexual attraction . . .*' Darwin, as usual, has displayed his prescience in this quotation, for it is the visual signalling of reproductive status in female baboons that best meets the criteria for an intersexually selected signal.

Many authors have described the sexual swellings of baboons and chimpanzees (see Dixson, 1998 for review). There is a cyclic variation in swelling size, and males appear to be more interested in females at the time of maximum swelling, with more dominant males mating at this time. Let me take this phenomenon and analyse it with respect to the criteria given above. If sexual swelling is a visual signal of reproductive state, then there should be variation in the signal over time; these variations should correlate with some measure of female ovulatory activity, and there should be sexual arousal in males at the peak of the cycle.

Demonstration of these points requires some experimental manipulation, and Craig Bielert deserves the major credit for his experimental demonstration of sexual swellings in chacma baboons (*Papio ursinus*) as true visual sexual signals (see Table 4.2 for summary). First, he demonstrated that females with normal ovarian cycles undergo changes in sex-skin swelling and that males were more likely to copulate with females during the period of swelling. Male copulation rates declined rapidly 2–3 days before the decline in sex-skin swellings, an adaptive behaviour since insemination must occur 24–48 hours prior to ovulation if it is to be successful (Bielert, 1982).

Subsequently, using ovariectomised females Bielert showed that injections of oestrogen would produce fully developed sexual swellings similar to those of hormonally cycling females. Moreover, males copulated with these ovariectomised females after they received hormonal treatment and had displayed sexual swellings. Bielert and his colleagues also noted that after copulatory tests with females, male baboons masturbated frequently, and the resulting seminal emissions were observable on the floor below the male cages (Bielert *et al.*, 1981).

Was direct copulation needed to produce masturbation? In another study, ovariectomised females were treated with oestrogen again, but males did not have any chance to copu-

Table 4.2 *Experimental paradigm for showing sexual swellings that are sexually selected.*

1. Intact, non-pregnant female baboons show variation in sex-skin swelling.
2. Males copulate with females more during swelling (suggesting reproductive function).
3. Injection of oestrogen to ovariectomised females produces swellings similar to normal females (showing swellings are oestrogen dependent).
4. Males copulate with ovariectomised females injected with oestrogen (suggesting males are using swellings as a cue).
5. Males demonstrate sexual arousal (masturbation) to sexually swollen females (copulation not needed).
6. Increasing doses of oestrogen produce increasing swelling in ovariectomised females (showing swelling size depends on levels of oestrogen).
7. Male arousal correlates with oestrogen dose and swelling size (showing males attend to this too).
8. Males show more arousal to intact females with supernormal swelling (but only in early follicular phase when conception is not possible; showing variation in females leads to variation in male response).
9. Females with very high doses of oestrogen do not elicit male arousal when visual barrier is present (ruling out odour and vocal cues).
10. Untreated ovariectomised females wearing prosthetic sex swelling elicit more male arousal than when not wearing prosthesis (showing sex swelling alone and not behavioural changes due to oestrogen are involved).

Source: Bielert *et al.*, 1981.

late with the females. Cages were arranged so that each individual male could see the oestrogen-treated, ovariectomised females, and again male masturbation rates were directly correlated with female swellings (Bielert *et al.*, 1981). However, the oestrogen treated females were possibly producing a variety of other signals: olfactory and vocal signals as well as proceptive behaviours, and we need to consider these factors as potential cues. In addition, there may be something abnormal about using ovariectomised females, even though the oestrogen doses resulted in serum oestrogen levels and swellings that were similar to those of normal ovulating females.

Bielert's next study involved presenting the males with females that were undergoing normal menstrual cycles. As serum oestrogen levels increased so did the sexual swelling, and immediately after ovulation the swelling deflated rapidly. Male masturbation rates increased as the female swelling increased and when the sex-skin swelling deflated, there was a decline in male masturbation (Bielert, 1986). Thus, masturbation rates closely tracked sexual swellings, but a systematic study of oestrogen doses and the effects on female swelling and male response was needed. Bielert injected the ovariectomised females with a range of doses: 1, 5, 10 and 25 μg of oestrogen per day and found a graded response in terms of both size of sexual swelling and male masturbatory response. The lowest dose produced no swelling and no male responses. The 5 μg dose produced a small swelling but no male response, and the 10 μg and 25 μg doses produced increasing swelling sizes with concomitant increases in male masturbation. Bielert and Anderson (1985) further demonstrated this correlation between sexual skin swelling and male sexual arousal. They found that some females naturally had larger swellings than others, and they compared the sexual reaction of males to these supernormal females versus the normal females. Males showed greater masturbatory responses to the females with supernormal swellings, but responded in the early follicular stage of the ovulatory cycle (a time when conception is not possible). However, there were no observable differences in the behaviour of females, suggesting, but not proving, that the differences in male response were due to the swellings and not different behavioural cues. Thus, there is natural variation in sex-skin swellings among female baboons; the variation within a female correlates with her time of ovulation, and male sexual arousal directly correlates with these natural changes.

Although all these studies suggest that sexual swellings provide sexually selected visual cues, there are still several alternative hypotheses. Oestrogen, either changing naturally across the cycle or injected into ovariectomised females, is likely to have many effects other than simply altering the sex skin. There may be changes in olfactory stimuli, vocalisations and behaviour that occur along with changes in sex-skin swelling. Two more studies were needed. Bielert and van der Walt (1982) placed a visual barrier around the females so that males could still hear and smell the females, but there were no visual cues. Ovariectomised females were injected with 50 μg of oestrogen, twice the dose needed to mimic the normal swelling of sex skins. Despite this high dose of oestrogen, the males never masturbated when the females could not be seen. Thus, auditory and olfactory cues were ruled out.

But there is still a potential confounder. Behavioural cues of female solicitation have not been separated from the cues of sexual swelling. In a final study, Girolami and Bielert (1987) created a prosthetic sexual swelling that could be strapped on to an ovariectomised female. The ovariectomised female received no oestrogen, and thus the only variable to differ was the presence or absence of a swollen sex skin. Males masturbated only when the female was wearing the prosthetic sex skin and never responded when she was observed without the swelling.

I have presented these experiments in some detail because they illustrate the care that must be taken to demonstrate that the hypothesised cue is, in fact, the cue involved, and that other variables are not influencing the behaviour. The first four criteria for sexual selection have now been met: sexual swellings are dimorphic; they vary within a population; and males react more strongly to larger swellings than to smaller ones, showing that they can discriminate among variations. Furthermore, the signals vary as a function of reproductive state, and males are more likely to mate (or masturbate) when swellings are maximal. Other potential cues have been ruled out. There remains the final criterion of whether reproductive success is at all affected by variation in sex-skin swelling.

Both Nunn (1999), and Stallman and Froehlich (2000) review several different functional hypotheses concerning sexual swellings – honest communication, paternity confusion, paternity confidence and paternal investment, protection, incitement of pre-copulatory male–male competition, post-copulatory sexual selection and sensory exploitation. These authors ruled out a few hypotheses but suggested that empirical evidence was lacking in support of most remaining hypotheses. Pagel (1994), in contrast, developed a theoretical model that considered benefits to males and concluded that honest advertisement of female fitness was the best hypothesis for explaining sex-skin swellings.

Domb and Pagel (2001) appear to have provided empirical support for Pagel's hypothesis. Studying a population of olive baboons (*Papio cynocephalus anubis*) living at Gombe National Park in Tanzania, they measured the length, width and depth of sexual swellings of females during ovulatory cycles, and simultaneously kept track of male interest in females, as well as aggression received by males when consorting with females. Based on a long-term database of female reproductive history, they correlated swelling size with a variety of parameters. Swelling length (although not width or depth) correlated negatively with age at first conception, meaning that females with longer swellings became reproductive at an earlier age ($r = -0.61$). There were also significant positive correlations between swelling size and number of offspring per year ($r = 0.58$), number of surviving offspring per year ($r = 0.68$) and proportion of all offspring surviving ($r = 0.59$). The correlations given above were significant and are also controlled for female age and rank. Thus female swelling size appears to provide an honest signal of female fitness. Males consorting with females with larger swellings also received more aggression from other males ($r = 0.82$) suggesting a potential of larger swellings for incitement of male–male competition.

However, Zinner *et al.* (this volume) have challenged these results with a re-analysis of the Domb and Pagel data, looking at several additional factors such as female size and age, troop differences in feeding and reproductive success, and sex ratio within each group. Zinner *et al.* found that, after correcting for these factors, the correlation between swelling size and reproductive success became non-significant (although still positive). Zinner *et al.* suggest that female swelling size may be an indicator of phenotypic quality but not genetic quality, and further suggest that the increased male aggression over females with large swellings, might be an artefact of increased aggression in those troops with more male-biased sex ratios. However, if swelling size is related to increased competition among males, then there is a close functional parallel between sexual swellings and the copulatory vocalisations studied by Semple and colleagues.

Despite the critique by Zinner *et al.*, the Domb and Pagel study illustrates the kind of data needed to show that variation in a signal can lead to variation in reproductive success. Field data are essential to document reproductive success. However, field data alone are not sufficient. To demonstrate unambiguously that the swelling itself and not some other aspect of behaviour or communication was critical, and to demonstrate conclusively that male reproductive effort was influenced by the swelling, required the lengthy series of controlled experiments by Bielert and his colleagues described above. Neither field studies nor captive studies alone are sufficient for a complete documentation of sexual selection.

Although this discussion has focused on sexual swellings as intersexual communication signals, the swellings also indicate intrasexual selection as well. If swelling does communicate female quality, then one would expect competition among females to produce ever larger swellings, and if swellings incite competition among males, then females' swellings are directly affecting intrasexual selection among males. Although conceptual distinctions between intrasexual and intersexual selection are important, in fact the same signal might be effective simultaneously in both types of sexual selection.

CHEMICAL SIGNALS: EVIDENCE FOR INTRASEXUAL AND INTERSEXUAL SELECTION

Chemical signals have been studied primarily in prosimians and Callitrichid monkeys, although there is emerging evidence that chemical signals change throughout the cycle of a human female, with potential consequences in regulating ovulation in other women (Stern & McClintock, 1998), as well as providing cues of ovulation to men (Singh & Bronstad, 2001). This is another example of signals that might be affected by both intrasexual and intersexual selection. There has been extensive documentation that chemical signals provide information about species, sex, population and individual identity, but chemical signals also provide information useful for mate choice and reproduction (Epple, 1985).

In pygmy marmosets (*Cebuella pygmaea*), Converse *et al.* (1995) reported that although female pygmy marmosets did not scent-mark more often during the periovulatory period, males showed increased investigation of female scents at this time. Ziegler *et al.* (1993) developed a behavioural bioassay for cotton-top tamarin (*Saguinus oedipus*) scents. They collected daily scent marks from a donor female throughout her ovulatory cycle and presented these marks to pairs of tamarins. Males demonstrated a significant increase in erections and also increased mounting of their own mates on days when they received odours from the periovulatory period of the donor female, compared with odours from the same female collected at other times in the cycle. Subsequently, Washabaugh and Snowdon (1998) tested cotton-top tamarins with odours of unfamiliar females that were either reproductively active or reproductively suppressed. Both males and females discriminated between reproductive and non-reproductive unfamiliar females and part of the response involved increased sexual solicitation of the male by the female and increased rates of mounting by the males. A study of common marmosets (*Callithrix jacchus*) also found that males discriminated between odours of periovulatory and anovulatory females (Smith & Abbott, 1998). These studies suggest that chemical signals might play a role in sexual selection.

Heymann (2003) has argued that scent-marking is a sexually selected cue in Callitrichines based on variation in male parental care among species. The role of caregivers, other than the mother, appears more pronounced in tamarins compared with marmosets. Two recent studies on captive cotton-top tamarins have documented male weight loss of up to 10 per cent during the time that males care for infants (Sanchez *et al.*, 1999; Achenbach & Snowdon,

2002). In both field and captive populations of cotton-top tamarins infant survival does not reach 100 per cent until there are five caregivers in the group (Snowdon, 1996), and in moustached tamarins (*Saguinus mystax*) Garber *et al.* (1984) reported greater reproductive success in groups with at least two males. These findings contrast with studies of marmosets and lion tamarins (*Leontopithecus rosalia*) where helpers (the male and others) do not appear so essential. Heymann (2003) also notes female sexual dimorphism in rates of scent-marking in *Saguinus* species in contrast to other Callitrichines, and argues that because male *Saguinus* invest so heavily in parental care, they benefit more from careful mate choice, and that females engage in significantly more scent-marking to communicate reproductive potential to mates.

There is also evidence of intrasexual competition among females with reports of infanticide in wild common marmosets (Digby, 1995; Lazaro-Perea *et al.*, 2000). Scent marks from dominant reproductive females inhibit ovulation in subordinate female saddlebacks (*Saguinus fuscicollis*) and cotton-top tamarins (Epple & Katz, 1984; Savage *et al.*, 1988). Thus, for several reasons, female chemical cues are good candidates for sexually selected signals for both intrasexual and intersexual selection. Even in species such as common marmosets, where both sexes scent-mark at equal rates, there are still sex differences in the quality of odours since males can discriminate between periovulatory and anovulatory females (Smith & Abbott, 1998).

Thus we have considerable evidence of sexual dimorphism in tamarin scent-marking rates, and likely dimorphism in chemical signals, if not marking rates, in marmosets. We also have evidence that males can discriminate between reproductively active versus anovulatory or reproductively inhibited females, and can discriminate odours from an ovulating female versus odours from the same female at times when she is not ovulating. There is variation in the signal both among females in a population and within a female over her reproductive cycle. Do males demonstrate different sexual responses to these variations in signals? Although it has been hypothesised that males make use of olfactory cues in mating, no direct tests have been done.

To answer this question we have made use of a new non-invasive methodology for studying brain function in monkeys (Ferris *et al.*, 2001). Based on work by Dixson and colleagues (Kendrick & Dixson, 1986; Dixson & Lloyd, 1988; 1989) two brain areas – the anterior hypothalamus and the medial preoptic areas – are known to be involved in sexual arousal in both sexes, and in copulatory behaviour in

male marmosets. Lesions to these areas reduce copulatory behaviour, and produce less following behaviour of females, as well as less anogenital investigation. If olfactory cues from females do produce sexual arousal in males, then we would predict increased neuronal activation in both the anterior hypothalamus and preoptic areas in male marmosets presented with female odours, and we would expect greater activity in response to odours of periovulatory females than to odours from non-reproductive females.

We used functional magnetic resonance imaging (fMRI), a non-invasive method for visualising brain activity, so that the same subjects could be tested repeatedly over time. The method produces greater temporal and spatial resolution than traditional lesioning methods, but has less resolution than single-unit recording. Because the same subjects can be tested repeatedly, and the entire brain can be scanned, variability is minimised among subjects and one can learn much more from a few animals than would be possible with previous methods. The fMRI method works by placing the subject in a very high magnetic field with a coil close to the area being imaged. Deoxygenated haemoglobin disrupts the local magnetic field; thus, with enhanced neural activity in an area, the increased capillary flow and corresponding increase in oxygenated haemoglobin lead to a change in the magnetic field. This change can be detected, leading to a signal that in our study ranged from 3 to 15 per cent above baseline activity.

We habituated four male common marmosets to head and body restraint, initially anaesthetising them for placement in the apparatus. After the animals were positioned in the restraining apparatus, they were administered an antidote to the anaesthesia and were fully recovered in 12 to 30 minutes. At this time, we began collecting anatomical data using the 9.4 Tesla research magnet at the Center for Magnetic Resonance Research at the University of Minnesota, about 400 km away. (We brought the monkeys and their cage mates with us from the Wisconsin National Primate Center.) After the anatomical data were gathered, we then presented monkeys in various sequences with a baseline vehicle control, with odours from ovariectomised females and with odours gathered at the time of ovulation from normally cycling females. We gathered data first from 7 minutes of baseline, followed by 7 minutes of odour presentation, followed by a 10-minute post-stimulus data collection. Finally, another anatomic series was collected. At the end of a sequence of imaging, the monkeys were removed from the magnet and restraint, rewarded with a highly preferred food and returned to their cage mate.

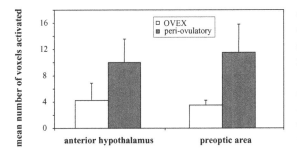

Fig. 4.1   Mean number of voxels significantly activated compared to vehicle control in response to scent marks from ovariectomised females (OVEX) and normal females at the time of ovulation, for the anterior hypothalamus and medial preoptic area of male common marmosets (based on data from Ferris *et al.*, 2001).

There are some important controls needed. First, at the high magnetic field used even slight movement could create artefacts, but subtraction of anatomical data from the beginning and end of an imaging session showed no evidence of movement. Second, we measured cortisol levels before and after imaging to monitor stress. While we found cortisol levels to be a mean of 100 μg/dl higher after imaging than before, this change is the same as that found when animals with their cage mates are moved to a new cage. This change is the equivalent of a mild social stress.

The results were clear. Looking only at the anterior hypothalamus and medial preoptic area, we observed increases in activity in response to both odours. That is, neural activity was significantly above baseline levels (Fig. 4.1). But, there was significantly more activation in each area in response to odours from periovulatory females compared to odours from ovariectomised females (Ferris *et al.*, 2001). Two of the four males were virgins and had never copulated before, but their activity was the same as that of sexually experienced males. Thus fMRI provided evidence that brain areas associated with sexual arousal and with copulation were activated in common marmosets by female scent marks in the absence of any other cues. Furthermore, odours from females of different reproductive value (ovulating versus ovariectomised) had differential activational effects. We do not have the field data to determine whether variation in quality of female odours is directly related to differential reproductive success. However, there is evidence for high reproductive skew among female Callitrichids owing to reproductive inhibition of subordinate females, and, as noted, in captivity males can discriminate between the odours of unfamiliar reproductively active versus reproductively suppressed females. The differential response to ovulating versus ovariectomised females suggests the likelihood that variation in female chemical cues is related to reproductive success.

Although I have described so far the responses in areas involved in sexual interest and copulation, the virtue of fMRI is that we also have data on activation patterns in other parts of the brain, such as the dopamine arousal and reward pathways, and the cortical and subcortical areas known to be involved in emotional responding in humans and other animals. Thus we can go well beyond questions of sexual arousal to see what other brain systems are activated by odours.

SEXUAL SELECTION OR SEXUAL LEARNING?

So far I have been reviewing cues that are thought to have been selected to promote intersexual mating or mediate intrasexual competition, but are the cues that human or non-human animals use purely the ones that might result from sexual selection? Goldfoot *et al.* (1976), in following up on work by Michael and Keverne (1970; Michael *et al.*, 1971) on putative vaginal pheromones in rhesus monkeys, found several results at odds with sexually selected odours. Males would mate preferentially with specific females, including some that were ovariectomised. As mentioned above, Goldfoot *et al.* (1976) found that males showed as much interest in females with artificial odours of green peppers as females with natural vaginal odours, and males often showed more interest in females that had recently mated and still had semen from a different male. In a review of his work, Goldfoot wrote:

> If a male has learned previously to associate copulatory success with visual and odour qualities of his partner, the peripheral cues would be expected to increase sexual arousal . . . In addition to these stimulus conditions, the male must assess the female's social condition since from previous experience he might have learned that he will not be allowed to copulate if he is within view of a dominant and aggressive male.
>
> (Goldfoot, 1982, p. 424)

Are we primates sexual automata led to mate choice and mating behaviour based on sexually selected cues, or are we instead sensitive to a variety of signals both unlearned (selected) and learned? It seems unlikely that we are sexual automata, yet there have been no studies demonstrating sexual conditioning in non-human primates. This is all the more strange when sexual conditioning has been demonstrated in many other species: in rats (*Rattus rattus*) (e.g. Kippin *et al.*, 1998; Pfaus *et al.*, 2001); in gerbils (*Meriones unguiculatus*) (Villarreal & Domjan, 1998); in Japanese quail (*Coturnix japonica*; e.g. Koksal & Domjan, 1998); and even in fish, the blue gurami (*Trichogaster trichopterus*; Hollis *et al.*, 1989). Sexual conditioning can even lead to male rats displaying preferences for odours of decaying flesh (Pfaus *et al.*, 2001).

Two studies have shown greater reproductive success after conditioning. In blue guramis, males that learn a cue that predicts the arrival of a female inhibit their normal territorial aggressive behaviour, spawn more quickly with the female, and produce more young than males who did not learn a signal to predict the arrival of females (Hollis *et al.*, 1997). Japanese quail placed in a novel chamber, where they had previously learned to expect to copulate with a female released greater volumes of semen and greater numbers of spermatozoa than unconditioned males, although there was no change in serum testosterone (Domjan *et al.*, 1998). Both of these studies demonstrate a significant reproductive advantage to those males that can learn cues to anticipate the arrival of a female.

My colleague Pamela Tannenbaum recently completed a study of sexual conditioning in male marmosets to an arbitrary odour – lemon. In our previous fMRI studies we included a lemon odour as one of the stimulus conditions both to provide a novel olfactory stimulus as a control and to prepare for the conditioning study. The four males from the fMRI studies were tested in their home cages with their partners absent. They were tested for 6 days of pre-conditioning with a nest box placed inside the home cage. Then followed 12 conditioning trials, interspersed with 12 control trials. The nest box contained a periovulatory female in each trial (both conditioning and control), and after 7 minutes of exposure to a lemon-scented wood block, the female was released from the nest box. On interspersed control trials, no scent was on the wooden block, and the female was not released from the nest box. All males copulated on each lemon conditioning trial, and three of the males produced anticipatory erections in response to the odour prior to the release of the female. Over the 12 trials, males spent significantly more time close to the nest box on lemon trials than they did on either control trials or in the pre-conditioning tests. This increase in approach to the nest box only in the lemon condition is all the more striking, because on the control trials the nest boxes still contained an ovulating female who had all of the natural periovulatory olfactory cues as on

the lemon trials. Male marmosets can become conditioned to arbitrary stimuli that predict an opportunity to copulate with females. The final step is to see if the previously neutral lemon odours now produce activation of the anterior hypothalamus and preoptic areas. We have completed the imaging but do not have the data analyses completed at time of going to press.

With respect to chemical signals produced by female marmosets, we have considerable evidence that these cues may be sexually selected stimuli: there is variation in quality with reproductive value, the odours lead directly to neuronal activation in the brain areas involved in sexual investigation and copulation in males, and the brain responses are found even in virgin males with no prior copulatory experience. But the same monkeys also can be readily conditioned to a completely arbitrary odour when that odour predicts an opportunity to copulate.

Although there do appear to be some communication signals that may be sexually selected, it may be best to consider these signals as biasing factors rather than the determinants of mate choice. For primates, human and non-human, as well as for Japanese quail, gerbils, rats and blue guramis, there is more to successful reproduction than simply responding to sexually selected cues. Although I might be initially attracted to a woman with the 'correct' breast-to-waist-to-hip ratios, a symmetric face and all of the other hypothesised sexually selected cues, I will quickly learn if she is intelligent or not, if she is emotionally stable, and many other things that should be more important in my reproductive decisions than mere appearance. It is important to keep this in mind in any discussion of sexual selection.

Sexual learning also suggests a mechanism for maintaining variation in sexually selected traits. Different individuals have different developmental histories and experiences. A mate choice that is appropriate for one may not be appropriate for another: an obvious example is avoiding breeding with one's kin no matter how high their quality may be. Sexual conditioning provides a mechanism for choosing an individually appropriate mate and therefore can serve to maintain variance of sexually selected traits within a population.

SEXUAL SELECTION AND FEMALE TRAITS

It is interesting that the best examples of intersexually selected communication signals (vocal, visual and olfactory) in primates seem to be those produced by females. We have long assumed that mate choice only mattered to females and not to males. In many primates, however, there are significant costs to males – either in the increased aggression and competition as seen in the baboons in Domb and Pagel (2001), or in the costs of paternal care in the marmosets and tamarins (Sanchez et al., 1999; Achenbach & Snowdon, 2002). Therefore, males should be carefully evaluating females.

As noted earlier, mate choice may be more difficult to study in primates than in other species. The best-documented examples of sexual selection in communication signals are in species with brief mating bouts (such as frogs) and where new pairs are formed each breeding season (as in many birds). Primate social groups tend to be much more stable, and female mate-choice decisions can be formed on the basis of information gathered over a long time. Although male primates also have a long time period for evaluating females, they also need to detect when it is time to mate, and in polygynous groups to evaluate the social risks of potential aggression in attempting to mate. Male primates may have as much at stake as females in mate-choice decisions, and thus intersexual selection may be acting as much or more on female traits.

At the same time, female primates are also competitive. This is most evident in Callitrichids where there is high reproductive skew among females and where, typically, only one female in a group will breed while suppressing other females. When this suppression fails, competing females will kill each other's offspring (Digby, 1995; Lazaro-Perea et al., 2000). Female competition is emerging as an important variable in other species as well (e.g. Pusey et al., 1999). Male and female primates both have much at stake in reproductive decisions and we need to consider more egalitarian models of mate choice and sexual selection.

SUMMARY AND CONCLUSIONS

I have presented the criteria necessary for determining whether a signal is sexually selected or not; criteria that require an understanding of both proximate mechanisms and ultimate causes, and an integration of experimental and field studies. Using these criteria I have suggested that there is little evidence (so far) of intersexual selection in vocal signals, although the loud or long calls of some male primates may function as intrasexually selected signals. Sexual swellings of female baboons provide the most complete documentation needed for intersexual signals in primates owing to a set of comprehensive experiments and field data. Chemical signals appear to function both in intrasexual competition among females and as intersexual attractants for males in both Callitrichid and human primates. However, sexual conditioning to arbitrary cues has been demonstrated in a wide variety of non-primate species leading, in some cases, to increased reproductive success. I outlined the first sexual conditioning study in primates and suggested that conditioning to specific cues may function to provide individuals with their best mates, and thus maintain variation of cues within a population. The strongest evidence, so far, for intersexual selection of traits is observed in female primates, suggesting that male mate choice and female competition may be as important as male competition and female mate choice. This review leads to the following conclusions:

(1) The study of sexual selection in primates needs more critical examination of the assumptions made and the data gathered. Strong inference models used in other scientific disciplines must be applied to the study of sexual selection.

(2) We have differing levels of evidence for different communication modalities on the relative contribution of signals to intrasexual versus intersexual selection. Vocal signals appear most involved in male–male competition or avoidance; sexual swellings in female attraction of good mates, and chemical signals are effective in both mate attraction and female–female competition.

(3) Because most primates live in stable, long-lasting social groups, pressures for direct sexually selected communication cues may be less than in species with ephemeral mating groups or frequent pairings. Primates are likely to accumulate information about competitors and mates from many sources over a longer time frame.

(4) As a result of long-term information gathering, sexual conditioning to many potential cues is likely to be more important than direct automatic responses to communication cues that appear to be sexually selected. If different individuals are conditioned to different cues, variation in signals can be maintained in a population, despite sexual selection.

(5) The data suggest that intersexual selection is as strong if not stronger on female primates than on males. Thus we need to re-evaluate our models of female mate choice and male competition to include male mate choice and female competition as equally important.

acknowledgements

My own research presented here has been part of a team effort involving Toni Ziegler, Kate Washabaugh, Cristina

Lazaro-Perea, Pamela Tannenbaum, Craig Ferris, Jean King, Nancy Schultz-Darken, Timothy Duong, Reinhold Ludwig and David Olson. I thank Susan Alberts and Karen Strier for valuable discussions concerning the ideas developed here, and the editors and two reviewers for critical feedback. The research is supported by United States Public Health Service (USPHS) grants MH35215 to Charles T. Snowdon and Toni E. Ziegler by MH 58700 grants to Craig Ferris and Jean King and by an RR00167 grant to the Wisconsin Regional Primate Research Center.

REFERENCES

Achenbach, G. G. & Snowdon, C. T. 2002. Costs of caregiving: weight loss in captive adult male cotton-top tamarins (*Saguinus oedipus*) following the birth of infants. *International Journal of Primatology*, **23**, 179–89.

Bielert, C. 1982. Experimental examination of baboon (*Papio ursinus*) sex stimuli. In *Primate Communication*, ed. C. T. Snowdon, C. H. Brown & M. R. Petersen. New York, NY: Cambridge University Press, pp. 373–95.

1986. Sexual interactions between captive adult male and female chacma baboons as related to the female's menstrual cycle. *Journal of Zoology, London*, **209**, 521–36.

Bielert, C. & Anderson, C. M. 1985. Baboon sexual swellings and male response: a possible operational mammalian supernormal stimulus and response interaction. *International Journal of Primatology*, **6**, 377–93.

Bielert, C. & van der Walt, L. A. 1982. Male chacma baboon (*Papio ursinus*) sexual arousal: mediation by visual cues from female conspecifics. *Psychoneuroendocrinology*, **7**, 31–48.

Bielert, C., Howard-Tripp, M. E. & van der Walt, L. A. 1981. Environmental and social factors influencing seminal emission in chacma baboon (*Papio ursinus*). *Psychoneuroendocrinology*, **5**, 287–303.

Borgia, G. 1985. Bower quality, number of decorations and mating success of male satin bowerbirds (*Ptilonorhynchus violaceus*): an experimental analysis. *Animal Behaviour*, **33**, 266–71.

Carlson, A. A. & Isbell, L. A. 2001. Causes and consequences of single-male and multimale mating in free-ranging patas monkeys, *Erythrocebus patas*. *Animal Behaviour*, **62**, 1047–58.

Converse, L. A., Carlson, A. A., Ziegler, T. E. & Snowdon, C. T. 1995. Communication of ovulatory state to mates by female pygmy marmosets (*Cebuella pygmaea*). *Animal Behaviour*, **49**, 615–21.

Cords, M. 1987. Forest guenons and patas monkeys: male–male competition in one-male groups. In *Primate Societies*, ed. B. B. Smuts, D. L. Cheney, R. M. Seyfarth, R. W. Wrangham & T. T. Struhsaker. Chicago, IL: University of Chicago Press, pp. 98–111.

Darwin, C. 1871. *The Descent of Man and Selection in Relation to Sex*. London: Murray.

1876. Sexual selection as related to monkeys. *Nature*, **15**, 18–9.

Deputte, B. L. 1982. Duetting in male and female songs of the white-cheeked gibbon (*Hylobates concolor lucogenys*). In *Primate Communication*, ed. C. T. Snowdon, C. H. Brown & M. R. Petersen. New York, NY: Cambridge University Press, pp. 67–93.

Deputte, B. L. & Goustard, M. 1980. Copulatory vocalizations of female macaques (*Macaca fascicularis*): variability factors analysis. *Primates*, **21**, 83–99.

Digby, L. J. 1995. Infant care, infanticide and female reproductive strategies in polygynous groups of common marmosets (*Callithrix jacchus*). *Behavioral Ecology and Sociobiology*, **37**, 51–61.

Dixson, A. F. 1998. *Primate Sexuality: Comparative Studies of the Prosimans, Monkeys, Apes and Human Beings*. Oxford: Oxford University Press.

Dixson, A. F. & Lloyd, S. A. C. 1988. The hormonal and hypothalamic control of primate sexual behaviour. *Symposium of the Zoological Society of London*, **80**, 81–117.

1989. Effects of male partners on proceptivity in ovariectomized estradiol-treated marmosets (*Callithrix jacchus*). *Hormones and Behavior*, **23**, 211–20.

Domb, L. G. & Pagel, M. 2001. Sexual swellings advertise female quality in wild baboons. *Nature*, **410**, 204–6.

Domjan, M., Blesbois, E. & Williams, J. 1998. Pavlovian conditioning of sperm release. *Psychological Science*, **9**, 411–5.

Eisner, T. & Meinwald, J. 1995. The chemistry of sexual selection. *Proceedings of the National Academy of Science, USA*, **92**, 50–5.

Epple, G. 1985. Communication by chemical signals. In *Comparative Primate Biology*. Vol. 2A: *Behavior, Conservation, and Ecology*, ed. G. Mitchell & J. Erwin. New York, NY: Alan R. Liss, pp. 531–80.

Epple, G. & Katz, Y. 1984. Social influences on oestrogen excretion and ovarian cyclicity in saddle back tamarins (*Saguinus fuscicollis*). *American Journal of Primatology*, **6**, 215–27.

Ferris, C. F., Snowdon, C. T., King, J. A. *et al.* 2001. Functional imaging of brain activity in conscious monkeys responding to sexually-arousing cues. *NeuroReport*, **12**, 2231–6.

Fischer, J., Hammerschmidt, K., Cheney, D. L. & Seyfarth, R. M. 2002. Acoustic features of male baboon loud calls: influence of context, age and individuality. *Journal of the Acoustic Society of America*, **111**, 1465–74.

Garber, P. A., Moya, L. & Malaga, C. 1984. A preliminary study of the mustached tamarin (*Saguinus mystax*) in northeastern Peru: questions concerned with the evolution of a communal breeding system. *Folia Primatologica*, **42**, 17–32.

Gautier, J.-P. & Gautier, A. 1977. Communication in Old World Monkeys. In *How Animals Communicate*, ed. T. A. Sebeok. Bloomington: Indiana University Press, pp. 890–964.

Girolami, L. & Bielert, C. 1987. Female perineal swelling and its effects on male sexual arousal: an apparent sexual releaser in the chacma baboon (*Papio ursinus*). *International Journal of Primatology*, **8**, 651–61.

Goldfoot, D. A. 1982. Multiple channels of sexual communication in rhesus monkeys: role of olfactory cues. In *Primate Communication*, ed. C. T. Snowdon, C. H. Brown & M. R. Petersen. New York, NY: Cambridge University Press, pp. 413–28.

Goldfoot, D. A., Kravetz, M. A., Goy, R. W. & Freeman, S. K. 1976. Lack of effect of vaginal lavages and aliphatic acids on ejaculatory responses in rhesus monkeys. *Hormones and Behavior*, **7**, 1–27.

Green, S. M. 1981. Sex differences and age gradations in vocalizations of Japanese and lion-tailed monkeys (*Macaca fuscata* and *Macaca silenus*). *American Zoologist*, **21**, 165–83.

Gubernick, D. & Addington, R. L. 1994. The stability of social and mating preferences in the monogamous California mouse (*Peromyscus californicus*). *Animal Behaviour*, **47**, 559–67.

Heymann, E. W. 2003. Scent marking, paternal care and sexual selection in Callitrichines. In *Sexual Selection and Reproductive Competition in Primates: New Perspectives and Directions*, Special Topics in Primatology 4, ed. C. Jones. New York, NY: Wiley-Liss.

Hollis, K. L., Cadieux, E. L. & Colbert, M. M. 1989. The biological function of Pavlovian conditioning: a mechanism for mating success in the blue gurami (*Trichogaster tricopterus*). *Journal of Comparative Psychology*, **103**, 115–21.

Hollis, K. L., Pharr, V. L., Dumas, M. J., Britton, G. B. & Field, J. 1997. Classical conditioning provides paternity advantage for territorial male blue guramis (*Trichogaster tricopterus*). *Journal of Comparative Psychology*, **111**, 219–25.

Kendrick, K. M. & Dixson, A. F. 1986. Anteromedial hypothalamic lesions block proceptivity but not receptivity in the female common marmoset (*Callithrix jacchus*). *Brain Research*, **375**, 221–9.

Kippin, T. E., Talianakis, S., Schatterman, L., Bartholomew, S. & Pfaus, J. G. 1998. Olfactory conditioning of sexual behavior in the male rat (*Rattus norvegicus*). *Journal of Comparative Psychology*, **112**, 389–99.

Koksal, F. & Domjan, M. 1998. Observational conditioning of sexual behavior in the domesticated quail. *Animal Learning and Behavior*, **26**, 427–32.

Lande, R. 1980. Sexual dimorphism, sexual selection and adaptation in polygenic characteristics. *Evolution*, **34**, 292–307.

Lazaro-Perea, C. 2001. Intergroup interactions in wild common marmosets, *Callithrix jacchus*: territorial defence and assessment of neighbours. *Animal Behaviour*, **62**, 11–21.

Lazaro-Perea, C., Castro, C. S. S., Harrison, R. *et al.* 2000. Behavioral and demographic changes following the loss of the breeding female in cooperatively breeding marmosets. *Behavioral Ecology and Sociobiology*, **48**, 137–46.

Michael, R. P. & Keverne, E. B. 1970. Primate sex pheromones of vaginal origin. *Nature*, **225**, 84–5.

Michael, R. P., Keverne, E. B. & Bonsall, R. W. 1971. Pheromones: isolation of male sexual attractant from a female primate. *Science*, **172**, 964–6.

Mitani, J. C. 1985a. Location-specific responses of gibbons (*Hylobates muelleri*) to male songs. *Zeitschrift für Tierpsychologie*, **70**, 219–24.

1985b. Gibbon song duets and intergroup spacing. *Behaviour*, **92**, 59–96.

1985c. Sexual selection and male orangutan long calls. *Animal Behaviour*, **33**, 272–83.

1988. Male gibbon (*Hylobates agilis*) singing behavior: natural history, song variations and function. *Ethology*, **79**, 177–94.

Nunn, C. L. 1999. The evolution of exaggerated sexual swellings in primates: the graded signal hypothesis. *Animal Behaviour*, **58**, 229–46.

Pagel, M. 1994. The evolution of conspicuous oestrous advertisement in Old World monkeys. *Animal Behaviour*, **47**, 1333–41.

Petrie, M. 1994. Improved growth and survival of offspring of peacocks with more elaborate trains. *Nature*, **371**, 598–9.

Pfaus, J. G., Kippin, T. E. & Centeno, S. 2001. Conditioning and sexual behavior: a review. *Hormones and Behavior*, **40**, 291–321.

Pusey, A., Williams, J. & Goodall, J. 1999. The influence of dominance rank on reproductive success in female chimpanzees. *Science*, **277**, 828–31.

Ryan, M. J. & Rand, A. S. 1990. The sensory basis of sexual selection for complex calls in the tungara frog, *Physalaemus pustulosis* (sexual selection for sensory exploitation). *Evolution*, **44**, 305–14.

Sanchez, S., Pelaez, F., Gil-Burmann, C. & Kaumanns, W. 1999. Costs of infant carrying in the cotton-top tamarins (*Saguinus oedipus*). *American Journal of Primatology*, **48**, 99–111.

Savage, A., Ziegler, T. E. & Snowdon, C. T. 1988. Sociosexual development, pair bond formation and mechanisms of fertility suppression in female cotton-top tamarins (*Saguinus oedipus oedipus*). *American Journal of Primatology*, **14**, 345–59.

Searcy, W. A. & Andersson, M. B. 1986. Sexual selection and the evolution of song. *Annual Review of Ecology and Systematics*, **17**, 507–33.

Semple, S. 1998. The function of Barbary macaque copulation calls. *Proceedings of the Royal Society of London, Series B*, **265**, 287–91.

Semple, S. & McComb, K. 2000. Perception of female reproductive state from vocal cues in a mammal species. *Proceedings of the Royal Society of London, Series B*, **267**, 707–12.

Semple, S., McComb, K., Alberts, S. & Altmann, J. 2002. Information content of female copulation calls in yellow baboons. *American Journal of Primatology*, **56**, 43–56.

Seyfarth, R. M. & Cheney, D. L. 1997. Some general features of vocal development in nonhuman primates. In *Social Influences on Vocal Development*, ed. C. T. Snowdon & M. Hausberger. Cambridge: Cambridge University Press, pp. 249–73.

Singh, D. & Bronstad, P. M. 2001. Female body odour is a potential cue to ovulation. *Proceedings of the Royal Society of London, Series B*, **268**, 797–801.

Smith, T. E. & Abbott, D. H. 1998. Behavioral discrimination between circumgenital odor from peri-ovulatory dominant and anovulatory female common marmosets (*Callithrix jacchus*). *American Journal of Primatology*, **46**, 265–84.

Snowdon, C. T. 1996. Infant care in cooperatively breeding species. *Advances in the Study of Behavior*, **25**, 643–89.

Stallman, R. R. & Froehlich, J. W. 2000. Primate sexual swellings as coevolved signal systems. *Primates*, **41**, 1–16.

Stern, K. & McClintock, M. K. 1998. Regulation of ovulation by human pheromones. *Nature*, **392**, 177–9.

van Schaik, C. P., Assink, P. & Salafsky, N. 1992. Territorial behavior in Southeast Asian langurs: resource defense or mate defense? *American Journal of Primatology*, **26**, 233–42.

Villarreal, R. & Domjan, M. 1998. Pavlovian conditioning of social-affiliative behavior in the Mongolian gerbil (*Meriones unguiculatus*). *Journal of Comparative Psychology*, **112**, 26–35.

Waser, P. M. 1977. Individual recognition, intragroup cohesion and intergroup spacing: evidence from sound playback to forest monkeys. *Behaviour*, **60**, 28–74.

1982. The evolution of male loud calls. In *Primate Communication*, ed. C. T. Snowdon, C. H. Brown & M. R. Petersen. New York, NY: Cambridge University Press, pp. 117–45.

Washabaugh, K. & Snowdon, C. T. 1998. Chemical communication of reproductive status in female cotton-top tamarins (*Saguinus oedipus oedipus*). *American Journal of Primatology*, **45**, 337–49.

Whitehead, J. M. 1987. Vocally mediated reciprocity between neighbouring groups of mantled howler monkeys, *Alouatta palliata palliata*. *Animal Behaviour*, **35**, 1615–27.

Wilkinson, G. S. & Reillo, P. R. 1994. Female choice response to artificial selection on an exaggerated trait in a stalk-eyed fly. *Proceedings of the Royal Society of London, Series B*, **255**, 1–6.

Ziegler, T. E., Epple, G., Snowdon, C. T. *et al.* 1993. Detection of chemical signals of ovulation in the cotton-top tamarin, *Saguinus oedipus*. *Animal Behaviour*, **45**, 313–22.

5 • Sexual selection and exaggerated sexual swellings of female primates

DIETMAR P. ZINNER
Department of Behaviour and Ecology
German Primate Centre
Göttingen, Germany

CAREL P. VAN SCHAIK
Department of Biological Anthropology and Anatomy
Duke University
Durham, NC, USA

CHARLES L. NUNN
Department of Biology
University of California
Davis, CA, USA

PETER M. KAPPELER
Department of Behaviour and Ecology
German Primate Centre
Göttingen, Germany

INTRODUCTION

Females of several species of Old World monkeys and apes exhibit enlarged perineal swellings that include the skin of the anogenital region and rump (see Fig. 5.1). Swellings are normally produced by adult females undergoing ovarian activity and they have stimulated evolutionary biologists since Darwin (1876) to think about their adaptive value and the evolutionary mechanisms responsible for their origin and maintenance. Given the association between sexual swellings and mating activity, it seems likely that some aspect of sexual selection is responsible for the evolution of this exaggerated trait. However, even today the functional significance of exaggerated swellings, as well as the processes responsible for their evolution, remain controversial (Dixson, 1983, 1998; Pagel, 1994, 1995; Radwan, 1995; Wiley & Poston, 1996; Nunn, 1999a; Stallmann & Froehlich, 2000; Domb & Pagel, 2001; Nunn *et al.*, 2001; Domb & Pagel, 2002; Zinner *et al.*, 2002; Snowdon, this volume).

In this chapter, we explore the role of sexual selection in the evolution of exaggerated sexual swellings. Because sexual swellings are associated with mating behaviour and competition among males for access to females, sexual selection has figured prominently among hypotheses for this exaggerated trait. Hypotheses have incorporated the two primary components of sexual selection, sometimes within the same explanation. For example, the best-male hypothesis (Clutton-Brock & Harvey, 1976) states that swellings stimulate male–male competition, improving the ability of females to identify and mate with the highest-quality males.

In recent years, our understanding of the theoretical basis for sexual selection has increased. Based on these advances, two hypotheses about the evolution of sexual swellings have attracted attention in the 1990s. First, the reliable-indicator hypothesis (Pagel, 1994) proposes that sexual swellings involve a reversal of sexual selection, with males choosing females based on characteristics of their swellings that reliably signal female fecundity and ability to rear surviving offspring (Domb & Pagel, 2001). This hypothesis provides a mechanism for how swellings stimulate male competition under the best-male hypothesis: males are likely to compete if swelling size indicates aspects of female reproductive quality (Pagel, 1994).

Second, the graded-signal hypothesis (Nunn, 1999a) incorporates intersexual conflict, which has been recognised as a third form of sexual selection (Parker, 1979; Smuts & Smuts, 1993; Arnqvist & Rowe, 1995; Clutton-Brock & Parker, 1995; Gowaty, 1996, 1997; Holland & Rice, 1998; Nunn & van Schaik, 2000). Due to intersexual conflict over control of reproduction (Trivers, 1972; van Schaik *et al.*, this volume), males are expected to evolve traits that manipulate female reproductive decisions, such as mate choice, and females are expected to evolve traits that resist these manipulations. Under the graded-signal hypothesis, swelling size indicates the probability of ovulation and enables females to balance the benefits of (1) confusing paternity to reduce the risk of infanticide (a form of sexual coercion) and (2) biasing paternity, which increases paternal care and protection and may allow females to obtain the benefits of 'good genes'. Hence, the graded-signal hypothesis proposes that sexual

Sexual Selection in Primates: New and Comparative Perspectives, ed. Peter M. Kappeler and Carel P. van Schaik. Published by Cambridge University Press. © Cambridge University Press 2004.

swellings represent a morphological adaptation of females to mitigate reproductive restrictions set by antagonistic male mating strategies.

Following background details on the characteristics and distribution of sexual swellings and their relationship to ovulation, we briefly review previous hypotheses for this exaggerated female trait. This review will emphasise the important links between previous hypotheses for exaggerated sexual swellings and more recent hypotheses that incorporate reversals of sexual selection or intersexual conflict. In fact, benefits that are proposed to accrue under the more recent hypotheses include many of the benefits proposed by previous hypotheses, but within an updated sexual selection framework. We then focus our attention on the reliable indicator and graded-signal hypotheses. Throughout our review, we highlight the important gaps in our understanding of exaggerated sexual swellings. Hence, the goal of our chapter is to synthesise information on empirical patterns and theoretical advances in sexual selection to understand better the function and evolutionary origins of this exaggerated trait.

BACKGROUND

Moderate swelling and pinkness of external genitalia can be observed during the periovulatory phase in many female mammals, including most primate species (Dixson, 1983; Sillén-Tullberg & Møller, 1993). An analogous structure around the female cloaca was described in a bird, the alpine accentor (*Prunella collaris*), during the mating period (Davies *et al.*, 1996; Nakamura, 1998). The changes of the external genitalia of primates are oestrogen-dependent and appear, at least in prosimians, in close temporal association with ovulation (Jolly, 1967; Dixson, 1983). In females of a number of catarrhine species, however, these changes are not limited to the external genitalia (Fig. 5.1). Depending on the species, they extend to the circumanal, subcaudal and paracallosal regions (Dixson, 1998). Once fully exaggerated, they can make up as much as 14 per cent of female body mass (chacma baboon, *Papio ursinus*: Bielert & Busse, 1983; the value of 25 per cent for red colobus (*Piliocolobus preussi*) provided by Struhsaker (1975) is probably exaggerated, H.-J. Kuhn, personal communication). The exaggerated swelling is primarily an oestrogen-dependent oedema of the tissues involved, with mostly extracellular water retention, although some intracellular retention may occur (Krohn & Zuckermann, 1937; Aykroyd & Zuckermann, 1938; Zuckermann & Parkes, 1939).

Fig. 5.1   Female hamadryas baboon (*Papio hamadryas hamadryas*) with maximum sexual swelling. (Photo: D. Zinner.)

Sexual swellings are not permanent, but rather fluctuate in size and coloration during the ovarian cycle, with maximum size lasting for as much as two thirds of the cycle (Rowell, 1972; Dixson, 1983; Dahl, 1986). Endocrinological studies suggest that exaggerated swellings in female primates reflect changes in oestrogen and progesterone secretion during the menstrual cycle (Dixson, 1983). Their increase during the follicular phase is correlated with the increase of oestrogen levels, and their detumescence in the luteal phase with rising progesterone levels. These relationships are probably causal: ovariectomy causes the swelling to decrease, an effect that is reversed by application of oestradiol (Saayman, 1970; Dixson & Herbert, 1977).

The distribution of exaggerated swellings among primates has been documented previously and examined comparatively. Prosimians show some reddening and swelling of the external genitalia (Dixson, 1983), and slight changes of the external genitalia also occur in some platyrrhines, such as Callitrichids (Sicchar & Heymann, 1992) and howling monkeys (Glander, 1980). But these changes are not equivalent in degree to the exaggerated swellings found exclusively among Old World monkeys and apes. Within catarrhines, several distantly related genera exhibit swellings (e.g. *Pan* and *Piliocolobus*), whereas some closely related genera do not (e.g. *Pan* and *Gorilla*). According to phylogenetic reconstructions (Nunn, 1999a), it is likely that exaggeration of swellings evolved independently at least three times in the cercopithecines, African colobines and the great apes. Within the cercopithecines, it is also likely that two losses of swellings have occurred, in the macaques and the guenons.

Comparative analyses also revealed that all taxa that exhibit swellings live in multi-male social systems, including the multi-level systems of some baboons and the fission–fusion systems of chimpanzees and bonobos. Other factors appear to be involved, however, because females of some catarrhine species that live in multi-male groups lack exaggerated sexual swellings, e.g. the vervet monkey, *Chlorocebus aethiops* (Clutton-Brock & Harvey, 1976; Sillén-Tullberg & Møller, 1993; Nunn, 1999a). Nonetheless, all species that exhibit exaggerated sexual swellings are found in multi-male social units of one kind or another, and most hypotheses concerning the adaptive value of this trait have based their line of argument on this relationship.

More recently, van Schaik *et al.* (1999) demonstrated an additional effect of breeding seasonality on the distribution of sexual swellings. Their analyses revealed that exaggerated swellings are more common in multi-male species that do not breed seasonally. In macaques, for example, seasonally breeding species show no exaggerated swellings (e.g. the *sinica* group), whereas species that breed year-round exhibit the largest swellings. This association is a statistical one, and thus some exceptions exist to the general pattern. For example, Barbary macaque females (*Macaca sylvanus*) produce striking, exaggerated swellings despite the seasonality of their breeding (Kuester & Paul, 1984) and the same may hold true for the talapoin monkey (*Miopithecus talapoin*: Rowell & Dixson, 1975).

SWELLINGS AND OVULATION

Swellings do signal changes in the hormonal status of females, although the correlation between swelling size and the probability of ovulation is likely to be lower than once thought. The high levels of oestrogen, which peak at mid-cycle, cause the pituitary to produce a surge of luteinising hormone that stimulates ovulation and results in maximum swelling. After ovulation the follicle forms the corpus luteum, which secretes progesterone and induces rapid detumescence of the swelling (Dixson, 1998). In only a few studies has ovulation been documented directly using laparoscopy (Wildt *et al.*, 1977; Shaikh *et al.*, 1982). In these studies, however, oestradiol and FSH peaks exhibited a strong correlation with ovulation, allowing the use of changes in hormone concentrations to detect ovulation. In particular, ovulation has most likely occurred when one finds a peak in oestradiol followed by a rise in progesterone.

Endocrinological and experimental studies have indicated that the temporal relationship between changes in sexual swellings and ovulation may be variable among species (Nunn, 1999a). There is a tendency for ovulation to coincide with maximum sexual swelling in mangabeys (*Cercocebus atys lunulatus*: Aidara *et al.*, 1981; *Cercocebus torquatus atys*: Whitten & Russel, 1996); macaques (*Macaca tonkeana*:

Thierry *et al.*, 1996; *Macaca mulatta*: Bielert *et al.*, 1976); baboons (*Papio* spp.: Wildt *et al.*, 1977; Shaikh *et al.*, 1982); bonobos (*Pan paniscus*: Heistermann *et al.*, 1996); and chimpanzees (*Pan troglodytes*: Deschner *et al.*, 2003). In baboons, for example, the probability of ovulation is highest on the last day of maximum swelling and the first day of sex-skin detumescence (Gillman & Gilbert, 1946; Hendrickx & Kraemer, 1969; Wildt *et al.*, 1977) but ovulation was also detected several days before and after this period. In an attempt to find the optimum time for mating in baboons under laboratory conditions, Gillman and Gilbert (1946) and Hendrickx and Kraemer (1969) found that the probability of conception was highest between two and seven days prior to detumescence, normally during the maximum swelling phase. A more extreme case was found in captive bonobos, with endocrinological studies revealing that swelling size and ovulation are only weakly associated. Ovulation was detected from 16 days before and up to 7 days after the first day of detumescence. In only 20 out of 32 cycles among 14 females did ovulation occur within five days before detumescence (Heistermann *et al.*, 1996; Reichert *et al.*, 2002). In species that signal ovulation but lack the exaggerated swellings that are the focus of this chapter, the temporal relation of slight swellings, or reddening of the genitals, and ovulation seems to be more precise. In gorillas, for example, ovulation (determined by LH peaks) always occurred exactly one day after peak labial tumescence (Nadler *et al.*, 1979).

Hence, the probability of ovulation, and thus its predictability for males, seems to vary among species. A striking example is the comparison between bonobos and chimpanzees (Fig. 5.2). In captive bonobos maximum swelling lasts for an average of 15.1 days (SD 7.1, 14 females with 32 swellings) and no clear peak in ovulation probability is detectable. In free-ranging chimpanzees from the Taï Forest, however, average duration of maximum swelling is 10.9 days (SD 3.0, 12 females with 33 swellings) and the probability of ovulation rises on day seven of maximum swelling to almost 28 per cent. Days seven to nine of a maximum swelling have a 58 per cent chance of ovulation in chimpanzees.

The longer and the more variable the duration of the maximum swelling phase, the more difficult it is for males to predict the day of ovulation. In some species maximum swelling phase is comparatively short. In *Cercocebus atys lunulatus*, for example, maximum swelling lasts only two to three days (Aidara *et al.*, 1981), and swellings in this species may provide more reliable cues of ovulation because inter- and intra-individual variability are also small. In other species, such as captive bonobos, maximum size can be maintained for

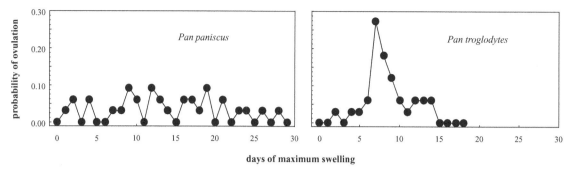

days of maximum swelling

Fig. 5.2   Probability of ovulation in relation to first day of maximum swelling period (day 0) of captive bonobos (*Pan paniscus*) and free-ranging chimpanzees (*Pan troglodytes*) from Taï Forest (probability calculation based on data from Heistermann *et al.*, 1996; Reichert *et al.*, 2002; Deschner *et al.*, 2003).

several weeks (Heistermann *et al.*, 1996; Reichert *et al.*, 2002), thereby forming not a single peak but a plateau that varies markedly among individuals and even within individuals over time. In captive hamadryas baboons (*Papio hamadryas hamadryas*), the maximum swelling phase ranges from 1 to 20 days (mean 5.7 ± 3.4, 327 swellings of 14 females; D. Zinner, unpublished data), in wild chimpanzees from 6 to 18 days (mean 10.9 ± 3.0, 33 swellings of 12 females; Deschner *et al.*, 2003), and in captive bonobos from 3 to 30 days (mean 15.1 ± 7.1, 32 swellings of 14 females; Heistermann *et al.*, 1996; Reichert *et al.*, 2002). The degree to which slight changes in swelling size occur within the maximum swelling phase, and the ability of males to discern such differences, remain unclear and require further study.

Despite the general association between swellings and ovulation, swellings also occur during times in which ovulation is not possible. They occur during anovulatory cycles (M. Heistermann, personal communication; Wasser, 1996). More strikingly, they also occur during pregnancy (Hrdy & Whitten, 1987). For example, Wallis and Goodall (1993) report that up to 25 per cent of swellings in chimpanzees occur during gestation. In some species, such as red colobus (Harth, 1978) and mangabeys (Gordon *et al.*, 1991), sexual swellings continue for several weeks after conception or occur during the first part of gestation. In some cases these non-ovulatory swellings seem to be situation-dependent. For example, situation-dependent receptivity may occur during encounters with strange males (Hrdy & Whitten, 1987), such as the take-over of a group by a new male. In this case, swellings may occur even when females are unlikely to be near ovulation, or when females have dependent offspring,

normally causing lactational amenorrhea (e.g. hamadryas baboons: Swedell, 2000; Zinner & Deschner, 2000).

MALE ALLOCATION OF MATING EFFORT IN RELATION TO SWELLING SIZE

The information content of swellings is further indicated by male responses to swelling size and coloration. In species without exaggerated sexual swellings, males frequently approach and sniff the anogenital region of females in prooestrus (prosimians: Schilling, 1979; Kappeler, 1988, 1998; New World monkeys: Converse *et al.*, 1995). Limited evidence suggests that olfactory cues play a significant role in the control of female attractiveness in prosimians and New World monkeys (Dixson, 1998). Males of species that show an exaggerated swelling also frequently approach, manually inspect and smell the perineal area of females, suggesting that these males may also use olfactory signals to monitor female reproductive condition (Hrdy & Whitten, 1987).

However, in contrast to New World primates and prosimians, the importance of olfactory signals as a sexual attractant in catarrhines remains questionable (Goldfoot *et al.*, 1976; Goldfoot, 1981; Dixson, 1998; Snowdon, this volume). Furthermore, females are not passive in establishing sexual relationships. Significant changes in the behaviour of females occur during the follicular phase. The frequencies of proceptive and receptive behaviour increase and they approach and present to males more often (Hrdy & Whitten, 1987). Such behavioural changes are probably also used by males as indicators of readiness to conceive (ovulation) and males intensify their sexual investment accordingly.

Although olfactory and behavioural cues may play ancillary roles in signalling female fertility in species with sexual swellings (Michael & Keverne, 1968; Dixson *et al.*, 1973), experiments using models of sexual skins demonstrated that male baboons responded to visual cues of receptivity alone (Bielert & Anderson, 1985; Bielert & Girolami, 1986; Girolami & Bielert, 1987; Bielert et al., 1989; reviewed in Snowdon, this volume). Males are strongly attracted to females during the stage of maximum turgescence of their sexual skin. Adult and dominant males, in particular, consort and mate preferentially during this period (*Macaca tonkeana*: Thierry et al., 1996; *Papio hamadryas ursinus*: Saayman, 1970; *Papio hamadryas hamadryas*: Bernhardt, 1993; *Macaca nigra*: Dixson, 1977; reviewed in Nunn, 1999a). Intra-male competition and wounding also peak during phases of maximum swelling (Nunn, 1999a), indicating that males obviously intensify their reproductive investment in such periods. Subadult and subordinate males, however, often tend to mate outside the period of maximum swelling (Nunn, 1999a), possibly because this is the only time they have access to females. Limited evidence suggests that males of some species can distinguish fertile from non-fertile swellings, particularly in sooty mangabeys, in which alpha males have been shown to discriminate between 'normal' and post-conception swellings that appear to be virtually identical to observers (Gordon et al., 1991; Gust, 1994). Whether this ability depends on the characters of the swelling, olfactory stimuli, female behaviour or on long-term experience of the male with the females in the group is not clear.

Males generally respond to sexual swellings as if they are indicators of female fertility, and increase their sexual efforts according to the degree of swelling. The advantage for males is obvious. Males that recognise swellings and mate preferentially when swelling size is at its maximum increase their chances of siring infants. However, the problem for males is that their ability to detect ovulation in females is imperfect (van Schaik et al., 2000), so that, whenever mating is polyandrous, complete concentration of paternity on the dominant male is less likely.

SEXUAL SELECTION AND SEXUAL SWELLINGS

EARLY HYPOTHESES FOR THE FUNCTION OF EXAGGERATED SEXUAL SWELLINGS

Based on features of exaggerated swellings, their distribution and their relation to certain mating systems, it has commonly been hypothesised that sexual selection plays a major role in the evolution of swellings (e.g. Darwin, 1876; Clutton-Brock & Harvey, 1976; Hrdy, 1981; Pagel, 1994; Dixson, 1998; Nunn, 1999a; Stallmann & Froehlich, 2000). Dixson

(1998) proposed that, in general, an adaptive process similar to the classical 'Fisherian runaway process' (Fisher, 1930) is responsible for the exaggeration of sexual swellings. Here, we review the most commonly cited hypotheses (see Hrdy & Whitten, 1987; Dixson, 1998; Nunn, 1999a). To varying degrees, these hypotheses assume that males compete for matings, swellings indicate fertility, and benefits accrue to females according to their reproductive goals of raising high-quality offspring.

(1) Hamilton (1984) proposed the 'obvious ovulation' or 'paternity assurance' hypothesis, in which swellings pinpoint ovulation precisely. Paternity certainty will be very high for the mating male. Thus, it is proposed that this male will subsequently provide care to the infant, to the benefit of the female. However, given the promiscuous mating behaviour among catarrhines (e.g. Hrdy & Whitten, 1987; Dixson, 1998), as well as error in the timing of ovulation, as discussed above, this hypothesis lacks empirical support.

(2) The 'many-males' hypothesis argues that swellings function to increase a female's opportunity to mate with several different males within a cycle and among cycles (Hrdy, 1981; Hrdy & Whitten, 1987). This promiscuity serves to dilute paternity and to create paternity illusions when swellings occur outside ovulatory cycles. As a result, females may reduce the risk of infanticide and also obtain protection from more males (Hrdy, 1981; Hrdy & Whitten, 1987; for discussion see van Schaik et al., this volume). Indeed, promiscuous mating is known to reduce the risk of infanticide (e.g. comparative studies: van Schaik et al., 1999, 2000), and the opportunities for widespread promiscuity in multi-male groups are part of the explanation for the higher rates of infanticide usually observed in single-male group contexts. Moreover, swellings during postpartum lactational amenorrhoea following take-overs have been observed in hamadryas baboons (Zinner & Deschner, 2000). Interbirth intervals are not shortened in this case, however, so that the swellings do not speed up reproduction. Zinner and Deschner (2000) interpret these situation-dependent swellings as 'deceptive' swellings that falsely signal receptivity. These observations are consistent with the many-males hypothesis, and difficult to interpret otherwise. However, it is not clear why exaggerated swellings are needed to achieve these effects. Elements of the many-males hypothesis will therefore be used in one of the current hypotheses discussed below, the

graded-signal hypothesis. Additional benefits of some secondary hypotheses may fall under the graded-signal hypothesis. For example, by increasing promiscuity, females are likely to be inseminated by males that are most successful in sperm competition. In general, mating promiscuity will be required for benefits acquired through cryptic female choice (Eberhard, 1996; Birkhead & Kappeler, this volume). Thus, Dixson and Mundy (1994) proposed that swellings function in cryptic female choice by lengthening the female reproductive tract and making it more difficult for males to inseminate females. Under this proposal, swelling size may also influence penile morphology and make it even more costly for males to mate. Additionally, such morphological changes may eventually support reproductive isolation of populations and may lead to speciation, as proposed for Sulawesi macaques by Stallmann and Babo (1996).

(3) The 'best-male' hypothesis argues that swellings function to attract males and provoke male–male interference competition within a group for sexual access to the receptive female, thus increasing the chances that the female will acquire desirable traits for her offspring, including competitive ability (Clutton-Brock & Harvey, 1976). An extension of the many-males and best-male hypotheses was proposed by Dixson (1983), in the 'distant-male' hypothesis. Females advertise receptivity to males outside the group with their swellings to increase the number of possible mates, either to enhance the chances for female mate choice or to reduce the number of possibly infanticidal males. Although exaggerated swellings are likely to attract males from farther afield, and this effect is no doubt part of the functional explanation, the distant-male hypothesis is unlikely to provide a complete explanation for their existence and distribution.

Thus, many species with multi-male groups, including some with large and dispersed groups, do not show exaggerated swellings (see above), whereas attraction of outsider males is most common in single-male groups (Cords, 2000), in which females lack exaggerated sexual swellings. Wiley and Poston (1996) provide a theoretical framework for understanding the best-male hypothesis. They argue that it would be sufficient if swellings demonstrate that females are fertile, and hence indicate that this is the optimum time for males to compete and to mate. Within this framework of indirect mate choice, exaggerated signals may emerge, and such signals need not be restricted to visual sensory modalities. Thus, depending on the species, other signals, such as odours (Keverne, 1982; Hrdy & Whitten, 1987), vocalisations (copulation calls: Cox

& LeBoeuf, 1977; Hauser, 1990; Oda & Masataka, 1992; but see O'Connell & Cowlishaw, 1994) or behavioural displays may be used to enhance indirect female mate choice. In contrast to Wiley and Poston (1996), Pagel (1994) argues that a small swelling would be sufficient to incite male–male competition because the evolutionary stable strategy (ESS) is for males to compete for receptive females whenever they are fertile. Therefore, the exaggeration of the swelling requires additional explanation; in particular, it requires a mechanism linking swelling size to male–male competition. We turn to this 'reliable-indicator' hypothesis next.

THE RELIABLE-INDICATOR HYPOTHESIS

BACKGROUND TO THE RELIABLE-INDICATOR HYPOTHESIS

In the traditional view, intrasexual selection is stronger in males than in females because male reproductive success is limited by access to choosy females, who typically invest more in their offspring and have slower potential reproductive rates than males (Darwin, 1871; Bateman, 1948; Trivers, 1972; Clutton-Brock & Parker, 1992). Among mammals, parental investment is generally highly biased toward females through the constraints and costs of gestation and lactation (Williams, 1966). As a result, exaggerated and conspicuous secondary sexual traits, such as costly displays, bright colours, and decorative ornaments or weapons, are usually found in males (Darwin, 1871; Andersson, 1994).

Under certain conditions, however, this pattern may be reversed, with the result that competition among females exceeds that found in males and males are the choosy sex (e.g. Reynolds & Szekely, 1997; Jones et al., 2000). More important, even in species in which males also exhibit intense competition, mate choice and intrasexual competition can occur in both sexes (Johnstone et al., 1996; Gowaty, this volume), so that female ornaments are theoretically expected under a wide range of conditions. Specifically, if males provide substantial paternal investment or if males or their sperm become limiting resources to females, females should compete for males or matings. Moreover, if there are large differences in female fecundity, or if mating effort reduces a male's chances to fertilise other females, males may evolve preferences for high-quality females and females may evolve traits that indicate female quality (Andersson, 1994). The few cases in species with conventional sex roles in which females advertise good genes or males prefer more ornamented females come from birds. For example, in barn owls

(*Tyto alba*), females indicate quality by the degree of spottiness of their plumage (Roulin *et al.*, 2000), and in bluethroats (*Luscinia s. svecica*), males prefer more ornamented females (Amundsen *et al.*, 1997).

In mammals, a complete sex-role reversal is not expected because of the constraints set by basic mammalian reproductive biology; only females are able to carry a foetus to term and subsequently nurse the offspring. Nonetheless, because the exaggerated sexual swellings found in some female primates are such a strikingly prominent trait, it is reasonable to investigate whether sexual swellings of female primates represent a reversal of sexual selection. The reliable-indicator hypothesis proposes that sexual swellings are involved in female–female competition for the 'best' male, and male mate choice for high-quality females (Pagel, 1994). By providing a mechanism underlying male intrasexual competition, the reliable-indicator hypothesis can be viewed as an extension of the best-male hypothesis (Clutton-Brock & Harvey, 1976). This hypothesis is intriguing because it involves a rare example of sexual selection in females analogous to the genetic indicator (good genes) mechanism of sexual selection in males of other species (Fisher, 1915; Williams, 1966; Zahavi, 1975; Hamilton & Zuk, 1982), such as the elongated tails of widowbirds or peacocks (Andersson, 1982; Petrie *et al.*, 1991; Petrie, 1994).

The basic assumption of the reliable-indicator hypothesis is that a female's ultimate reproductive goal is to reproduce with the 'best' male, and that a female that can increase the level of competition among males will mate with 'better' males than will other females. Under this hypothesis, sexual swellings are honest indicators of female quality because they are costly to produce. Males should be choosy because contest competition raises the mating costs for males in multimale groups; hence, males should regulate their mating effort according to the 'quality' of the female. However, theoretical and empirical questions have been raised concerning the reliable-indicator hypothesis (Radwan, 1995; Wiley & Poston, 1996; Nunn, 1999a; Nunn *et al.*, 2001; Zinner *et al.*, 2002). In what follows, we outline the key predictions of this hypothesis and evaluate empirical support for each prediction.

TESTING THE RELIABLE-INDICATOR HYPOTHESIS

It has been demonstrated that sexual swellings indicate female hormonal status and that males use the size of the swelling to allocate mating effort. The crucial issues for testing the reliable-indicator hypothesis concern the nature of competition among females, variation in female quality, and male responses to this variation. The reliable-indicator hypothesis makes six main predictions.

(1) Sexual swellings should evolve in situations in which female–female competition for males or matings is most intense. Only when access to mates limits reproductive success of females should females evolve mechanisms of intrasexual competition for mates, such as exaggerated swellings. When female–female competition for mates or matings is relaxed, sex-specific female structures, such as swellings, might instead evolve as exaggerated mechanisms of indirect mate choice (Wiley & Poston, 1996).

(2) To be honest and reliable signals, swellings should be costly, with larger swellings carrying greater costs than smaller ones.

(3) High-quality females should have the largest swellings.

(4) Swelling size signals enduring quality of individual females; hence, the variance of swelling size within females should be less than the variance among females.

(5) Males prefer females with larger swellings and compete more intensely for them.

(6) Females with larger swellings conceive more often from high-quality males, such as those with the longest period of alpha tenure, than females with smaller swellings.

Testing these assumptions and predictions is difficult because the necessary data are scarce or costly to collect. Recent advances have been made, however, using phylogenetic comparisons (Nunn *et al.*, 2001) and collection of original data from wild baboons (Domb, 2000; Domb & Pagel, 2001). First, in phylogenetic comparative tests (Nunn *et al.*, 2001), no association was found between the presence of exaggerated swellings and female mating competition (prediction 1). Female mating competition was measured using the adult sex ratio, female canine size, and expected female mating synchrony (Fig. 5.3).

In captive hamadryas baboons, the probability of conception was negatively correlated with the number of synchronous-cycling females, pointing to a condition in which female mating competition may occur (Zinner *et al.*, 1994). This species is unusual in the context of exaggerated swellings, however, because it is characterised by a higher-level social organisation. In fact, the general absence of exaggerated swellings in single-male groups, except when embedded in higher-level multi-male societies, suggests that female competition for mates is less severe in species with swellings. Thus, in species in which females exhibit exaggerated swellings and reside in multi-male groups, females can (and do) mate with multiple males while in oestrus.

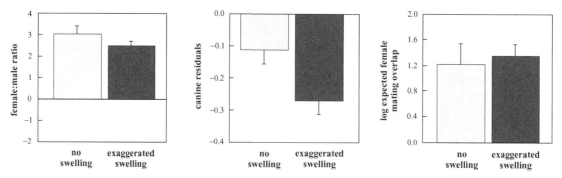

Fig. 5.3   Indicators for the potential of female–female reproductive competition in species with and without exaggerated sexual swellings. Female:male sex ratio ($+1\ SE$), mean canine residuals ($-1SE$) and mean expected female mating overlap ($+1\ SE$). Under the reliable-indicator hypothesis one would expect increased mating competition among females, as measured by a higher female:male sex ratio, larger canines in females and an increased mating overlap. Two of the three variables show patterns opposite to expectations of the reliable-indicator hypothesis (for details see Nunn et al., 2001).

Second, the reliable-indicator hypothesis assumes that swellings are costly (prediction 2). It seems plausible that exaggerated swellings pose some costs on females (Nunn, 1999a). At present, however, the magnitude of these costs is speculative. Highly turgescent swellings are probably prone to injuries and often show cuts, which may lead to infections. Whether females with conspicuous swellings suffer from a higher predation risk than females in other reproductive states is unknown. Maximum swellings can increase female body mass considerably, and hence possibly travel costs. Even if swellings are costly, however, this is not definitive evidence for the reliable-indicator hypothesis. Any of the possible functions for this trait should provide benefits, such as a reduction in infanticide risk, that offset costs of producing swellings.

In a pioneering study, Domb (2000) and Domb and Pagel (2001) tested predictions (3 and 5). By studying the population of olive baboons at Gombe, these authors examined the relationship between sexual swellings, female fitness measurements and male interest scores. They found positive correlations among the vertical length of swellings and the number of offspring born and surviving per year, and they discovered a negative correlation with age at first conception (prediction 3, controlling for age and family rank of the female). Furthermore, they found a positive correlation between the length of a swelling and the amount of aggres-

sion a male received while in consort with a female. The authors interpreted this result as showing increased competition among males for females with longer swellings (prediction 5). Other measures of swelling size, such as width, were not significant in the statistical tests.

However, Zinner et al. (2002) pointed to several methodological problems with the empirical study, and in re-analysis of the Gombe data it became clear that variation among females in fertility is not reflected in swelling size independently of other variables. Female body height explained a significant portion of variation in swelling size in their data (Domb, 2000), with taller females having longer swellings. Furthermore, data from five baboon groups were pooled. These groups show significant differences in swelling length, in age at first conception and number of surviving offspring per year. The groups also differed substantially in food availability, because of variable access to a garbage dump, and fish laid out for drying by local villagers (Domb, 2000). Differences in food availability are known to affect dramatically female fertility (Fa & Southwick, 1988) and swelling characteristics. There is growing evidence that the nutritional status of females may influence the size and duration of sexual swellings (Macaca fuscata: Mori et al., 1997; food-enhanced Macaca sylvanus in Gibraltar: U. Möhle, personal communication). Although such condition dependence is common to indicator mechanisms of sexual selection, in this case it reflects a characteristic of the group (i.e. access to garbage) rather than individual, genetically based differences. In fact, females that fail to conceive because of genetic or physiological defects often have larger swellings (Hausfater, 1975; de Waal, 1982; Wasser, 1983; Pfeiffer et al., 1985; Goodall, 1986; for a compilation see Nunn et al., 2001), possibly because they are able to sequester more resources for themselves rather than investing in offspring growth and development. The groups at Gombe also differed in their adult sex ratio (1.3 to 2.2 females/male, Domb, 2000), which may influence the male competitive regime (Clutton-Brock &

Harvey, 1977). When female height and group membership are also controlled statistically (Zinner *et al.*, 2002), swelling length is no longer a significant predictor of female quality (prediction 3).

Male preference for females with longer swellings is also not unambiguously supported by the data from the Gombe baboons (prediction 5; Zinner *et al.*, 2002). First, sex ratio was ignored in the analysis, as noted above. Second, periods of time when more than one female had a swelling were excluded from the analysis (Domb, 2000). These are in fact the times when male choice of females is most relevant to testing the hypothesis.

The data from Gombe are not sufficient to demonstrate that swelling size is an enduring quality signal of females (prediction 4). Only a few repeated measures of female swelling size are available, precluding an analysis of whether differences among females in swelling size are consistent. However, even in the small sample of repeated measures, the intra-individual variance of swelling length was as large as the variation found among females. Thus, swelling size may provide little information on individual differences across cycles (Zinner *et al.*, 2002).

No data are available for testing prediction 6. In the scenario of Domb and Pagel (2001), high-quality females with larger swellings should gain by reproducing with the male that out-competes all other males. In many cases, this might be the dominant male of the group. In fact, a female preference for dominant males is one of the most often reported findings from primate studies on female mate choice (Small, 1989), but there is also evidence that females choose males based on other criteria (Manson, 1995), such as friendship, novelty or genetic compatibility (see Gangestad & Thornhill, this volume). Moreover, an even more striking finding is that female primates prefer to mate with multiple males, rather than seeking copulations with one, or a few 'best' males (Manson, 1992; Bercovitch, 1995; Paul, 2002).

In summary, the reliable-indicator hypothesis is plausible, and the theoretical model underlying exaggerated sexual swellings has motivated new research on the function of this possibly costly trait. This hypothesis also fits with the observation that males cue into the size and colour of female swellings (Bielert & van der Walt, 1982; Bielert & Anderson, 1985). However, there is as yet no convincing empirical support for the reliable-indicator hypothesis, and the data are consistent with alternative hypotheses. Moreover, even if such support will emerge in the future, we still need additional explanations for the occurrence of exaggerated swellings during non-fertile periods, such as pregnancy or lactation, and for their highly limited taxonomic distribution.

THE FEMALE DILEMMA AND THE GRADED-SIGNAL HYPOTHESIS

BACKGROUND: INTERSEXUAL CONFLICT

Male and female reproductive strategies are often antagonistic, resulting in intersexual conflict (Smuts & Smuts, 1993; Clutton-Brock & Parker, 1995; Rice, 1996). Due to patterns of sexual dimorphism in body and canine size (Clutton-Brock *et al.*, 1977; Plavcan & van Schaik, 1994; Plavcan, 2001), males of many anthropoid primate taxa have the potential to increase their reproductive success through sexual coercion, including harassment, forced copulation, prolonged mate guarding and infanticide (Hrdy, 1974; Smuts & Smuts, 1993; Clutton-Brock & Parker, 1995). These strategies have been incorporated in sexual selection theory as mechanisms additional to intrasexual contest competition and mate choice (Andersson & Iwasa, 1996). Sexual coercion is costly to females; thus, females are expected to evolve counter-strategies. An evolutionary arms race emerges between the sexes, and many aspects of male and female reproductive behaviour and physiology can be seen as resulting from intersexual conflict (van Schaik *et al.*, this volume).

Sexual coercion probably varies according to the mating system, with the important distinction being whether males are able to monopolise access to a female or a group of females (single-male groups), or whether competition for females takes place among males in multi-male groups (Kappeler, 1999). In single-male groups, the major threat that females face involves infanticide, which happens mainly in the context of group take-overs. In multi-male groups, infanticide remains a threat to females, particularly in the context of rank reversals (e.g. Palombit *et al.*, 1997; Weingrill, 2000). Thus, females who rely on protection from one male may be at risk from other males in the group. In multi-male groups, females can also acquire protection from one or more dominant males. In such groups, however, females have another viable option for reducing infanticide risk: they can spread both paternity and paternity estimates among males, reducing the probability that any particular male in the group will kill her infant (van Schaik *et al.*, this volume).

These considerations suggest that females in multi-male groups face the dual need of biasing and confusing paternity. On the one hand, females benefit from concentrating paternity in one or a few dominant males, thereby securing protection for their offspring (Palombit *et al.*, 1997). They may also obtain good genes from more competitive males (Clutton-Brock & Harvey, 1976; Pagel, 1994). These benefits would be most effectively acquired if females signalled receptivity with an honest signal that pinpointed the timing of ovulation (Nunn, 1999a). On the other hand, females benefit from confusing paternity to reduce the risk of infanticide (Hrdy, 1981, 1988; van Schaik *et al.*, 1999, 2000). This benefit would be achieved most effectively by concealing ovulation and mating with as many males in the group as possible (Nunn, 1999a). In addition to confusing paternity, mating with multiple males may incite sperm competition (Harvey & May, 1989), enabling females to identify males that are successful in sperm competition.

These goals of biasing and confusing paternity in multi-male groups through sexual behaviour (Hrdy & Whitten, 1987) appear to be at odds with one another; hence, this situation has been called the female dilemma (Nunn, 1999a; van Schaik *et al.*, 1999, 2000, this volume). For example, females may reduce the ability of a single male to monopolise sexual access by synchronising their receptive periods (oestrus) within and between menstrual cycles (see Nunn, 1999a, b; van Schaik *et al.*, 1999), or by extending the length of such periods, all of which result in greater overlap among females. Moreover, as the receptive period increases, it becomes more costly for a male to monopolise a female during the entire time that she is potentially fertile.

The mechanism underlying these female counter-strategies is economic: females manipulate male behaviour by altering the costs and benefits of mating. This works because competition for access to receptive females involves risks and costs in multi-male groups, for example involving constraints on foraging activity (Alberts *et al.*, 1996). Thus, males are expected to compete selectively based on the fitness returns of mating. Opportunity costs also arise because mating with one female when she is potentially fertile reduces a male's ability to mate with other females. The typical pattern emerging from the few studies of male mate choice in primates is consistent with expectations. Male primates mostly prefer females that indicate proximity of ovulation, e.g. show a maximum swelling (Dixson, 1983), and they prefer older, experienced females or those that are high ranking (Anderson, 1986; Keddy-Hector, 1992). Thus, females can manipulate male behaviour by changing the incentives for males to compete, as these benefits are embedded in the costs of male mating effort. These economic considerations form the conceptual framework that underlies the graded-signal hypothesis.

THE GRADED-SIGNAL HYPOTHESIS

Under the graded-signal hypothesis, exaggerated sexual swellings alleviate the female dilemma that arises from intersexual conflict in multi-male groups (Nunn, 1999a). Under this hypothesis, sexual swellings constitute graded signals that are probabilistic signals of females' readiness to conceive (ovulation). The gradual change in size and long duration of the exaggerated swellings enable females to manipulate male behaviour by altering the costs and benefits of mate guarding, so that dominant or 'best' males tend to mate-guard only at peak swelling. Hence, these competitive males obtain high chances to sire offspring and are more likely to protect the female's offspring. Outside peak swelling, the probability of ovulation decreases but is still greater than zero. Thus, lower-ranking males who mate outside maximum swelling may sire offspring, reducing the risk that these males will kill infants at a later time. The increase in the duration of receptivity makes it less likely that a single male will monopolise access over the female's entire duration of oestrus, especially because doing so will often result in lost mating opportunities with other females in the group. Because females in the group will vary in swelling size and thus probability of ovulation, competitive males should focus their mating effort on females that are maximally swollen, switching to other females as their swellings increase in size.

The graded-signal hypothesis combines aspects of all three hypotheses that were discussed above. In particular, by concentrating paternity among competitive males who can monopolise the female when she is most likely to conceive, swellings may provide benefits involving offspring care and protection (the obvious ovulation hypothesis) and good genes (the best-male hypothesis). By retaining a probability of ovulation outside peak swelling, as the swelling increases or decreases in size, exaggerated sexual swellings allow females to mate with multiple, less competitive males when ovulation is less likely, but still possible. This allows females to reduce the overall level of infanticide risk (the many-males hypothesis). In terms of infanticide, the end result is a reduction in overall risk within the group, but also one or more males who are willing to protect the infants from attacks should they occur from new immigrants or floater males (see van Schaik *et al.*, this volume).

If swellings indicate the probability of ovulation, then males that can distinguish variation in swelling size will have an advantage. Exaggeration of the signal will possibly strengthen this effect, because among the acceptable signals, sexual selection should favour those that most effectively stimulate the recipients: that is, intense, persistent, or otherwise conspicuous signals (Ryan, 1990). Exaggeration of the signal also allows for the best discrimination of the probability of ovulation, and may arise as a consequence of indirect selection (Wiley & Poston, 1996).

Many previous studies are consistent with aspects of the graded-signal hypothesis, but no empirical tests have been conducted that focus directly on this hypothesis. For example, studies of friendships among baboon males and females reveal that certain males are willing to protect infants from infanticide (Palombit *et al.*, 1997), mating effort is costly for males (e.g. Alberts *et al.*, 1996), and females of a variety of species with exaggerated sexual swellings are able to mate with multiple males and thus reduce their risk of infanticide (Hrdy & Whitten, 1987; Nunn, 1999a; van Schaik *et al.*, 1999). Finally, signals in other sensory modalities should also be graded signals of ovulation, as demonstrated for copulation calls in baboons (O'Connell & Cowlishaw, 1994) and Barbary macaques (Semple *et al.*, 2002).

Comparative analysis would seem a worthwhile avenue for testing the graded-signal hypothesis. However, the graded-signal hypothesis was in fact generated by using much of what we know about the phylogenetic distribution of exaggerated swellings and their ecological and social correlates (Nunn, 1999a), and evidence used to generate a hypothesis cannot be used subsequently to test it. For example, species with exaggerated sexual swellings exhibit a longer duration of receptivity, the swelling indicates the probability of ovulation, and swelling size increases gradually (Nunn, 1999a).

One comparative pattern, involving seasonality of breeding and the presence of exaggerated sexual swellings, does support the graded-signal hypothesis over the reliable-indicator hypothesis (Nunn, 1999a). An important prediction of the graded-signal hypothesis is that swellings increase overlap among females and make it difficult for males to monopolise matings over the entire duration of oestrus. This effect should be needed most in species that are non-seasonal breeders, because in seasonal breeders females are expected to exhibit increased overlap simply by chance as more females come into oestrus during a limited period of time (Ridley, 1986; Nunn, 1999b). Thus, the graded-signal hypothesis can account for the association between non-seasonal breeding and the presence of exaggerated sexual swellings, while the reliable-indicator hypothesis makes the opposite prediction:

swellings should evolve when competition is greatest, as expected in seasonally breeding primate troops (see above; cf. Nunn *et al.*, 2001).

EVOLUTIONARY ORIGINS OF EXAGGERATED SEXUAL SWELLINGS: THE CHASE-AWAY MODEL

The graded-signal hypothesis focuses on the adaptive function of sexual swellings in the context of intersexual conflict, but we must also consider the origins of this trait to understand better its exaggeration. A slightly modified version of the 'chase-away' model of sexual selection, proposed by Holland and Rice (1998), may account for the evolution of exaggerated sexual swellings from an ancestrally small signal of fertility, although it was not developed with this question in mind. This model also stresses intersexual conflict, and within this framework the evolution of exaggerated swellings is based on cyclically antagonistic, rather than reinforcing, coevolution between the sexes.

The chase-away model is based on 'sensory exploitation' (Ryan, 1990), in which the effective signal creates a sensory trap to manipulate behaviour in the signaller's favour (West-Eberhard, 1984). The model assumes that pre-existing sensory biases of males select females to evolve an initial, rudimentary trait that enhances their attractiveness to males. In the case of swellings, the initial stage of the evolutionary process involves the small swellings that males use to identify potentially fertile females in prosimians and other primate species. The signal could be used by females to increase their attractiveness to males by extending and enlarging the swelling, thus enabling females to manipulate male behaviour as proposed under the many-male hypothesis and graded-signal hypothesis. Individual males may then mate in a suboptimal manner, for example by allocating too much time, sperm or energy to mating. Advantages for the female could be an earlier exhaustion of the male, thereby increasing her chances to mate with additional males. If females use the swelling to manipulate males, males are expected to evolve counter-adaptations. Males are trapped by the fact that swelling size indicates the probability of ovulation. If males evolve a certain resistance against the swellings this, in turn, would select females to evolve even more pronounced swellings to overcome the threshold of males, and a cyclic antagonistic coevolution ensues.

Several observations of male and female sexual behaviour appear to be consistent with the assumptions of the chase-away model. Males are often reluctant to mate, in spite of female solicitation of mating (e.g. Sommer *et al.*, 1992), and females that fail to conceive sometimes show the largest and most persistent swellings (Hausfater, 1975; de Waal, 1982; Wasser, 1983; Pfeiffer *et al.*, 1985; Goodall, 1986). Less fertile females, such as adolescents, show large swellings, whereas fully adult females show small swellings or no swellings, as in several macaque species (Anderson & Bielert, 1994).

SYNTHESIS

Several factors make it difficult to evaluate sexual-selection-based explanations for the function of exaggerated swellings. First, our general understanding of the role of sexual selection in primate evolution is limited compared to other taxa (Paul, 2002; Snowdon, this volume), particularly with regard to mate choice and intersexual conflict. For example, little is known about the function of bright ornaments among males, such as the brightly coloured face of the mandrill or the red head of the uakari. We also lack an understanding of the role of olfaction in mate choice by either sex, while understanding that female mate choice in primates is complicated by cases of female mating promiscuity and the possibility of cryptic female choice (Eberhard, 1996; Birkhead & Kappeler, this volume). First, in the case of sexual swellings, our understanding is further complicated by the cyclical nature of the signal: rather than being a fixed trait with striking variation among individuals, within-individual variation is more remarkable, with the sex skin reddening and swelling over the course of the female cycle. Moreover, variance within a female in swelling size is comparable to variance found among females (e.g. Zinner *et al.*, 2002).

Second, hypotheses about the functional significance of swellings and the underlying evolutionary mechanisms have not been separated clearly, thus creating a mixture of explanatory levels. On the functional level, the analysis of sexual swellings in female primates includes fundamental questions about the ultimate reproductive goals of females, such as mating with the 'best' male (Clutton-Brock & Harvey, 1976), mating with many males (Hrdy, 1981) or attraction of distant males (Dixson, 1983). In terms of evolutionary mechanisms, Fisherian runaway processes (Dixson, 1998) and sensory exploitation (Stallmann & Froehlich, 2000) have been proposed. For most of these hypotheses, conclusive tests have not been conducted.

Finally, because sexual swellings have evolved multiple times in independent lineages of Old World primates, it is possible that they evolved for different reasons or are maintained because of secondary functions, such as social advantages, that may conceal their original function (Kummer, 1968; Kuhn, 1972; Colmenares & Gomendio, 1988; Wrangham, 2002). A striking example of this is the case of swellings in juvenile males of some colobus monkey species (Hill, 1952; Wickler, 1967; Kuhn, 1972). These male signals may function to appease adult males. Hence, they are consistent with a sensory exploitation process operating in males following evolution of the trait in females. More generally, it is possible that the function of swellings in female chimpanzees, for example, differs from the function in female colobus monkeys, while loss of swellings may be caused by different environmental and social factors from those responsible for evolutionary gains. The generality of most comparative patterns, however, provides hope that a single explanation will account for most origins of this exaggerated sexual trait.

Nonetheless, three points are clear concerning the function and evolution of exaggerated sexual swellings.

(1) Exaggerated swellings evolved from small swellings that existed in close physiological and temporal association with ovulation (Dixson, 1998).

(2) Exaggerated swellings appear to be probabilistic signals of the timing of ovulation (Nunn, 1999a). Although they indicate ovulation, they do not pinpoint ovulation exactly; instead, the highest probability of ovulation occurs at peak swelling in most species, and males allocate their mating effort accordingly. The exact benefits that females obtain from these graded signals require further testing, but patterns are consistent with those predicted by intersexual conflict.

(3) The currently available data provide no clear evidence that swellings additionally indicate female quality as predicted by the reliable-indicator hypothesis, and many comparative patterns run counter to this hypothesis when it is examined in isolation from other benefits that females may acquire from this exaggerated trait.

The specific function of swellings still remains obscure. In fact, support exists for several hypotheses, and this trait may have evolved by multiple evolutionary mechanisms. However, incorporating intersexual conflict as a driving force in the evolution of exaggerated sexual swelling explains most of the features of this trait, such as its relationship to the hormonal status of females, the long duration of the signal, its presence in non-seasonal breeders that live in multi-male mating systems, and perhaps its limited taxonomic distribution (see van Schaik *et al.*, 2000, this volume). Swellings therefore address the female dilemma arising in multi-male groups. Other hypotheses based solely upon chase-away selection or a reversal of sexual selection are unable to account for these patterns.

FUTURE DIRECTIONS

New conceptual frameworks are needed better to understand the evolution of exaggerated sexual swellings. Zinner *et al.* (2002) distinguish among the levels at which swellings may operate, and they note that hypotheses should identify the level at which swellings provide benefits to females. Three levels of analysis are possible. First, swellings may indicate the optimal time for males to compete during a given female's oestrous cycle. Under this hypothesis, swelling size is a *within-cycle* indicator of the probability of conception (e.g. the graded-signal hypothesis). Second, swellings may also signal cycle-to-cycle variability in the probability that a given female will conceive. Swelling size then would be a *within-female* indicator of the probability of conception across cycles. This hypothesis has been found to have some support in a study of chimpanzee oestrous cycles (M. Emery & P. Whitten, personal communication), but would require that males are capable of remembering such differences. Finally, swellings may indicate differences among females in their abilities to conceive and raise offspring. At this level, swellings may signal enduring differences *among females* in female quality (the best-male hypothesis, the reliable-indicator hypothesis).

Future work on the function and evolution of sexual swellings requires an interdisciplinary approach and quantitative measures of female swelling size and male reaction. Ongoing studies (e.g. Sulawesi macaques: R. Stallmann; Barbary macaques: U. Möhle; chimpanzees: T. Deschner) are measuring changes in swelling size with the help of digital video recording, similar to the pioneering protocol of Domb (2000). Using these methods, changes in swelling morphology *within cycles–within females, between cycles–within females* and *among females* should be documented quantitatively and examined in relation to the hormonal status of the respective females. Behavioural data on sexual activity, male interest and male–male competition, and genetic data on reproductive success would complete such a study. Hormonal and genetic information can be inferred by non-invasive methods from faecal samples.

Finally, we need to incorporate new theoretical concepts within the study of sexual swellings, such as antagonistic coevolution (Holland & Rice, 1998) and indirect mate choice

(Wiley & Poston, 1996). More rigorous modelling, using mathematical tools and simulation approaches, is needed to investigate the underlying functions of exaggerated sexual swellings (e.g. Pagel, 1994). We also need to be clear about the evolutionary transitions leading to exaggerated sexual swellings. A working hypothesis is as follows. Slight changes in the morphology of the external genitalia of female primates were needed at ovulation to prepare the genitals for copulation. Males used these changes as an indicator of ovulation. In multi-male groups, males that were able to detect ovulation by visual cues may have had a reproductive advantage, particularly when faced with choices about how to optimise the allocation of mating effort. Females therefore could use swellings to manipulate males by altering the costs and benefits of mate guarding.

SUMMARY AND CONCLUSIONS

Females of several species of Old World monkeys and apes exhibit enlarged perineal or sexual swellings that include the skin of the anogenital region and rump. Swelling characters, such as size and colour, co-vary with hormonal status of females. Depending on the species, sexual swellings indicate probability of ovulation, more or less reliably. However, the functional significance of exaggerated swellings, as well as the processes responsible for their evolution, remain controversial. Given the association between sexual swellings and mating activity, it seems likely that some aspect of sexual selection is responsible for the evolution of this exaggerated trait. We reviewed previous hypotheses and focused on two more recent ones: the reliable-indicator hypothesis (Pagel, 1994) and the graded-signal hypothesis (Nunn, 1999a). We also considered how new theoretical concepts apply to the study of sexual swellings, such as antagonistic coevolution (Holland & Rice, 1998) and indirect mate choice (Wiley & Poston, 1996).

The reliable-indicator hypothesis involves a reversal of sexual selection. Based on existing comparative and field data, however, we found no support for the hypothesis that swellings indicate female quality. The graded-signal hypothesis considers intersexual conflict as a third form of sexual selection. This conflict arises from antagonistic reproductive goals of males and females. Although further research is needed, we find that existing evidence is most consistent with the graded-signal hypothesis, including the relationship between hormonal status of females and swelling size, the long duration of the signal, the presence of swellings in non-seasonal breeders that live in multi-male mating systems, and the limited taxonomic distribution of exaggerated swellings to Old World monkeys and apes.

ACKNOWLEDGEMENTS

We thank many colleagues for stimulating discussions over the years on the topic of sexual swellings, among them the members of the Department of Behaviour and Ecology and the Department of Reproductive Biology at the German Primate Centre, also Tobias Deschner and Susan Alberts. C. L. Nunn was funded during the writing of this chapter by a National Science Foundation (NSF) Postdoctoral Research Fellowship in Biological Informatics.

REFERENCES

Aidara, D., Badawi, M., Tahiri-Zagret, C. & Robyn, C. 1981. Changes in concentrations of serum prolactin, FSH, oestradiol and progesterone and of the sex skin during the menstrual cycle in the mangabey monkeys (*Cercocebus atys lunulatus*). *Journal of Reproduction and Fertility*, **62**, 475–81.

Alberts, S. C., Altmann, J. & Wilson, M. L. 1996. Mate guarding constrains foraging activity of male baboons. *Animal Behaviour*, **51**, 1269–77.

Amundsen, T., Forsgren, E. & Hansen, L. T. T. 1997. On the function of female ornaments: male bluethroats prefer colourful females. *Proceedings of the Royal Society London, Series B*, **264**, 1579–86.

Anderson, C. M. 1986. Female age: male preferences and reproductive success in primates. *International Journal of Primatology*, **7**, 305–26.

Anderson, C. M. & Bielert, C. F. 1994. Adolescent exaggeration in female catarrhine primates. *Primates*, **35**, 283–300.

Andersson, M. 1982. Female choice selects for extreme tail-length in a widowbird. *Nature*, **229**, 818–20.

1994. *Sexual Selection*. Princeton, NJ: Princeton University Press.

Andersson, M. & Iwasa, Y. 1996. Sexual selection. *Trends in Ecology and Evolution*, **11**, 53–9.

Arnqvist, G. & Rowe, L. 1995. Sexual conflict and arms races between the sexes – a morphological adaptation for control of mating in female insects. *Proceedings of the Royal Society London, Series B*, **261**, 123–7.

Aykroyd, O. E. & Zuckermann, S. 1938. Factors in sexual-skin oedema. *The Journal of Physiology, London*, **94**, 13–25.

Bateman, A. J. 1948. Intra-sexual selection in *Drosophila*. *Heredity*, **2**, 349–68.

Bercovitch, F. B. 1995. Female cooperation, consortship maintenance, and male mating success in savanna baboons. *Animal Behaviour*, **50**, 137–49.

Bernhardt, C. 1993. Änderungen im Verhalten adulter Mantelpavianweibchen (*Papio hamadryas hamadryas*) in Abhängigkeit vom Östruszustand. Diploma thesis, University of Erlangen–Nürnberg.

Bielert, C. & Anderson, C. M. 1985. Baboon sexual swellings and male response: a possible operational mammalian supernormal stimulus and response interaction. *International Journal of Primatology*, **6**, 377–93.

Bielert, C. & Busse, C. 1983. Influences of ovarian hormones on the food intake and feeding of captive and wild female chacma baboons (*Papio ursinus*). *Physiology and Behavior*, **30**, 103–11.

Bielert, C. & Girolami, L. 1986. Experimental assessments of behavioral and anatomical components of female chacma baboon (*Papio ursinus*) sexual attractiveness. *Psychoneuroendocrinology*, **11**, 75–90.

Bielert, C. & van der Walt, L. A. 1982. Male chacma baboon (*Papio ursinus*) sexual arousal: mediation by visual cues from female conspecifics. *Psychoneuroendocrinology*, **7**, 31–48.

Bielert, C., Czaja, J. A., Eisele, S. *et al.* 1976. Mating in the rhesus monkey (*Macaca mulatta*) after conception and its relationship to oestradiol and progesterone levels throughout pregnancy. *Journal of Reproduction and Fertility*, **46**, 179–87.

Bielert, C., Girolami, L. & Jowell, S. 1989. An experimental examination of the colour component in visually mediated sexual arousal of the male chacma baboon (*Papio ursinus*). *Journal of Zoology, London*, **219**, 569–79.

Clutton-Brock, T. H. & Harvey, P. H. 1976. Evolutionary rules and primate societies. In *Growing Points in Ethology*, ed. P. Bateson & R. Hinde. Cambridge: Cambridge University Press, pp. 195–237.

1977. Primate ecology and social organisation. *Journal of Zoology, London*, **183**, 1–39.

Clutton-Brock, T. H. & Parker, G. A. 1992. Potential reproductive rates and the operation of sexual selection. *The Quarterly Review of Biology*, **67**, 437–56.

1995. Sexual coercion in animal societies. *Animal Behaviour*, **49**, 1345–65.

Clutton-Brock, T. H., Harvey, P. H. & Rudder, B. 1977. Sexual dimorphism, socionomic sex ratio and body weight in primates. *Nature*, **269**, 797–800.

Colmenares, F. & Gomendio, M. 1988. Changes in female reproductive condition following male take-overs in a colony of hamadryas and hybrid baboons. *Folia Primatologica*, **50**, 157–74.

Converse, L. J., Carlson, A. A., Ziegler, T. E. & Snowdon, C. T. 1995. Communication of ovulatory state to mates by female pygmy marmosets, *Cebuella pygmaea*. *Animal Behaviour*, **49**, 615–21.

Cords, M. 2000. The number of males in guenon groups. In *Primate Males: Causes and Consequences of Variation in Group Composition*, ed. P. M. Kappeler. Cambridge: Cambridge University Press, pp. 84–96.

Cox, C. R. & LeBoeuf, B. J. 1977. Female incitation of male competition: a mechanism in sexual selection. *The American Naturalist*, **111**, 317–35.

Dahl, J. F. 1986. Cyclic perineal swelling during the intermenstrual intervals of captive female pygmy chimpanzees (*Pan paniscus*). *Journal of Human Evolution*, **15**, 369–85.

Darwin, C. 1871. *The Descent of Man, and Selection in Relation to Sex*. London: John Murray.

1876. Sexual selection in relation to monkeys. *Nature*, **15**, 18–19.

Davies, N. B., Hartly, I. R., Hatchwell, B. J. & Langmore, N. E. 1996. Female control of copulations to maximise male help: a comparison of polygynandrous alpine accentors, *Prunella collaris*, and dunnocks, *P. modularis*. *Animal Behaviour*, **51**, 27–47.

Deschner, T., Heistermann, M., Hodges, J. K. & Boesch, C. 2003. Timing and probability of ovulation in relation to sex skin swelling in wild West African chimpanzees, *Pan troglodytes verus*. *Animal Behaviour*, in press.

de Waal, F. B. M. 1982. *Chimpanzee Politics*. New York, NY: Harper & Row.

Dixson, A. F. 1977. Observations on the displays, menstrual cycles and sexual behaviour in the 'black ape' of Celebes (*Macaca nigra*). *Journal of Zoology, London*, **182**, 63–84.

1983. Observations on the evolution and behavioral significance of 'sexual skin' in female primates. *Advances in the Study of Behavior*, **13**, 63–106.

1998. *Primate Sexuality: Comparative Studies of the Prosimians, Monkeys, Apes and Human Beings*. Oxford: Oxford University Press.

Dixson, A. F. & Herbert, J. 1977. Gonadal hormones and sexual behavior in groups of adult talapoin monkeys (*Miopithecus talapoin*). *Hormones and Behavior*, **8**, 141–54.

Dixson, A. F. & Mundy, N. I. 1994. Sexual behavior, sexual swelling and penile evolution in chimpanzees (*Pan troglodytes*). *Archives of Sexual Behavior*, **23**, 267–80.

Dixson, A. F., Everitt, B. J., Herbert, J., Rugma, S. M. & Scruton, D. M. 1973. Hormonal and other determinants of sexual attractiveness and receptivity in rhesus and talapoin monkeys. In *Primate Reproductive Behavior*, ed. C. H. Phoenix. Basel: Karger, pp. 36–63.

Domb, L. G. 2000. Sexual swellings in wild baboons (*Papio cynocephalus anubis*) at Gombe National Park, Tanzania. Ph.D. thesis, Harvard University, Cambridge, MA.

Domb, L. G. & Pagel, M. 2001. Sexual swellings advertise female quality in wild baboons. *Nature*, **410**, 204–6.

2002. Significance of primate sexual swellings. Reply to Zinner *et al. Nature*, **420**, 143.

Eberhard, W. G. 1996. *Female Control: Sexual Selection by Cryptic Female Choice*. Princeton, NJ: Princeton University Press.

Fa, J. E. & Southwick, C. H. (eds.) 1988. *Ecology and Behavior of Food-Enhanced Primate Groups*. New York, NY: Alan R. Liss.

Fisher, R. A. 1915. The evolution of sexual preference. *Eugenics Review*, **7**, 184–92.

1930. *The Genetical Theory of Natural Selection*. Oxford: Clarendon Press.

Gillman, J. & Gilbert, C. 1946. The reproductive cycle of the chacma baboon with special reference to the problems of menstrual irregularities as assessed by the behavior of the sex skin. *The South African Journal of Medical Sciences, Biological Supplement*, **11**, 1–54.

Girolami, L. & Bielert, C. 1987. Female perineal swelling and its effects on male sexual arousal: an apparent sexual releaser in the chacma baboon (*Papio ursinus*). *International Journal of Primatology*, **8**, 651–61.

Glander, K. E. 1980. Reproduction and population growth in free-ranging mantled howling monkeys. *American Journal of Physical Anthropology*, **53**, 25–36.

Goldfoot, D. A. 1981. Olfaction, sexual behavior, and the pheromone hypothesis in rhesus monkeys: a critique. *American Zoologist*, **21**, 153–64.

Goldfoot, D. A., Kravetz, M. A., Goy, R. W. & Freeman, S. K. 1976. Lack of effect of vaginal lavages and aliphatic acids on ejaculatory responses in rhesus monkeys: behavioral and chemical responses. *Hormones and Behavior*, **7**, 1–27.

Goodall, J. 1986. *The Chimpanzees of Gombe*. Cambridge, MA: Harvard University Press.

Gordon, T. P., Gust, D. A., Busse, C. D. & Wilson, M. E. 1991. Hormones and sexual behavior associated with postconception swelling in the sooty mangabey (*Cercocebus torquatus atys*). *International Journal of Primatology*, **12**, 585–97.

Gowaty, P. A. 1996. Battles of the sexes and origins of monogamy. In *Partnership in Birds*, ed. J. M. Black, pp. 21–52. Oxford: Oxford University Press.

1997. Sexual dialectics, sexual selection, and variation in mating behavior. In *Feminism and Evolutionary Biology: Boundaries, Intersections, and Frontiers*, ed. P. A. Gowaty, pp. 351–84. New York: Chapman & Hall.

Gust, D. A. 1994. Alpha-male sooty mangabeys differentiate between females' fertile and their postconception maximal swellings. *International Journal of Primatology*, **15**, 289–301.

Hamilton, W. D. & Zuk, M. 1982. Heritable true fitness and bright birds: a role for parasites? *Science*, **218**, 384–7.

Hamilton, W. J. 1984. Significance of paternal investment by primates to the evolution of male–female associations. In *Primate Paternalism*, ed. D. M. Taub. New York, NY: Van Nostrand, pp. 309–35.

Harth, I. 1978. Zyklusabhängige Veränderungen der Geschlechtsorgane weiblicher *Procolobus badius*. Ph.D. thesis, University of Frankfurt.

Harvey, P. H. & May, R. M. 1989. Out for the sperm count. *Nature*, **337**, 508–9.

Hauser, M. 1990. Do chimpanzee copulatory calls incite male–male competition? *Animal Behaviour*, **39**, 596–7.

Hausfater, G. 1975. Dominance and reproduction in baboons: a quantitative analysis. *Contributions to Primatology*, **7**, 1–150. Basel: Karger.

Heistermann, M., Möhle, U., Vervaecke, H., van Elsacker, L. & Hodges, J. K. 1996. Application of urinary and fecal steroid measurements for monitoring ovarian function and pregnancy in the bonobo (*Pan paniscus*) and evaluation of perineal swelling patterns in relation to endocrine events. *Biology of Reproduction*, **55**, 844–53.

Hendrickx, A. G. & Kraemer, D. C. 1969. Observation of the menstrual cycle, optimal mating time, and pre-implantation embryos of the baboons, *Papio anubis* and *Papio cynocephalus*. *Journal of Reproduction and Fertility, Supplement*, **6**, 119–28.

Hill, W. C. O. 1952. The external and visceral anatomy of the olive colobus monkey (*Procolobus verus*). *Proceedings of the Zoological Society, London*, **122**, 127–86.

Holland, B. & Rice, W. R. 1998. Perspective: chase-away sexual selection – antagonistic seduction versus resistance. *Evolution*, **52**, 1–7.

Hrdy, S. B. 1974. Male–male competition and infanticide among the langurs (*Presbytis entellus*) of Abu, Radjasthan. *Folia Primatologica*, **22**, 19–58.

1981. *The Woman that Never Evolved*. Cambridge, MA: Harvard University Press.

1988. The primate origins of human sexuality. In *The Evolution of Sex*, ed. R. Bellig & G. Stevens. San Francisco, CA: Harper & Row, pp. 101–36.

Hrdy, S. B. & Whitten, P. L. 1987. Patterning of sexual activity. In *Primate Societies*, ed. B. B. Smuts, D. L. Cheney, R. W. Seyfarth, R. W. Wrangham & T. T. Struhsaker. Chicago, IL: University of Chicago Press, pp. 370–84.

Johnstone, R. A., Reynolds, J. D. & Deutsch, J. C. 1996. Mutual mate choice and sex differences in choosiness. *Evolution*, **50**, 1382–91.

Jolly, A. 1967. Breeding synchrony in wild *Lemur catta*. In *Social Communication among Primates*, ed. S. A. Altman. Chicago, IL: University of Chicago Press, pp. 3–14.

Jones, A. G., Rosenqvist, G., Berglund, A., Arnold, S. J. & Avise, J. C. 2000. The Bateman gradient and the cause of sexual selection in a sex-role-reversed pipefish. *Proceedings of the Royal Society London, Series B*, **267**, 677–80.

Kappeler, P. M. 1988. A preliminary study of olfactory behavior of captive *Lemur coronatus* during the breeding season. *International Journal of Primatology*, **9**, 135–46.

1998. To whom it may concern: transmission and function of chemical signals in *Lemur catta*. *Behavioral Ecology and Sociobiology*, **42**, 411–21.

1999. Primate socioecology: new insights from males. *Naturwissenschaften*, **86**, 18–29.

Keddy-Hector, A. C. 1992. Mate choice in non-human primates. *American Zoologist*, **32**, 62–70.

Keverne, E. B. 1982. Olfaction and the reproductive behaviour of nonhuman primates. In *Primate Communication*, ed. C. T. Snowdon, C. H. Brown & M. R. Petersen. Cambridge: Cambridge University Press, pp. 396–412.

Krohn, P. L. & Zuckermann, S. 1937. Water metabolism in relation to the menstrual cycle. *The Journal of Physiology, London*, **88**, 369–87.

Kuester, J. & Paul, A. 1984. Female reproductive characteristics in semifree-ranging Barbary macaques (*Macaca sylvanus* L. 1758). *Folia Primatologica*, **43**, 69–83.

Kuhn, H. J. 1972. On the perineal organ of male *Procolobus badius*. *Journal of Human Evolution*, **1**, 371–8.

Kummer, H. 1968. *Social Organization of Hamadryas Baboons. A Field Study*. Chicago, IL: University of Chicago Press.

Manson, J. H. 1992. Measuring female mate choice in Cayo Santiago rhesus macaques. *Animal Behaviour*, **44**, 405–16.

1995. Female mate choice in primates. *Evolutionary Anthropology*, **3**, 192–5.

Michael, R. P. & Keverne, E. B. 1968. Pheromones and the communication of sexual status in primates. *Nature*, **218**, 746–9.

Mori, A., Yamaguchi, N., Watanabe, K. & Shimizu, K. 1997. Sexual maturation of female Japanese macaques under poor nutritional conditions and food-enhanced perineal swelling in Koshima troop. *International Journal of Primatology*, **18**, 553–80.

Nadler, R. D., Graham, C. E., Collins, D. C. & Gould, K. G. 1979. Plasma gonadotropins, prolactin, gonadal steroids, and genital swelling during the menstrual cycle of lowland gorillas. *Endocrinology*, **105**, 290–6.

Nakamura, M. 1998. Multiple mating and cooperative breeding in polygynous alpine accentors. I. Competition among females. *Animal Behaviour*, **55**, 259–75.

Nunn, C. L. 1999a. The evolution of exaggerated sexual swellings in primates and the graded signal hypothesis. *Animal Behaviour*, **58**, 229–46.

1999b. The number of males in primate social groups: a comparative test of the socioecological model. *Behavioral Ecology and Sociobiology*, **46**, 1–13.

Nunn, C. L. & van Schaik, C. P. 2000. Social evolution in primates: the relative role of ecology and intersexual conflict. In *Infanticide by Males and Its Implications*, ed. C. P. van Schaik & C. H. Janson. Cambridge: Cambridge University Press, pp. 388–412.

Nunn, C. L., van Schaik, C. P. & Zinner, D. 2001. Do exaggerated sexual swellings function in female mating competition in primates? A comparative test of the reliable indicator hypothesis. *Behavioral Ecology*, **5**, 646–54.

O'Connell, S. M. & Cowlishaw, G. 1994. Infanticide avoidance, sperm competition and mate choice: the function of copulation calls in female baboons. *Animal Behaviour*, **48**, 687–94.

Oda, R. & Masataka, N. 1992. Functional significance of female Japanese macaque copulatory calls. *Folia Primatologica*, **58**, 146–9.

Pagel, M. 1994. The evolution of conspicuous oestrus advertisement in Old World monkeys. *Animal Behaviour*, **47**, 1333–41.

1995. Sexual selection and oestrus advertisement: a reply to Radwan. *Animal Behaviour*, **49**, 1401–2.

Palombit, R. A., Seyfarth, R. M. & Cheney, D. L. 1997. The adaptive value of 'friendships' to female baboons: experimental and observational evidence. *Animal Behaviour*, **54**, 599–614.

Parker, G. A. 1979. Sexual selection and sexual conflict. In *Sexual Selection and Reproductive Competition in Insects*, ed. M. S. Blum & N. A. Blum. New York, NY: Academic Press, pp. 123–66.

Paul, A. 2002. Sexual selection and mate choice. *International Journal of Primatology*, **23**, 877–904.

Petrie, M. 1994. Improved growth and survival of offspring of peacocks with more elaborate trains. *Nature*, **371**, 598–9.

Petrie, M., Halliday, T. & Sanders, C. 1991. Peahens prefer peacocks with elaborate trains. *Animal Behaviour*, **41**, 323–31.

Pfeiffer, G., Kaumanns, W. & Schwibbe, M. H. 1985. A female 'defeated leader' in hamadryas baboons. A case study. *Primate Report*, **12**, 18–26.

Plavcan, J. M. 2001. Sexual dimorphism in primate evolution. *Yearbook of Physical Anthropology*, **44**, 25–53.

Plavcan, J. M. & van Schaik, C. P. 1994. Canine dimorphism. *Evolutionary Anthropology*, **2**, 208–14.

Radwan, J. 1995. On oestrus advertisement, spite and sexual harassment. *Animal Behaviour*, **49**, 1399–400.

Reichert, K. E., Heistermann, M., Hodges, J. K., Boesch, C. & Hohmann, G. 2002. What females tell males about their reproductive status: are morphological and behavioural cues reliable signals of ovulation in bonobos (*Pan paniscus*)? *Ethology*, **108**, 1–18.

Reynolds, J. & Szekely, T. 1997. The evolution of parental care in shorebirds: life histories, ecology and sexual selection. *Behavioral Ecology*, **8**, 126–34.

Rice, W. R. 1996. Sexually antagonistic male adaptation triggered by experimental arrest of female evolution. *Nature*, **381**, 232–4.

Ridley, M. 1986. The number of males in a primate troop. *Animal Behaviour*, **34**, 1848–58.

Roulin, A., Jungi, T. W., Pfister, H. & Dijkstra, C. 2000. Female barn owls (*Tyto alba*) advertise good genes. *Proceedings of the Royal Society London, Series B*, **267**, 937–41.

Rowell, T. E. 1972. Female reproductive cycles of the talapoin monkey (*Miopithecus talapoin*). *Folia Primatologica*, **28**, 188–202.

Rowell, T. E. & Dixson, A. F. 1975. Changes in social organisation during the breeding season of wild talapoin monkeys. *Journal of Reproduction and Fertility*, **43**, 419–34.

Ryan, M. J. 1990. Sexual selection, sensory systems and sensory exploitation. *Oxford Surveys in Evolutionary Biology*, 7, 157–95.

Saayman, G. S. 1970. The menstrual cycle and sexual behaviour in a troop of free-ranging chacma baboons (*Papio ursinus*). *Folia Primatologica*, 12, 81–110.

Schilling, A. 1979. Olfactory communication in prosimians. In *The Study of Prosimian Behavior*, ed. G. A. Doyle & R. D. Martin. New York, NY: Academic Press, pp. 461–542.

Semple. S., McComb, K., Alberts, S. C. & Altmann, J. 2002. Information content of female copulation calls in yellow baboons. *American Journal of Primatology*, 56, 43–56.

Shaikh, A. A., Celaya, C. L., Gomez, I. & Shaikh, S. A. 1982. Temporal relationship of hormonal peaks to ovulation and sex skin deturgescence in the baboon. *Primates*, 23, 444–52.

Sicchar, L. A. & Heymann, E. W. 1992. Preliminary observations on external signs of oestrus in moustached tamarins, *Saguinus mystax*, Callitrichidae. *Laboratory Primate Newsletter*, 31, 4–6.

Sillén-Tullberg, B. & Møller, A. P. 1993. The relationship between concealed ovulation and mating systems in anthropoid primates: a phylogenetic analysis. *American Naturalist*, 141, 1–25.

Small, M. 1989. Female choices in nonhuman primates. *Yearbook of Physical Anthropology*, 32, 103–27.

Smuts, B. B. & Smuts, R. W. 1993. Male aggression and sexual coercion of females in nonhuman primates and other mammals: evidence and theoretical implications. *Advances in the Study of Behavior*, 22, 1–63.

Sommer, V., Srivastava, A. & Borries, C. 1992. Cycles, sexuality, and conception in free-ranging langurs (*Presbytis entellus*). *American Journal of Primatology*, 28, 1–27.

Stallmann, R. R. & Babo, N. 1996. The telltale tail: a possible mechanism of specific mate recognition in Sulawesi macaques. *Abstracts, 16th IPS Conference, Madison*, #175.

Stallmann, R. R. & Froehlich, J. W. 2000. Primate sexual swellings as coevolved signal systems. *Primates*, 41, 1–16.

Struhsaker, T. T. 1975. *The Red Colobus Monkey*. Chicago, IL: University of Chicago Press.

Swedell, L. 2000. Two takeovers in wild hamadryas baboons. *Folia Primatologica*, 71, 169–72.

Thierry, B., Heistermann, M., Aujard, F. & Hodges, J. K. 1996. Long-term data on basic reproductive parameters and evaluation of endocrine, morphological, and behavioral measures for monitoring reproductive status in a group of semifree-ranging tonkean macaques (*Macaca tonkeana*). *American Journal of Primatology*, 39, 47–62.

Trivers, R. L. 1972. Parental investment and sexual selection. In *Sexual Selection and the Descent of Man, 1871–1971*, ed. B. Campbell. Chicago, IL: Aldine, pp. 136–79.

van Schaik, C. P., van Noordwijk, M. A. & Nunn, C. L. 1999. Sex and social evolution in primates. In *Comparative Primate Socioecology*, ed. P. C. Lee. Cambridge: Cambridge University Press, pp. 204–31.

van Schaik, C. P., Hodges, J. K. & Nunn, C. L. 2000. Paternity confusion and the ovarian cycles of female primates. In *Infanticide by Males and Its Implications*, ed. C. P. van Schaik & C. H. Janson. Cambridge: Cambridge University Press, pp. 361–87.

Wallis, J. & Goodall, J. 1993. Anogenital swelling in pregnant chimpanzees of Gombe National Park. *American Journal of Primatology*, 31, 89–98.

Wasser, S. K. 1983. Reproductive competition and cooperation among female yellow baboons. In *Social Behavior of Female Vertebrates*, ed. S. K. Wasser. New York, NY: Academic Press, pp. 349–90.

1996. Reproductive control in wild baboons measured by fecal steroids. *Biology of Reproduction*, 55, 393–9.

Weingrill, T. 2000. Infanticide and the value of male–female relationships in mountain chacma baboons. *Behaviour*, 137, 337–59.

West-Eberhard, M. J. 1984. Sexual selection, competitive communication and species-specific signals in insects. In *Insect Communication*, ed. T. Lewis. New York, NY: Academic Press, pp. 283–324.

Whitten, P. L. & Russel, E. 1996. Information content of sexual swellings and fecal steroids in sooty mangabeys (*Cercocebus torquatus atys*). *American Journal of Primatology*, 40, 67–82.

Wickler, W. 1967. Socio-sexual signals and their intraspecific imitation among primates. In *Primate Ethology*, ed. D. Morris. London: Weidenfeld and Nicolson, pp. 69–147.

Wildt, D. E., Doyle, U., Stone, S. C. & Harrison, R. M. 1977. Correlation of perineal swelling with serum ovarian hormone levels, vaginal cytology and ovarian follicular development during the baboon reproductive cycle. *Primates*, 18, 261–70.

Wiley, R. H. & Poston, J. 1996. Perspective: indirect mate choice, competition for mates, and coevolution of the sexes. *Evolution*, 50, 1371–81.

Williams, G. C. 1966. *Adaptation and Natural Selection: A Critique of Some Current Evolutionary Thoughts.* Princeton, NJ: Princeton University Press.

Wrangham, R. W. 2002. The cost of sexual attraction: is there a trade-off in female *Pan* between sex appeal and received coercion? In *Behavioural Diversity in Chimpanzees and Bonobos*, ed. C. Boesch, G. Hohmann & L. F. Marchant. Cambridge: Cambridge University Press, pp. 204–15.

Zahavi, A. 1975. Mate selection – a selection for handicap. *Journal of Theoretical Biology*, **53**, 205–14.

Zinner, D. & Deschner, T. 2000. Sexual swellings in female hamadryas baboons after male take-overs: 'deceptive' swellings as a possible female counter-strategy against infanticide. *American Journal of Primatology*, **52**, 157–68.

Zinner, D., Schwibbe, M. H. & Kaumanns, W. 1994. Cycle synchrony and probability of conception in female hamadryas baboons, *Papio hamadryas. Behavioral Ecology and Sociobiology*, **35**, 175–83.

Zinner, D., Alberts, S. C., Nunn, C. L. & Altmann, J. 2002. Significance of primate sexual swellings. *Nature*, **420**, 142–3.

Zuckerman, S. & Parkes, A. S. 1939. Observations on the secondary characters in monkeys. *Journal of Endocrinology*, **1**, 430–9.

6 • Female multiple mating and genetic benefits in humans: investigations of design

STEVEN W. GANGESTAD
Department of Psychology
University of New Mexico
Albuquerque, NM, USA

RANDY THORNHILL
Department of Biology
University of New Mexico
Albuquerque, NM, USA

INTRODUCTION

William James, one of the founders of scientific psychology (see James, 1890), tells a personal story in which he awoke from a dream one night with a flash of insight. Wanting not to forget it, he scribbled down, in his half-wakened state, the insight and went back to bed. In the morning he recalled having this revelation but not its content, and excitedly went to read his recording. Disappointed, he found these words: *Higamus, hogamus, women are monogamous; Hogamus, higamus, men are polygamous* (Kitcher, 1987).

Almost certainly, James would not have been able to anticipate that, 100 years later, the whole question of female monogamy or its absence, polyandry, would become one of the most fascinating topics in behavioural biology.

A recent paper published in *Animal Behaviour* (Zeh & Zeh, 2001, p. 1051) claimed that behavioural ecology is in the process of undergoing a paradigm shift, with 'the traditional concept of the choosy, monogamous female increasingly giving way to the realisation that polyandry is pervasive in natural populations' even when males invest substantially in offspring. One form of polyandry that has received much attention is extra-pair copulation (EPC) – sex that a female with a social mate has with a male who is not the social mate. The data showing a mean extra-pair paternity rate of 10–15 per cent in socially monogamous birds (with some rates as high as 70 per cent) are highly familiar (Birkhead & Møller, 1995; Petrie & Kempenaers, 1998). We know of only two published studies that used DNA or blood markers to estimate the extra-pair paternity rate in human populations. One found a rate of about 1 per cent in a Swiss sample (Sasse *et al.*, 1994). Another estimated a rate of 12 per cent in Monterrey, Mexico, with a rate of 20 per cent for a subsample in a low income bracket (Cerda-Flores *et al.*, 1999; see also McIntyre & Sooman, 1991; Baker & Bellis, 1995; Hill & Hurtado, 1996; Beckerman *et al.*, 1998). The extra-pair paternity rate hence appears to vary across human groups but, in some circumstances, can be appreciable.

Direct as well as genetic benefits have been offered as explanations of polyandry through EPC. Although historically more controversial, genetic benefits have received much attention in recent years (see Jennions & Petrie, 2000, for a review). Three types of genetic benefits should be distinguished:

(1) *Intrinsically good genes*: a female seeks an extra-pair sire whose genes have additive effects on offspring fitness and hence affect offspring fitness independent of maternal phenotypic and genotypic features.
(2) *Compatible and, hence, good genes*: a female seeks an extra-pair sire whose genes match well with the mother's, to enhance offspring fitness; that is, the sire has good genes from the chooser's standpoint, but may not offer good genes for other females.
(3) *Diverse genes*: a female seeks an extra-pair sire who will produce an offspring different from other offspring of the mother, thereby increasing total fitness through bet-hedging; for example, by diversifying a brood's self-recognition components of the immune system, a female may lower the probability that the entire brood will be wiped out by a single epidemic.

As noted by Zeh and Zeh (2001), any appreciable rate of extra-pair paternity may have vast ramifications. A variety of inter- and intragenomic conflicts result, including conflicts between the sexes, conflicts between offspring and mother, conflicts between offspring within a brood, and conflicts between paternal and maternal contributions to a foetus's genome. These conflicts can fuel escalating coevolutionary processes that profoundly influence social interactions as

Sexual Selection in Primates: New and Comparative Perspectives, ed. Peter M. Kappeler and Carel P. van Schaik. Published by Cambridge University Press. © Cambridge University Press 2004.

well as sexual selection (e.g. Rice, 1996; Rice & Holland, 1998; Arnqvist *et al.*, 2000; Gavrilets *et al.*, 2001; Arnqvist & Rowe, 2002; Birkhead & Pizzari, 2002). Hence, documenting the nature, causes and conditions under which polyandry occurs within a species importantly informs an understanding of mating and sexual reproduction for that species in a wide-ranging way.

Since the 1990s, we have collaborated on a research program concerning human sexual selection. One main, overarching goal has been to inquire about the role of good genes in sexual selection in humans. We have a working premise that paternal investment has been important in human evolutionary history, that females have benefited by selecting mates able and willing to provide material benefits such as nutrition and protection, and that selection pressures due to these benefits have importantly influenced the evolution of human sexuality and male–female relations (see Kaplan *et al.*, 2000).

Studies of human foragers indicate that their total energy expenditure is near the high extreme of living primates (Leonard & Robertson, 1997), and analyses of *Homo erectus* females suggest that their energetic costs during pregnancy and lactation would have been very substantial (Aiello & Key, 2002). These high energy costs were probably only possible with a diet heavily dependent on food items with high caloric density. Human forager diets contrast sharply with those of our nearest ancestors. Whereas human foragers obtain about 30–80 per cent of their calories from vertebrate meat, hunted foods account for only 2 per cent of the chimpanzee diet. By contrast, collected foods (e.g. fruits, leaves) account for 95 per cent of the caloric intake of chimpanzees but just 8 per cent of the diet of human foragers (Kaplan *et al.*, 2000). With the expansion of the African savannah between 2.5 and 1.5 million years ago, early humans may have entered a feeding niche in which animal meat became a primary source of food, which both required and provided a high-quality diet (Leonard & Robertson, 1997). This niche appears to have selected for a variety of characteristics that distinguish humans from their relatives: an extended juvenile period of growth and learning (e.g. Leigh, 2001; Bock, 2002); investment in mortality reduction leading to an extended lifespan during which capital investments in learning pay off (e.g. Kaplan *et al.*, 2000; Leigh, 2001); new and extended forms of sociality permitting and fostering cooperative hunting and other alliances (e.g. Geary & Flinn, 2002); increased brain size in response to both ecological and social demands.

This adaptive shift also had implications for the hominid sexual selection system. Means by which the high energetic costs of reproduction for females could be lowered became particularly profitable, leading to the evolution of bi-parental care and a cooperative division of labour of the sexes (e.g. Aiello & Key, 2002). In human foraging groups, men subsidise the diets of females and juveniles through hunting to a considerable degree. Without this, human interbirth intervals would be much higher (Kaplan *et al.*, 2000). Offspring with access to paternal investment are advantaged in traditional societies (Hill & Hurtado, 1996; Hagen *et al.*, 2001), as some paternal activities are directed to benefit offspring (see also Geary, 2000). Paternal investment led to increased pay-offs to mate guarding, which may be at least as important as paternal investment in the evolution of predominantly monogamous marriage in human groups (Marlowe, 2000).

But might selection in humans also have shaped individuals to prefer and choose mates who provide genetic benefits to offspring? Indicators of genetic benefits might co-vary with ability to provide material benefits and, hence, preferences for genetic and material benefits might be difficult to disentangle in the context of women's choice of long-term, investing partners. Indeed, Hawkes *et al.* (2001; see also Hawkes & Bird, 2002) argue that, because meat-sharing is widespread, the benefits of hunting to men, and marriage to a good hunter for women, are not because of enhanced nutritional benefits to offspring. Rather, they argue, good hunting functions as a male competitive display. These indicators may be more easily disentangled in the context of women's preferences for men as short-term and, particularly, extra-pair mates. Specifically, do women possess features that evolved, at least in part, because of genetic benefits garnered through extra-pair sex in ancestral populations? (Because an evolved adaptation is a product of past direct selection for a function, the question of whether EPC by women is currently adaptive or currently advances women's reproductive success (RS) is a distinct one. An evolved adaptation may be currently non-adaptive and even maladaptive because the current ecological setting in which it occurs differs from the evolutionary historical setting that was the selection favouring it; cf. Reeve and Sherman, 1993).

The criterion we have used to identify features that evolved to have a specific function is special design. A feature exhibits special design for a particular function if it proficiently performs the function; moreover, it is difficult to imagine any alternative evolutionary process that would have led to the feature or its details other than selection for that function. Evidence for special design is evidence that adaptation has been at work, but more precisely, evidence that the organism has been shaped by *particular* selection pressures (Williams, 1966; Thornhill, 1990, 1997; Symons, 1992). Crudely put, the special design of eyes and wings for sight and flight, respectively, are evidence for historical selection for sight and flight. Women are historical documents in that the features they possess resulted from past evolutionary processes. The special design of certain *psychological* features in modern women may testify to the existence of particular selection pressures in ancestral populations (see Thornhill, 1997; Andrews *et al.*, 2003; Thornhill & Gangestad, 2003).

In the current paper, we review several literatures. First, we consider the role of sexual selection for good genes in general, with particular attention to intrinsically good genes. Second, we examine the potential role of sexual selection for good genes through extra-pair mating in non-human species characterised by bi-parental care of offspring. Third, we describe empirical work that suggests that sexual selection for genetic benefits has effectively shaped particular features in humans, and consider potential alternative explanations.

GOOD-GENES SEXUAL SELECTION: THEORETICAL CONSIDERATIONS

Intrinsically good-genes sexual selection (hereafter, GGSS) involves evolution of (a) preferences in one sex for traits of the opposite sex because of intrinsic genetic benefits to offspring advertised by the preferred traits, and (b) changes in the level of the traits that advertise genetic benefits in the chosen sex as a result of the preferences. Despite long-standing evidence that females in a number of species prefer particular males even when males appear not to contribute anything other than genes to offspring (see Andersson, 1994), GGSS has been a highly controversial topic since it was first explicitly discussed by Trivers (1972).

The problem of adequate additive genetic variance in fitness. Historically, one major reason to doubt the good-genes process concerns whether the amount of genetic variance in fitness renders choice for good genes profitable. Fisher's (1930) fundamental theorem of natural selection states that directional selection on a trait reduces its additive genetic variance. Fitness, by definition under directional selection, should thus have its additive genetic variation exhausted, 'which creates a serious difficulty for the good-genes hypothesis' (Charlesworth, 1987, p. 22). Naturally, if there is no additive genetic fitness variation in a pool of potential mates, there is no reason to choose one mate over any other for additive genetic benefits.

One measure of the genetic variation in a trait is the additive genetic coefficient of variation (CV_A), the square root of the trait's genetic variance standardised by the phenotypic mean multiplied by 100 (to remove decimal places). This measure is evolutionarily meaningful because Fisher's fundamental theorem states that the expected proportional change in a trait in a single generation due to selection (and hence the rate of evolution owing to selection) is a function of the square of this value (without multiplication times 100). Hence, the larger the CV_A, the greater the potential rate of evolution of the trait. To address the issue of whether fitness traits have little genetic variance empirically, Houle (1992) and Pomiankowski and Møller (1995) compared CV_As of different sorts of traits of many organisms: ordinary morphological traits, a subset of these known to be under stabilising selection, and fitness components such as longevity and fecundity. They found that, not only do fitness traits *not* have smaller genetic variance; they actually have substantially *greater* CV_As. Specifically, life history traits such as fecundity and survivorship typically have CV_As in the range of 10–20, and the CV_A of fitness itself is indirectly estimated to be between 10 and 30 (Burt, 1995). By contrast, the CV_As of ordinary morphological traits and those under stabilising selection average about 5 – only 20–40 per cent that of life history traits.

Fisher (1930) himself provided the framework for understanding how genetic variance in fitness could be maintained despite strong selection on it. Populations are generally characterised by a shifting equilibrium. Each generation, those individuals who successfully reproduce have a greater mean fitness and less variable fitness than the entire population. Within each generation, however, various processes degrade the fitness of progeny relative to their parents, including harmful changes in the environment (e.g. changes in pathogens' ability to parasitise their hosts; Hamilton, 1982; Hamilton & Zuk, 1982) and deleterious changes in the organism itself due to mutation. These degrading processes not only reduce fitness but also reintroduce variation in fitness. At a stable equilibrium, natural selection and degrading processes balance out so that the population fitness mean and variance remain unchanged. Hence, degrading processes not only ensure a level of maladaptation in populations; they also prevent populations from becoming uniform, so that some individuals inevitably have lower fitness than others (Burt, 1995).

The extent to which the variation in fitness is due to mutation alone as opposed to changes in the environment (e.g. host–pathogen coevolution) is open to debate. Based on empirical parameter estimates such as the mutation rate and the strength of selection against mutations, Charlesworth and Hughes (1998) estimated the CV_A of fitness due to mutation alone to be about 8 for *Drosophila*, in which the number of new mutations per genome per generation is approximately 1. Charlesworth (1990) had earlier estimated a value of about 17. (In this model, the vast majority of deleterious mutations that exist in the population at any given time have been in the pool for multiple generations. Whereas the per genome mutation rate is assumed to be about 1, an average of at least 20 mutations per individual (perhaps 200 in humans) are assumed to reside in the population because, given weak selection against them, they have not yet been eliminated by selection. For an overview, see Lynch *et al.* (1999).) Perhaps anywhere from half to nearly all of the genetic variance in fitness in *Drosophila*, then, may be due to mutation. More precise estimates must await further research. Whether due to mutation or changing environments, the abundant genetic variation in fitness in natural populations should generally be thought of as differences in health or 'condition', broadly defined (see Rowe & Houle, 1996; see also Thornhill & Gangestad, 1999b), though it should be noted that individuals of superior condition need not have lower mortality rates. Under certain circumstances, it may pay them to allocate much greater effort to mating – at a cost to mortality, somatic maintenance and health – than individuals of poorer condition and, ultimately, they may actually have higher mortality rates (see Kokko, 1998).

Additional factors in GGSS. GGSS requires at least two conditions beyond genetic variance in fitness. First, genetic variation in condition must co-vary with traits that serve to indicate genetic benefits. Grafen (1990) and Getty (1999) note that condition should, in theory, co-vary with any trait for which (a) the marginal costs of allocating effort into the development of the trait vary with condition (Grafen, 1990), and/or (b) the marginal benefits of allocating effort into the development of the trait vary with condition (Getty, 1999).

Even in the absence of any sexual selection for a trait, any number of traits could fulfil the conditions. For example, males in better condition may invest more effort into the development of traits useful in male–male competition (e.g. size) due to the energetic costs of the traits, the socially mediated costs of competition, or the fact that they may derive greater benefits from these investments owing to greater longevity and, hence, period of use. Female preferences for such indicators of condition could potentially evolve because

the traits advertise genetic benefits to offspring, and subsequently drive evolution of increased allocation of effort to the indicators (Pomiankowski, 1988; Iwasa *et al.*, 1991; Iwasa & Pomiankowski, 1994).

For females to drive indicator evolution, they must be in a position to observe male indicators and, hence, features that evolve as honest advertisements of genetic benefits should be reliably detected by females. For instance, song rate, song repertoire, or other acoustic features of male singing (e.g. amplitude) appear to have evolved to signal male quality in a number of bird species (e.g. Hasselquist *et al.*, 1996; Hasselquist, 1998; Møller *et al.*, 1998; Forstmeier *et al.*, 2002), perhaps because they are readily discerned by females at a distance. In other species, visual cues or behavioural performance (e.g. in relation to other males) serve as reliably detected, valid cues of quality.

Second, for female preferences to evolve, they must not have costs that outweigh their benefits. Some costs are costs of mating. Preference for some mates over others can lead to increased search costs (mating and re-mating times), which can be considerable in low-density, highly dispersed populations. Other costs may be due to pleiotropic effects of preferences on other traits (e.g. the ways in which they compromise other functions of the sensory system, such as food choices and abilities to avoid predators).

Except when particularly high, the costs on mating appear insufficient to prevent or greatly hinder GGSS. Indeed, Houle and Kondrashov (2002) modelled GGSS and found, under a wide range of reasonable estimates of genetic variation in fitness, co-variation between indicator traits and condition, and mating costs of preferences, that preferences readily evolved and often caused extreme exaggeration of indicator traits. By contrast, Kirkpatrick (1996) and Kirkpatrick and Barton (1997) found that GGSS is present but weak when the costs of female preferences increase with exaggeration of the male display due to pleiotropy. The nature of the costs of female preferences hence appears to be a key factor affecting the strength of GGSS. Houle and Kondrashov (2002) assumed that selection on cognitively complex organisms can separate naturally selected aspects of sensory and cognitive systems and thereby lead to minimal costs of preference due to pleiotropy (e.g. Wagner & Altenberg, 1996). But empirical data directly addressing this point are not available at this time.

In sum, a variety of aspects of GGSS – the genetic variance in fitness, the evolution of costly indicator traits of fitness, and the costs of female preference – have been subject to intense debate. Despite the history of controversy, at present there is no clearly established theoretical reason to doubt that GGSS can be a powerful evolutionary force. Naturally, if future research demonstrates considerable costs to female preferences due to pleiotropy, there would be new reason to doubt the strength of GGSS.

EMPIRICAL DATA ON GGSS

In order to demonstrate GGSS of female preference on male traits directly, it must be shown that:

(1) females prefer males who possess specific traits;
(2) these males produce offspring with greater than average fitness components; and
(3) the apparent advantages of choice for preferred males cannot be due to other reasons (e.g. material benefits provided by males, co-variation between female preference and condition)

As some factors are difficult to control precisely, even in experimental investigations (e.g. degree of female investment contingent on male features, see Burley, 1986; Gil et al., 1999), few studies rule out all possible alternatives. Later, we describe a programme of research on one species (collared flycatchers) that appears to demonstrate GGSS. (For an excellent example of study that rules out alternatives and thereby shows that a preference by sierra dome female spiders for large males increases viability of offspring, see Watson, 1998.)

Møller and Alatalo (1999) reviewed 22 studies in which preferred male features were correlated with offspring survivorship. They estimated a mean correlation coefficient of 0.12 (i.e. about 1.5 per cent of the variance in offspring survivorship was accounted for by the male trait), which was significantly different from zero. The effect was moderated by taxon (at least for those species examined, birds showed larger effects than other organisms), reproductive skew (as might be expected, greater skew was associated with greater effects), and publication date (for unknown reasons, earlier papers showed larger effects than more recent papers). Though the overall effect is modest, two points should be noted:

(1) the effects of female choice on offspring RS may be partly or largely mediated through mating success rather than survivorship;
(2) the heritability of offspring survivorship is fairly small itself (typically less than 30 per cent; Charlesworth, 1987;

Houle, 1992), which, in combination with the fact that male indicator traits may share only 20 per cent of their variance with heritable condition and still effectively mediate GGSS (Houle & Kondrashov, 2002), means that a few per cent of the variance in offspring survivorship accounted for is all one can theoretically expect.

None of the species studied in work reviewed by Møller and Alatalo (1999) was a primate; GGSS in non-human primates remains undemonstrated.

GGSS IN HUMANS?

In humans, evidence that good-genes sexual selection has operated on females' preferences has been indirect. Researchers have asked whether women prefer male traits that in ancestral conditions may well have been associated with genetic benefits to offspring. Traits preferred in modern environments due to GGSS may not be associated with viability in current environments. The extraordinary health care and life style changes that have occurred in the past century, for instance, may have altered substantially any correlations between phenotypic features and viability.

A candidate trait that may well have been fitness-related in ancestral environments is developmental instability – the imprecise expression of developmental design due to genetic and environmental perturbations. These perturbations importantly include mutations and pathogens and, hence, factors that contribute to genetic variation in fitness. Because of its conceptual link to maladaptation (see Møller, 1999), developmental instability became a focal point of research on sexual selection in the 1990s (Møller & Swaddle, 1997; Møller & Thornhill, 1998a).

The primary measure of developmental instability used in biology is fluctuating asymmetry (FA) – absolute asymmetry in bilateral traits owing to random errors in the development of the two sides. In research conducted at the University of New Mexico, we measured a number of asymmetries in humans – of the ears, elbows, wrists, ankles, feet, fingers (see, for instance, Gangestad et al., 1994; Thornhill & Gangestad, 1994, 1999a; Thornhill et al., 1995; Furlow et al., 1997, 1998; Gangestad & Thornhill, 1997a, b, 1998; Thoma et al., 2002). The asymmetry that exists in these traits is very small, the mean being 1–2 mm, so small that one cannot detect it reliably through normal social interaction. Hence, the asymmetries we measure cannot serve as cues by which individuals assess others' developmental imprecision. The reason we measure them, then, is because they purportedly are markers of underlying developmental imprecision, which substantially affects the overall phenotypic fitness of individuals.

As developmental perturbations can be caused by both environmental factors (e.g. pathogens, toxins, nutritional stress) and genetic factors (e.g. mutations, resistance to pathogens, resistance to toxins), there is much controversy about the extent to which FA is a marker of genetic health (for a recent review, see Lens *et al.*, 2002). A large number of studies have examined the heritability of FA on single traits in a wide variety of organisms. Heritabilities are generally estimated to be small and, in fact, average just 0.03 (Fuller & Houle, 2003; see also van Dongen, 2000). Taken at face value, these results may appear to indicate that variation in FA largely reflects environmental rather than genetic variation in developmental perturbations across individuals. Examination of the data in light of recent modelling, however, reveals that this simple interpretation may be wrong. Model-based interpretation of data indicates that a single trait's FA is typically a very poor indicator of underlying developmental instability, with less than 10 per cent of the variance in a single trait's FA owing to systematic differences in developmental instability (the remaining 90-plus per cent is owing to the stochastic nature of developmental error and is not informative about an individual's propensity to develop imprecisely (Gangestad & Thornhill, 1999, 2003b; see also Houle, 2000). Only about 8 per cent of the variance in the FA of the human traits that we measure reflects true differences across individuals in developmental instability (Gangestad *et al.*, 2001). If single-trait FA has a typical heritability of 0.03, then results actually suggest that underlying developmental instability has a moderate heritability: $0.03/0.08 \cong 0.4$. Because small differences in the true heritability of a single trait's FA can have sizeable effects on estimation of the heritability of underlying developmental instability, no strong conclusions about the heritability of developmental instability can be drawn at this time (Fuller & Houle, 2003).

If FA taps developmental instability only weakly, the question of how it can serve as a reasonable measure of it arises. Here, the critical point is that a *single* trait's FA weakly taps developmental instability. When multiple traits' asymmetries are added to yield a multi-trait index of FA, the resulting measure can be a reasonable valid measure of underlying developmental instability. For instance, we estimate that when our ten individual traits' asymmetries are aggregated into a total index, with each trait's FA owing just 8 per cent variance to developmental instability, the resulting measure owes about 45 per cent of its variance to develop-

mental instability – that is, it correlates approximately 0.65 (i.e. $\sqrt{0.45}$) with developmental instability (see Gangestad & Thornhill, 1999, 2003b).

At this point, we have a working hypothesis that our aggregate measure of FA taps an important component of heritable developmental health in humans, one that seems reasonable given both theory and available data. None the less, definitive evidence is not available at time of going to press.

Several years ago, we began asking whether FA predicts number of sex partners in college students, similar to associations examined in other species. We have now studied over 500 college students of each sex. In sum, we find that, in this population, men's FA does, whereas women's FA does not reliably do so (see Gangestad & Thornhill, 1999 for an overview; see also Thornhill & Gangestad, 1994; Gangestad & Thornhill, 1997b). Recently, we estimated (with Kevin Bennett; Gangestad *et al.*, 2001) the correlation between men's developmental instability and number of sex partners using latent structural equation modelling, which takes each individual trait's FA as an independent marker of underlying developmental instability. The estimated correlation was −0.4 to −0.5 (with size and age controlled), a sizeable effect.

We have also collaborated with two anthropologists, Mark Flinn and Rob Quinlan, in work in a rural village on the Caribbean island of Dominica, and there too we find that FA predicts number of romantic partners in men (Gangestad *et al.*, 2003b unpublished). In that study we measured number of romantic partners by asking other villagers, not target individuals themselves. The correlation between FA and men's number of partners was about −0.4.

We suspect that female preferences for symmetry per se have little if anything to do with the causal process that drives these associations. FA is *our* measure of developmental health. It correlates with a variety of physical and behavioural features that may mediate these associations between FA and number of sex partners. Consider the following examples:

(1) In Dominica, FA negatively predicts men's peer status, as assessed through interviews with men. More symmetrical men are seen to be better coalition partners than less symmetrical men. Female preferences for men with favourable peer status could lead more symmetrical men to have more romantic partners (Gangestad *et al.*, 2003b).

(2) US college men who have greater body symmetry have more masculine faces, as assessed by a variety of sexually dimorphic facial dimensions (Gangestad & Thornhill, 2003a). Even the association between facial symmetry

and attractiveness may be largely mediated by other facial features (Scheib *et al.*, 1999).

(3) Simpson *et al.* (1999) had US college men interviewed for a potential lunch date with an attractive female. As part of the interview, each man was asked to tell the woman, as well as a male competitor – someone else she was purportedly interviewing – why he should be chosen for the lunch date over the other. More symmetrical men were more likely to engage in direct intrasexual competitive tactics – directly comparing themselves with the other and stating that they were the better choice. Other research shows that more symmetrical men are described as less willing to back down from threats to their status (Gangestad & Thornhill, 1997a) and, perhaps as a result, more likely to get into physical fights, particularly those that they themselves escalated into a fight (Furlow *et al.*, 1998; Manning & Wood, 1998).

(4) In college samples, symmetrical men appear to be more muscular and vigorous (Gangestad & Thornhill, 1997a) and have lower basal metabolic rates (Manning *et al.*, 1997).

(5) Research in Dominica and on college students indicates that more symmetrical individuals are more intelligent (e.g. Furlow *et al.*, 1997; Gangestad *et al.*, 2003b; also Thornhill & Møller, 1997; Yeo *et al.*, 2000; Prokosch *et al.* (submitted for publication); Thoma *et al.*, 2002).

(6) Symmetrical men appear to have more attractive voice qualities (Hughes *et al.*, 2002).

Again, we suspect that these features (or some set of them) mediate FA's association with sexual history, partly because of female preferences for them (intrasexual competitive advantages may also play some role). Because these features may also carry with them some material benefits, however, preferences for them do not constitute special design evidence for good-genes sexual selection. A study of romantically involved college couples suggested that more symmetrical men actually invest *less*, not more, in their romantic relationships (e.g. they give less time to them, sexualise other women more, and are more dishonest to their partners; Gangestad & Thornhill, 1997a). Nonetheless, although symmetrical men may often invest less in their relationships in general, they possibly provide specific forms of material benefits to their partners (e.g. physical protection, which more symmetrical men are judged better able to provide; Gangestad & Thornhill, 1997a). Broad preferences for traits associated with purported heritable fitness markers probably cannot, by themselves, demonstrate special design for

obtaining genetic benefits. Design may be found in the particular contexts in which preferences are expressed, which is why we began to examine preferences pertaining to extra-pair mating.

GGSS AND EXTRA-PAIR MATING

In many species of birds, males and females bi-parentally invest in offspring. Males and females form cooperative social mateships that may last seasonally or for many years. In these systems, multiple mating can take a particular form. Individuals copulate with their social or 'in-pair' mates, with whom they share responsibilities for caring for young, but may also copulate with one or more 'extra-pair' mates, with whom they do not typically share parental responsibilities. There are asymmetries in the male and female extra-pair mating. Whereas female extra-pair mating may result in investment by her in-pair mate in offspring not his own, male extra-pair mating often leads to investment by his extra-pair mate's mate in offspring not his own.

A spectacular empirical discovery is the well-known fact that extra-pair paternity is common in many species of birds, accounting for, on average, 10–15 per cent of offspring and not uncommonly exceeding 25 per cent of them (Birkhead & Møller, 1995; Petrie & Kempenaers, 1998). As noted earlier, a number of benefits, both direct and indirect, have been proposed to account for female EPC, one being intrinsically good genes for offspring. The hypothesis that females engage in EPCs because they have obtained intrinsically good genes states that:

(1) When males and females assort into breeding pairs, many females are inevitably paired with males who provide near-average or below-average intrinsic genetic benefits to offspring.

(2) Females therefore may be able to increase the genetic benefits to offspring by mating with an extra-pair male who possesses indicators of high levels of genetic benefits. The potential benefits that females may accrue through extra-pair mating are partly an inverse function of the level of genetic benefits provided by a mate.

(3) Females may also suffer costs through extra-pair mating. They may incur search costs to the extent that they look for potential extra-pair mates.

Because selection should favour male motivation, ability to detect extra-pair mating in their partners and a strategy of paternal investment that allocates care contingent on likelihood of paternity, females may also suffer direct costs of withheld paternal investment in their offspring. Finally, extra-pair mating generates half-siblings within a family and thereby can increase sibling–sibling conflict, which carries fitness costs. A number of predictions follow:

(1) Males who are the EPC partners of females ought to, on average, possess markers of intrinsically good genes. The greater (RS) of males who possess markers of intrinsically good genes may hence be largely mediated through enhanced extra-pair paternity (EPP) rather than greater success of offspring with their in-pair mates.
(2) Males particularly susceptible to having their mates engage in EPCs may tend to lack markers of intrinsically good genes. Females with such males may particularly increase the fitness of offspring through EPC and, therefore, be most willing to pay its costs. This prediction may not straightforwardly apply when males guard their mates contingent on their own features, such that males who lack markers of genetic benefits impose greater costs on their female mates.
(3) Males particularly successful at obtaining EPC partners may allocate greater effort to extra-pair mating effort at a cost of allocation to other activities, and hence may engage in less parenting effort than males less successful at obtaining these partners. Even if more efficient at obtaining resources relevant to parenting (e.g. more efficient at feeding), then, males who possess markers of genetic benefits may actually provide fewer material benefits to offspring.
(4) Because 'good-genes' males may provide no more material benefits to offspring, female preferences for social partners and extra-pair mates differentially weight indicators of genetic benefits. Females may not prefer (or, in certain systems, may even disfavour) markers of genetic benefits in social, in-pair partners, even when strongly preferring them in EPC mates.
(5) Offspring sired by extra-pair males should possess greater mean viability than offspring sired by in-pair males, and males sired by extra-pair fathers should possess greater mean breeding success than their counterparts sired by in-pair fathers.
(6) If females or males can bias the sex ratio of offspring, a male-biased sex ratio is expected among the offspring produced by females that mate with males with high intrinsic genetic quality, owing to the particular edge their sons have in competition for matings. Hence, offspring sired by males favoured as extra-pair mates should tend to have a male-biased sex ratio.

In a number of systems, the evidence supports the good-genes explanation of extra-pair mating. Behavioural or direct DNA fingerprinting data indicate that males who are solicited as extra-pair mates or who are responsible for extra-pair fertilisations possess features that distinguish them from other males in a number of species: attractive zebra finch males (Houtman, 1992); dusky warblers who sing for longer periods of time at high amplitude (Forstmeier et al., 2002); black-capped chickadees who are more socially dominant than in-pair males (Otter et al., 1998); great reed warblers with a more extensive song repertoire (Hasselquist et al., 1996); collared flycatchers who have a more extensively developed condition-dependent secondary sexual character (a larger forehead patch; Sheldon et al., 1997); barn swallows who have a higher song rate (Møller et al., 1998). In some instances, researchers have linked preferred traits to fitness components: extensive singing at high frequencies is associated with male longevity in dusky warblers (Forstmeier et al., 2002); song repertoire of a father predicts the post-fledging survival rate of offspring in great reed warblers (Hasselquist et al., 1996; Hasselquist, 1998); offspring sired by male collared flycatchers with a larger forehead patch are in better condition than their half-siblings within the same brood (Sheldon et al., 1997). Females in some species (e.g. Bullock's orioles; Richardson & Burke, 1999) favour older males as EPC partners, perhaps because age itself is a marker of viability (see Brooks & Kemp, 2001). Female bearded tits invite extra-pair males to chase them more often when fertile, and Hoi (1997) proposed that they do so to incite male–male competition as a means of assessing relative genetic quality. In species in which the extra-pair paternity rate is high (yielding relatively many opportunities for males to obtain extra-pair matings), more attractive males tend to engage in less parental effort, in accordance with the good-genes hypothesis (Møller & Thornhill, 1998b).

At the same time, studies in a variety of species in which extra-pair mating is common have yielded negative or equivocal evidence for intrinsic genetic benefits: e.g. razorbills (Wagner, 1992); hooded warblers (Stutchbury et al., 1997); sedge warblers (Buchanan & Catchpole, 2000). Some researchers have argued that genetic benefits achieved in particular species are better understood in terms of compatible genes (e.g. in bluethroats: Johnsen et al., 2000, 2001: in pied flycatchers: Ratti et al., 1995) or diverse genes (e.g. in great tits: Otter et al., 2001; see also Lubjuhn et al., 1999) rather than intrinsically good genes. Different genetic and material benefits may account for the evolution of female extra-pair mating in different species. For reviews, see Petrie and Kempenaers (1998) and Jennions and Petrie (2000).

GGSS AND EXTRA-PAIR MATING IN THE COLLARED FLYCATCHER

One population in which extensive investigation has yielded an impressive array of convergent evidence that females obtain intrinsic genetic benefits through extra-pair mating is the group of collared flycatchers (*Ficedula albicollis*) on the island of Gotland, just off the east coast of Sweden. As noted above, size of a male secondary sex character – a white forehead patch – is condition-dependent (e.g. compromised by infectious disease; Gustafsson *et al.*, 1994). It possesses substantial heritability (∼ 0.55) in favourable environments and low to moderate heritability (∼ 0.25) in poor rearing environments (Qvarnström, 1999a), and is sexually selected due to two mating advantages. First, males with large patches are paired to females that breed earlier. Patches are used as plumage displays in male–male contests over territories, patch size predicts the winners of these contests, and winners obtain females more quickly than losers (Pärt & Qvarnström, 1997). Indeed, experimental enlargement of a male's patch leads him to be more willing to engage in contests over territories and hence increase this component of mating effort. Second, males with large patches sire more offspring through EPCs. (The EPP rate in this population is about 15 per cent; Sheldon & Ellegren, 1999.) The effect of patch size on reproductive success mediated by extra-pair paternity is substantially greater than the effect mediated by early mating, accounting for 65–90 per cent of the total selective advantage (Sheldon & Ellegren, 1999).

Not only do males with small forehead patches obtain fewer extra-pair matings, their in-pair mates preferentially engage in EPCs (Michl *et al.*, 2002). In an attempt to provide females opportunity and motivation to mate with multiple males, Sheldon *et al.* (1999) removed males from their nests for two days, which, as expected, greatly increased the rate of multiple paternity of the subsequent brood. When the removed male had a large forehead patch, there was lower probability of a replacement male appearing. As a result, removed males with a larger forehead patch fathered more young than males with smaller patches.

Females tend to engage in EPCs selectively within the middle part of their fertile period and after their last copulation with their in-pair mates. Hence, despite just 1.33 extra-pair copulations per cuckolding female, Michl *et al.* (2002) measured the ratio of extra-pair to in-pair sperm stored to

be about 5 : 1. Through timing of copulations, then, females appear to bias paternity strongly in favour of extra-pair males.

There is a clearly established basis for females to prefer the sires of their offspring to possess large patches: offspring sired by collared flycatcher males with a larger forehead patch are in better condition than their half-siblings within the same brood (Sheldon *et al.*, 1997), which indicates a genetic correlation between male patch size and offspring viability. In addition, because male patch size is itself somewhat heritable, male mating success rates associated with patch size should also be heritable.

Males with larger patches do not invest in greater parental effort than others. In fact, because these males engage in greater mating effort among early breeders, they are in worse condition when feeding begins (Qvarnström, 1999b). Experimental manipulations of paternal effort (through increased brood size) reduces patch size in the following year (Gustafsson *et al.*, 1995; Griffith & Sheldon, 2001) and hence males face trade-offs between parental and mating effort. Males with experimentally enlarged patches engage in less parental effort (lower feeding rates), presumably owing at least in part to the larger negative impact on their potential mating success the following year. Yet, because their mates increase feeding, their offspring do not suffer lower parental effort overall (Qvarnström, 1997). Perhaps because of trade-offs between genetic quality and parental investments (both their partners' and their own), early breeding females do not demonstrate a preference for mating with males with large patches. When they breed late, males with large patches engage in greater parental care than when they breed early, purportedly because most females have not only mated but also reproduced and, therefore, male mating effort has relatively low pay-offs even for large-patched males (Qvarnström, 1999b; Qvarnström *et al.*, 2000). Interestingly, then, late-breeding females, whose reproductive success is enhanced when their mates have large patches, *do* show a clear preference for these males.

Finally, Ellegren *et al.* (1996) found that paternal size of the forehead patch predicts the sex ratio of offspring: males with large forehead patches father relatively more males. Experimental manipulations of brood size demonstrate a trade-off between quantity and quality of males, as measured by patch size (Gustafsson *et al.*, 1995). These experimental manipulations also affect the sex ratio of offspring in a predictable manner: under conditions favouring male quality, relatively more males are produced (Ellegren *et al.*, 1996).

In sum, data clearly demonstrate that the size of the male forehead patch is partly heritable and under sexual selection, as mediated through timing of breeding (as affected by ability to compete for territories) and, more importantly, through extra-pair paternity. Whereas the former effect is attributable to differences in male competitive abilities, the latter effect is at least partly due to female choice; females possess adaptive, *special design* for seeking EPCs with large-badged males, biasing paternity in favour of their extra-pair mates, and possibly altering sex-ratio dependence on sire features. Males apparently possess design for advertising quality through the condition-dependent trait of forehead patch size, as well as conditionally altering their relative allocation of effort to mating and parenting based on their own trait level (or correlated response evoked by their trait level by other males and females). The evidence for current sexual selection clarifies why we observe these design features if we assume that selection pressures now were also effective forces of phenotypic (physiological and psychological/behavioural) evolution in ancestral populations of collared flycatchers, i.e. that the selective environment has not changed. *Crucial* to the argument that sexual selection led to the evolution of female strategies for seeking EPCs, however, is the evidence for design. Were that evidence lacking, the evidence for current selection pressures would in no way by itself demonstrate that those pressures have been effective forces in the evolution of collared flycatchers.

GGSS AND EXTRA-PAIR MATING IN HUMANS

Female EPC is not a rare occurrence in humans. During a face-to-face interview with a large random sample of married women in the USA, 15 per cent admitted to having extra-marital sex (Laumann *et al.*, 1994). Because of response bias, that value may be underestimated (for a review of questionnaire data, yielding a mean estimated rate of female marital infidelity of about 30 per cent, see Thompson, 1983). In our study of 201 romantically involved university women of an average age of 20 years and who had been in their current relationship an average of about two years, 13 per cent reported having an EPC in the current relationship and 17 per cent in any relationship. Furthermore, 6 per cent of a sample of British women (albeit far from a random sample) with one primary partner reported their *last* act of intercourse to be an EPC (Bellis & Baker, 1990). As noted earlier, EPC apparently results in an appreciable rate of extra-pair paternity in at least some populations (e.g. the low-income subpopulation of Monterrey, Mexico has a rate of about 20 per cent;

Cerda-Flores *et al.*, 1999), though only rare extra-pair paternity in others (e.g. a 1 per cent rate in a Swiss population; Sasse *et al.*, 1994).

Female EPC may be a relatively common occurrence now. But was it sufficiently common in small ancestral populations of humans or pre-hominid primates to be an effective selective force of evolution? Evidence suggests yes, and perhaps the best evidence comes from design features of men rather than women. Men, but not women, can be duped about parentage as a result of EPC, leading to the unknowing investment in another man's offspring. Men show a rich diversity of mate guarding and anti-cuckoldry tactics ranging from sexual jealousy, vigilance, monopolising a mate's time, pampering a mate, threatening a mate with harm if she shows interest in other men, and adjusting ejaculate size to defend against the mate's insemination by a competitor (e.g. Buss, 1988; Wilson & Daly, 1992; Baker & Bellis, 1995; Shackelford *et al.*, 2002). Some mate guarding tactics appear to be conditional, such that men guard mates of high fertility status (young or not pregnant) more intensely than ones of low-fertility status (older or pregnant) (Flinn, 1987) and hence appear not to be caused by general male–male competitive strivings but rather concern for fidelity of a primary social mate (see Buss, 2000).

The functional design and, hence, adaptation in men's sexual proprietariness (Wilson & Daly, 1992) is itself evidence for female EPC in human evolutionary history, indicating that female EPC was of sufficient importance to have generated selection on males in the deep-time past, shaping male features that appear to function as anti-cuckoldry tactics (Buss, 2000). Nonetheless, these features are not evidence that women evolved functional design due to the benefits of EPCs. Female EPC may be forced or unforced and, if the latter, could arise from incidental by-products of adaptations the functions of which do not pertain to obtaining benefits through EPC. Male sexual propriety would be expected to evolve in response to appreciable rates of female EPC, no matter its cause.

EVIDENCE FOR DESIGN IN WOMEN'S EPC

To demonstrate that female EPC is at least partly attributable to adaptations the beneficial effects of which were achieved specifically through EPC, one must establish design features of behaviour, desires and preferences that affect EPC and that strongly point to a particular function. That is, one must show that physiological or psychological features affecting

EPC possess special design for achieving particular beneficial effects through EPC.

For the remainder of this chapter, we emphasise one specific set of features of women's sexuality that, when considered collectively, provisionally indicate special design for acquiring genetic benefits through EPC: changes in women's sexual interests and preferences across the menstrual cycle. Suppose ancestral females could have secured a better genetic complement for their offspring from extra-pair males at the expense of the RS of their primary social partners, who impose a cost on such female behaviour when they observe it (e.g. by divestment of the female or her offspring). If the benefits and costs of specific cognitive, emotional and overt behavioural responses varied across contexts ancestrally, selection is expected to have acted to shape psychological adaptations that produce responses contingent on context – that is, these psychological adaptations will give condition-dependent outputs.

Because women obtain genetic benefits from mating only when at fertile times in the menstrual cycle (approximately six to seven days per month; Wilcox et al., 1995) but could pay the costs of EPC throughout the cycle, selection for obtaining genetic benefits through EPC should be expected to have shaped female interest in and pursuit of men who possess indicator traits of high genetic quality, such that the interest and pursuit changes across the cycle: maximal mid-cycle when genetic benefits can be obtained and lower at other times. As this prediction is not readily derived from alternative viewpoints, evidence for a set of such features consistent with good-genes sexual selection constitutes at least provisional evidence for such an account of EPC behaviour.

SHIFTS IN WOMEN'S EPC INTEREST ACROSS THE CYCLE

The first such demonstration of a menstrual cycle effect came from a large survey of UK women (Bellis & Baker, 1990). Whereas the frequency of in-pair sex was relatively evenly distributed across the cycle, women's EPC occurred more often on high-fertility days (Bellis & Baker, 1990). Indeed, the peak rate of EPCs (occurring about the tenth day of the cycle) was about 2.5 times the minimal rate of EPCs (occurring the last week of the cycle) (Baker & Bellis, 1995).

Research by Gangestad et al. (2002) indicates that this pattern is at least partly due to changes in female sexual interest in extra-pair partners. Participants were 51 young women (31 with primary romantic partners) not using hormone-based contraceptives. They filled out questionnaires about their sexual interests and fantasies at two points in their menstrual cycle: once within five days before or one day after a luteinising hormone (LH) surge, which is a period of high fertility with the surge corresponding roughly to the peak (Jöchle, 1973; Baker & Bellis, 1995; Wilcox et al., 1995), and ovulation occurring approximately 24–48 hours after; and once during the luteal phase (low conception risk). The study yielded three salient findings. First, on average, the effect of fertility status on women's sexual attraction to and fantasy about *primary* partners was small and statistically insignificant. Second, the same women's attraction to and fantasy about men *other than primary partners* increased substantially during the fertile phase of the cycle, and to a degree significantly greater than increases in women's attraction to primary partners. Third, some analyses suggested that the second effect increases in strength with how close our questionnaire session occurred to actual ovulation.

SHIFTS IN WOMEN'S PREFERENCE FOR THE SCENT OF SYMMETRY

The hypothesis that EPC by women has been directly selected because it resulted in genetic benefits to offspring suggests more. Naturally, women should not show increased interest to extra-pair men in general when fertile; they should show increased interest in men who exhibit indicators of genetic benefits. Hence, women's preferences should change across the cycle (Gangestad & Thornhill, 1998). These preference shifts should be particularly strong when women evaluate men as short-term sex partners (i.e. men's 'sexiness'), as opposed to men as long-term relationship partners (Penton-Voak et al., 1999). The study by Gangestad et al. (2002a) on female sexual interests in extra-pair partners does not address the question of *which* extra-pair men women find most attractive mid-cycle. But other research does.

As discussed earlier, symmetrical men tend to report a greater number of sexual partners and possess a number of traits that women find attractive in sexual partners. We first tested the prediction that women's preferences change across the menstrual cycle because they could have garnered genetic benefits from extra-pair partners mid-cycle by looking at olfactory preferences – specifically, whether women prefer the scent of symmetrical men more strongly when fertile than not. Male scent strongly affects women's attraction to them, perhaps more so than their looks (e.g. Regan & Berscheid, 1995; Herz & Cahill, 1997). We have now performed three studies, and all confirm the prediction (Gangestad & Thornhill, 1998; Thornhill & Gangestad, 1999a; Thornhill et al.,

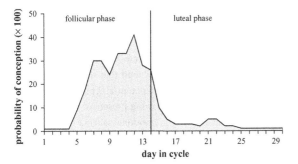

Fig. 6.1 Women's actuarial probability of conception as a function of day of the cycle. From Jöchel (1973). Total $N > 1800$.

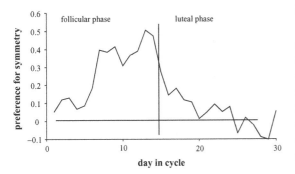

Fig. 6.2 Women's preference for the scent of symmetry as a function of day of the cycle. Data combined from Gangestad and Thornhill (1998), Thornhill and Gangestad (1999a), and Thornhill *et al.* (2003). Total $N = 141$ women. Each day represents a 3-day moving average to smooth the curve (due to low N for some days).

2003). In all studies, 40–100 men wore T-shirts for two nights. They were instructed to avoid eating certain foods (e.g. spicy foods), smoking, sleeping with someone else, and sex. On the two days following the second night, normally ovulating (non-pill-using) women rated the shirts' scent on scales of pleasantness and sexiness, combined to form an attractiveness rating. For each woman, we computed preference for symmetry as the slope of the regression of their ratings on men's symmetry.

Figure 6.1 is based on data presented by Jöchle (1973) showing how fertility risk varies across the cycle. Figure 6.2 shows preference for the scent of symmetry as a function of the day of the cycle (3-day moving averages) for all 141 women in our studies. As shown, women prefer the scent of symmetrical men only during the fertile period. Peak preference matches peak conception risk, which occurs on average on day 12 (Jöchle, 1973). Another way to look at the data is

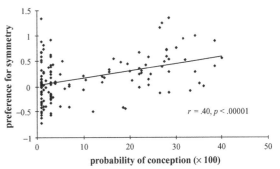

Fig. 6.3 Women's preference for the scent of symmetry as a function of estimated conception risk. Data combined from Gangestad and Thornhill (1998), Thornhill and Gangestad (1999a), and Thornhill *et al.* (2003). Total $N = 141$ women.

to convert the day of the cycle to conception risk, estimated from the Jöchle data (Fig. 6.3). The correlation with preference for symmetry is 0.4. Rikowski and Grammer (1999) replicated this finding in Austria. As these studies used an imperfect measure of fertility status (as opposed to detection of LH in urine), the true association between fertility status and the preference for the scent of symmetry is probably stronger.

We do not know yet what chemical substance in men's sweat is associated with symmetry (presuming such a chemical exists) and mediates this finding. Theoretical reasons suggest that it may be androgen-derived, as testosterone may be elevated in men who exert greater mating effort (e.g. Gray *et al.*, 2002), and sweat glands in human skin have high levels of 5-alpha reductase activity, which converts the weak androgen testosterone into a powerful androgen, dihydrotestosterone (Luuthe *et al.*, 1994). The fact that more symmetrical men have more masculine features provides some empirical reasons to suspect that they have increased testosterone (or are more sensitive to its effects), at least during critical periods of adolescence. Hence, one possibility is that the chemical mediator of ovulating women's preference for the scent of symmetrical men is androgen-derived (Thornhill & Gangestad, 1999a), but no direct evidence for this claim exists. Future research may explore this hypothesis.

The cues in scent that signal developmental stability may co-vary with other cues of developmental stability. In one study, normally ovulating women (from New Mexico) rated the attractiveness of photos of men from Dominica. Women found the men whose bodies had been measured as symmetrical more attractive when they were more fertile (as estimated from their day in the cycle) than when they were outside the fertile phase (Thornhill & Gangestad, 2003). It is not known which facial features associated with symmetry are particularly preferred in mid-cycle, though the research we discuss below suggests one possibility: men's facial hormone markers.

SHIFTS IN PREFERENCES FOR MALE FACIAL MASCULINITY

Following the initial finding that women's olfactory preferences change across the cycle, researchers have begun to look for other preference shifts. One fruitful line of research that may be relevant to the claim that more testosteronised men are particularly attractive mid-cycle was initiated by Ian Penton-Voak and his colleagues. Through computerised technology, multiple faces can be digitised and morphed to create average faces. Perrett *et al.* (1998) used this methodology to produce an image corresponding to an average male face and an image corresponding to an average female face. By blending or exaggerating the differences between these two faces, they were able to create an array of male faces that varied from relatively androgynous to hypermasculine. When asked which face they found most attractive, Scottish and Japanese women, on average, actually tended to choose a face somewhat more feminine than the average male face (see also Swaddle & Reierson, 2002). Penton-Voak *et al.* (1999) hypothesised that women's preferences would change across the cycle, however, such that the face they find most attractive when fertile is more masculine than the one they most prefer when outside the fertile phase. Four different studies have confirmed their prediction: two in the UK (Penton-Voak *et al.*, 1999, Study 2; Penton-Voak & Perrett, 2000), one in Japan (Penton-Voak *et al.*, 1999, Study 1), and one in the US and Austria (Johnston *et al.*, 2001).

In one of their studies, Penton-Voak *et al.* (1999) added an important twist. They reasoned that, if preference-shifts toward markers of genetic benefits mid-cycle were selected because they would have led women to conditionally seek genetic benefits through extra-pair mating mid-cycle (or be more likely to seek such benefits), then the shift in women's preferences should pertain to their evaluation of men as short-term sex partners, not as long-term, investing primary mates. Hence, in one study Penton-Voak *et al.* specifically asked women to evaluate men's attractiveness in both of these relationship contexts. A significant fertility-status × relationship-context interaction effect emerged. Follow-up analyses showed that, as predicted, fertility status affected ratings of men's attractiveness as a sex

partner only, not their attractiveness as a long-term, investing partner.

Further evidence that women's responses to male faces change across the menstrual cycle was found by Oliver-Rodriguez *et al.* (1999). They found that the magnitude of the P300 response of the evoked potential (a positive potential around 300 milliseconds following presentation of a stimulus, which co-varies with the emotional salience of the stimulus) of women in the fertile cycle phase correlated with their rating of male facial attractiveness, but not their ratings of female facial beauty. During the infertile phase, women's P300 responses were undifferentiated and co-varied with both male and female attractiveness judgments.

Men's facial masculinity is affected by androgen production during development, which, like symmetry, may be a signal of better condition during development (Thornhill & Gangestad, 1993; Thornhill & Møller, 1997; Thornhill & Gangestad, 1999b). Indeed, as we noted earlier, male facial masculinity positively co-varies with body symmetry and, hence, these traits tap a common underlying factor (Gangestad & Thornhill, 2003a).

Men who have relatively masculine faces are perceived by people to be men who will invest in a mateship less than will a man with lower facial masculinity (Perrett *et al.*, 1998; Johnston *et al.*, 2001), and there is evidence that possibly related attributions (e.g. how agreeable or warm a man is) are accurate (Berry & Wero, 1993; Graziano *et al.*, 1997). Women's preference for men with less masculinised or slightly feminised faces during the infertile period may thus function to secure certain material benefits from men who are willing to invest as a result of their relatively low genetic quality. Fertile women's preference for symmetrical men and men with highly masculine features suggests that women are willing to trade-off between heritable benefits from mate choice and willingness to provide material benefits in mate choice.

SHIFTS IN PREFERENCES FOR MEN'S BEHAVIOURAL DISPLAYS

Research has also begun to address whether women's preferences for men's behavioural displays shift across the cycle. As noted earlier, Simpson *et al.* (1999) found that more symmetrical men tended to exhibit more intrasexually competitive displays in a situation in which men competed for a potential lunch date with an attractive woman. Gangestad *et al.* (2003a, unpublished) asked whether women are more attracted to men who exhibit these and related displays when in the fertile phase of their cycles. The behaviours of men being interviewed by women for a lunch date were coded for a host of verbal and non-verbal qualities. Through principal components analysis of these codes, two major dimensions along which men's performance varied were identified; 'social presence', marked by a man's composure, his direct eye contact and lack of downward gaze, as well as a lack of self-deprecation, and emphasis that he's a 'nice guy'; and 'direct intrasexual competitiveness', marked by a man's explicit derogation of his competitor and statements to the effect that he is the better choice, as well as not being as obviously agreeable.

Based on the working assumption that a man's social presence and intrasexual competitiveness may both signal broadly defined developmental health (e.g. Simpson *et al.*, 1999), women were predicted, when fertile, to prefer particularly men who exhibited these behaviours, and particularly as short-term mates – sex partners. That is, a pattern of preferences that parallels the findings of Penton-Voak *et al.* (1999) for facial masculinity was expected. Normally ovulating women were shown one-minute segments of the videotapes and asked to rate the attractiveness of men as a long-term partner as well as a short-term or sex partner. Each woman's ratings (for each attractiveness dimension) were regressed on each of the two male behavioural traits to obtain measures of preference for each behavioural trait – the slope of the regression line. Because physical attractiveness accounts for much of the variance in the attractiveness ratings and the study focused on variance caused by what the men did and said, men's physical attractiveness was statistically controlled for in these regressions (though this procedure did not affect the results). Women's fertility risk was estimated from day of the cycle.

As predicted, a significant fertility-risk × relationship-context interaction emerged. Women's preferences for social presence and direct intrasexual competitiveness in sex partners and long-term mates were quite similar when they were in the low fertility phase of their cycles (early follicular and luteal). Their sex-partner and long-term mating preferences diverged, however, in the fertile phase of their cycles, such that women particularly preferred these behavioural traits in sex partners at this time. Separate analysis of the preferences showed a reliable effect of fertility status on women's sex-partner preferences for the behavioural traits, but a small and statistically unreliable effect on the long-term mating preferences. Analysis of the two traits separately yielded similar findings. Again, this fertility risk × relationship context effect directly parallels what Penton-Voak *et al.* (1999) found with facial masculinity – the difference here being that these

effects concern preferences for behavioural displays rather than facial traits.

SUMMARY: CYCLE EFFECTS ON WOMEN'S PREFERENCES

Women's preferences shift across the cycle in a number of ways. They particularly prefer the scent and faces of more symmetrical men when fertile. The face they find most attractive when fertile is more masculine than the face they most prefer when not fertile. They prefer more assertive, intrasexually competitive displays when fertile than when not. Furthermore, evidence indicates that their preferences when evaluating men as sex partners (i.e. their sexiness) is particularly affected; evidence shows that their evaluations of men as long-term partners shift little, if at all.

These effects were predicted based on the theory that ancestral women may have benefited by obtaining genetic benefits for offspring through extra-pair mating with men exhibiting markers of genetic benefits, but at some cost. Selection should be expected to have shaped their preferences for markers of genetic benefits to be most pronounced when women could obtain these benefits – that is, when in the fertile phase of their cycle – and less pronounced outside the fertile phase. Women's preferences for male traits hence change across the cycle in ways predicted, if they have adaptations designed for obtaining genetic benefits for offspring through sex with partners other than long-term social partners, and symmetry and associated traits do tap some variation associated with additive genetic benefits.

It stands to reason that these or related shifts in women's preferences for male attributes partly drive changes in women's interest in men other than primary partners (Gangestad *et al.*, 2002). That is, we suspect that:

(1) when fertile, women experience heightened sexual attraction to men who possess certain characteristics, which only a minority of women's male partners possess;

(2) this leads women to experience heightened sexual attraction to men other than primary partners. Taken together, the evidence indicates that, when fertile, women's interest in men other than primary partners is not highly generalised, but rather is highly selective.

A crucial prediction has not yet been examined: women's enhanced interest in extra-pair men should in fact be conditionally influenced by the characteristics of their primary mates. Women with men who possess features of symmetry, facial masculinity and associated behavioural displays should not be particularly interested in men other than primary partners when fertile. The factors that moderate menstrual cycle effects should be explored in future research.

MENSTRUAL CYCLE EFFECTS ON MALE MATE GUARDING

In addition to examining shifts in women's interests across the cycle, Gangestad et al. (2002) also examined shifts in men's mate guarding. If women's sex with an extra-pair partner has historically been both more likely and, in reproductive terms, more costly to male partners, selection may have favoured male counter-adaptations to attend to their mates more in mid-cycle, using the minimal cues of fertility status that are available. (e.g. Singh and Bronstad, 2001, found that men are more attracted to the scent of fertile women). We also asked women in this study to report their primary mate's mate-retention tactics. Our questionnaire measures two major dimensions, 'proprietariness' and 'attentiveness'. Women reported their partners to be higher on both when fertile. The subscale showing the largest effects was 'vigilance' – frequent checking up on partners. The men whose partners have increased attraction to and fantasy about other men mid-cycle showed a greater effect. How much women have increased attraction to their primary partners mid-cycle did not predict changes in men's proprietariness and attentiveness. This pattern of findings may reveal something about the conflicts of interest between the sexes.

OTHER DESIGN ELEMENTS

Though the menstrual-cycle effects on female sexuality and preferences suggest female adaptation for obtaining genetic benefits, demonstration of additional features designed to increase genetic benefits would obviously constitute stronger evidence. Evidence suggests that female orgasm may function to increase retention of sperm (Baker & Bellis, 1995). In Japanese macaques, females orgasm more frequently with dominant males, perhaps to bias paternity in favour of these males (Troisi & Carosi, 1998). In our work with Randall Comer, we found that the partners of symmetrical men tend to orgasm during sexual intercourse more frequently, perhaps for similar reasons (Thornhill et al., 1995; see also Møller et al., 1999). Using a different method of measuring symmetry, Montgomerie and Bullock (1999) did not replicate this finding. Shackelford et al. (1999) reported that women whose partners they rate as more physically attractive

than women who rate their partners as less attractive also say that they have more orgasms with their partners. Additional work on these associations, as well as the basic claim that orgasm functions to bias paternity, is needed (see also Birkhead & Kappeler, this volume).

As noted earlier, selection on males and female partners may favour the biasing of sex ratios of offspring such that men with indicators of genetic benefits produce more sons. Gangestad and Simpson (1990) found that men in the Kinsey sample (e.g. Kinsey et al., 1953) who reported a greater number of premarital sex partners (and who might hence have been found more attractive as short-term partners) tended to produce more sons. Amongst 34 fathers in a rural village in Dominica, more symmetrical men tended to have more sons (r with FA $= -0.52$) but not more daughters ($r = -0.27$). These correlations did not differ significantly from one another, however, and therefore this pattern is only suggestive (Gangestad et al., 2003b).

DO WOMEN SEEK COMPATIBLE GENES WHEN FERTILE?

Research on shifts in women's preferences across the cycle have generally been inspired by ideas about the benefits of obtaining *intrinsically* good genes: genes that would presumably benefit any female's offspring. Variation across males in their ability to provide such benefits could be maintained by mutation–selection balance or important coevolutionary processes such as host–parasite coevolution (e.g. Houle & Kondrashov, 2002). But what of other kinds of genetic benefits? Might women also have been designed by selection preferentially to seek compatible or diverse genes through extra-pair sex and, hence, have stronger preferences for indicators of these benefits mid-cycle? Might the effects that we have observed even be interpreted in terms of these benefits?

In one of our studies of scent preferences, we gathered data that partly address these issues. Specifically, we collected blood and typed each individual's major histocompatibility complex (MHC) alleles at three major loci, A, B, and DRβ. Claus Wedekind and colleagues (Wedekind et al., 1995; Wedekind & Füri, 1997) have found that individuals prefer the scent of others who possess dissimilar MHC alleles, which could be because of selection for a number of functions: inbreeding avoidance and producing disease-resistant offspring being two major ones. In either case, preference for MHC-dissimilar mates is preference for compatible genes. Jacob et al. (2002) found the opposite effect in normally ovulating women (they chose as most desired the scent of individuals who possess similar MHC alleles – in particular, alleles that women had received from their fathers), but used a different method for assessing preferences – they asked women to choose smells that they would want to spend the rest of their lives with – and this instruction set may not have probed women's assessment of scents' 'sexiness'.

Wedekind's studies did not assess cycle variation in these preferences. Together with Rob Miller, Glenn Scheyd, Julie McCollough and Melissa Franklin, we examined cycle variations in this preference (Thornhill *et al.*, 2003). We found no evidence for it; fertility status correlated just 0.03 with preference for MHC dissimilarity. (In fact, we found no overall preference for MHC dissimilarity in women, though we did find one in men.)

We also found that women preferred the scent of men who are relatively heterozygous at MHC loci. Two potentially important historical benefits might account for this preference. One is that heterozygotes may be healthier and hence better investors in offspring. Another is that heterozygotic mates produce a family with a given female that is more diverse in terms of MHC alleles than the families of homozygotic mates.

Within-family diversity of MHC may be important because if a pathogen adapts well to the MHC markers of one member of a family, different cell-surface markers for other family members may protect against a within-family epidemic (Tooby, 1982). Either of these benefits is particularly possible when the mate is a long-term partner with whom a female will have multiple offspring. It may matter little if a female has just one offspring with the chosen male. (A man passes on only one of his alleles at a particular locus to an offspring and hence his heterozygosity per se should not provide a genetic benefit to a single offspring sired by, say, an extra-pair partner.) Interestingly, then, we found a marginally significant trend for women to prefer the scent of heterozygotic males *outside* of the fertile period. Moreover, the pattern of preference for heterozygosity across the cycle significantly differed from the pattern of preference for symmetry. Possibly, women tend to prefer indicators relatively important in a long-term mate, particularly outside the fertile period. During the fertile period, preferences for indicators of good genes in sex partners increase, with corresponding dampening of preferences for traits particularly valuable in a long-term partner.

This pattern of findings suggests that it is not simply the case that all traits preferred by females are particularly preferred mid-cycle; that fertility status simply enhances existing preferences. Rather, it appears that only specific preferences are enhanced – perhaps those for features that ancestrally were indicators of genetic benefits. Preferences for features particularly important in long-term, investing mates may actually be more prominent outside the fertile period.

CYCLE VARIATIONS, BY-PRODUCTS AND ALTERNATIVE FUNCTIONS

We have suggested that the apparent design of menstrual cycle variations in women's interest in extra-pair men and preferences for men as short-term partners provides provisional evidence for historical selection for women to have engaged in extra-pair mating to obtain intrinsic genetic benefits. As already noted, additional research is needed to bolster further the argument for design.

An argument of special design requires that one consider alternative evolutionary scenarios and the plausibility that they could have led to the features suggestive of design. Female interest in extra-pair mating could be: (1) an incidental by-product of female adaptations, the functions of which do not specifically pertain to extra-pair mating; or (2) the product of selection for benefits of extra-pair mating other than obtaining intrinsic genetic benefits. To say that women possess features the function of which is to obtain genetic benefits is not to deny that female interest in extra-pair mating could not have other functions. Indeed, Hill and Hurtado (1996), in a study of Ache Indians, and Beckerman *et al.* (1998), in a study of the Barí, provide evidence that extramarital copulations by women enhance offspring survival, possibly as a result of the greater parental investment from multiple 'fathers'. We now consider possible alternative explanations for menstrual cycle variations in female interests and preferences.

Women's EPC shifts in sexual interest occur as by-products of women's sexual desire directly selected because the desire achieved successful mating with a long-term mate(s). This hypothesis is inconsistent with a number of lines of evidence:

(1) Women's expressed desires for traits of short-term, extra-pair partners focus on physical attractiveness, sexiness and sensuality, whereas women's expressed desires for more committed relationships target, to a much greater extent, a man's resources and status (Buss & Schmitt, 1993; Buss, 2000; Greiling & Buss, 2000).

(2) Menstrual cycle shifts are more marked for preferences concerning short-term partners rather than long-term

partners (Penton-Voak *et al.*, 1999; Gangestad *et al.*, 2003a).

(3) As noted above, at least one characteristic (MHC heterozygosity) that women should prefer in long-term mates shows a different pattern of menstrual cycle variation, and may even be heightened when women are outside the fertile phase.

(4) On average, women's increased sexual interest in non-partner men at mid-cycle does not seem to be accompanied, in general, by an expression of lust toward the primary partner (Gangestad *et al.*, 2002).

Women's extra-pair sex occurs as a side-effect of men's sexually selected adaptation for obtaining many mates. According to this hypothesis, women engage in extra-pair sex for the same reason men do and the sexes have similar sexual–psychological adaptation affecting EPC. Symons (1979) and Geary (1998) review the many lines of evidence demonstrating that women are not designed to seek partner variety per se. Moreover, women's changes in mate preferences across the menstrual cycle and their actual EPC behaviour of preferring symmetrical EPC partners indicate that extra-pair mating in women is associated with a high degree of mate choice and is inconsistent with women simply pursuing a variety of sex partners through EPC.

Women seek extra-pair sex with men who are physical protectors rather than sires with good genes. Ethnographic data indicate that protection by a pair-bond mate may reduce sexual coercion by other males (Smuts & Smuts, 1993; Thornhill & Palmer, 2000). Possibly, women prefer short-term mates who are symmetrical and possess masculine features because these men can provide physical protection. Indeed, although symmetrical men appear to invest less time in and are less faithful to their primary relationship partners, they are perceived to be better able to provide physical protection (Gangestad & Thornhill, 1997a); masculine features (robustness, muscularity) may function similarly. As protection may be a form of male investment that minimally interferes with a man's access to partners other than his pair-bond mate (because his protective ability is attractive to women in general and, compared to investment in the form of time, honesty and sexual exclusivity, competes with male pursuit of additional partners to a lesser degree), female preference for male protection is not necessarily an explanation that is exclusive of good-genes mate choice; women may obtain benefits of protection by selecting masculine, symmetrical sex partners. Nonetheless, it is not obvious how the hypothesis that women seek protection from extra-pair mates leads to the expectation that women will prefer indicators of ability to provide

protection mid-cycle. Hence, this hypothesis does not appear sufficient by itself to account for menstrual cycle preference shifts.

Female multiple mating functions to give many males in the social group a probability of paternity, thereby reducing the likelihood that subsequent offspring will be injured or killed by males (Hrdy, 1999). This hypothesis too is difficult to reconcile with observed menstrual cycle shifts. It would appear to predict either that women will pursue more EPCs at infertile than at fertile cycle points (assuming that females desire to give the primary partner the highest probability of conception) or that they pursue EPCs equally across the cycle (assuming sire genetic quality is unimportant).

Female extra-pair mating functions to obtain food or protection for offspring from males from some paternity prospect. This hypothesis similarly should predict that women will pursue more EPCs at infertile than at fertile cycle points or that they pursue EPCs equally across the cycle and, hence, is inconsistent with observed menstrual cycle effects.

Female preferences for extra-pair mates target men who provide adequate sperm and, hence, increase the probability of successful conception rather than intrinsic genetic benefits. This hypothesis can account for women's preferences for symmetric and testosteronised men as short-term partners at mid-cycle if, in fact, these men are more fertile due to higher-quality sperm. In support of this hypothesis, Manning *et al.* (1998) found positive associations of body symmetry with ejaculate size and sperm motility in a sample of men attending a fertility clinic. These effects appeared to be largely due to a subsample of men who produced no sperm owing to obstructive lesions between the seminiferous tubules and the orifice of the ejaculatory duct. As the proportion of men in general who have such developmental defects may be very low, it is not clear that selection for low-FA men as short-term partners provides large reproductive benefits due to sperm quality. In fact, associations between symmetry, ejaculate size and sperm motility were not replicated in a sample of Boston men not selected for fertility problems (Ellis *et al.*, 2003). Indeed, the latter study found some evidence that more symmetrical men have *higher* proportions of sperm with particular kinds of abnormalities (abnormal heads).

Research by Pizzari *et al.* (2002) may be relevant to interpreting this finding, if it is indeed real. Dominant domesticated male chickens are consistently preferred as copulation partners by females. Yet dominant males actually produce lower-quality sperm. Possibly, expending the costs associated with producing high-quality sperm is more worthwhile for lower-quality males than for higher-quality males, as the former get fewer chances to inseminate females. At the present time, the evidence for the hypothesis that women seek partners with high-quality sperm at mid-cycle is not convincing, but it deserves attention in future work.

PHYLOGENETIC CONSIDERATIONS

Thus far, we have not discussed the phylogeny of menstrual cycle shifts in humans. In comparison with our closest relatives, there are two noteworthy features about women's cycles. First, human females do not have sexual swellings that advertise their reproductive condition. Bonobos and chimpanzees do and, hence, this feature possibly evolved in the hominid line (though ancestors that are shared with gorillas and orangutans may also have lacked signs of ovulation; see Sillén-Tullberg & Møller, 1993; Zinner et al., this volume). Second, humans have sex throughout the cycle. Chimpanzees do not and, in the wild, male–female sex in bonobos is largely specific to the period of swellings (Stanford, 1998).

The selective pressures that led to concealed ovulation remain a matter of debate. Several theories have been put forward:

(1) Concealed ovulation evolved to promote the bond between reproductive pairs and to keep an investing male close by, increasing the costs of male extra-pair mating effort (Alexander & Noonan, 1979).
(2) Concealed ovulation evolved to prohibit men from effectively guarding a female, hence increasing women's ability to choose sires (Benshoof & Thornhill, 1979).
(3) Concealed ovulation evolved to confuse paternity as a means to reduce infanticide (Hrdy, 1979).
(4) Because sexual swellings are themselves costly, relaxed selection for them may lead to their disappearance (such that ovulation is not adaptively 'concealed' – rather, it is simply not advertised; Burt, 1992; see also Pawlowski, 1999). One theory states that they act as graded signals that allow dominant males to copulate with females at peak fertility, while females mate promiscuously to confuse paternity (e.g. as in chimpanzees and bonobos; Burt, 1992; see also Snowdon, this volume; Zinner et al., this volume). Diminished benefits to females having dominant males that sire offspring, then, may have led to loss of advertised ovulation.

Comparative data indicate that, within anthropoid primates, loss of visual signs of ovulation has evolved more often in the context of non-monogamy than monogamy, offering little support to theories proposing that monogamy is the primary condition that predisposes to concealed ovulation (Sillén-Tullberg & Møller, 1993). At the same time, monogamy has more often evolved when signs of ovulation are absent than when they are present; perhaps, the absence of ovulatory signs facilitates the evolution of social monogamy.

If this pattern applies to humans, then loss of visual signs of ovulation may have evolved prior to the evolution of paternal investment (perhaps even in pre-hominid ancestors). The evidence for design we have described here suggests, however, that women's sexual interests and preferences were modified to change across the cycle once paternal investment did evolve. Concealed ovulation itself may have been exapted for the benefit of increasing sire choice (perhaps with secondary modification, e.g. due to an evolutionary arms race between male detection ability and female ability to conceal).

SUMMARY

The hypothesis that women possess adaptation to seek genetic benefits from partners other than primary social partners has led researchers to search for design features that would constitute tell-tale evidence of selection for this function – ones that are consistent with the function while also difficult to account for in terms of alternative evolutionary scenarios, whether producing the features as adaptations or by-products. The specific design features that research has focused on to date have concerned menstrual cycle variations in female mate preferences and sexual interests. This research has produced a wealth of findings indicating that female preferences and interests do change across the cycle in a variety of ways. They fit with the hypothesis that women are particularly attracted to male indicators of genetic benefits when fertile, and particularly when evaluating men as potential sex partners as opposed to long-term mates. Alternative explanations appear, at this time, inadequate to account for the totality of the evidence. Nonetheless, identifying adaptation through special design is an onerous burden and, hence, additional empirical probing into the nature of female preferences and interests is needed before definitive conclusions can be reached.

ACKNOWLEDGEMENTS

We thank Peter Kappeler, Carel van Schaik and two anonymous reviewers for very helpful commentary on an earlier draft of this chapter. We are grateful to the Princeton Biomedical Corporation for their donation of luteinising hormone detection kits for use in our research. The work was in part supported by the National Science Foundation through Grant Award 0136023. We are grateful to Peter Kappeler for the invitation to present this material at the Göttinger conference.

REFERENCES

Aiello, L. C. & Key, C. 2002. Energetic consequences of being a *Homo erectus* female. *American Journal of Human Biology*, **14**, 551–65.

Alexander, R. D. & Noonan, K. 1979. Concealment of ovulation, parental care, and human social evolution. In *Evolutionary Biology and Human Behavior: An Anthropological Perspective*, ed. N. A. Chagnon & W. Irons. North Scituate, MA: Duxbury, pp. 402–35.

Andersson, M. 1994. *Sexual Selection*. Princeton, NJ: Princeton University Press.

Andrews, P. W., Gangestad, S. W. & Matthews, D. 2003. Adaptationism – how to carry out the exaptationist program. *Behavioral and Brain Sciences*, in press.

Arnqvist, G. & Rowe, L. 2002. Antagonistic coevolution between the sexes in a group of insects. *Nature*, **415**, 787–9.

Arnqvist, G., Edvardsson, M., Friberg, U. & Nilsson, T. 2000. Sexual conflict promotes speciation in insects. *Proceedings of the National Academy of Sciences*, USA, **97**, 10460–4.

Baker, R. R. & Bellis, M. A. 1995. *Human Sperm Competition: Copulation, Masturbation and Infidelity*. London: Chapman and Hall.

Beckerman, S., Lizarralde, R., Ballew, C. *et al.* 1998. The Bari partible paternity project: preliminary results. *Current Anthropology*, **39**, 164–7.

Bellis, M. A. & Baker, R. R. 1990. Do females promote sperm competition? Data for humans. *Animal Behaviour*, **40**, 997–9.

Benshoof, L. & Thornhill, R. 1979. The evolution of monogamy and loss of estrus in humans. *Journal of Social and Biological Structures*, **2**, 95–106.

Berry, D. S. & Wero, J. L. 1993. Accuracy of face perception: a view from ecological psychology. *Journal of Personality*, **61**, 497–503.

Birkhead, T. R. & Møller, A. P. 1995. Extra-pair copulation and extra-pair paternity in birds. *Animal Behaviour*, **49**, 843–8.

Birkhead, T. R. & Pizzari, T. 2002. Postcopulatory sexual selection. *Nature Reviews Genetics*, **3**, 262–73.

Bock, J. 2002. Learning, life history, and productivity: children's lives in the Okavango Delta, Botswana. *Human Nature*, **13**, 161–97.

Brooks, R. & Kemp, D. J. 2001. Can older males deliver the good genes? *Trends in Ecology and Evolution*, **16**, 308–13.

Buchanan, K. L. & Catchpole, C. K. 2000. Extrapair paternity in the sedge warbler *Acrosephalus schoenobaenus* as veiled by multilocus DNA fingerprinting. *Ibis*, **142**, 12–20.

Burley, N. 1986. Sexual selection for aesthetic traits in species with biparental care. *American Naturalist*, **127**, 415–45.

Burt, A. 1992. Concealed ovulation and sexual signals in primates. *Folia Primatologica*, **58**, 1–6.

1995. Perspective: the evolution of fitness. *Evolution*, **49**, 1–8.

Buss, D. M. 1988. From vigilance to violence: mate retention tactics in American undergraduates. *Ethology and Sociobiology*, **9**, 291–317.

2000. *Dangerous Passions*. New York, NY: Free.

Buss, D. M. & Schmitt, D. P. 1993. Sexual strategies theory: a contextual evolutionary analysis of human mating. *Psychological Review*, **100**, 204–32.

Cerda-Flores, R. M., Barton, S. A., Marty-Gonzalez, L. F., Rivas, F. & Chakraborty, R. 1999. Estimation of nonpaternity in the Mexican population of Nueveo Leon: a validation study with blood group markers. *American Journal of Physical Anthropology*, **109**, 281–93.

Charlesworth, B. 1987. The heritability of fitness. In *Sexual Selection: Testing the Alternatives*, ed. J. W. Bradbury & M. B. Andersson. New York, NY: Wiley, pp. 22–40.

1990. Mutation–selection balance and the evolutionary advantage of sex and recombination. *Genetical Research*, **55**, 199–221.

Charlesworth, B. & Hughes, K. A. 1998. The maintenance of genetic variation in life history traits. In *Evolutionary Genetics from Molecules to Morphology*, ed. R. S. Singh & C. B. Krimbas. Cambridge: Cambridge University Press, pp. 369–91.

Ellegren, H., Gustafsson, L. & Sheldon, B. C. 1996. Sex ratio adjustment in relation to paternal attractiveness in a wild bird population. *Proceedings of the National Academy of Sciences, USA*, **93**, 11723–8.

Ellis, P., Cristello, M. & Whitmeyer, J. 2003. FA and sperm quality in a sample of Boston men. *Evolution and Human Behavior*, in press.

Fisher, R. A. 1930. *The Genetical Theory of Natural Selection*. Oxford: Clarendon.

Flinn, M. 1987. Mate guarding in a Caribbean village. *Ethology and Sociobiology*, **8**, 1–28.

Forstmeier, W., Kempenaers, B., Meyer, A. & Leisler, B. 2002. A novel song parameter correlates with extra-pair paternity and reflects male longevity. *Proceedings of the Royal Society of London, Series B*, **269**, 1479–85.

Fuller, R. C. & Houle, D. 2003. Inheritance of developmental instability. In *Developmental Instability: Causes and Consequences*, ed. M. Polak, in press. New York, NY: Oxford University Press.

Furlow, B. F., Armijo-Prewitt, T., Gangestad, S. W. & Thornhill, R. 1997. Fluctuating asymmetry and psychometric intelligence. *Proceedings of the Royal Society of London, Series B*, **264**, 823–9.

Furlow, B. F., Gangestad, S. W. & Armijo-Prewitt, T. 1998. Developmental stability and human violence. *Proceedings of the Royal Society of London, Series B*, **266**, 1–6.

Gangestad, S. W. & Simpson, J. A. 1990. Toward an evolutionary history of female sociosexual variation. *Journal of Personality*, **58**, 69–96.

Gangestad, S. W. & Thornhill, R. 1997a. Human sexual selection and developmental stability. In *Evolutionary Social Psychology*, ed. J. A. Simpson & D. T. Kenrick. Mahwah, NJ: Lawrence Erlbaum Associates, pp. 169–95.

1997b. The evolutionary psychology of extrapair sex: the role of fluctuating asymmetry. *Evolution and Human Behavior*, **18**, 69–88.

1998. Menstrual cycle variation in women's preferences for the scent of symmetrical men. *Proceedings of the Royal Society of London, Series B*, **265**, 927–33.

1999. Individual differences in developmental precision and fluctuating asymmetry: a model and its implications. *Journal of Evolutionary Biology*, **12**, 402–16.

2003a. Facial masculinity and fluctuating asymmetry. *Evolution and Human Behavior*, in press.

2003b. Fluctuating asymmetry, developmental stability, and fitness: toward model-based interpretation. In *Developmental Instability: Causes and Consequences*, ed. M. Polak, in press. New York, NY: Oxford University Press.

Gangestad, S. W., Thornhill, R. & Yeo, R. A. 1994. Facial attractiveness, developmental stability, and fluctuating asymmetry. *Ethology and Sociobiology*, **15**, 73–85.

Gangestad, S. W., Bennett, K. L. & Thornhill, R. 2001. A latent variable model of developmental stability in relation to men's number of sex partners. *Proceedings of the Royal Society of London, Series B*, **268**, 1677–84.

Gangestad, S. W., Thornhill, R. & Garver, C. E. 2002. Changes in women's sexual interests and their partners' mate retention tactics across the menstrual cycle: evidence for shifting conflicts of interest. *Proceedings of the Royal Society of London, Series B*, **269**, 975–82.

Gavrilets, S., Arnqvist, G. & Friberg, U. 2001. The evolution of female mate choice by sexual conflict. *Proceedings of the Royal Society of London, Series B*, **268**, 531–9.

Geary, D. C. 1998. *Male, Female: The Evolution of Human Sex Differences*. Washington, DC: American Psychological Association.

2000. Evolution and proximate expression of human paternal investment. *Psychological Bulletin*, **126**, 55–77.

Geary, D. C. & Flinn, M. V. 2002. Sex differences in behavioural and hormonal response to social threat: commentary on Taylor *et al.* 2000. *Psychological Review*, **109**, 745–50.

Getty, T. 1999. Handicap signalling: when fecundity and mortality do not add up. *Animal Behaviour*, **56**, 127–30.

Gil, D., Graves, J., Hazon, N. & Wells, A. 1999. Male attractiveness and differential testosterone investment in zebra finch eggs. *Science*, **286**, 126–8.

Grafen, A. 1990. Biological signals as handicaps. *Journal of Theoretical Biology*, **144**, 517–46.

Gray, P. B., Kahlenberg, S. M., Barrett, E. S., Lipson, S. F. & Ellison, P. T. 2002. Marriage and fatherhood are associated with lower testosterone in males. *Evolution and Human Behavior*, **23**, 193–201.

Graziano, W. G., Jensen-Campbell, L. A., Todd, M. & Finch, J. F. 1997. Interpersonal attraction from an evolutionary perspective: women's reactions to dominant and prosocial men. In *Evolutionary Social Psychology*, ed. J. A. Simpson & D. T. Kenrick. Mahwah, NJ: Lawrence Erlbaum Associates, pp. 141–67.

Greiling, H. & Buss, D. M. 2000. Women's sexual strategies: the hidden dimension of short-term extra-pair mating. *Personality and Individual Differences*, **28**, 929–63.

Griffith, S. C. & Sheldon, B. C. 2001. Phenotypic plasticity in the expression of sexually selected traits: neglected components of variation. *Animal Behaviour*, **61**, 987–93.

Gustafsson, L., Nordling, D., Andersson, M. S., Sheldon, B. C. & Qvarnström, A. 1994. Infectious diseases, reproductive effort and the cost of reproduction in birds. *Philosophical Transactions of the Royal Society of London, Series B*, **346**, 323–31.

Gustafsson, L., Qvarnström, A. & Sheldon, B. C. 1995. Trade-offs between life history traits and a secondary sexual character in male collared flycatchers. *Nature*, **375**, 311–13.

Hagen, E. H., Hames, R. B., Craig, N. M., Lauer, M. T. & Price, M. E. 2001. Parental investment and child health in a Yanomamo village suffering short-term food stress. *Journal of Biosocial Science*, **33**, 503–28.

Hamilton, W. D. 1982. Pathogens as causes of genetic diversity in their host populations. In *Population Biology of*

Infectious Diseases, ed. R. M. Anderson & R. M. May. New York, NY: Springer-Verlag, pp. 269–96.

Hamilton, W. D. & Zuk, M. 1982. Heritable true fitness and bright birds: a role for parasites. *Science*, **218**, 384–7.

Hasselquist, D. 1998. Polygyny in great reed warblers: a long-term study of factors contributing to male fitness. *Ecology*, **79**, 2376–90.

Hasselquist, D., Bensch, S. & von Schantz, T. 1996. Correlation between male song repertoire, extrapair paternity and offspring survival in the great reed warbler. *Nature*, **381**, 229–32.

Hawkes, K. & Bird, R. B. 2002. Showing off, handicap signaling, and men's work. *Evolutionary Anthropology*, **11**, 58–67.

Hawkes, R., O'Connell, J. F. & Blurton Jones, N. G. 2001. Hunting and nuclear families: some lessons from the Hadza about men's work. *Current Anthropology*, **42**, 681–709.

Herz, R. S. & Cahill, E. D. 1997. Differential use of sensory information in sexual behavior as a function of gender. *Human Nature*, **8**, 275–86.

Hill, K. & Hurtado, A. M. 1996. *Ache Life History*. New York, NY: Aldine de Gruyter.

Hoi, H. 1997. Assessment of the quality of copulation partners in the monogamous bearded tit. *Animal Behaviour*, **53**, 277–86.

Houle, D. 1992. Comparing evolvability and variability of traits. *Genetics*, **130**, 195–204.

  2000. A simple model of the relationship between asymmetry and developmental stability. *Journal of Evolutionary Biology*, **13**, 720–30.

Houle, D. & Kondrashov, A. S. 2002. Coevolution of costly mate choice and condition-dependent display of good genes. *Proceedings of the Royal Society of London, Series B*, **269**, 97–104.

Houtman, A. M. 1992. Female zebra finches choose extra-pair copulations with genetically attractive males. *Proceedings of the Royal Society of London, Series B*, **249**, 3–6.

Hrdy, S. B. 1979. Infanticide among animals: a review, classification, and examination of the implications for reproductive strategies of females. *Ethology and Sociobiology*, **1**, 13–40.

  1999. *Mother Nature: A History of Mothers, Infants and Natural Selection*. New York, NY: Pantheon.

Hughes, S. M., Harrison, M. A. & Gallup, G. G. 2002. The sound of symmetry: voice as a marker of developmental instability. *Evolution and Human Behavior*, **23**, 173–80.

Iwasa, Y. & Pomiankowski, A. 1994. The evolution of mate preferences for multiple sexual ornaments. *Evolution*, **48**, 853–67.

Iwasa, Y., Pomiankowski, A. & Nee, S. 1991. The evolution of costly mate preferences. II. The 'handicap' principle. *Evolution*, **45**, 1431–42.

Jacob, S., McClintock, M.K., Zelano, B. & Ober, C. 2002. Paternally inherited alleles are associated with women's choice of male odor. *Nature Genetics*, **30**, 175–9.

James, W. 1890. *The Principles of Psychology*. London: Macmillan.

Jennions, M. D. & Petrie, M. 2000. Why do females mate multiply? A review of the genetic benefits. *Biological Reviews*, **75**, 21–64.

Jöchle, W. 1973. Coitus induced ovulation. *Contraception*, **7**, 523–64.

Johnsen, A., Andersen, V., Sunding, C. & Lifjeld, J. T. 2000. Female bluethroats enhance offspring immunocompetence through extra-pair copulations. *Nature*, **406**, 296–9.

Johnsen, A., Lifjeld, J. T., Andersson, S., Ornborg, J. & Amundsen, T. 2001. Male characteristics and fertilisation success in male bluethroats. *Behavior*, **138**, 1371–90.

Johnston, V. S., Hagel, R., Franklin, M., Fink, B. & Grammer, K. 2001. Male facial attractiveness: evidence for hormone mediated adaptive design. *Evolution and Human Behavior*, **23**, 251–67.

Kaplan, H. S., Hill, K., Lancaster, J. B. & Hurtado, A. M. 2000. A theory of human life history evolution: diet, intelligence, and longevity. *Evolutionary Anthropology*, **9**, 156–85.

Kinsey, A. C., Pomeroy, W. B., Martin, C. E. & Gebhard, P. H. 1953. *Sexual Behavior in the Human Female*. Philadelphia, PA: W. B. Saunders.

Kirkpatrick, M. 1996. Good genes and direct selection in the evolution of mating preferences. *Evolution*, **50**, 2125–40.

Kirkpatrick, M. & Barton, N. H. 1997. The strength of indirect selection on female mating preferences. *Proceedings of the National Academy of Sciences, USA*, **94**, 1282–6.

Kitcher, P. 1987. *Vaulting Ambition: Sociobiology and the Quest for Human Nature*. Cambridge, MA: MIT Press.

Kokko, H. 1998. Good genes, old age and life history trade-offs. *Evolutionary Ecology*, **12**, 739–50.

Laumann, E. O., Gagnon, J. H., Michael, R. T. & Michaels, S. 1994. *The Social Organization of Sexuality*. Chicago, IL: University of Chicago Press.

Leigh, S. R. 2001. Evolution of human growth. *Evolutionary Anthropology*, **10**, 223–36.

Lens, L., van Dongen, S., Kark, S. & Matthysen, E. 2002. Fluctuating asymmetry as an indicator of fitness: can we bridge the gap between studies? *Biological Reviews*, 77, 27–38.

Leonard, W. R. & Robertson, M. L. 1997. Comparative primate energetics and hominid evolution. *American Journal of Physical Anthropology*, 102, 265–81.

Lubjuhn, T., Strohbach, S., Brun, J., Gerken, T. & Epplen, J. T. 1999. Extra-pair paternity in great tits (*Parus major*): a long-term study. *Behavior*, 136, 1157–72.

Luuthe, V., Sugimoto, Y., Puy, L. *et al.* 1994. Characterization, expression, and immunohistochemical localization of 5-alpha-reductase activity in human skin. *Journal of Investigative Dermatology*, 102, 221–6.

Lynch, M., Blanchard, J., Houle, D. *et al.* 1999. Perspective: spontaneous deleterious mutation. *Evolution*, 53, 645–63.

Manning, J. T. & Wood, D. 1998. Fluctuating asymmetry and aggression in boys. *Human Nature*, 9, 53–65.

Manning, J. T., Koukourakis, K. & Brodie, D. A. 1997. Fluctuating asymmetry, metabolic rate and sexual selection in human males. *Evolution and Human Behavior*, 18, 15–21.

Manning, J. T., Scutt, D. & Lewis-Jones, D. I. 1998. Developmental stability, ejaculate size and sperm quality in men. *Evolution and Human Behavior*, 19, 273–82.

Marlowe, F. 2000. Paternal investment and the human mating system. *Behavioural Processes*, 51, 45–61.

McIntyre, S. & Sooman, A. 1991. Non-paternity and prenatal genetic screening. *Lancet*, 338, 869.

Michl, G., Torok, J., Griffith, S. C. & Sheldon, B. C. 2002. Experimental analysis of sperm competition mechanisms in a wild bird population. *Proceedings of the National Academy of Sciences, USA*, 99, 5466–70.

Møller, A. P. 1999. Asymmetry as a predictor of growth, fecundity and survival. *Ecology Letters*, 2, 149–56.

Møller, A. P. & Alatalo, R. V. 1999. Good-genes effects in sexual selection. *Proceedings of the Royal Society of London, Series B*, 266, 85–91.

Møller, A. P. & Swaddle, J. P. 1997. *Asymmetry, Developmental Stability and Evolution*. Oxford: Oxford University Press.

Møller, A. P. & Thornhill, R. 1998a. Bilateral symmetry and sexual selection: a meta-analysis. *American Naturalist*, 151, 174–92.

1998b. Male parental care, differential parental investment by females, and sexual selection. *Animal Behaviour*, 55, 1507–15.

Møller, A. P., Saino, N., Taramino, G., Galeotti, P. & Ferrario, S. 1998. Paternity and sexual signaling: effects of a secondary sexual character and song on paternity in the barn swallow. *American Naturalist*, 151, 236–42.

Møller, A. P., Gangestad, S. W. & Thornhill, R. 1999. Nonlinearity and the importance of fluctuating asymmetry as a predictor of fitness. *Oikos*, 86, 366–8.

Montgomerie, R. & Bullock, H. 1999. Fluctuating asymmetry and the human female orgasm. Paper presented at the annual meeting of the Human Behavior and Evolution Society, Salt Lake City, UT, June.

Oliver-Rodriguez, J. C., Guan, Z. & Johnston, V. S. 1999. Gender differences in late positive components evoked by human faces. *Psychophysiology*, 36, 176–85.

Otter, K., Ratcliffe, L., Michaud, D. & Boag, P. T. 1998. Do female black-capped chickadees prefer high-ranking males as extra-pair partners? *Behavioral Ecology and Sociobiology*, 43, 25–36.

Otter, K. A., Stewart, I. R. K., McGregor, P. K. *et al.* 2001. Extra-pair paternity among great tits *Parus major* following manipulation of male signals. *Journal of Avian Biology*, 32, 338–42.

Pärt, T. & Qvarnström, A. 1997. Badge size in collared flycatchers predicts outcome of male competition over territories. *Animal Behaviour*, 54, 893–9.

Pawlowski, B. 1999. Loss of oestrus and concealed ovulation in human evolution: the case against the sexual-selection hypothesis. *Current Anthropology*, 40, 257–75.

Penton-Voak, I. S. & Perrett, D. I. 2000. Female preference for male faces changes cyclically: further evidence. *Evolution and Human Behavior*, 21, 39–48.

Penton-Voak, I. S., Perrett, D. I., Castles, D. *et al.* 1999. Female preference for male faces changes cyclically. *Nature*, 399, 741–2.

Perrett, D. I., Lee, K. J., Penton-Voak, I. *et al.* 1998. Effects of sexual dimorphism on facial attractiveness. *Nature*, 394, 884–7.

Petrie, M. & Kempenaers, B. 1998. Extra-pair paternity in birds: explaining variation between species and populations. *Trends in Ecology and Evolution*, 13, 52–8.

Pizzari, T., Froman, D. P. & Birkhead, T. R. 2002. Pre- and post-insemination episodes of sexual selection in the fowl, *Gallus g. domesticus*. *Heredity*, 88, 112–16.

Pomiankowski, A. N. 1988. The evolution of mate preferences for male genetic quality. *Oxford Surveys in Evolutionary Biology*, 5, 136–84.

Pomiankowski, A. N. & Møller, A. P. 1995. A resolution of the lek paradox. *Proceedings of the Royal Society of London, Series B*, 260, 21–9.

Qvarnström, A. 1997. Experimentally increased badge size increases male competition and reduces male parental care in the collared flycatcher. *Proceedings of the Royal Society of London, Series B*, **264**, 1225–31.

1999a. Genotype-by-environment interactions in the determination of the size of a secondary sexual character in the collared flycatcher (*Ficedula albicollis*). *Evolution*, **53**, 1564–72.

1999b. Different reproductive tactics in male collared flycatchers signalled by size of secondary sexual character. *Proceedings of the Royal Society of London, Series B*, **266**, 2089–93.

Qvarnström, A., Part, T. & Sheldon, B. C. 2000. Adaptive plasticity in mate preference linked to differences in reproductive effort. *Nature*, **405**, 344–7.

Ratti, O., Hovi, M., Lundberg, A., Telegstrom, H. & Alatalo, R. V. 1995. Extra-pair paternity and male characteristics in the pied flycatcher. *Behavioral Ecology and Sociobiology*, **37**, 419–25.

Reeve, H. K. & Sherman, P. W. 1993. Adaptation and the goals of evolutionary research. *Quarterly Review of Biology*, **68**, 1–32.

Regan, P. C. & Berscheid, E. 1995. Gender differences in beliefs about the causes of male and female sexual desire. *Personal Relationships*, **2**, 345–58.

Rice, W. R. 1996. Sexually antagonistic male adaptation triggered by experimental arrest of female evolution. *Nature*, **381**, 232–4.

Rice, W. R. & Holland, B. 1998. The enemies within: intragenomic conflict, interlocus contest evolution (ICE), and the intraspecific Red Queen. *Behavioral Ecology and Sociobiology*, **41**, 1–10.

Richardson, D. S. & Burke, T. 1999. Extra-pair paternity in relation to male age in Bullock's orioles. *Molecular Ecology*, **8**, 2115–26.

Rikowski, A. & Grammer, K. 1999. Human body odour, symmetry and attractiveness. *Proceedings of the Royal Society of London, Series B*, **266**, 869–74.

Rowe, L. & Houle, D. 1996. The lex paradox and the capture of genetic variance by condition-dependent traits. *Proceedings of the Royal Society of London, Series B*, **263**, 1415–21.

Sasse, G., Muller, H., Chakraborty, R. & Ott, J. 1994. Estimating the frequency of nonpaternity in Switzerland. *Human Heredity*, **44**, 337–43.

Scheib, J. E., Gangestad, S. W. & Thornhill, R. 1999. Facial attractiveness, symmetry, and cues of good genes. *Proceedings of the Royal Society of London, Series B*, **266**, 1318–21.

Shackelford, T. K., Weekes-Shackelford, V. A., LeBlanc *et al.* 1999. Female coital orgasm and male attractiveness. *Human Nature*, **11**, 299–306.

Shackelford, T. K., LeBlanc, G. J., Weekes-Shakelford *et al.* 2002. Psychological adaptation to human sperm competition. *Evolution and Human Behavior*, **23**, 123–38.

Sheldon, B. C. & Ellegren, H. 1999. Sexual selection resulting from extrapair paternity in collared flycatchers. *Animal Behaviour*, **57**, 285–98.

Sheldon, B. C., Merila, J., Qvarnström, A., Gustafsson, L. & Ellegren, H. 1997. Paternal genetic contribution to offspring condition predicted by size of male secondary sexual character. *Proceedings of the Royal Society of London, Series B*, **264**, 297–302.

Sheldon, B. C., Davidson, P. & Lindgren, G. 1999. Mate replacement in experimentally widowed collared flycatchers (*Ficedula albicollis*): determinants and outcomes. *Behavioral Ecology and Sociobiology*, **46**, 141–8.

Sillén-Tullberg, B. & Møller, A. P. 1993. The relationship between concealed ovulation and mating systems in anthropoid primates: a phylogenetic analysis. *American Naturalist*, **141**, 1–25.

Simpson, J. A., Gangestad, S. W., Christensen, P. N. & Leck, K. 1999. Fluctuating asymmetry, sociosexuality and intrasexual competitive tactics. *Journal of Personality and Social Psychology*, **76**, 159–72.

Singh, D. & Bronstad, P. M. 2001. Female body odour is a potential cue to ovulation. *Proceedings of the Royal Society of London, Series B*, **268**, 797–801.

Smuts, B. B. & Smuts, R. W. 1993. Male aggression and sexual coercion of females in nonhuman primates and other mammals: evidence and theoretical implications. *Advances in the Study of Behavior*, **22**, 1–63.

Stanford, C. B. 1998. The social behavior of chimpanzees and bonobos. *Current Anthropology*, **39**, 399–419.

Stutchbury, B. J. M., Piper, W. H., Neudorf, D. L. *et al.* 1997. Correlates of extra-pair fertilization success in hooded warblers. *Behavioral Ecology and Sociobiology*, **40**, 119–26.

Swaddle, J. P. & Reierson, G. W. 2002. Testosterone increases the perceived dominance but not attractiveness of human males. *Proceedings of the Royal Society of London, Series B*, **269**, 2285–9.

Symons, D. 1979. *The Evolution of Human Sexuality*. Oxford: Oxford University Press.

1992. On the use and misuse of Darwinism in the study of human behavior. In *The Adapted Mind: Evolutionary Psychology and the Generation of Culture*, ed. J. Barkow,

L. Cosmides & J. Tooby. Oxford: Oxford University Press, pp. 137–59.

Thoma, R. J., Yeo, R. A., Gangestad, S. W., Lewine, J. D. & Davis, J. T. 2002. Fluctuating asymmetry and the human brain. *Laterality*, **7**, 45–58.

Thompson, A. P. 1983. Extramarital sex: a review of the research literature. *Journal of Sex Research*, **19**, 1–22.

Thornhill, R. 1990. The study of adaptation. In *Interpretation and Explanation in the Study of Behavior*, ed. M. Bekoff & D. Jamieson. Boulder, CO: Westview, pp. 31–62.

1997. The concept of an evolved adaptation. In *Characterizing Human Psychological Adaptations*, ed. M. Daly. London: Wiley, pp. 4–13.

Thornhill, R. & Gangestad, S. W. 1993. Human facial beauty: averageness, symmetry, and parasite resistance. *Human Nature*, **4**, 237–70.

1994. Fluctuating asymmetry and human sexual behavior. *Psychological Science*, **5**, 297–302.

1999a. The scent of symmetry: a human sex pheromone that signals fitness? *Evolution and Human Behavior*, **20**, 175–201.

1999b. Facial attractiveness. *Trends in Cognitive Sciences*, **3**, 452–60.

2003. Do women have evolved adaptation for extra-pair copulation? In *Evolutionary Aesthetics*, ed. E. Voland & K. Grammer. Hamburg: Springer-Verlag, pp. 341–68.

Thornhill, R. & Møller, A. P. 1997. Developmental stability, disease and medicine. *Biological Reviews*, **72**, 497–548.

Thornhill, R. & Palmer, C. T. 2000. *A Natural History of Rape: Biological Bases of Sexual Coercion*. Cambridge, MA: MIT Press.

Thornhill, R., Gangestad, S. W. & Comer, R. 1995. Human female orgasm and mate fluctuating asymmetry. *Animal Behaviour*, **50**, 1601–15.

Thornhill, R., Gangestad, S. W., Miller, R. *et al.* 2003. MHC, symmetry and body scent attractiveness in men and women (*Homo sapiens*). *Behavioral Ecology*.

Tooby, J. 1982. Pathogens, polymorphism, and the evolution of sex. *Journal of Theoretical Biology*, **97**, 557–76.

Trivers, R. L. 1972. Parental investment and sexual selection. In *Sexual Selection and the Descent of Man, 1871–1971*, ed. B. Campbell. Chicago, IL: Aldine, pp. 139–79.

Troisi, A. & Carosi, M. 1998. Female orgasm rate increases with male dominance in Japanese macaques. *Animal Behaviour*, **56**, 1261–6.

van Dongen, S. 2000. The heritability of fluctuating asymmetry: a Bayesian hierarchical model. *Acta Zoologica Fennici*, **37**, 15–23.

Wagner, G. P. & Altenberg, L. 1996. Perspective: complex adaptations and the evolution of evolvability. *Evolution*, **50**, 967–76.

Wagner, R. H. 1992. The pursuit of extra-pair copulations by female razorbills: how do females benefit? *Behavioral Ecology and Sociobiology*, **29**, 455–64.

Watson, P. J. 1998. Multi-male mating and female choice increase offspring growth in the spider *Neriene litigiosa* (*Linyphiidae*). *Animal Behaviour*, **55**, 387–403.

Wedekind, C. & Füri, S. 1997. Body odour preference in men and women: do they aim for specific MHC combinations or simply heterozygosity? *Proceedings of the Royal Society of London, Series B*, **264**, 1471–9.

Wedekind, C., Seebeck, T., Bettens, F. & Paepke, A. J. 1995. MHC-dependent mate preferences in humans. *Proceedings of the Royal Society of London, Series B*, **260**, 245–9.

Wilcox, A. J., Weinberg, C. R. & Baird, B. D. 1995. Timing of sexual intercourse in relation to ovulation. *New England Journal of Medicine*, **333**, 1517–21.

Williams, G. C. 1966. *Adaptation and Natural Selection: A Critique of Some Current Evolutionary Thought*. Princeton, NJ: Princeton University Press.

Wilson, M. & Daly, M. 1992. The man who mistook his wife for a chattel. In *The Adapted Mind: Evolutionary Psychology and the Generation of Culture*, ed. J. Barkow, L. Cosmides & J. Tooby. Oxford: Oxford University Press, pp. 289–326.

Yeo, R. A., Hill, D., Campbell, R., Vigil, J. & Brooks, W. M. 2000. Developmental instability and working memory ability in children: a magnetic resonance spectroscopy investigation. *Developmental Neuropsychology*, **17**, 143–59.

Zeh, J. A. & Zeh, D. W. 2001. Reproductive mode and the genetic benefits of polyandry. *Animal Behaviour*, **61**, 1051–63.

Part III
Sexual selection in action

7 • Sexual selection, behaviour and sexually transmitted diseases

CHARLES L. NUNN
Department of Biology
University of California
Davis, CA, USA

SONIA M. ALTIZER
Department of Environmental Studies
Emory University
Atlanta, GA, USA

INTRODUCTION

Factors that alter the contact structure of individuals within populations will influence the spread of parasites that are transmitted by direct contact (Anderson & May, 1991; Blower & McLean, 1991). Few cases illustrate this fundamental principle of epidemiology better than sexual selection and the spread of sexually transmitted diseases (STDs). Sexual selection involves variation in mating success mediated by male–male competition or female choice. By changing the structure of mating contacts within a population, sexual selection influences the spread of sexually transmitted infections. In particular, those individuals with the greatest mating success are at highest risk of contracting STDs, and will also contribute disproportionately to STD spread and persistence (Graves & Duvall, 1995; Thrall *et al.*, 2000). Moreover, promiscuity associated with sperm competition is predicted to increase both the spread and virulence of STDs (Thrall *et al.*, 1997). Therefore STDs may represent a substantial cost of sexual selection and non-monogamous mating behaviour (Thrall *et al.*, 2000).

Sexually transmitted diseases have been virtually ignored in studies of animal mating systems (Smith & Dobson, 1992; Lockhart *et al.*, 1996), but it is now possible to link epidemiological theory on STDs to patterns of infection in wild populations. In this chapter, we explore the consequences of sexual selection, for the spread of STDs in primates. We also examine behavioural defences to avoid infection, specifically addressing interactions between parasite fitness and host reproductive success. Primates represent an ideal system for studying STDs and sexual selection, because of the large amount of data available on their parasites and mating behaviour. It is important to note that, throughout this chapter, we apply the general term 'parasite' to any organism that lives in or on a host and utilises host resources, usually to the detriment of the host (thus including viruses, bacteria, protozoa and fungi, in addition to more traditionally defined helminths and arthropods).

First, we review evidence for STDs in non-human primates and the effects of STDs on their hosts. This is an important step, because if STDs are extremely rare or have only minor effects on host fitness, they are unlikely to represent a primary selective force influencing mating behaviour. Second, we summarise the results of a simulation study that examined the spread of STDs within populations, assuming high variance in male mating success, as expected under sexual selection (Thrall *et al.*, 2000). We test the key prediction of this model using data on prevalence of STDs in free-living primate populations. Third, we review a theoretical model that investigates the effects of mate choice on virulence evolution in STDs (Knell, 1999). This model suggests that mate choice to avoid STD infection will reduce the virulence of sexually transmitted parasites. Because the spread of STDs is sensitive to sexual selection, it might seem that male viability traits, such as those predicted under the Hamilton–Zuk (1982) model, should be particularly fine-tuned to STD infection. However, Knell's model provides the opposite conclusion and raises more general questions about behavioural counter-strategies to STDs. Thus, in the final section of this chapter we investigate the consequences of STDs for host behavioural defences by examining comparative patterns of putative behavioural counter-strategies in relation to STD risk (Nunn, 2003, in press). In so doing, we distinguish between pre-copulatory and post-copulatory defensive behaviours, and we examine the prediction that post-copulatory behaviours will be more effective than those performed prior to mating.

We use these diverse questions to illustrate the conceptual links between sexual selection, host behaviour and the spread of STDs. Moreover, we show that sufficient comparative data are emerging to test theoretical models, although our results must be considered initial explorations of these

Sexual Selection in Primates: New and Comparative Perspectives, ed. Peter M. Kappeler and Carel P. van Schaik. Published by Cambridge University Press. © Cambridge University Press 2004.

questions. A major goal of our chapter is to highlight the types of data needed for future tests. We also discuss the ways in which existing theoretical models should be extended to consider the spread of parasites in primate populations, particularly with regard to polygynandrous mating systems found in many primate species.

DO STDs OCCUR IN WILD PRIMATE POPULATIONS?

STDs typically have been viewed as a curious group of parasites rather than established entities with important selective effects on their hosts (Lockhart et al., 1996). In recent decades, this view has changed, primarily through our increased understanding of HIV in the context of the AIDS crisis (e.g. Garnett & Anderson, 1993; Schwartländer et al., 2000). It is now well established that the simian form of HIV, the simian immunodeficiency virus (SIV), is found in many Old World monkey and ape species and appears to show a high degree of strain specificity (e.g. Phillips-Conroy et al., 1994; Hahn et al., 2000; Santiago et al., 2002). As in humans, SIV is spread primarily through sexual contact, but the relative importance of other transmission routes (e.g. biting) may vary among species or environmental conditions (Nerrienet et al., 1998). Unlike HIV in humans, SIV is not known to cause severe immunodeficiency or mortality in its native hosts (Norley et al., 1999; Swanstrom & Wehbie, 1999).

Several reviews have examined the distribution of STDs across host species, but the most comprehensive review at time of going to press was conducted by Lockhart et al. (1996). These authors surveyed the veterinary, medical and parasitological literature and found evidence for over 200 STDs in 27 orders of hosts. Parasites that exhibit sexual transmission represented most major taxonomic groups, including viruses, bacteria, helminths, protozoa, fungi and arthropods. For primates, our own literature search on nearly 200 host species confirms the conclusion of Lockhart et al. (1996), that STDs are better studied in Old World monkeys. In many of the species in which STDs have been documented, individuals mate promiscuously, and none of these species is typically classified as monogamous. These phylogenetic and social correlates are not definitive, however, because surveys of STDs in wild populations may be biased toward group-living, promiscuous primates. In fact, wild populations of most primate species have not yet been examined for STD infection. As a consequence, we cannot conduct broad-scale comparative tests of STDs in primates, but analyses within subsets of taxa are possible, as illustrated below.

HOW COSTLY ARE STDs?

Parasites differ markedly in their effects on host fitness (i.e. virulence), and much of this variation is predicted to relate to transmission mode (Ewald, 1994; Herre, 1995; Sorci et al., 1997). This is particularly important for parasites with limited transmission routes (e.g. STDs), where negative effects on host survival or conspicuous signs of infection may reduce new transmission events below threshold conditions for establishment or persistence (Getz & Pickering, 1983; Thrall et al., 1993). As a result, some STDs, such as those caused by retroviruses, may simply hitchhike along with their hosts, causing little harm and resulting in few host counter-strategies (Norley et al., 1999). We acknowledge that many STDs are relatively benign in relation to other parasites. But recent analyses by Lockhart et al. (1996) and others (Smith & Dobson, 1992; Holmes et al., 1999) demonstrate at least three major costs of STDs: (1) A large proportion of STDs increase the risk of sterility in males or females. (2) STDs commonly exhibit vertical transmission, with severe consequences for offspring health. In humans, for example, syphilis causes congenital defects that are likely to reduce the survival and reproductive success of offspring (Radolf et al., 1999). Similarly detrimental effects are found in infants that contract papilloma virus, gonorrhoea, herpes and HIV from their infected mothers (see chapters in Holmes et al., 1999). (3) Relative to infectious disease transmitted by non-sexual contact, STDs commonly exhibit long infectious periods with low host recovery, failure to clear infectious organisms following recovery, or limited immunity to reinfection. This pattern arises because many sexually transmitted parasites (like some non-STDs) have mechanisms of 'hiding' from host immune defences (e.g. persisting in neural ganglia; Hoeprich et al., 1994). In fact, many STDs are impossible to eradicate from the body through the immune response alone, resulting in lifetime infections.

Many negative consequences of STD infection probably provide benefits to the parasites themselves, increasing the likelihood of invasion, transmission and persistence (see Lockhart et al., 1996). In mammals, for example, host infertility is likely to result in repeated cycling by females and may consequently increase their number of sexual contacts. Primates offer an important opportunity to test this hypothesis, because the frequency of infertile females within wild groups may exceed 10 per cent (Anderson, 1986). Similarly,

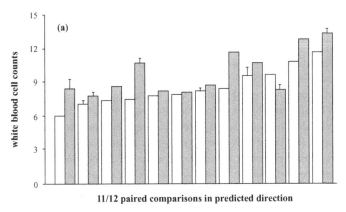

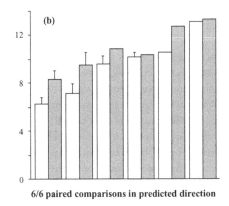

Fig. 7.1   White blood cell counts in primates (a) and carnivores (b). Bars show phylogenetically independent comparisons of less promiscuous (open bars) and more promiscuous (filled bars) species. Error bars are provided when comparisons involve phylogenetically weighted mean values for more than one species. The prediction to test is whether evolutionary transitions to increased promiscuity show increases in white blood cell counts. Results are significant at $P < 0.05$ (Nunn, 2002; Nunn *et al.*, 2003), with comparisons calculated using the BRUNCH method in the computer program CAIC (Purvis & Rambaut, 1995).

STDs that increase host mortality or possess short infectious periods are less likely to survive until the next breeding season, when contact is established with new, uninfected hosts (e.g. Thrall *et al.*, 1993). Thus, in addition to long infectious periods, STDs tend to produce less disease-induced mortality relative to other infectious diseases (Lockhart *et al.*, 1996).

Another approach to assessing the costs of STDs is to examine their effects on host defences. Nunn *et al.* (2000; Nunn, 2002) conducted comparative tests across a diverse assemblage of primates to assess whether baseline white blood cell (WBC) counts are associated with risk of STD infection. White blood cell counts among healthy, captive animals may indicate the capacity of innate immune cells (monocytes, granulocytes or natural killer cells) to respond quickly to parasite infection. Such generalised defences could be critical to STD prevention because antibody-mediated immune responses are unlikely to be effective in eradicating venereal diseases following establishment. Consistent with predictions that more promiscuous species of primates should experience increased risk of STD infection, species with greater promiscuity exhibited higher baseline WBC counts in phylogenetic comparative tests (Nunn *et al.*, 2000; Fig. 7.1). These results were upheld when

using different measures of promiscuity, after controlling for other sources of infection, such as terrestrial locomotion and sociality, when limiting the analysis to adult females or males only, and in an independent data set on WBC counts (M. J. Anderson, J. Hessel & A. F. Dixson, unpublished). Moreover, analyses were repeated in carnivores and found to be consistent with the patterns in primates, although it proved difficult to rule out confounding effects of sociality, life history and body size in carnivores (Nunn *et al.*, 2003, in review).

One mechanistic explanation for the association between promiscuity and primate immune defence involves the interaction between WBCs and sperm in the female reproductive tract. Immediately following copulation, massive numbers of WBCs flood the female reproductive tract and actively engulf sperm and seminal fluid (e.g. Phillips & Mahler, 1977; Pandya & Cohen, 1985; Barratt *et al.*, 1990). Neutrophils are a primary phagocytic WBC central to this process, a fact that is relevant because analysis of neutrophil counts provided the most consistent results in phylogenetic comparative tests in primates and carnivores (Nunn *et al.*, 2000; Nunn, 2002). Given that STDs are present in seminal fluid (Holmes *et al.*, 1999), a plausible interpretation is that active and immediate phagocytosis of ejaculate functions to reduce the risk of STD infection. It is important to note, however, that sperm destruction may involve other functions, including cryptic female choice (Eberhard, 1985).

Further research is needed to document the occurrence and epidemiology of STDs in wild animal populations. Information is also needed on the costs of infection for different STDs in their natural hosts. The limited information that is available demonstrates that STDs are present in non-human primates and can have a considerable impact on host fitness, thus representing an important selective force in wild populations.

SEXUAL SELECTION AND PATTERNS OF STD SPREAD

As noted above, STDs spread through populations via networks of sexual contacts. By affecting mating patterns within populations, sexual selection should be a major determinant of how STDs spread through animal populations (Thrall *et al.*, 2000). The spread of STDs will be influenced by either male–male competition or female choice, with the critical variable being the variance in mating contacts in the two sexes. The following questions therefore arise. Does sexual selection affect the prevalence of STDs within populations? Do males and females exhibit different infection rates? How do other host characteristics, such as dispersal and life-history parameters, influence the prevalence of STDs?

Several investigators have addressed these questions using epidemiological models. For example, a recent paper by Thrall *et al.* (2000) examined disease spread in the context of a polygynous mating system applicable to mammals. The authors used a simulation model to investigate the spread of STDs in males and females with respect to variance in male mating success, transfer of females among groups, and adult mortality rates. Variance in male mating success simply represents sexual selection, which is the focus of this chapter, but dispersal and mortality also are expected to modify the distribution of infections within and among groups. Dispersal is a key predictor of disease spread that is particularly relevant to primates because most (but not all) primate matings take place within social groups. Thus, greater dispersal increases the prevalence of infection in the entire population because this allows greater mixing among infected and uninfected sub-populations. Host mortality is important because it influences the duration of infectiousness: when an infected host dies, the parasite dies with it. Thus, higher host mortality is expected to reduce prevalence.

It is important to note that data required to estimate key variables in the model by Thrall *et al.* (2000) – reproductive skew among males, female dispersal and adult mortality – are available for many free-living animal populations. As applied to primates, however, the assumptions of the model do not fit all types of mating systems, with polygynandrous groups noticeably absent. To the extent that variance in mating success determines the spread of STDs, the model makes some useful predictions for primate groups in initial tests. This is because high variance in male mating success within polygynandrous groups may yield transmission dynamics similar to the polygamous systems described below.

Simulations by Thrall *et al.* (2000) were conducted by randomly assigning 'attractivity scores' to each of 250 males in a simulated population. These values were drawn from a log-normal distribution, with increasing variance in this distribution, reflecting increasing sexual selection. The same total number of females was assigned to these males in proportion to each male's attractivity code. Dispersal was accomplished by drawing a random subset of females to transfer to other groups, with assignment to groups proportional to male attractivity scores. Thus, female dispersal provides a means for the STD to spread outside polygynous groups and infect males in other groups. Mortality was simulated by eliminating a set number of males and females and replacing them with healthy individuals of the same sex. If a male died, his group of females was dissolved and reassigned to other males. Disease transmission was assumed to be a function of mating probabilities between healthy and infected hosts, number of copulations, and per-mating transmission probabilities. Thus, by systematically varying key parameters representing sexual selection, dispersal and life history, Thrall *et al.* (2000) examined the spread of STDs within and among mating groups.

Model simulations showed that increasing variance in male mating success resulted in higher prevalence for both males and females. In addition, STD prevalence tended to be higher in females than males, and this difference increased with greater variance in male attractiveness. An intuitive explanation for this outcome is that as sexual selection increases, a smaller percentage of males in the population actually mate, generating lower prevalence among males than females. The simulations also revealed that greater among-group dispersal increased population-wide prevalence, whereas increased mortality reduced overall prevalence. These simulation results make predictions that can be tested comparatively. We acquired published data on STD prevalence in wild primates, using only studies with prevalence data separated by sex among adult individuals (Table 7.1). Because quantitative data were not available on dispersal or adult mortality for each population, we focus here on the key prediction, namely that when variance in male mating opportunities is greater than variance in female mating opportunities (i.e. sexual selection on males), the prevalence of STDs should be higher in females than in males. The detailed information that we required was available for only two STDs, both of which are retroviruses: SIV and simian T-cell lymphoma/leukaemia virus (STLV).

Table 7.1 *Prevalence of two retroviruses, SIV and STLV–1, in free-living primates.*[a]

Species	Locality–group	Parasite	Prevalence (males)	Number of males	Prevalence (females)	Number of females	Reference
Cercopithecus aethiops	Saloum Delta National Forest, Senegal	SIV	0.64	14	0.95	21	Bibollet-Ruche et al., 1996
Cercopithecus aethiops	Samburu, Kenya	STLV	0.69	26	0.68	38	Drapcoli et al., 1986
Cercopithecus aethiops	Mosiro, Kenya	STLV	0.6	5	1.00	13	Drapcoli et al., 1986
Cercopithecus aethiops	Naivasha, Kenya	STLV	0.45	11	0.61	33	Drapcoli et al., 1986
Cercopithecus aethiops	Kimana, Kenya	STLV	0.78	18	0.67	15	Drapcoli et al., 1986
Cercopithecus aethiops	Awash National Park, Ethiopia	STLV	0.12	26	0	35	Drapcoli et al., 1986
Cercopithecus aethiops	Fathala Forest, Senegal, Group P	SIV	0.71	7	0.90	10	Galat-Luong et al., 1994b
Cercopithecus aethiops	Fathala Forest, Senegal, Group G	SIV	0.33	3	1.00	5	Galat-Luong et al., 1994b
Cercopithecus aethiops	Awash National Park, Ethiopia, 1990–93	SIV	0.76	37	0.97	33	Jolly et al., 1996
Cercopithecus aethiops	Awash National Park, Ethiopia, 1973	SIV	0.61	18	0.79	29	Jolly et al., 1996
Colobus guereza	Cameroon	SIV	0.27	11	0.44	9	Courgnaud et al., 2001
Erythrocebus patas	Saloum Delta National Forest, Senegal	SIV	0.14	7	0.07	14	Bibollet-Ruche et al., 1996
Erythrocebus patas	Fathala Forest, Senegal	SIV	0.17	6	0.09	11	Galat-Luong et al., 1994a
Macaca fascicularis	Indonesia	STLV	0.36	28	0.14	36	Ishikawa et al., 1987
Macaca fuscata	Japan	STLV	0.35	314	0.43	719	Hayami et al., 1984
Macaca fuscata	Nagano Prefecture, Japan	STLV	0.23	13	0.55	20	Ishida et al., 1983
Macaca maurus	Indonesia	STLV	0.28	18	0.43	14	Ishikawa et al., 1987
Papio ursinus	Northern and Eastern Transvaal	STLV	0.30	56	0.25	52	Botha et al., 1985
Macaca nemestrina	Southern Sumatra, Indonesia	STLV-1	0	5	0.15	55	Richards et al., 1998

[a] Information was found on two viruses that are known to exhibit some degree of sexual transmission. Prevalence and sample sizes for males and females in wild populations are shown separately. Many studies reported estimates of both seroprevalence (the presence of anti-viral antibodies as determined by enzyme-linked immunosorbent assay (ELISA) and western blot analysis) and the presence of virus using polymerase chain reaction (PCR) methods with virus-specific primers. ELISA and PCR analyses provided consistent results (e.g. Richards et al., 1998), probably because animals remained infected with these retroviruses for life (Lockhart et al., 1996). Although existing evidence strongly suggests that SIV and STLV are sexually transmitted in wild and captive primates, other close-contact transmission (allowing exchange of bodily fluids) is possible and the degree of sexual transmission may vary among host species (e.g. Georges-Courbet et al., 1996).

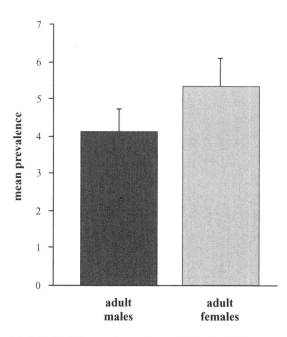

Fig. 7.2   Sex differences in prevalence of STDs in wild primates. Bars represent mean prevalence of two STDs in males and females, +1 *SE*. Data are from Table 7.1, and statistical results are provided in the text.

First, we tested whether prevalence was higher among adult than immature animals, an obvious expectation if the parasite is transmitted sexually. This prediction was consistently supported by all studies with necessary information (matched pairs $t_{17} = 7.95$, $P < 0.0001$, one-tailed; one study provided no data on immature animals and was therefore excluded). We then tested whether STD prevalence was greater in females than males, with analyses restricted to data from sexually mature adults. Again, the pattern was statistically significant (Fig. 7.2, matched pairs $t_{18} = 2.49$, $P = 0.011$, one-tailed). Twelve of the 19 studies showed the predicted pattern regarding sex differences.

These analyses represent an initial test of predictions, and prior to conducting more detailed analyses it is important to examine critically the assumptions and consider alternative explanations for our findings. First, we assumed that both SIV and STLV are transmitted sexually, but they also may be transmitted from infected mothers to infants (vertical transmission) or through aggressive interactions, such as biting (Nerrienet *et al.*, 1998). However, greater competition among males means that transmission via aggressive contacts is expected to bias patterns opposite to our findings,

suggesting that this is not an alternative explanation for the results. Second, we assumed that transmission probabilities are equal in males and females. Currently we lack quantitative data on the mechanisms of transmission for these STDs in their natural hosts, but retroviruses in humans often show higher transmission rates from males to females than vice versa (Alexander, 1990; Padian *et al.*, 1997). Differential transmission probabilities could be incorporated in future simulation studies to refine the comparative predictions. Finally, the predictions that we tested were generated from simulations of disease spread in polygynous mating systems, whereas many of the species in Table 7.1 are polygynandrous, in which more than one male breeds with a group of females. Violations of this assumption are likely to have minor consequences in the present case, however, because males within multi-male groups also show striking variance in mating success, limiting transmission opportunities to one or a few dominant males (Cowlishaw & Dunbar, 1991).

CAN STDs BE AVOIDED THROUGH MATE CHOICE?

One expected behavioural consequence of infection by an STD would involve increased host sexual activity or attractiveness of infected hosts. Although such modification of host sexual behaviour would increase STD transmission, few examples involving direct manipulation by parasites have been documented (but see Møller, 1993). Dourine, a sexually transmitted trypanosome of horses, is one parasite thought to increase sexual activity of stallions (Thrall *et al.*, 1997), thus increasing the probability of sexual transmission. Other evidence suggests that STD infection increases the duration of oestrus in cows (bovine genital campylobacteriosis; Roberts, 1979). However, examples are fragmentary and inconsistent (Webberley *et al.*, 2002), and more detailed studies are needed.

Because sexual reproduction offers an important mechanism for disease spread and may even be influenced by infection status, it is pertinent to ask whether animals can identify infected individuals and avoid mating with them. Symptoms such as visible lesions, sores, discharge around the genitalia or olfactory cues may provide evidence of infection. Infection cues might work both ways, with females inspecting males and males inspecting females. If STDs influence the expression of secondary sexual characteristics, then parasite-mediated sexual selection may play a role in female avoidance of infected partners (Hamilton & Zuk, 1982), in addition to more general contagion-avoidance mechanisms (Able, 1996).

Application of either framework requires that STDs produce reliable indicators of infection so that potential partners can identify infected individuals. We know of no studies that have examined mate choice in relation to STD infections in vertebrates. In two species of beetle, however, potential partners showed no evasion of mates that were infected with sexually transmitted mites (Abott & Dill, 2001; Webberley et al., 2002).

In humans, STDs (such as genital herpes) are notorious for producing unpleasant symptoms that could be detected by potential mating partners. Despite this dogma, many human STDs are more frequently characterised by limited symptoms or, in the case of viruses, asymptomatic shedding (Holmes et al., 1999). The most prominent example of this is HIV, in which infection status (prior to advanced stages) can be determined only through medical tests that detect the virus or host antibodies in blood or other bodily fluids. The same absence of obvious symptoms is likely to characterise many non-human STDs (Lockhart et al., 1996).

It might seem puzzling that STDs produce few severe or notable signs of infection, as compared to other diseases. A recent theoretical model by Knell (1999) sheds light on this issue by considering how STD virulence interacts with host mating success. Virulence has many definitions (Bull, 1994; Ebert, 1994; Ewald, 1994; Read, 1994; Herre, 1995). In the context of Knell's (1999) model, virulence reflects the degree to which the parasite produces symptoms in infected hosts. This is a reasonable assumption, given that increased virulence is likely to drain resources from the host, which then becomes unavailable for investment in sexual ornamentation and courtship displays. Knell's (1999) model shows that mate choice is unlikely to evolve as an effective mechanism for avoiding STDs, making parasite avoidance through secondary sexual characteristics or contagion-avoidance mechanisms less likely (Hamilton & Zuk, 1982; Able, 1996). The reason for this is that reproductive success of an STD is correlated with partner exchange and successful matings of infected hosts. Therefore, virulent parasites that produce outward signs of infection will experience decreased transmission because they provide conspicuous cues for choosy members of the opposite sex to avoid infected mates. Thus, Knell's (1999) model predicts that STDs will be less virulent, which is a general pattern that emerges when comparing STDs to non-sexually transmitted parasites (see Lockhart et al., 1996). To state the conclusion differently, a host and its sexually transmitted parasites have congruent interests in facilitating host mating success, favouring low virulence among STDs.

BEHAVIOURAL COUNTER-STRATEGIES TO STDs

The above discussion raises the issue of behavioural counter-strategies more generally. Based on Knell's (1999) model and the evolutionary importance of mating success, we propose that pre-copulatory behaviours that limit exposure to STD infection will arise less frequently in natural populations than behaviours performed after copulation. In this section, we review recent comparative results (Nunn, 2003) that examine a range of STD-avoidance behaviours available to primates before and after mating.

Figure 7.3 illustrates the basic process of dispersal and invasion for any kind of parasite. A parasite faces two main barriers, or defences, imposed by the host: behavioural counter-strategies to avoid exposure, and physical or immune defences (including both innate and acquired immunity). The order of events can vary, but behavioural mechanisms commonly are viewed as the first line of defence. An important point we wish to emphasise is that host behaviour to avoid exposure prior to mating is likely to have other reproductive costs, and these costs may outweigh their benefits. Three examples support this point. First, individuals of either sex could reduce STD risk by limiting their number of mating partners (Loehle, 1995; Thrall et al., 1997), but this is likely to be costly in terms of reproductive success. For a female, failure to mate multiply may increase the risk of infanticide committed by males with whom she did not mate while fertile (Hrdy, 1979; Hausfater & Hrdy, 1984; van Schaik & Janson, 2000). For males, missed mating opportunities directly impact reproductive success by reducing the number of offspring potentially sired. Second, individuals could reduce the number and duration of copulatory bouts with each mating partner (e.g. Hooper et al., 1978; Sheldon, 1993; Thrall et al., 1997). However, copulatory patterns in primates are likely to have been shaped by selection to increase fertilisation success, for example by sperm competition (Dixson, 1998); hence, manipulating these parameters is likely to be costly. Finally, simulations by Thrall et al. (2000) show that males who are more successful in sexual selection are more likely to be infected with an STD. Thus, females could avoid mating with successful males to reduce STD risk (Graves & Duvall, 1995). However, these females would be sacrificing the direct and indirect benefits of mating with successful males.

In addition to pre-copulatory behaviours, animals possess post-copulatory anti-parasite behaviours that may exhibit fewer reproductive trade-offs in the context of sexual

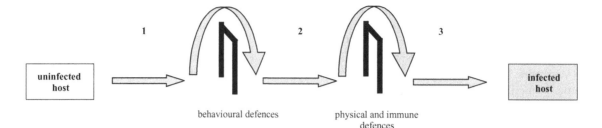

behavioural defences   physical and immune
                                    defences

Fig. 7.3   Parasite transmission. Generalised steps required for successful infection of susceptible hosts: (1) encounter with infected host or infectious material, (2) exposure of susceptible tissues to parasites; and (3) successful invasion of host and evasion of immune defences. A parasite must therefore overcome two main lines of host defences in transmission and invasion: behavioural defences that involve avoiding contagion or overcoming infection (e.g. Richards *et al.*, 1998), probably because animals remained infected with these retroviruses for life (Lockhart *et al.*, 1996); and physical and immune defences to prevent and eliminate infections. Although existing evidence strongly suggests that SIV and STLV are sexually transmitted in wild and captive primates, other close-contact transmission (allowing exchange of bodily fluids) is possible and the degree of sexual transmission may vary among host species (e.g. Georges-Courbot *et al.*, 1996), overcoming infection, and defences that provide physical barriers to infection and the immune response.

selection. In what follows, we consider two post-copulatory behaviours: oral self-grooming of the genitalia and post-copulatory urination. Oral self-grooming of the genitalia is common in many mammals after mating, and has long been known to occur in prosimians (e.g. *Lemur catta*: Jolly, 1966). It has now been established that saliva has anti-bacterial and anti-viral properties (Baron *et al.*, 2000), and grooming in rats has been shown to reduce transmission of STDs (Hart *et al.*, 1987). In many human societies, genital washing is practised before and after sex (Donovan, 2000a, b). Urination also is a human folk remedy for prevention of STDs (Donovan, 2000b) and is commonly practised after 'risky' sex (Hooper *et al.*, 1978). This behaviour may be more effective for males than females because the urethra is the primary site of infection for some STDs in males (Holmes *et al.*, 1999), whereas females have a larger mucosal area for STD infection that cannot be as effectively 'flushed' by urinating.

Because data on oral–genital grooming and urination among primate species are not readily available from the literature, Nunn (2003) compiled data using an email survey of primatologists. These data represent the best available information at this time and are unlikely to be systemati-

cally biased. Replies were received from 77 primatologists, including individuals that work with wild, zoo and laboratory populations of primates. Many respondents provided information on more than one species, and for 21 primate species, two to four responses were available, allowing assessment of data quality. Because inconsistencies were found for some questions, analyses were conducted first with variable responses coded as behaviour absent for a species and, second, as the behaviour present.

We predicted that post-copulatory oral–genital grooming would be more common among promiscuous species of primates. Surprisingly, this prediction was not supported. We estimated promiscuity in two ways – as relative testes mass and the duration of oestrus (Nunn *et al.*, 2000). In phylogenetic analyses of male and female oral–genital grooming, grooming was unrelated to either variable after incorporating variation among survey respondents (i.e. we found a nearly even mixture of positive and negative results, $P = 0.04$ to 0.78 in sensitivity tests; no results were significant after correcting for multiple comparisons; Nunn, 2003). It is unlikely that the data are too fragmentary or variable for testing the hypothesis, because further analysis revealed a clear phylogenetic signal in the frequency of post-copulatory grooming among species (Fig. 7.4). Most prosimians show very stereotyped genital grooming after mating, whereas this stereotyped behaviour is largely absent in anthropoids. The other major phylogenetic group showing oral–genital grooming is the callitrichids, where both sexes of all species in the database were reported to groom their genitals orally after mating, in at least one response to the survey. While several other species were reported to exhibit the behaviour, not all replies were consistent within species or genera (see Fig. 7.4). These results highlight two points. First, prosimians and callitrichids may be key clades for testing the hypothesis observationally or experimentally. Interestingly, both clades exhibit marked variation in mating behaviour and flexible mating systems (Goldizen, 1988; Kappeler, 2000). Second, these analyses suggest that body size may be a correlate of oral-grooming behaviour because callitrichids and

(a)

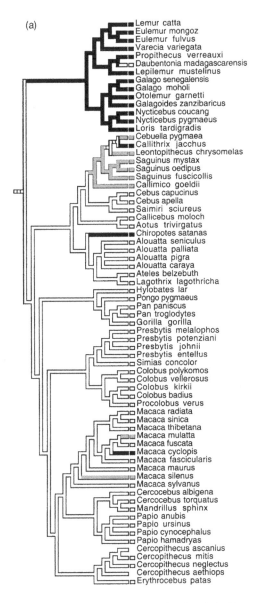

(b)

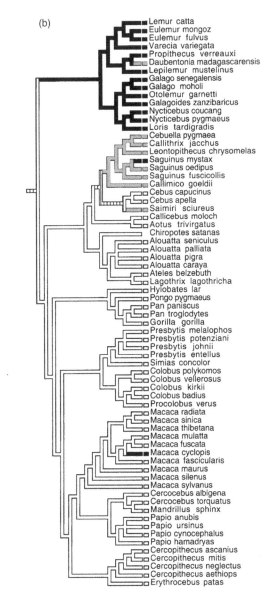

Fig. 7.4 Phylogenetic distribution of oral–genital self-grooming in male (a) and female (b) primates following copulation. Dark shading indicates presence of the trait, white shading indicates absence, and grey shading indicates conflicting responses or intermediate values of the trait (see Nunn, 2003). Species with no boxes indicate that no information was available. The phylogeny was taken from Purvis (1995).

prosimians are among the smallest-bodied primates. One possible explanation is that larger-bodied species experience physical limitations in reaching their genitalia for oral grooming.

Comparative tests using data on the occurrence of post-copulatory urination also failed to support predictions that this behaviour is associated with the risk of STD infection. It should be noted, however, that many respondents expressed uncertainty in their answers to these questions, reflecting that few observers systematically record urination in relation to mating behaviour. With this limitation in mind, survey responses indicated that very few species exhibit the behaviour, with only 2–6 per cent of species ($n = 53$) exhibiting post-copulatory urination by males, and only 5–13 per cent of species ($n = 55$) exhibiting post-copulatory urination by females. Rather than being involved in STD prevention, many respondents noted that urination was part of scent-marking behaviour, as reported for some species (e.g. Robinson, 1979; Boinski, 1992). Moreover, post-copulatory urination was not significantly related to quantitative measures of promiscuity involving relative testes mass or the duration of oestrus (Nunn, 2003). A secondary prediction from models of disease risk is that post-copulatory urination should be more common in males than females (see above). In fact, the behaviour was slightly more common in females than males.

These results must be considered preliminary, but thus far it appears that primates fail to exhibit post-copulatory behaviours that would prevent STD transmission. Nunn (2003) also examined patterns of genital inspection prior to copulation, but found no correlations between inspection and STD risk factors. Despite these negative results, some variables, such as oral–genital grooming, showed strong phylogenetic signals (Fig. 7.4), suggesting that the data are not so 'noisy' that patterns are undetectable. Further research is needed to identify the causes of variation within particular primate clades and to understand the absence of behavioural defences to STDs in other species.

SUMMARY AND CONCLUSIONS

STDs clearly exist in free-living primate populations and are likely to be costly to infected hosts. Although STDs have been relatively under-studied in comparison to other parasites, many examples of primate STDs exist, and a growing number of studies have monitored prevalence in wild primate populations. From what we know of STDs in humans and domesticated animals, there is good reason to expect that STDs impact reproductive success in wild primates. This is borne out in comparative studies of immune-system parameters in relation to host promiscuity and infection risk (Nunn et al., 2000, in review; Nunn, 2002). Moreover, our analysis of prevalence indicates that STDs are distributed in primate populations as expected from theoretical models of parasite transmission and sexual selection (Thrall et al., 2000).

Based on these lines of evidence, it is surprising that animals have not been reported to show behavioural defences to STDs, including mate choice to avoid infected partners and post-copulatory grooming and urination. As illustrated by Knell's (1999) model and primate options to reduce STD risk, pre-copulatory behavioural defences are expected to show trade-offs with other reproductive activities. This is made clear by the case of monogamy in primates, which is expected to be an effective defence against STDs in humans (Immerman, 1986) and other animals (Loehle, 1995; Thrall et al., 2000). In many of these 'monogamous' species, however, extra-pair matings have been observed (Callicebus moloch: Mason, 1966; Hylobates syndactylus: Palombit, 1994; Hylobates lar: Reichard, 1995). One interpretation of these observations is that immediate reproductive benefits from multiple mating outweigh the risks of STD infection, even in presumably monogamous species.

More generally, male and female behaviour indicates that STD risk is of secondary importance relative to other selective pressures operating on mating success. Females mate polyandrously to reduce infanticide risk (van Schaik & Janson, 2000) and, for similar reasons, they prefer novel males, though risking infection with STDs acquired from other social groups. Males prefer females of intermediate age that have already produced offspring, as these females have high reproductive value (see Anderson, 1986). Both sets of decisions by males and females are expected to increase exposure to STDs by increasing the number of partners and mating events.

These results represent an initial exploration of the consequences of sexual selection for the spread of STDs, and the distribution of behavioural defences across primate species. Many important questions remain to be tested both within and among natural populations of primates, including:

(1) Is male mating success associated with the risk of acquiring STDs?
(2) How does STD infection influence the expression of sexually selected traits?
(3) How do age, sex and social status correlate with STD risk?
(4) What symptoms do STDs cause in wild primates, and to what extent do STDs impact host reproductive success?
(5) Are there medicinal plants that reduce the risk of acquiring sexually transmitted diseases, as suggested for non-sexually transmitted parasites (e.g. Huffman et al., 1997)?

Finally, sexual transmission is only one transmission route for parasites, and it may not be equally effective in all mating systems (Thrall et al., 1997, 2000). We therefore need to investigate the ecological and evolutionary effects of other parasite transmission modes, including trade-offs between sexual and non-sexual transmission (Thrall et al., 1998). It seems likely that variation in host behaviour affects the selective advantages conferred to parasites with different transmission strategies, although very few studies have quantified transmission in free-living populations. With increasing data on parasites and phylogenies, it will become increasingly possible to answer these broad evolutionary questions and link them to the behavioural ecology of hosts.

ACKNOWLEDGEMENTS

For helpful discussion or comments on the manuscript, we thank Sarah Hrdy, Peter Kappeler, Monique Borgerhoff Mulder, Mary Poss, Carel van Schaik and two anonymous reviewers.

C. Nunn thanks the respondents to the email survey, whose names can be found at http://faculty.virginia.edu/charlienunn. This research was supported through a grant from the National Center for Ecological Analysis and Synthesis (NCEAS) to S. Altizer and C. Nunn, and a National Science Foundation (NSF) Postdoctoral Research Fellowship in Biological Informatics to C. Nunn.

REFERENCES

Abbot, P. & Dill, L. M. 2001. Sexually transmitted parasites and sexual selection in the milkweed leaf beetle, *Labidomera clivicollis. Oikos*, **92**, 91–100.

Able, D. J. 1996. The contagion indicator hypothesis for parasite-mediated sexual selection. *Proceedings of the National Academy of Sciences, USA*, **93**, 2229–33.

Alexander, N. J. 1990. Sexual transmission of human immunodeficiency virus: virus entry into the male and female genital tract. *Fertility and Sterility*, **54**, 1–18.

Anderson, C. M. 1986. Female age: male preference and reproductive success in primates. *International Journal of Primatology*, **7**, 305–26.

Anderson, R. M. & May, R. M. 1991. *Infectious Diseases of Humans: Dynamics and Control*. Oxford: Oxford University Press.

Baron, S., Singh, I., Chopra, A., Coppenhaver, D. & Pan, J. Z. 2000. Innate antiviral defenses in body fluids and tissues. *Antiviral Research*, **48**, 71–89.

Barratt, C. L. R., Bolton, A. E. & Cooke, I. D. 1990. Functional significance of white blood cells in the male and female reproductive tract. *Human Reproduction*, **5**, 639–48.

Bibollet-Ruche, F., Galat-Luong, A., Cuny, G. *et al.* 1996. Simian immunodeficiency virus infection in a patas monkey (*Erythrocebus patas*): evidence for cross-species transmission from African green monkeys (*Cercopithecus aethiops sabaeus*) in the wild. *Journal of General Virology*, **77**, 773–81.

Blower, S. M. & McLean, A. R. 1991. Mixing ecology and epidemiology. *Proceedings of the Royal Society of London, Series B*, **245**, 187–92.

Boinski, S. 1992. Olfactory communication among Costa Rican squirrel monkeys: a field study. *Folia Primatologica*, **59**, 127–36.

Botha, M. C., Jones, M., Deklerk, W. A. & Yamamoto, N. 1985. Spread and distribution of human T-cell leukemia-virus type-I-reactive antibody among baboons and monkeys in the northern and eastern Transvaal. *South African Medical Journal*, **67**, 665–8.

Bull, J. J. 1994. Virulence. *Evolution*, **48**, 1423–37.

Courgnaud, V., Pourrut, X., Bibollet-Ruche, F. *et al.* 2001. Characterization of a novel simian immunodeficiency virus from guereza colobus monkeys (*Colobus guereza*) in Cameroon: a new lineage in the nonhuman primate lentivirus family. *Journal of Virology*, **75**, 857–66.

Cowlishaw, G. & Dunbar, R. I. M. 1991. Dominance rank and mating success in male primates. *Animal Behavior*, **41**, 1045–56.

Dixson, A. F. 1998. *Primate Sexuality: Comparative Studies of the Prosimians, Monkeys, Apes, and Humans*. Oxford: Oxford University Press.

Donovan, B. 2000a. The repertoire of human efforts to avoid sexually transmissible diseases: past and present. Part 1: strategies used before or instead of sex. *Sexually Transmitted Infections*, **76**, 7–12.

  2000b. The repertoire of human efforts to avoid sexually transmissible diseases: past and present. Part 2: strategies used during or after sex. *Sexually Transmitted Infections*, **76**, 88–93.

Dracopoli, N. C., Turner, T. R., Else, J. G. *et al.* 1986. STLV-I antibodies in feral populations of east-african vervet monkeys (*Cercopithecus aethiops*). *International Journal of Cancer*, **38**, 523–9.

Eberhard, W. G. 1985. *Sexual Selection and Animal Genitalia*. Cambridge, MA: Harvard University Press.

Ebert, D. 1994. Virulence and local adaptation of a horizontally transmitted parasite. *Science*, **265**, 1084–6.

Ewald, P. W. 1994. *Evolution of Infectious Disease*. Oxford: Oxford University Press.

Galat-Luong, A., Bibollet-Ruche, F., Pourrut, X. *et al.* 1994a. Social organization and SIV seroepidemiology of a patas monkey population in Senegal. *Folia Primatologica*, **63**, 226–8.

Galat-Luong, A., Galat, G., Bibollet-Ruche, F. *et al.* 1994b. Social structure and SIVagm prevalence in two groups of green monkeys, *Cercopithecus aethiops sabaeus*, in Senegal.

In *Current Primatology*. Vol. 3: *Behavioral Neuroscience, Physiology and Reproduction*, ed. J. R. Anderson, J. J. Roeder, B. Thierry & N. Herrenschmidt. Strasbourg: Université Louis Pasteur, pp. 259–62.

Garnett, G. P. & Anderson, R. M. 1993. No reason for complacency about the potential demographic impact of AIDS in Africa. *Transactions of the Royal Society of Tropical Medicine and Hygiene*, **87**, 19–22.

Georges-Courbot, M. C., Moisson, P., Leroy, E. *et al.* 1996. Occurrence and frequency of transmission of naturally occurring simian retroviral infections (SIV, STLV, and SRV) at the CIRMF Primate Center, Gabon. *Journal of Medical Primatology*, **25**, 313–26.

Getz, W. M. & Pickering, J. 1983. Epidemic models: thresholds and population regulation. *American Naturalist*, **121**, 892–8.

Goldizen, A. W. 1988. Tamarin and marmoset mating systems: unusual flexibility. *Trends in Ecology and Evolution*, **3**, 36–40.

Graves, B. M. & Duvall, D. 1995. Effects of sexually-transmitted diseases on heritable variation in sexually selected systems. *Animal Behaviour*, **50**, 1129–31.

Hahn, B. H., Shaw, G. M., de Cock, K. M. & Sharp, P. M. 2000. AIDS as a zoonosis: scientific and public health implications. *Science*, **287**, 607–14.

Hamilton, W. D. & Zuk, M. 1982. Heritable true fitness and bright birds: a role for parasites? *Science*, **218**, 384–7.

Hart, B. J., Korinek, E. & Brennan, P. 1987. Postcopulatory genital grooming in male rats: prevention of sexually transmitted infections. *Physiology and Behavior*, **41**, 321–5.

Hausfater, G. & Hrdy, S. B. 1984. *Infanticide: Comparative and Evolutionary Perspectives*. New York, NY: Aldine de Gruyter.

Hayami, M., Komuro, A., Nozawa, K. *et al.* 1984. Prevalence of antibody to adult T-cell leukemia virus-associated antigens (ATLA) in Japanese monkeys and other non-human primates. *International Journal of Cancer*, **33**, 179–83.

Herre, E. A. 1995. Factors affecting the evolution of virulence: nematode parasites of fig wasps as a case study. *Parasitology*, **111**, S179–91.

Hoeprich, P. D., Jordan, M. C. & Ronald, A. R. 1994. *Infectious Diseases: A Treatise of Infectious Processes*. Philadelphia, PA: J. B. Lippincott.

Holmes, K. K., Sparling, P. F., Mardh, P.-A. *et al.* 1999. *Sexually Transmitted Diseases*. New York, NY: McGraw Hill.

Hooper, R. R., Reynolds, G. H., Jones, O. G. *et al.* 1978. Cohort study of venereal disease. I: The risk of gonorrhea transmission from infected women to men. *American Journal of Epidemiology*, **108**, 136–44.

Hrdy, S. B. 1979. Infanticide among animals: a review, classification, and examination of the implications for the reproductive strategies of females. *Ethology and Sociobiology*, **1**, 13–40.

Huffman, M. A., Gotoh, S., Turner, L. A., Hamai, M. & Yoshida, K. 1997. Seasonal trends in intestinal nematode infection and medicinal plant use among chimpanzees in the Mahale Mountains, Tanzania. *Primates*, **38**, 111–25.

Immerman, R. S. 1986. Sexually transmitted disease and human evolution: survival of the ugliest? *Human Ethology Newsletter*, **4**, 6–7.

Ishida, T., Yamamoto, K., Kaneko, R., Tokita, E. & Hinuma, Y. 1983. Seroepidemiological study of antibodies to adult T-cell leukemia virus-associated antigen (ATLA) in free-ranging Japanese monkeys (*Macaca fuscata*). *Microbiology and Immunology*, **27**, 297–301.

Ishikawa, K., Fukasawa, M., Tsujimoto, H. *et al.* 1987. Serological survey and virus isolation of simian T-cell leukemia/lymphotropic virus Type-I (STLV-I) in nonhuman primates in their native countries. *International Journal of Cancer*, **40**, 233–9.

Jolly, A. 1966. *Lemur Behavior*. Chicago, IL: University of Chicago Press.

Jolly, C. J., Phillips-Conroy, J. E., Turner, T. R., Broussard, S. & Allan, J. S. 1996. SIVagm incidence over two decades in a natural population of Ethiopian grivet monkeys (*Cercopithecus aethiops aethiops*). *Journal of Medical Primatology*, **25**, 78–83.

Kappeler, P. M. 2000. Causes and consequences of unusual sex ratios among lemurs. In *Primate Males: Causes and Consequences of Variation in Group Composition*, ed. P. M. Kappeler. Cambridge: Cambridge University Press, pp. 55–63.

Knell, R. J. 1999. Sexually transmitted disease and parasite-mediated sexual selection. *Evolution*, **53**, 957–61.

Lockhart, A. B., Thrall, P. H. & Antonovics, J. 1996. Sexually transmitted diseases in animals: ecological and evolutionary implications. *Biological Reviews*, **71**, 415–71.

Loehle, C. 1995. Social barriers to pathogen transmission in wild animal populations. *Ecology*, **76**, 326–35.

Mason, W. A. 1966. Social organization of the South American monkey, *Callicebus moloch*: a preliminary report. *Tulane Studies in Zoology*, **13**, 23–8.

Møller, A. P. 1993. A fungus infecting domestic flies manipulates sexual behavior of its host. *Behavioral Ecology and Sociobiology*, **33**, 403–7.

Nerrienet, E., Amouretti, X., Mullertrutwin, M. C. *et al.* 1998. Phylogenetic analysis of SIV and STLV Type I in mandrills (*Mandrillus sphinx*): indications that intracolony transmissions are predominantly the result of male-to-male aggressive contacts. *AIDS Research and Human Retroviruses*, **14**, 785–96.

Norley, S., Beer B., Holzammer S., Zur Megede, J. & Kurth, R. 1999. Why are natural hosts of SIV resistant to AIDS? *Immunology Letters*, **66**, 47–52.

Nunn, C. L. 2002. A comparative study of leukocyte counts and disease risk in primates. *Evolution*, **56**, 177–90.

2003. Behavioural defences against sexually transmitted diseases in primates. *Animal Behaviour*, in press.

Nunn, C. L., Gittleman, J. L. & Antonovics, J. 2000. Promiscuity and the primate immune system. *Science*, **290**, 1168–70.

2003. A comparative study of white blood cell counts and disease risk in carnivores. *Proceedings of the Royal Society of London, Series B. Biological Sciences*, **270**, 347–56.

Padian, N. S., Shiboski, S. C., Glass, S. O. & Vittinghoff, E. 1997. Heterosexual transmission of human immunodeficiency virus (HIV) in Northern California: results from a ten-year study. *American Journal of Epidemiology*, **146**, 350–7.

Palombit, R. A. 1994. Extra-pair copulations in a monogamous ape. *Animal Behaviour*, **47**, 721–3.

Pandya, I. J. & Cohen, J. 1985. The leukocytic reaction of the human uterine cervix to spermatozoa. *Fertility and Sterility*, **43**, 417–21.

Phillips, D. M. & Mahler, S. 1977. Leukocyte emigration and migration in the vagina following mating in the rabbit. *Anatomical Record*, **189**, 45–60.

Phillips-Conroy, J. E., Jolly, C. J., Petros, B., Allan, J. S. & Desrosiers, R. C. 1994. Sexual transmission of SIV(agm) in wild grivet monkeys. *Journal of Medical Primatology*, **23**, 1–7.

Purvis, A. 1995. A composite estimate of primate phylogeny. *Philosophical Transactions of the Royal Society of London, Series B*, **348**, 405–21.

Purvis, A. & Rambaut, A. 1995. Comparative analysis by independent contrasts (CAIC): an Apple Macintosh application for analysing comparative data. *Computer Applications in the Biosciences*, **11**, 247–51.

Radolf, J. D., Sanchez, P. J., Schulz, K. F. & Murphy, F. K. 1999. Congenital syphilis. In *Sexually Transmitted Diseases*, ed. K. K. Holmes, P. F. Sparling, P.-A. Mardh, S. M. Lemon, W. E. Stamm, P. Piot & J. N. Wasserheit. New York, NY: McGraw-Hill, pp. 1165–89.

Read, A. F. 1994. The evolution of virulence. *Trends in Microbiology*, **2**, 73–6.

Reichard, U. 1995. Extra-pair copulation in a monogamous gibbon (*Hylobates lar*). *Ethology*, **100**, 99–112.

Richards, A. L., Giri, A., Iskandriati, D. *et al.* 1998. Simian T-lymphotrophic virus type-I infection among wild-caught Indonesian pig-tailed macaques (*Macaca nemestrina*). *Journal of Acquired Immune Deficiency Syndromes and Human Retrovirology*, **19**, 542–5.

Roberts, L. 1979. Bovine venereal campylobacteriosis (vibriosis) in north east Scotland. *Veterinary Record*, **105**, 295–6.

Robinson, J. G. 1979. Correlates of urine washing in the wedge-capped capuchin *Cebus nigrivittatus*. In *Vertebrate Ecology in the Northern Neotropics*, ed. J. F. Eisenberg. Washington, DC: Smithsonian Institution Press, pp. 137–43.

Santiago, M. L., Rodenburg, C. M., Kamenya, S. *et al.* 2002. SIVcpz in wild chimpanzees. *Science*, **295**, 465.

Schwartländer, B., Garnett, G., Walker, N. & Anderson, R. 2000. AIDS in a new millennium. *Science*, **289**, 64–7.

Sheldon, B. C. 1993. Sexually-transmitted disease in birds: occurrence and evolutionary significance. *Philosophical Transactions of the Royal Society of London, Series B*, **339**, 491–7.

Smith, G. & Dobson, A. P. 1992. Sexually transmitted diseases in animals. *Parasitology Today*, **8**, 159–66.

Sorci, G., Møller, A. P & Boulinier, T. 1997. Genetics of host–parasite interactions. *Trends in Ecology and Evolution*, **12**, 196–200.

Swanstrom, R. & Wehbie, R. 1999. The biology of HIV, SIV and other lentiviruses. In *Sexually Transmitted Diseases*, ed. K. K. Holmes, P. F. Sparling, P.-A. Mardh, S. M. Lemon, W. E. Stamm, P. Piot & J. N. Wasserheit. New York, NY: McGraw-Hill, pp. 215–29.

Thrall, P. H., Antonovics, J. & Bever, J. D. 1997. Sexual transmission of disease and host mating systems: within-season reproductive success. *American Naturalist*, **149**, 485–506.

Thrall, P. H., Antonovics, J. & Dobson, A. P. 2000. Sexually transmitted diseases in polygynous mating systems: prevalence and impact on reproductive success. *Proceeding of the Royal Society of London, Series B*, **267**, 1555–63.

Thrall, P. H., Antonovics, J. & Hall, D. W. 1993. Host and
pathogen coexistence in vector-borne and venereal
diseases characterized by frequency-dependent
disease transmission. *American Naturalist*, **142**, 543–52.

Thrall, P. H., Antonovics, J. & Wilson, W. G. 1998. Allocation
to sexual vs. non-sexual disease transmission. *American
Naturalist*, **151**, 29–45.

van Schaik, C. P. & Janson, C. H. 2000. *Infanticide by Males
and Its Implications*. Cambridge: Cambridge University
Press.

Webberley, K. M., Hurst, G. D. D., Buszko, J. & Majerus,
M. E. N. 2002. Lack of parasite-mediated sexual selection
in a ladybird/sexually transmitted disease system. *Animal
Behaviour*, **63**, 131–41.

8 • Mating conflict in primates: infanticide, sexual harassment and female sexuality

CAREL P. VAN SCHAIK, GAURI R. PRADHAN &
MARIA A. VAN NOORDWIJK
Department of Biological Anthropology and Anatomy
Duke University
Durham, NC, USA

INTRODUCTION

In a variety of mammals and a few birds, newly immigrated or newly dominant males are known to attack and kill dependent infants (Hausfater & Hrdy, 1984; Parmigiani & vom Saal, 1994; van Schaik & Janson, 2000). Hrdy (1974) was the first to suggest that this bizarre behaviour was the product of sexual selection: by killing infants they did not sire, these males advanced the timing of the mother's next oestrus and, owing to their new social position, would have a reasonable probability of siring this female's next infant. Infanticide would therefore be one of the most dramatic expressions of sexual conflict (Smuts & Smuts, 1993; Gowaty, 1997, this volume).

Although this interpretation, and indeed the phenomenon itself, has been hotly debated for decades (e.g. Dolhinow, 1977; Boggess, 1984; Bartlett *et al.*, 1993; Sussman *et al.*, 1995), on balance, this hypothesis provides a far better fit with the observations on primates than any of the alternatives (cf. van Schaik, 2000a). First, several detailed studies showed that the males never attacked or killed their own offspring (Borries *et al.*, 1999; Soltis *et al.*, 2000), in accordance with the more anecdotal information compiled from all directly observed cases of infanticide in the wild (van Schaik, 2000a). Second, several large-scale studies have estimated that the time gained by the infanticidal male amounts to 25 per cent, 26 per cent and 32 per cent of the mean interbirth interval (Crockett & Sekulic, 1984; Sommer, 1994; Borries, 1997). Third, in most cases (e.g. in 78 per cent of 49 directly observed cases in the wild; van Schaik, 2000a), these males subsequently gained mating access to the female and had above-average chances of siring the next infant because of their dominant status. Because males rarely, if ever, suffer injuries during infanticidal attacks, and because there is no evidence that committing infanticide leads to reduced tenure length, one can safely conclude that, on average, infanticide is an adaptive male strategy.

The curious taxonomic concentration of observations of infanticide by males in primates, carnivores and sciurognath rodents can be explained by the fact that these radiations all share a peculiar life-history feature. In all of them, females that lose their dependent offspring are ready to conceive sooner than they otherwise would – the key benefit to the male – due to the long period of infant dependence relative to the duration of gestation (van Schaik, 2000b).

Hrdy (1979) was also the first to argue that if we accept that infanticide by males is an adaptive phenomenon we should also ask about evolved counter-strategies. As she put it recently: 'it should logically follow that infanticide must have acted as an important selection pressure shaping the behaviour and reproductive physiologies of mothers as well the protective responses by fathers and other relatives' (Hrdy, 2000a). Soon after her proposal (Hrdy, 1979), female sexual behaviour became the focus of scrutiny from the perspective of infanticide reduction (e.g. Hrdy & Whitten, 1987; Small, 1993; van Schaik *et al.*, 1999), but association with protector males has received attention as well (van Schaik & Dunbar, 1990; Smuts & Smuts, 1993; cf. Wrangham, 1979).

In this chapter, we extend the work on female sexuality in primates in relation to the risk of infanticide. First, after reviewing the basic logic and empirical evidence for sexual counter-strategies against infanticide risk, we examine in more detail how female sexual behaviour can be effective in reducing attack tendencies by unlikely sires while still securing protection from more likely sires. Second, we turn to sexual harassment,[1] i.e. male aggression targeted against sexually active ('oestrous') females, and show how it can be

---

[1] We follow the use of 'harassment' as in Smuts and Smuts (1993), which differs from that in Clutton-Brock and Parker (1995).

Sexual Selection in Primates: New and Comparative Perspectives, ed. Peter M. Kappeler and Carel P. van Schaik. Published by Cambridge University Press. © Cambridge University Press 2004.

viewed as an expression of mating conflict between the female and the dominant male. Because harassment is remarkably concentrated in catarrhine primates, we then examine sexual behaviour and physiology in that radiation, relative to species vulnerable to infanticide in other primate radiations. We conclude that various features of catarrhine sexuality can plausibly be understood as responses to this harassment in the evolutionary arms race, although this topic requires much additional work.

MALES AS PROTECTORS

Before discussing aspects of female sexual behaviour as a counter-strategy against infanticide, some attention to the association with likely sires is needed in order to explain why female sexual behaviour tends to lead to protection by one, or sometimes more, likely sires. Protection of infants by likely sires is made possible by the year-round male–female association found in virtually all primates in which females carry their infants (van Schaik & Kappeler, 1997). At least in multi-male groups, likely sires tend to be in close spatial proximity with infants (Janson, 1986; Paul et al., 2000), and numerous reports indicate that these males actually defend the infants against attacks by other males (Borries et al., 1999; review: van Schaik, 2000a).

Infanticide often happens when the former dominant male, the most likely sire of most infants even in multi-male groups (reviews in Cowlishaw & Dunbar, 1991; Paul, 1997; van Noordwijk & van Schaik, this volume), is eliminated or incapacitated. Infanticidal attacks could have been provoked by his experimental removal, either in the wild (Sugiyama, 1966) or (quite frequently, and inadvertently) in captivity (e.g. Angst & Thommen, 1977), or by natural demographic processes (e.g. Steenbeek, 1996). In all these situations infanticide by the new male is commonly observed. Conversely, when one interprets the social context of all directly observed cases of infanticide in wild primates that occurred spontaneously, the great majority (85 per cent of 55 cases: van Schaik, 2000a) are associated with a change in the dominant position in the group, which involves the ousting or weakening of the former dominant. Relative take-over rate (corrected for variation in interbirth interval) also explains much of the variation in infanticide rates in well-studied populations (Janson & van Schaik, 2000).

Thus, dominant males are effective protectors of infants as long as they are not ousted or incapacitated. Females should therefore be expected to mate preferentially with these powerful males whenever females are confronted by males who are unlikely sires.

FEMALE SEXUAL BEHAVIOUR AS A COUNTER-STRATEGY TO INFANTICIDE

THEORY

Hrdy (1974, 1979) hypothesised that female sexual behaviour in primate species vulnerable to infanticide has been modified to reduce the risk of infanticide by males. The basis for the argument is that primate males, like those of the majority of mammals, do not recognise their offspring (possibly as a result of female-driven evolution), and therefore must rely on indirect indicators of paternity probability. These indicators have been studied experimentally in rodents (vom Saal, 1984; Perrigo & vom Saal, 1994), but not in primates, where the indicators have to be pieced together by analysing individual cases (reviewed in van Schaik, 2000a). Primate males are thought to use rules of thumb such as the quality of their sexual experience with the female (i.e. mating frequency relative to her attractivity – her 'stimulus value' to the male) weighted for her mating frequency with other males, the interval between matings and birth, and perhaps the continuity of association between male and female. We assume that natural selection has favoured males that use those rules of thumb that yield the closest average match with the probability of paternity.

Conceptually, we can distinguish two kinds of matings by females that may reduce the risk of infanticide. First, by mating polyandrously in potentially fertile periods, females can reduce the concentration of paternity in the dominant male, and spread some of it to other males, so that long-term average paternity probabilities will be somewhat below 1 for the dominant and somewhat above 0 for the subordinates. Second, by mating during periods of non-fertility (e.g. mating during pregnancy; situation-dependent receptivity: Hrdy, 1979), a female may be able to manipulate the assessment by the various males of their paternity chances, although she obviously cannot change the actual paternity values allocated to the various males. This distinction is theoretically useful, but the players are probably quite unaware of it, with males merely responding to visual, olfactory and behavioural stimuli emanating from the female that create variation in her attractivity (Snowdon, this volume; Zinner et al., this volume).

If male behaviour depends on their estimate of paternity, this sexual behaviour can be effective. Hrdy (1979, 1997) reasoned that where males have a low, but non-zero probability, they would refrain from attacking the infant, whereas they would defend it when the estimate is higher.

EMPIRICAL EVIDENCE

The basic prediction is that females that are vulnerable to infanticide by males should be actively polyandrous whenever potentially infanticidal males are present in the mating pool (i.e. the sexually mature males in the social unit or nearby with which the female can mate, in principle). There is ample evidence that primate females in vulnerable species actively pursue polyandrous matings and that they often engage in matings when fertilisation is unlikely or impossible (Small, 1993; Manson, 1995; summarised in van Schaik *et al.*, 1999). Indeed, females often target low-ranking or peripheral males reluctant to mate in the presence of dominant central males, especially during pregnancy (e.g. Watts, 1991; Wallis & Bettinger, 1993; Gust, 1994).

There are two sources of more direct empirical evidence to assess whether these derived features of primate sexual behaviour are indeed an evolutionary response to vulnerability to infanticide: (1) direct sexual responses by females to changes in group composition or male status; and (2) broader comparisons of sexual behaviour between taxa that are or are not vulnerable to infanticide by males.

In species vulnerable to infanticide, females often respond to changes in the male cohort of a group with immediate proceptivity, and effectively solicit matings with the new(or newly dominant) male (Struhsaker & Leland, 1985; Cords, 1988; Sommer *et al.*, 1992; Swedell, 2000), even showing rapid development of sexual swellings, in species that have them (Stein, 1984; Colmenares & Gomendio, 1988; Zinner & Deschner, 2000). In various catarrhine primates in which multiple males temporarily enter a group ('male influx'), mating periods (duration of oestrus) are relatively long compared to periods without male influxes (e.g. Cords, 1984, 1988; Takahata *et al.*, 1994). Similarly, in several species with both single-male and multi-male groups, female mating periods are longer in multi-male groups (Brockman, 1999; Heistermann *et al.*, 2001).

Interspecific comparisons provide similar support: van Noordwijk and van Schaik (2000) found a clear trend toward more polyandrous mating among primate and carnivore species vulnerable to infanticide, relative to those that are not vulnerable. Post-conception mating, while infrequently reported, was also concentrated in those orders of mammals where infanticide is to be expected based on their life history. Within primates, post-conception matings are found predominantly, and perhaps exclusively, in species vulnerable to infanticide (van Schaik *et al.*, 1999). However, one prediction was not upheld: in most other mammals no systematic trend toward longer mating periods in species

vulnerable to infanticide was apparent (van Noordwijk & van Schaik, 2000). We will later show that this prediction is only expected where males are able to force matings.

A different source of evidence for the effectiveness of sexual behaviour in reducing the risk of infanticide is the lower rate of infanticide in multi-male groups, when controlling for the effect of take-over of dominance (Janson & van Schaik, 2000). To some extent this reduction is obviously due to male protection, because in multi-male groups defeated dominants tend to remain in the group, at least for a while (e.g. van Noordwijk & van Schaik, 1988; Perry, 1998; Borries, 2000). However, sexual strategies are implicated as well because we occasionally see protection of the infant by other resident males (e.g. Borries *et al.*, 1999), or absence of attacks by the new dominant who was a long-term resident and had mated before with the mothers (see below).

A PROBLEM

A remarkable aspect of infanticide is that – especially in multi-male groups – only a small-to-moderate proportion of the infants typically ends up getting killed. While this high probability of survival probably has multiple sources, it is reasonable to attribute at least some of it to the female's sexual behaviour. Yet, the latter's effectiveness is somewhat surprising as paternity is a constant-sum game, in that each infant can have only one sire and that the long-term average probabilities of fertilisation of all the players must add up to 1. A female therefore faces a considerable challenge. On the one hand, by raising the long-term paternity probability of subordinate males through their mating behaviour, she reduces the risk of infanticide because these males are less likely to attack the infant when they become dominant. On the other hand, this behaviour must reduce the paternity probability of the dominant male, and hence make it less likely that the dominant will defend the infant against males entering from the outside (cf. Symons, 1982). It is therefore not immediately obvious that sexual behaviour could ever achieve an optimal balance and reduce overall risk.

Here, we will present a novel explanation for the effectiveness of sexual behaviour. Doing so, however, requires that we first determine the conditions in which natural selection favours infanticide by males (cf. van Schaik, 2000a).

WHEN IS INFANTICIDE FAVOURED BY NATURAL SELECTION?

In a species in which infanticide advances the female's next conception, and in a situation in which a male can be confident that $p = 0$ (i.e. he never mated with the female, where p is the probability of having sired the existing infant), infanticide is obviously an advantageous strategy, provided it can be committed at low cost. However, if the male has a mating history with the female, a more quantitative prediction is needed. In order to develop this prediction, we compare the expected mean number of offspring sired by a dominant male during his period of dominance (tenure) under two scenarios: with and without infanticide upon assuming the dominant position.

Denote the effective tenure period of a non-infanticidal dominant male as T, i.e. the period between the conception of the first infant sired during the new tenure to the end of the male's tenure. If the regular interbirth interval is t_n, and the interbirth interval following infanticide is t_i, the time gained by an infanticidal male is $t_n - t_i$. Thus, the effective tenure of an infanticidal male is $T + t_n - t_i$. Now we can calculate the benefit of the two strategies (B_n and B_i, for non-infanticide and infanticide benefit, respectively). These benefits in terms of expected number of infants are (cf. van Schaik (2000a), who overlooked infants sired by non-infanticidal males before becoming dominant):

$$B_n = P + \frac{T}{t_n}P$$

$$B_i = -p + \frac{T + t_n - t_i}{t_n}P$$

where P is the probability of siring infants (with any particular female) during tenure, assuming that this probability is constant, regardless of whether the male commits infanticide. Then, the net benefit of committing infanticide is:

$$B_i - B_n = -p + \frac{T + t_n - t_i}{t_n}P - \frac{T}{t}P$$

$$= -p + \frac{t_n - t_i}{t_n}P$$

Thus $B_i - B_n$ is positive (given that $t_n > t_i$) if

$$\frac{t_n - t_i}{t_n}P > p \qquad (8.1)$$

Inequality (8.1) ignores any costs to infanticide (see van Schaik (2000a) for discussion). In the average primate species, the maximum $(t_n - t_i)/t_n$ is around 0.5, which is attained when a newborn is killed. In this case $P > 2p$. As the infant gets older, P has to increase to make infanticide advantageous to the male. For the observed mean values of $(t_n - t_i)/t_n$, between 0.25 and 0.32 (see above), infanticide is advantageous if P is greater than approximately 3–4p.

OPTIMUM MALE DECISIONS

Inequality (8.1) shows when infanticide is expected, but we cannot assume that males have perfect estimates of the relevant parameters. We must therefore translate these criteria into decision-making rules for males. Some cases are simple. First, if the newly dominant male never mated with the female, and $p = 0$, the rule is easy. Indeed, infanticide is commonly seen after a new male immigrant takes over a group (e.g. Steenbeek, 1996; van Schaik, 2000a). Second, if the male has a sexual history with the female but his estimate of p is very small, infant age matters because the benefits decrease as infants get older (cf. Crockett & Sekulic, 1984; Sommer, 1994; summarised in van Schaik, 2000a, Fig. 2.1).

In multi-male groups, however, a male has generally mated with the female before, and optimum decision-making may be more difficult. The male is forced to use indirect indicators based on his sexual history with the female (see above) to produce estimates that will tend to be highly imprecise, except when the male could monopolise most matings with sexually highly attractive females or when he could get only very few matings when she was not very attractive.

Frequent polyandrous mating during periods of ovarian activity and periods of patent infertility (e.g. pregnancy) could have two consequences. First, it may lead to an increase in the estimated paternity probability (henceforth p'), especially if the males are not fully informed about the female mating activity with other males. Thus, especially by mating frequently at times of non-fertility, females may manage to increase p' of especially low-ranking males, producing a sum of these estimates greater than 1 (obviously, the actual paternity probabilities still add up to 1).

Second, frequent polyandry should lead to great uncertainty of each male's p'. One must assume that a higher quantity of matings can compensate to some extent for lower quality (i.e. the female was less attractive). We suspect that non-dominant males have very imprecise estimates of their chances of paternity. In a situation of high uncertainty as to the value of p', it may be impossible to find optimum decision-making algorithms.

A newly dominant male may decide to choose the option that maximises mean fitness. If uncertainty over the value of p' approaches ignorance, his best guess may be that $p' = 0.5$, and it is easy to show that not killing the infant is on average the best option (see Table 8.1). However, since a given male actually did or did not father the infant, in a case like this, a newly dominant male may actually maximise his fitness pay-off by avoiding costly mistakes (cf. Resnik, 1987, p. 28), i.e. minimise the risk of losing a large portion

Table 8.1 *The pay-offs of decisions made by newly dominant males with respect to infants, depending on whether the male had sired the infant or not (see text for explication of variables).*

Decision	Father	Not father
Kill	$-1 + \dfrac{T + t_n - t_i}{t_n} P$	$\dfrac{T + t_n - t_i}{t_n} P$
Not kill	$\dfrac{T}{t_n} P$	$\dfrac{T}{t_n} P$

Table 8.2 *The pay-offs of decisions made by currently dominant males with respect to infants, depending on whether the male had sired this infant or not.*

Decision	Father	Not father
Protect	$s - c$	$-c$
Not protect	$-s$	0

of fitness (cf. risk avoidance in foraging: Stephens & Krebs, 1986).

A newly dominant male can make two kinds of mistakes: (1) he can kill an infant that he had actually sired before becoming dominant, thus losing one infant from his total number produced; and (2) he can refrain from killing an infant that he did not sire, thus losing time to the next conception of the infant's mother. These two errors have different costs attached to them. Table 8.1 presents the pay-offs of the two possible male decisions (kill vs. not to kill), under two different conditions (male actually fathered the infant vs. did not). The cost of the mistake, i.e. the male killing his own infant, relative to the optimum tactic of refraining from killing it, is the difference between killing it and not killing it:

$$C_1 = -1 + \frac{T + t_n - t_i}{t_n} P - \frac{T}{t_n} P$$
$$= -1 + \frac{t_n - t_i}{t_n} P$$

Thus the cost of killing his own infant (C_1) is in the range $-1 \leq C_1 \leq 0$ (these are in infant units).

On the other hand, if the male did not sire the infant, the cost of not killing the infant (relative to the optimum tactic of killing it) is:

$$C_2 = \frac{T}{t_n} P - \frac{T + t_n - t_i}{t_n} P = -\frac{t_n - t_i}{t_n} P$$

The cost of not killing another male's infant (C_2) is also in the range $-1 \leq C_2 \leq 0$.

Committing the first error (killing one's own infant) has greater costs than committing the second error (not killing some other male's infant) if $(t_n - t_i/t_n)P < 0.5$. Thus, in all realistic conditions, killing one's own infant is costlier than not killing another male's infant. Hence, if female mating tactics have confused paternity estimates to the point of near-ignorance, a newly dominant male will do better to avoid the more costly error, and should thus refrain

from infanticide. This effect of deceptive female matings could explain why infanticide is not always seen in conditions where it might be expected.

We assume that the dominant males usually have less uncertainty concerning their decision whether to protect or not to protect an infant, because their estimates of paternity will tend to be close to 1. However, if they also face considerable uncertainty, verging on ignorance, they should also avoid making the costlier mistake. Table 8.2 provides the pay-offs of the dominant male's decisions for the two possible states (father vs. non-father). The cost of not protecting an infant that the male actually sired is:

$$C_3 = -s - (s - c) = -2s + c$$

(where $s = $ change in probability of infant survival owing to male protection, and $c = $ cost of protection, both expressed in infant units), whereas the cost of protecting an infant that he did not sire is:

$$C_4 = -c - 0 = -c.$$

For $s > c$ (a very reasonable assumption), the decision not to protect the infant when the dominant male is the actual sire is the costlier one. Thus, if a dominant male is so uncertain of his paternity as to be virtually ignorant of its value (i.e. if his estimate is close to 0.5), then his best decision is to protect the infant when it is at risk, up to a point. It also suggests, however, that males facing higher costs of protection, i.e. less powerful or injured males, are less likely to protect infants, or to protect with lesser intensity, even if they have mated extensively with the mother. As in the case of the newly dominant male, this approach leads to the same conclusion as the one that maximises his mean fitness (assuming that $s > c$).

CONCLUSIONS FROM THE MALE-DECISION MODEL

The main conclusion from this analysis is that female polyandry and mating during non-fertile periods serve to raise the estimated paternity probability (p') values for all males involved, and make their sum exceed 1. Polyandry may also confuse p' to the point that the males' best course of action is to refrain from killing infants and to protect them if they have mated extensively, even if they actually did not sire them. Thus, female sexual behaviour may serve to overcome the constant-sum nature of the paternity game.

Unfortunately, developing convincing tests of these ideas is not easy, especially because incomplete infanticidal attacks by newly dominant males and the reduced rates of infanticide in multi-male species can have multiple sources, as noted above. Nonetheless, we believe it is meaningful to translate the conditions in which natural selection should favour one action over another into the actual decision-making processes of animals. However, if the interpretation advanced here is correct, a clear prediction follows. It is in the female's interest to keep individual males guessing as to the extent to which other males have also mated with her: the lower the perception of that frequency the higher a male's p', his estimate of his own paternity chances, should be. Hence, females should be likely to mate discreetly, especially with subordinate males. We will develop the same prediction from a consideration of mating conflict (see below).

To reach this conclusion we have assumed that the female can bias the values of these estimates in directions favourable to her, implying that natural selection was unable to equip males with better assessment rules than the ones they currently have. Thus, female sexuality may have been designed to withhold potentially useful information from males. Physiological work done from this perspective might be rewarding.

SEXUAL HARASSMENT AS AN EXPRESSION OF MATING CONFLICT

Mating conflict between the sexes is now recognised as an intrinsic part of sexual selection (Hammerstein & Parker, 1987). One aspect of it concerns intersexual conflict over the identity of each individual's mates. Especially in some mammals, this kind of mating conflict has found expression in sexual coercion of females by stronger males. Smuts and Smuts (1993) defined sexual coercion and considered two components: infanticide and sexual harassment. Many students of primate behaviour have reported harassment or aggressive restriction of movements of sexually active (oestrous) females by males, especially high-ranking ones, sometimes to the point that the female is injured or even killed (Smuts & Smuts, 1993; chimpanzees: Goodall, 1986, p. 452; Matsumoto-Oda & Oda, 1998; macaques: Chapais, 1983; Huffman, 1987, 1992; Manson, 1992, 1994; Soltis,

1999; hamadryas baboons: Kummer, 1995). Although infanticide is seen by some as an extreme form of harassment (Smuts & Smuts, 1993), the two are not often treated as being directly interrelated phenomena. Here, we develop a model to show that at least some of the sexual harassment in primates is directly linked to females' attempts to be polyandrous.

THEORY

Assume a multi-male group of a primate species in which females are vulnerable to infanticide by males. One male is dominant and will guard a female when she is in oestrus, but other males are around and also interested in mating with the female. It is in the female's interest to dilute the paternity chances of the dominant male in order to reduce the risk of infanticide by mating polyandrously with other males in the group, or occasionally even in an adjacent group, in case one of them takes over top dominance or ousts the current dominant (Hrdy, 1979; Hrdy & Whitten, 1987; Small, 1993; van Schaik et al., 1999; Soltis et al., 2000; Heistermann et al., 2001). One might expect that the dominant male would also benefit from reducing the infant's risk of infanticide.

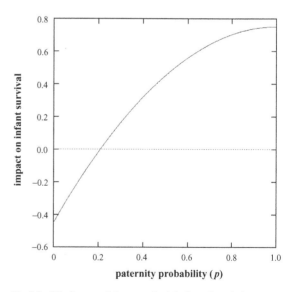

Fig. 8.1   The impact of the paternity (p) of a male on infant survival. The relationship has the form $g(p) = k - A(-1)^m \times (p - 1)^m$; the parameters chosen are $k = 0.75$, $A = 1.2$, $m = 2$. As the male's paternity increases, the impact varies from attack to tolerance to protection. At $p = 0$, the intercept on the y-axis is $(k - A)$, which is the maximum negative impact by a non-sire on the survival of the infant.

However, we will now show that the dominant male's probability of paternity that maximises his fitness is higher than that preferred by the female, and that he is therefore expected to attempt to prevent matings by the female with other males. Thus a male is not only in competition with other males, he may also have a conflict of interest with his mate(s).

A male's probability of paternity, p, is the long-term average proportion of infants sired by a male in similar conditions. We assume that the value of p is related to a male's assessment of it, although this relationship may be imprecise (see previous section). In general, a male's attitude toward the infant is a function of p, so that with increasing values of p, the male changes from attack (if given an opportunity to do so and if prospects for future mating access to the female exist), to indifference or tolerance, and finally to an increasingly strong tendency toward protection (Hrdy, 1979). We can represent the impact on infant survival of these changing attitudes as $g(p)$, with negative effects at low p and increasingly positive effects as p increases (see Fig. 8.1). We expect $g(p)$ to increase monotonically with p on domain [0, 1] and saturate at $p = 1$.

The function in Fig. 8.1 can be expressed as:

$$g(p) = k - A(-1)^m (p - 1)^m \qquad (8.2)$$

where m (integer, $m > 1$) is a shape parameter that determines how fast $g(p)$ rises and how soon it saturates as p increases, and k and A are positive constants (such that $A > k$). The value of k is the maximum positive impact of the likely sire on the infant's survival, whereas A can be seen as the maximum negative impact of a non-sire (the Y-intercept in Fig. 8.1 is at $k - A$). The assumption of $m > 1$ is critical for the result below but, in making it, we follow previous analyses of male parental care (Harada & Iwasa, 1996). It implies that, as p increases, the costs of male protection efforts are also likely to rise due to increased competition with other activities or increased risk of injury, leading to a relative slowdown in investment in, and thus effectiveness of, protection (in other words, one must assume protection efforts to saturate).

The question is at what value of *the dominant male's p* (here called q to avoid confusion) the fitness of the dominant male and of the female are maximised. In general, female fitness, F_F, is maximised when the infant's survival is maximised, assuming there are no other major effects on fitness, such as variation in male intrinsic genetic quality or relatedness to the female. On the other hand, the dominant male's fitness, F_{DM}, is maximised when $q \times F_F$ is maximised, whereas the fitness of a subordinate male (F_{SM}, or more precisely the highest-ranking among them) is maximised when

$(1 - q) \times F_F$ is maximised. We will now develop expressions for F_F, F_{DM} and F_{SM}.

We assume that infant survival is a function of q, but weighted for the dominant male's effective power $(1 - \varepsilon)$, where the parameter ε estimates the maximum strength of the strongest other male in the group or the vicinity (i.e. within the female's potential pool of mates), as well as a function of $(1 - q)$, i.e. the strongest subordinate male's paternity, p, but weighted for his effective power. This yields:

$$F_F = g(q)(1 - \varepsilon) + g(1 - q)\varepsilon \qquad (8.3)$$

$$F_{DM} = q\{F_F\} = q\{g(q)(1 - \varepsilon) + g(1 - q)\varepsilon\} \quad (8.4)$$

$$F_{SM} = (1 - q)\{g(q)(1 - \varepsilon) + g(1 - q)\varepsilon\} \qquad (8.5)$$

The strength parameter ε is best interpreted as the likelihood that this male will successfully challenge the current dominant in the near future (and thus also protect the infant in the future), hence $0 \leq \varepsilon \leq 1$. Its complement, $1 - \varepsilon$, represents the probability that the dominant male will be able to withstand challenges to his dominant position, and hence also his ability to protect the infant against infanticidal attacks by any of these males (or yet others).

Figure 8.2 illustrates the resulting mating conflict: the dominant male's fitness is maximised at a higher value of q compared to the value of q at which the female's fitness is

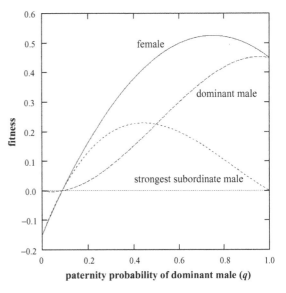

Fig. 8.2   The fitness of the female, dominant male and the strongest subordinate male as a function of the paternity q of the dominant male (Eqs. 8.2–8.5). The optimal fitness for each of the three is at different values of q, which cannot be satisfied simultaneously, reflecting a conflict.

maximised. We did not find analytical solutions to Eqs. 8.3–8.5 satisfying all values of m. However, for $m = 2$, the value of q at which the female's fitness is maximised ($q = \hat{q}_F$) is at $1 - \varepsilon$; in other words, she is expected to favour matings by other males in proportion to their relative strength. Still assuming $m = 2$, F_{DM}, the dominant male's fitness, is maximised at

$$q = \hat{q}_{DM} = (1 - \varepsilon) \left\{ \frac{2}{3} + \frac{1}{3} \sqrt{4 + \frac{3k}{A(1 - \varepsilon)^2} - \frac{3}{(1 - \varepsilon)}} \right\}$$

$$(8.6)$$

or at $q = 1$, whichever is the smaller.

It can easily be shown that $\hat{q}_{DM}$ is greater than $1 - \varepsilon$ (in other words, there is a conflict of interest between the female and the dominant male), provided the factor $(k/A) > \varepsilon(1 - \varepsilon)$. The factor $\varepsilon(1 - \varepsilon)$ reaches its maximum value at $\varepsilon = 0.5$, which means that if $(k/A) > 0.25$ the conflict is guaranteed for all values of ε; for $(k/A) < 0.25$ the mating conflict might still exist but is not guaranteed, since it will depend on the value of ε. For values of $m > 2$, we found mating conflict in all numerical solutions that we attempted for realistic values of m, k and A (i.e. for all $m \geq 2$, and k, A such that $(k/A) > 0.25$).[2]

Of course, this result raises the question as to the interpretation of $(k/A) > 0.25$. This inequality will often hold because protection is against infanticidal attacks; hence, if protection were totally effective, $k = A$. Since infanticide most often happens when the dominant male is eliminated or incapacitated, this implies that normally k is close to A, i.e. $k \gg 0.25A$. Hence, for all realistic ranges of values of all three parameters, there will be a mating conflict between the dominant male in the group or neighbourhood and the fertile female.

As implied by Fig. 8.2, a mating conflict exists between the female and the subordinate/peripheral males as well, because the dominant male's paternity at her maximum fitness, $\hat{q}_F$, is higher than that of the best subordinate male (the graph shows his optimum in terms of the dominant male's paternity; his own optimum is the complement of that value). However, the behavioural expression of this conflict is usually pre-empted by mating competition between the dominant and the subordinate males, forcing the latter to mate much less than they would otherwise do. Hence, the

subordinates never reach the zone in which females would prefer to mate less with them. Indeed, the subordinate males reach their optimum paternity, p, at a value much less than 1, because the dominant male is expected to attack infants obviously sired by them. On the other hand, when dominant males are not around, for instance because females are rather solitary, they too are expected to harass oestrous females.

In our equations we only incorporated the role of one subordinate male, the one most likely to succeed in challenging the dominant male in the near future. If only one, clearly stronger, young subordinate male resides in the group, this approach is acceptable, because one expects females to recognise such males. For instance, male baboons about to rise in dominance rank have different behavioural styles and endocrine profiles from others of similar rank, well before their actual rise (Virgin & Sapolsky, 1997). However, if there are more such males (and females recognise them as such), our treatment is conservative, and mating conflict between the female and the dominant male is even more intense. For instance, if two males have a reasonable chance of upsetting the dominant male (i.e. with $\varepsilon > 0$), the female's fitness will maximise at $\hat{q}_F = 1 - 2\varepsilon$.

PREDICTIONS AND EVIDENCE

A basic, albeit trivial, prediction is that, if no other males are in the mating pool, i.e. $\varepsilon = 0$, both the dominant male and the female will reach their optimum fitness at $q = 1$. Hence, in the absence of other males, no mating conflict is expected, and thus no harassment by the dominant male. When other males are present, however, various predictions can be made. We develop them here and also offer a preliminary evaluation of their fit with the primate literature.

(1) Harassment by dominant males

Observations of harassment of oestrous females, especially by high-ranking males, inspired the development of the model. Its fundamental prediction is that the conflict of interest between the female and the dominant male may find expression in a behavioural conflict if the male has the means to coerce the female, e.g. by coercive mate-guarding, and the female will try to escape in order to mate with other males (which, according to this model, she is likely to do).[3] This

---

[2] For the unrealistic case of $m = 1$ (see above), there is no mating conflict, with both sexes reaching their highest fitness at $q = 1$.

[3] Note that harassment of mating pairs by juveniles (for example, Drukker et al., 1991) and females (Linn et al., 1995) also occurs. Such harassment is of course not explained by the mating-conflict model.

is a strong prediction because it contrasts with the naïve expectation that the males most likely to force matings will be non-preferred, and hence generally subordinate.

One might object that attacking the female instead of the rival male of a mating or consorting pair is simply the least risky option for a dominant male, and is also likely to prevent the female from mating with a third male while he is engaged in fighting. However, there is abundant evidence for harassment of fertile females by males when the female is not actually mating or even near another male (see Smuts & Smuts, 1993; and see below). Moreover, if dominant males are not near the fertile female, lower-ranking males are also expected to employ harassment. Thus, in both chimpanzees (*Pan troglodytes*) and orangutans (*Pongo pygmaeus*), possessive mate-guarding and forced matings, respectively, by such non-dominant males are commonly observed (Tutin, 1979; Mitani, 1985; Goodall, 1986, pp. 457–64; Schürmann & van Hooff, 1986; Fox, 1998).

(2) Female polyandry in relation to the number of males

The mating-conflict model indicates that as the effective power of the dominant male $(1 - \varepsilon)$ declines, the conflict of interest between the dominant male and the female will increase (see Fig. 8.3). One common source of the reduced power of the dominant is an increased number of males in the mating pool, because this probably increases the strength of the strongest among them, and perhaps also because the larger number itself may wear out the dominant male, either directly or because coalitions are more likely (Bercovitch, 1989; Noë & Sluijter, 1990).

The model predicts that if the female wins the conflict, the concentration of paternity in the top-ranking male will decline as the number or strength of rivals increases. The female may achieve this outcome by mating more polyandrously during fertile cycles or by mating after conception.

The mating-conflict model thus offers an amendment to the explanation for the distribution of paternities over the available males provided by the Priority-of-Access (PoA) model (Altmann, 1962). According to the PoA model, a dominant male excludes other males from mating as long as there is only one female near ovulation. Lack of absolute concentration of paternity in dominant males would be owing to a reduction in male mating monopoly at times of overlap of female oestrous periods or high intruder pressure. The mating-conflict model claims that this pattern is due to

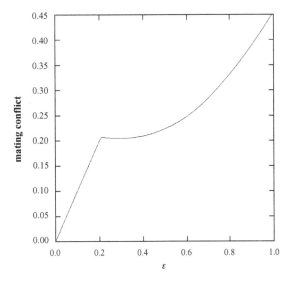

Fig. 8.3   The magnitude of the mating conflict ($\hat{q}_{DM} - \hat{q}_F$, see the text) between the dominant male and the female as a function of the strength of the strongest subordinate male (ε). The conflict between the dominant male and the female clearly increases with ε, albeit not monotonically.

active female polyandry, whereas PoA assumes females do not actively seek polyandry. The PoA model also assumes that male–male aggression serves only to monopolise access to females, and is therefore consistent with attacks on mating pairs, but not with male aggression targeted at the female in particular (see also van Noordwijk & van Schaik, this volume). To distinguish between the two models, detailed data on (changes in) male-dominance relations as well as matings and their timing (i.e. the number of females sexually active simultaneously) are needed.

If challenger males from within the group are more likely to succeed than recent immigrants, as is often found (Henzi & Lucas, 1980; Cheney, 1983; van Noordwijk & van Schaik, 1985, 2001; Robinson, 1988; Perry, 1998), a subsidiary prediction follows: females may prefer to live in multi-male groups, all other things being equal, because the 'insider' males pose less of an infanticide risk provided that they are granted a share of the matings. This idea has been suggested before but remains hard to test (cf. van Schaik, 1996; Nunn & van Schaik, 2000). However, our analysis here suggests that even the dominant male may find it in his interest to tolerate some unrelated subordinate males, because his optimum paternity may be slightly less than 1, assuming he can come close enough to this value in reality.

(3) Female polyandry in relation to potential change of male-dominance relations

The effective power of the dominant male may be reduced by increased strength of one or more of the subordinate males, increasing the risk of an effective challenge. If a female can recognise that the current top-ranking male is likely to be defeated before the birth of her infant, she should selectively decrease her matings with him. This scenario assumes that the top rank among males is acquired through challenge and not by succession, as in some macaque species with large groups and seasonal breeding (reviewed in van Noordwijk & van Schaik, this volume).

In many primate species, females are known to attempt to break away from the monopolisation of the dominant males and actively attempt to mate with subordinate or peripheral males (Hrdy, 1981; Small, 1993). The primate literature contains a few reports that females are more actively polyandrous when the dominance situation among the high-ranking males is not stable (Samuels et al., 1984; van Noordwijk, 1985; Janson, fide Manson, 1995; Manson et al., 1997; Alberts et al., 2003) or when the single male in the group is weak (Agoramoorthy & Hsu, 2000).

A more refined prediction is that we expect females to attempt to mate preferentially with those subordinate or peripheral males most likely to challenge successfully the current dominant male in the future. Hence, oestrous females should not randomly seek matings with males other than the dominant but show distinct preferences, which should be linked to the target males' prospects for future dominance. In several populations young maturing males are known to rise rapidly in rank and take over top rank, surpassing several males over a period of only a few months (e.g. Cheney, 1983; van Noordwijk & van Schaik, 1985, 1988, 2001; Hamilton & Bulger, 1990; Virgin & Sapolsky, 1997; Soltis et al., 2000). Since such challenges by maturing males are rather predictable, we expect the future top-ranking male (if already present in the group) to have a larger share of the matings than expected for his current dominance position. At least one study to date seems to confirm this point: Smith (1994) reports that for a large captive group of rhesus macaques, high male siring-success, attributed to female choice, preceded a rise to high dominance rank for young males.

(4) Surreptitious mating with subordinate males

The non-zero value of p that maximises the subordinate male's fitness leads to the prediction that it is in the interest of both the female and the subordinate male to mate as inconspicuously as possible in order to prevent the dominant male from adjusting his p estimate down, and withhold protection from the infant. We therefore expect that matings between females and subordinate males tend to take place out of sight of the dominant male, e.g. at the periphery or away from the group, and should less often be accompanied by calls. In order to distinguish this pattern from general mating competion, it should even happen if the dominant male is in visual contact but too far away to attack the pair effectively. We can further predict that inconspicuous mating with subordinates is even found in species without effective male harassment of females.

Despite a serious lack of quantitative data, it has been noted for several species that matings between females and subordinate males tend to occur rather surreptitiously (Pan troglodytes: Tutin, 1979; Goodall, 1986; Macaca fuscata: Huffman, 1992; M. mulatta: Berard et al., 1994; M. arctoides: Nieuwenhuijsen et al., 1986; M. sylvanus: Paul, 1989). One study focusing on matings in concealed places noted that lower-ranking males and their female partners were involved in matings outside visual contact with the rest of the group (M. fascicularis: Gygax, 1995), as predicted.

In many species, females give a specific copulation call, to which other males respond by looking at the pair (van Noordwijk, 1985) or even attacking them (Oda & Masataka, 1995). In at least one study a tendency was found for matings with subordinate males to be quiet (M. thibetana: Zhao, 1993), although another study found no effect of male identity on female calling-tendency (M. fascicularis: van Noordwijk, 1985). Hence, renewed examination of patterns in discreet matings may be worthwhile even in species without effective male harassment (as long as infanticide by males poses a risk).

SEXUAL HARASSMENT AND REPRODUCTIVE PHYSIOLOGY

THE DISTRIBUTION OF SEXUAL HARASSMENT

The model presented above confirms the existence of a conflict of interest between the breeding female and the group's males. The conflict with the dominant male(s) is most likely to be expressed because this male (or males) will try to monopolise the female most of the time, thus preempting any conflict with the other males. The question, of course, is who wins this conflict? Whenever male harassment occurs, females can only win at serious costs. Coercive mate-guarding makes it more difficult, and more costly, for the female to seek the matings with subordinate and peripheral males needed to achieve the distribution of p' values optimal to her. Harassment is more common where females are less powerful, both physically and socially (Smuts & Smuts, 1993). Giving in to coercion makes the female more vulnerable to infanticide, and if this risk is sufficiently increased, we expect that natural selection has produced physiological or behavioural tendencies in females to reduce the dominant male's monopolisation potential, provided that their costs do not exceed the gains of reduced infanticide rates.

Since it is not a priori clear what forms these female counter-strategies can take, we will first examine the taxonomic distribution of sexual harassment of sexually active (oestrous) females in primates, building on existing reviews (especially Smuts & Smuts, 1993; Dixson, 1998), and then search for derived reproductive features in the taxa with harassment.

In order to capture the variation in male behaviours directed at sexually active or 'oestrous' females, we propose the following categories of sexual harassment (defined as aggression by sexually mature males against sexually active or 'oestrous' females):

(1) No sexual harassment, nor any attempts, reported.
(2) Sexual harassment attempts reported, but only in the direct mating context and mostly ineffective, i.e. the female wards off the male or counter-attacks, and the male is unable to prevent the female from moving away or mating with others.
(3) Effective sexual harassment is observed in both the direct mating context and of oestrous females in general, as evidenced by coercive mate-guarding and physical attacks on the oestrous female followed by submission. Male behaviours sometimes include bites that result in wounding or even death, forced matings, and attacks on females when mating with other males.

The second category is needed because in several species (including many non-primate mammals), males show aggression toward females in the mating context that is not accompanied by any attempts at coercive mate-guarding. The presence of these behaviours suggests that aggression may be an integral component of mating in these species. The extent to which these attempts at harassment would constitute *de facto* harassment depends on the degree to which they stop females from achieving the preferred degree of polyandry. The literature is understandably vague on this,

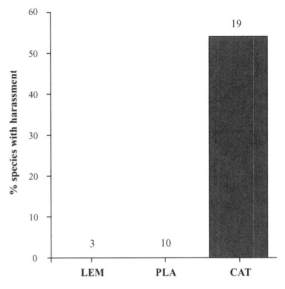

Fig. 8.4   The incidence of effective sexual harassment (which may include injury of females and forced matings) among primate species vulnerable to infanticide by males in three radiations (LEM = Lemuroidea; PLA = Platyrrhini – New World primates; CAT = Catarrhini – Old World primates). Based on literature review (starting with Smuts & Smuts, 1993), available from the authors upon request. Number of species with information indicated above columns.

but it is our impression that females can still mate with other males. In any case, it is useful to keep this second category separate from the category of effective harassment, where males also attack oestrous females outside the direct mating context, or force matings.

For the present purpose, we limit our review to species vulnerable to infanticide. A species is considered vulnerable if infanticide by males is reported for it, or if it has a life history that makes the females of the species vulnerable to such attacks, or both (van Noordwijk & van Schaik, 2000; van Schaik, 2000b). The advantage of this definition is that we need not rely only on reports of infanticide, which tend to be rare. As required, all species known to have infanticide are also predicted by the life-history measure to be vulnerable.

Although data on harassment are still very incomplete, a few clear patterns emerge. Consistent with earlier compilations, we see no evidence for effective harassment in *Lemuroidea* (lemurs) or in *Platyrrhini* (New World primates), but many reports for the *Catarrhini* (Old World primates; Fig. 8.4). Examples of adult females being able systematically to elicit submission from adult males are most

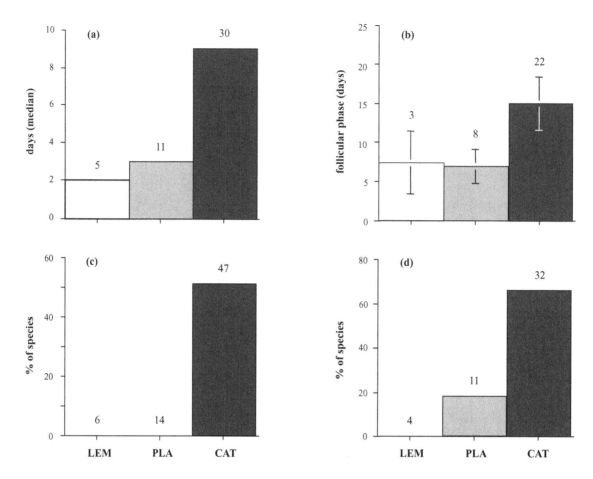

Fig. 8.5   Variation among species vulnerable to infanticide by males in three primate radiations (see Fig. 8.4 for abbreviations) for a variety of sexual features: (a) the median duration of the mating (oestrous) period ($P < 0.001$); (b) the mean duration of the follicular phase (in days) of the ovarian cycle ($P < 0.01$); (c) the percentage of species showing exaggerated sexual swellings ($P < 0.001$); (d) the percentage of species with mating calls by females ($P < 0.01$). Tested with Kruskal–Wallis one-way Anova. Numbers of species indicated above columns: (a), (c), and (d) based on data compiled in van Schaik *et al.*, 1999; (b) on data compiled by van Schaik *et al.*, 2000.

commonly found in lemurs, in several monomorphic New World primates and in the few pair-living Old World primates (Kappeler, 1993; Strier, 1994). In all these species, males attack rival males but we see, at most, only attempts at harassment of females, e.g. in lemurs (Pereira & Weiss, 1991; Sauther, 1991; Brockman, 1999), and females often counter-attack and are able to choose their mates (Richard, 1992). Likewise, only in some New World primates do we

see evidence of coalitions of males being required to inspect a female's sexual state (in *Saimiri oerstedi*, *Ateles*, *Lagothrix*: Boinski, 1987; Smuts & Smuts, 1993). Remarkably, effective sexual harassment is absent among New World primates, even the more dimorphic ones: for instance, female *Alouatta palliata* successfully rebuff attempts at forced matings by males (Jones, 1985).

Conversely, effective sexual harassment by males is reported for many Old World primate species. In this lineage, we also encounter many records of males generally harassing females, or females requiring coalitions in order to defend themselves against harassing males (Smuts, 1987; Smuts & Smuts, 1993). However, by no means all Old World primate species show sexual harassment by males.

Our sample is as yet too incomplete to allow analysis of the interspecific pattern in sexual harassment in relation to the known risk factors (larger male body size or weaponry, lack of female allies: Smuts & Smuts, 1993). Lemurs are largely monomorphic (Kappeler, 1991) and New World primates are far less dimorphic overall than Old World primates (e.g. van Schaik *et al.*, 2000). However, additional factors are probably involved because clearly dimorphic New World primate species and even some highly dimorphic Old World species (e.g. *Erythrocebus patas*) fail to produce clear evidence of sexual harassment. Hence, as yet unidentified additional factors are also involved in shaping interspecific variation in the occurrence and intensity of sexual harassment.

FEMALE COUNTER-STRATEGIES TO MALE SEXUAL HARASSMENT

Having found that effective sexual harassment of females by males is limited to Old World primates, we employ a two-step procedure to identify derived features of sexuality in Old World primates that possibly represent counter-strategies to the greater risk of sexual harassment. The first step is a systematic comparison of female sexuality in species vulnerable to infanticide in the three primate radiations compared above, in order to see in which respect Old World primates stand out. The second step is to conduct comparisons within the Old World primates with variation in coercive abilities or infanticide risk.

The mating period (oestrous period, i.e. period between first and last mating) within an ovarian cycle is at least three times longer in the average Old World primate than in the other radiations (Fig. 8.5a). Moreover, in Old World primates, but not in others, the mating period increases with the number of males that a female ends up mating with (Fig. 8.6), consistent with the expectation that only in some Old World primates would females find it difficult to escape from the monopolisation of the dominant male in order to mate with other males.

The long mating periods in Old World primates are made possible by a change in the ovarian cycle, in that their follicular phases are about twice as long as in the other radiations (Fig. 8.5b). Comparisons within the Old World primates suggested that follicular phases are longer where the sexes are more dimorphic in body size or weaponry; specifically, we found evidence for correlated evolution between the length of the follicular phase and the degree of canine dimorphism (van Schaik *et al.*, 2000), implying that the factor explaining variation among radiations is also at work within the Old World primates lineage. The long periods of sexual activity shown by Old World primate females require either that sperm remains viable in the female reproductive tract for long periods of time, or that ovulation is less tightly linked to the visual, olfactory or behavioural signals than usually

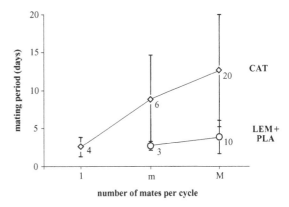

Fig. 8.6 Mean (±s.d.) duration of the mating ('oestrous') period in species vulnerable to infanticide by males in relation to the degree of female polyandry, in those vulnerable to male sexual harassment (CAT) and those not vulnerable (LEM and PLA). Degree of polyandry: 1 – mating with a single male in > 90 per cent of cycles; m – mating with multiple males in 10–50 per cent of cycles; and M – mating with multiple males in more than 50 per cent of the cycles (cf. van Noordwijk & van Schaik, 2000). Sample sizes (number of species) indicated at each point. Data on mating period taken from van Schaik *et al.* (1999). Rank correlation is significant (P < 0.01) for catarrhines.

assumed, or both (cf. Martin, 1992). There is no work in support of the former possibility, but recent work has supported the second idea. Nunn (1999) and van Schaik *et al.* (2000) review endocrinological work showing that ovulation is only poorly linked to these signals and thus should be rather unpredictable for males. Recent work by Heistermann *et al.* (2001) on hanuman langurs confirmed this prediction, showing that ovulation can take place with approximately equal probability at any time in the variable period of female sexual activity and that males do not recognise the female's time of ovulation, either in terms of behaviour or the degree to which paternity is concentrated in the dominant.

If unpredictable timing of ovulation was favoured by natural selection because it allowed the female to have longer mating periods of variable length, then there should be less need for it among the species that, while vulnerable to infanticide, are not vulnerable to male sexual harassment, i.e. lemurs and New World primates. There are as yet no detailed studies of the kind done on langurs for any of these species. However, we can get an indication of the extent to which ovulation is unpredictable by examining the variability in the duration of the follicular phase of the ovarian cycle (in most species, males can recognise when the luteal phase is well under way; e.g. Dixson, 1998). Data on variation in

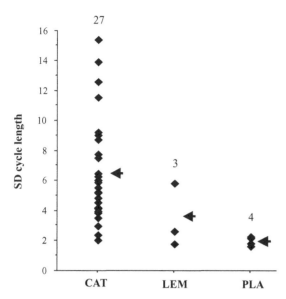

Fig. 8.7   The standard deviation (SD) of the length of the ovarian cycle in species vulnerable to infanticide in three primate radiations (see Fig. 8.4). Number of species indicated above columns. Arrows indicate averages. Tested with Kruskal–Wallis one-way Anova (P < 0.01).

follicular phase length are scarce, but because the variability in the follicular phase exceeds that in the luteal phase (on average by about a factor of two: see compilation in van Schaik *et al.*, 2000), variability in the total duration of the ovarian cycle may provide a rough indication of unpredictable ovulation. Figure 8.7 shows that the standard deviation of the length of the ovarian cycle in our sample (based on Hayssen *et al.*, 1993 and a compilation by K. Hodges & U. Möhle, unpublished) is indeed higher for Old World primates than it is for the two other radiations (including only species vulnerable to infanticide). The differences among them are significant (Kruskal–Wallis test, H [2] = 11.37; P < 0.01), and in the expected direction.

Old World primates differ from the other radiations in two more aspects of sexual behaviour: sexual swellings and copulation calls. The first has generated much interest from Darwin (1876) onwards (e.g. Dixson, 1983; Hrdy, 1997; Nunn, 1999): females of several species have exaggerated sexual swellings (Fig. 8.5c). Among Old World primates these swellings are found only in species that are actually or potentially polyandrous (cf. Clutton-Brock & Harvey, 1976), and among them predominantly in non-seasonal breeders (van Schaik *et al.*, 1999). The restriction to Old World primates and the predominance among non-seasonal breeders

is consistent with the hypothesis that these swellings are needed where the risk of harassment by dominant males is particularly serious (females in seasonal breeders can move more freely from male to male because they tend to be sexually active simultaneously). Thus, as argued by Nunn's (1999) graded-signal hypothesis, sexual swellings function not only to attract the dominant male(s) during their maximum size (when the probability of ovulation is highest), but through their exaggeration they can also attract lower-ranking or peripheral males when the dominant males are less attracted owing to the costs to these males of mate guarding for too long. For further discussion of exaggerated swellings see Zinner *et al.* (this volume).

We also find that females' mating or copulation calls are more common among Old World primates (Fig. 8.5d). Mating calls are given during or after ejaculation. They are not to be confused with 'oestrous calls' given when the female is receptive before actually mating (generally interpreted as alerting all males within hearing distance to her condition), or with 'distress' calls given by female northern elephant seals or chickens when mounted by a subordinate male or at a time when the female is not receptive, often followed by interference by a higher-ranking male (Cox & LeBoeuf, 1977; Pizzari & Birkhead, 2000). Mating calls may alert other males that the female has mated, and may thus attract them (O'Connell & Cowlishaw, 1994; van Schaik *et al.*, 1999). Mating calls may be a graded signal, the quality of which varies throughout the cycle (Semple & McComb, 2000) and thus somehow indicates the probability of ovulation, much like swellings (the distribution of mating calls across species largely mirrors that of swellings: van Schaik *et al.*, 1999). Females may also use mating calls tactically, calling only or more often with particular males than with others (see above). More detailed field studies and comparisons between species, including several New World primates, may help us determine whether their function is, as proposed here, linked to the reduction of male sexual harassment.

Thus we found strong evidence for a link between the distribution of physiological, morphological and behavioural characteristics of female sexuality with the occurrence of harassment by males who are physically stronger.

LINEAGE DIFFERENCES IN SEXUALITY

The significant differences between radiations in various features serve to show the existence of a grade shift. Explaining grade shifts is inherently difficult because many factors have changed in parallel. For instance, among Old World primates, we see, relative to the two other radiations, the presence of terrestrial adaptations, larger mean body size (relative to extant species in the other lineages), larger mean group size, a general reduction in reliance on olfactory communication, and greater tendencies toward folivory (relative to New World primates), etc. It is of course entirely possible that one or a combination of these factors facilitated the evolution of longer follicular phases or exaggerated swellings in Old World primates. However, the variation *within* the lineage, with respect to the duration of follicular phases in relation to male harassment potential, to the duration of mating period in relation to potential for polyandry, and to the conditions in which exaggerated swellings occur, suggests that subsequent evolution within the lineage in these traits did correlate with variation in harassment potential. Thus, variation within the lineage is in the same direction as that between the lineages. We therefore believe that the idea that the combination of male harassment and infanticide risk facilitated the evolution of these derived sexual features is a viable working hypothesis, worthy of further evaluation.

DISCUSSION AND PROSPECTS

It should be stressed that we do not propose that primate sexuality is moulded by infanticide risk alone: there are numerous selective factors on what is perhaps the behaviour most tightly linked to fitness, and thus most directly responsive to natural selection (Dixson, 1998). Birkhead and Kappeler (this volume) discuss additional hypotheses for female polyandry in primates, but these are not necessarily incompatible with the infanticide-avoidance function.

Nonetheless, the aim of this chapter has been to evaluate Hrdy's (1979) predictions concerning the impact of infanticide risk on sexual behaviour and reproductive physiology in primates. The original predictions are upheld but a link to sexual harassment can also be added to the list. Our preliminary exploration suggests that harassment has impacted female reproductive behaviour and physiology but, as always, the role of comparative work is best regarded as complementary to focused observational or experimental studies. The studies used to evaluate the hypotheses were not designed with these ideas in mind, and confounding factors can thus rarely be excluded. We therefore make the following suggestions for future work.

More detailed studies of female sexual behaviour could incorporate the role of social dynamics in the optimal balance between paternity concentration in dominants and dilution or manipulation of its assessment. For example, if new dominants are always recent immigrants, very little paternity dilution is expected, but if new dominants are always long-term residents, much more paternity dilution is expected. Strong tests would correct for the expectation, based on the Priority-of-Access model. As to male decision making it will be difficult to improve on the behavioural monitoring of Borries et al. (1999) on hanuman langurs, although additional work on different species would be worthwhile. With respect to mating conflict, detailed studies such as those of Gygax (1995) on the tendency of particular individuals or pairs to engage in matings in concealed locations or on the females' tendency to give copulation calls can be conducted.

Future work on female reproductive physiology could focus on three issues. First, detailed work is needed on the sexual behaviour of females in species in the different radiations along the lines of the study by Heistermann et al. (2001), where social and sexual behaviour, female ovarian state and paternity are all recorded simultaneously. Second, we should study (if necessary by experimental manipulation) the impact of male representation in the social unit on female reproductive physiology and behaviour in species of different lineages. Third, more fine-grained comparative work on the relationship between sexual harassment and possible coercion indicators in Old World primates, as well as detailed studies of highly dimorphic New World primates, may help us understand the distribution of male sexual harassment in relation to female reproductive physiology.

SUMMARY AND CONCLUSIONS

We explored the hypothesis that the vulnerability of females to infanticide by males has affected female sexual behaviour and given rise to sexual harassment by males in many primates. After establishing the adaptive nature of infanticidal behaviour for males who attack infants they did not father, we briefly reviewed sexual counter-strategies by females (polyandrous mating, mating during pregnancy). We then addressed two main issues. First, because paternity distribution is a constant-sum game, a female faces a considerable challenge. We argued that the female's sexual behaviour serves to produce paternity estimates among the various males that add up to more than 1. Moreover, paternity uncertainty may force males into avoiding the costliest mistakes, making them refrain from attacking infants they probably did not sire and making them protect those they may have sired. Second, we proposed an explanation that links infanticide with the common observation that high-ranking males often harass oestrous females. We developed a model that shows that a mating conflict exists between the female and the dominant male(s), and examined several detailed predictions about mating behaviour. Mating conflict can find expression in harassment if males can coerce females. This model led to several predictions, and a preliminary evaluation showed a good fit with existing data on primate sexual behaviour. We also noted that effective sexual harassment among primates is probably limited to Old World primates, in which it may have produced an arms race, leading to further changes in female sexual behaviour and physiology in this lineage, including longer mating periods, longer follicular phases with more unpredictable ovulation, exaggerated sexual swellings and perhaps more extensive female copulation vocalisations.

ACKNOWLEDGEMENTS

We thank Keith Hodges and Ulrike Möhle for allowing us to use their compilation of data on primate ovarian cycles; Helen Zayac for compiling many of the harassment references; and Tim Birkhead, Patty Gowaty, Sarah Hrdy, Peter Kappeler, Rebecca Lewis, and Sagar Pandit for helpful comments.

REFERENCES

Agoramoorthy, G. & Hsu, M. J. 2000. Extragroup copulation among wild red howler monkeys in Venezuela. *Folia Primatologica*, **71**, 147–51.

Alberts, S. C., Watts, H. & Altmann, J. 2003. Queuing and queue jumping: long-term patterns of dominance rank and mating success in male savannah baboons. *Animal Behaviour*, **65**, 821–40.

Altmann, S. A. 1962. A field study of the sociobiology of the rhesus monkey, *Macaca mulatta*. *Annals of the New York Academy of Sciences*, **102**, 338–435.

Angst, W. & Thommen, D. 1977. New data and a discussion of infant killing in Old World monkeys and apes. *Folia Primatologica*, **27**, 198–229.

Bartlett, T. Q., Sussman, R. W. & Cheverud, J. M. 1993. Infant killing in primates: a review of observed cases with specific reference to the sexual selection hypothesis. *American Anthropologist*, **95**, 958–90.

Berard, J. D., Nürnberg, P., Epplen, J. T. & Schmidtke, J. 1994. Alternative reproductive tactics and reproductive success in male rhesus macaques. *Behaviour*, **127**, 177–201.

Bercovitch, F. B. 1989. Body size, sperm competition, and determinants of reproductive success in male savanna baboons. *Evolution*, **43**, 1507–21.

Boggess, J. 1984. Infant killing and male reproductive strategies in langurs *Presbytis entellus*. In *Infanticide: Comparative and Evolutionary Perspectives*, ed. G. Hausfater & S. B. Hrdy. New York, NY: Aldine de Gruyter, pp. 283–310.

Boinski, S. 1987. Birth synchrony in squirrel monkeys, *Saimiri oerstedi*. *Behavioral Ecology and Sociobiology*, **21**, 393–400.

Borries, C. 1997. Infanticide in seasonally breeding multimale groups of Hanuman langurs, *Presbytis entellus* in Ramnagar South Nepal. *Behavioral Ecology and Sociobiology*, **41**, 139–50.

2000. Male dispersal and mating season influxes in Hanuman langurs living in multi-male groups. In *Primate Males: Causes and Consequences of Variation in Group Composition*, ed. P. M. Kappeler. Cambridge: Cambridge University Press, pp. 146–58.

Borries, C., Launhardt, K., Epplen, C., Epplen, J. T. & Winkler, P. 1999. Males as infant protectors in Hanuman langurs, *Presbytis entellus*, living in multimale groups: defence pattern, paternity, and sexual behavior. *Behavioral Ecology and Sociobiology*, **46**, 350–6.

Brockman, D. K. 1999. Reproductive behavior of female *Propithecus verreauxi* at Beza Mahafaly, Madagascar. *International Journal of Primatology*, **20**, 375–98.

Chapais, B. 1983. Reproductive activity in relation to male dominance and the likelihood of ovulation in rhesus monkeys. *Behavioral Ecology and Sociobiology*, **12**, 215–28.

Cheney, D. L. 1983. Proximate and ultimate factors related to the distribution of male migration. In *Primate Social Relationships: An Integrated Approach*, ed. R. A. Hinde. Oxford: Blackwell Scientific Publications, pp. 241–9.

Clutton-Brock, T. H. & Harvey, P. H. 1976. Evolutionary rules and primate societies. In *Growing Points in Ethology*, ed. P. P. G. Bateson & R. A. Hinde. Cambridge: Cambridge University Press, pp. 195–237.

Clutton-Brock, T. H. & Parker, G. A. 1995. Sexual coercion in animal societies. *Animal Behaviour*, **49**, 1345–65.

Colmenares, F. & Gomendio, M. 1988. Changes in female reproductive condition following male take-overs in a colony of hamadryas and hybrid baboons. *Folia Primatologica*, **50**, 157–74.

Cords, M. 1984. Mating patterns and social structure in red-tail monkeys, *Cercopithecus ascanius*. *Zeitschrift für Tierpsychologie*, **64**, 313–29.

1988. Mating systems of forest guenons: a preliminary review. In *A Primate Radiation: Evolutionary Biology of the African Guenons*, ed. A. Gautier-Hion, F. Bourliere, J. P. Gautier & J. Kingdon. Cambridge: Cambridge University Press, pp. 323–39.

Cowlishaw, G. & Dunbar, R. I. M. 1991. Dominance rank and mating success in male primates. *Animal Behaviour*, **41**, 1045–56.

Cox, C. R. & LeBoeuf, B. J. 1977. Female incitation of male competition: a mechanism in sexual selection. *American Naturalist*, **111**, 317–35.

Crockett, C. M. & Sekulic, R. 1984. Infanticide in red howler monkeys, *Alouatta seniculus*. In *Infanticide: Comparative and Evolutionary Perspectives*, ed. G. Hausfater & S. B. Hrdy. New York, NY: Aldine de Gruyter, pp. 173–91.

Darwin, C. 1876. Sexual selection in relation to monkeys. *Nature*, **15**, 18–19.

Dixson, A. F. 1983. Observations on the evolution and behavioral significance of 'sexual skin' in female primates. *Advances in the Study of Behavior*, **13**, 63–106.

1998. *Primate Sexuality: Comparative Studies of the Prosimians, Monkeys, Apes, and Human Beings*. Oxford: Oxford University Press.

Dolhinow, P. J. 1977. Normal monkeys? *American Scientist*, **65**, 266.

Drukker, B., Nieuwenhuijsen, K., van der Werff ten Bosch, J. J., van Hooff, J. A. R. A. M. & Slob, A. K. 1991. Harassment of sexual interactions among stumptail macaques, *Macaca arctoides*. *Animal Behaviour*, **42**, 171–82.

Fox, E. A. 1998. The function of female mate choice in the Sumatran orangutan (*Pongo pygmaeus abelii*). Ph.D. thesis, Duke University.

Gowaty, P. A. 1997. Sexual dialectics, sexual selection, and variation in mating behavior. In *Feminism and Evolutionary Biology: Boundaries, Intersections, and Frontiers*, ed. P. A. Gowaty. New York, NY: Chapman & Hall, pp. 351–84.

Goodall, J. 1986. *The Chimpanzees of Gombe*. Cambridge, MA: Harvard University Press.

Gust, D. A. 1994. Alpha-male sooty mangabeys differentiate between females' fertile and their postconception maximal swellings. *International Journal of Primatology*, **15**, 289–302.

Gygax, L. 1995. Hiding behaviour of long-tailed macaques, *Macaca fascicularis*. I. Theoretical background and data on mating. *Ethology*, **101**, 10–24.

Hamilton, W. J. & Bulger, J. B. 1990. Natal male baboon rank rises and successful challenges to resident alpha males. *Behavioral Ecology and Sociobiology*, **26**, 357–62.

Hammerstein, P. & Parker, G. A. 1987. Sexual selection: games between the sexes. In *Sexual Selection: Testing the Alternatives*, ed. J. W. Bradbury & M. B. Andersson. New York, NY: John Wiley, pp. 119–42.

Harada, Y. & Iwasa, Y. 1996. Female mate preference to maximize paternal care: a two-step game. *American Naturalist*, **147**, 996–1027.

Hausfater, G. & Hrdy, S. B. 1984. *Infanticide: Comparative and Evolutionary Perspectives*. New York, NY: Aldine de Gruyter.

Hayssen, V., van Tienhoven, A. & van Tienhoven, A. 1993. *Asdell's Patterns of Mammalian Reproduction: A Compendium of Species-Specific Data*. Ithaca, NY: Comstock Publishing Associates.

Heistermann, M., Ziegler, T., van Schaik, C. P. *et al.* 2001. Loss of oestrus, concealed ovulation and paternity confusion in free-ranging Hanuman langurs. *Proceedings of the Royal Society of London, Series B*, **268**, 2445–51.

Henzi, S. P. & Lucas, J. W. 1980. Observations on the inter-troop movement of adult vervet monkeys, *Cercopithecus aethiops*. *Folia Primatologica*, **33**, 220–35.

Hrdy, S. B. 1974. Male–male competition and infanticide among the langurs *Presbytis entellus* of Abu, Rajasthan. *Folia Primatologica*, **22**, 19–58.

1979. Infanticide among animals: a review, classification, and examination of the implications for the reproductive strategies of females. *Ethology and Sociobiology*, **1**, 13–40.

1981. *The Woman that Never Evolved*. Cambridge, MA: Harvard University Press.

1997. Raising Darwin's consciousness: female sexuality and the prehominid origins of patriarchy. *Human Nature*, **8**, 1–49.

2000a. Foreword. In *Infanticide by Males and Its Implications*, ed. C. P. van Schaik & C. H. Janson. Cambridge: Cambridge University Press, pp. xi–xiv.

Hrdy, S. B. & Whitten, P. L. 1987. Patterning of sexual activity. In *Primate Societies*, ed. B. B. Smuts, D. L. Cheney, R. M. Seyfarth, R. W. Wrangham & T. T. Struhsaker. Chicago, IL: University of Chicago Press, pp. 370–84.

Huffman, M. A. 1987. Consort intrusion and female mate choice in Japanese macaques, *Macaca fuscata*. *Ethology*, **75**, 221–34.

1992. Influences of female partner preference on potential reproductive outcome in Japanese macaques. *Folia Primatologica*, **59**, 77–88.

Janson, C. H. 1986. The mating system as a determinant of social evolution in capuchin monkeys (*Cebus*). In *Proceedings of the Xth International Congress of Primatology*, ed. J. Else & P. C. Lee. Cambridge: Cambridge University Press, pp. 169–79.

Janson, C. H. & van Schaik, C. P. 2000. The behavioral ecology of infanticide by males. In *Infanticide by Males and Its Implications*, ed. C. P. van Schaik & C. H. Janson. Cambridge: Cambridge University Press, pp. 469–94.

Jones, C. B. 1985. Reproductive patterns in mantled howler monkeys: estrus, mate choice, and copulation. *Primates*, **26**, 130–42.

Kappeler, P. M. 1991. Patterns of sexual dimorphism in body weight among prosimian primates. *Folia Primatologica*, **57**, 132–46.

1993. Female dominance in primates and other mammals. In *Perspectives in Ethology*, vol. 10, ed. P. P. G. Bateson, N. Thompson & P. H. Klopfer. New York, NY: Plenum Press, pp. 143–58.

Kummer, H. 1995. *In Quest of the Sacred Baboon: A Scientist's Journey*. Princeton, NJ: Princeton University Press.

Linn, G. S., Mase, D., Lafrançois, D., O'Keeffe, R. T. & Lifshitz, K. 1995. Social and menstrual cycle phase influences on the behavior of group-housed *Cebus apella*. *American Journal of Primatology*, **35**, 41–57.

Manson, J. H. 1992. Measuring female mate choice in Cayo Santiago rhesus macaques. *Animal Behaviour*, **44**, 405–16.

1994. Male aggression: a cost of female mate choice in Cayo Santiago rhesus macaques. *Animal Behaviour*, **48**, 473–5.

1995. Primate consortships: a critical review. *Current Anthropology*, **38**, 353–74.

Manson, J. H., Perry, S. & Parish, A. R. 1997. Nonconceptive sexual behavior in bonobos and capuchins. *International Journal of Primatology*, **18**, 767–86.

Martin, R. D. 1992. Female cycles in relation to paternity in primate societies. In *Paternity in Primates: Genetic Tests and Theories*, ed. R. D. Martin, A. F. Dixson & E. J. Wickings. Basel: Karger, pp. 238–74.

Matsumoto-Oda, A. & Oda, R. 1998. Changes in the activity budget of cycling female chimpanzees. *American Journal of Primatology*, **46**, 157–66.

Mitani, J. C. 1985. Mating behaviour of male orangutans in the Kutai Game Reserve, Indonesia. *Animal Behaviour*, **33**, 392–402.

Nieuwenhuijsen, K., de Neef, K. J. & Slob, A. K. 1986. Sexual behaviour during ovarian cycles, pregnancy and lactation in group-living stump-tail macaques, *Macaca arctoides*. *Human Reproduction*, **1**, 159–69.

Noë, R. & Sluijter, A. A. 1990. Reproductive tactics of male savanna baboons. *Behaviour*, **113**, 117–70.

Nunn, C. L. 1999. The evolution of exaggerated sexual swellings in primates and the graded signal hypothesis. *Animal Behaviour*, **58**, 229–46.

Nunn, C. L. & van Schaik, C. P. 2000. Social evolution in primates: the relative roles of ecology and intersexual conflict. In *Infanticide by Males and Its Implications*, ed. C. P. van Schaik & C. H. Janson. Cambridge: Cambridge University Press, pp. 388–420.

O'Connell, S. & Cowlishaw, G. 1994. Infanticide avoidance, sperm competition and mate choice: the function of copulation calls in female baboons. *Animal Behaviour*, **48**, 687–94.

Oda, R. & Masataka, N. 1995. Function of copulatory vocalizations in mate choice by females of Japanese macaques, *Macaca fuscata*. *Folia Primatologica*, **64**, 132–9.

Parmigiani, S. & vom Saal, F. S. (eds.) 1994. *Infanticide and Parental Care*. New York, NY: Harwood.

Paul, A. 1989. Determinants of male mating success in a large group of Barbary macaques, *Macaca sylvanus*, at Affenberg Salem. *Primates*, **30**, 461–76.

1997. Breeding seasonality affects the association between dominance and reproductive success in non-human male primates. *Folia Primatologica*, **68**, 344–9.

Paul, A., Preuschoft, S. & van Schaik, C. P. 2000. The other side of the coin: infanticide and the evolution of affiliative male–infant interactions in Old World primates. In *Infanticide by Males and Its Implications*, ed. C. P. van Schaik & C. H. Janson. Cambridge: Cambridge University Press, pp. 269–92.

Pereira, M. E. & Weiss, M. L. 1991. Female mate choice, male migration, and the threat of infanticide in ring-tailed lemurs. *Behavioral Ecology and Sociobiology*, **28**, 141–52.

Perrigo, G. & vom Saal, F. S. 1994. Behavioral cycles and the neural timing of infanticide and parental behavior in male house mice. In *Infanticide and Parental Care*, ed. S. Parmigiani & F. S. vom Saal. Chur, Switzerland: Harwood Academic Publishers, pp. 365–96.

Perry, S. 1998. Male–male social relationships in wild white-faced capuchins, *Cebus capucinus*. *Behaviour*, **135**, 139–72.

Pizzari, T. & Birkhead, T. R. 2000. Female feral fowl eject sperm of subdominant males. *Nature*, **405**, 787–9.

Resnik, M. D. 1987. *Choices: An Introduction to Decision Theory*. Minneapolis, MN: University of Minnesota Press.

Richard, A. 1992. Aggressive competition between males, female-controlled polygyny and sexual monomorphism in a Malagasy primate, *Propithecus verreauxi*. *Journal of Human Evolution*, **22**, 395–406.

Robinson, J. G. 1988. Demography and group structure in wedge-capped capuchin monkeys, *Cebus olivaceus*. *Behaviour*, **104**, 202–32.

Samuels, A., Silk, J. B. & Rodman, P. S. 1984. Changes in the dominance rank and reproductive behaviour of male bonnet macaques, *Macaca radiata*. *Animal Behaviour*, **32**, 994–1003.

Sauther, M. L. 1991. Reproductive behavior of free-ranging *Lemur catta* at Beza Mahafaly Special Reserve, Madagascar. *American Journal of Physical Anthropology*, **84**, 463–77.

Schürmann, C. L. & van Hooff, J. A. R. A. M. 1986. Reproductive strategies of the orang-utan: new data and a reconsideration of existing sociosexual models. *International Journal of Primatology*, **7**, 265–87.

Semple, S. & McComb, K. 2000. Perception of female reproductive state from vocal cues in a mammal species. *Proceedings of the Royal Society of London, Series B*, **267**, 707–12.

Small, M. F. 1993. *Female Choices: Sexual Behavior of Female Primates*. Ithaca, NY: Cornell University Press.

Smith, D. G. 1994. Male dominance and reproductive success in a captive group of rhesus macaques, *Macaca mulatta*. *Behaviour*, **129**, 225–42.

Smuts, B. B. 1987. Gender, aggression, and influence. In *Primate Societies*, ed. B. B. Smuts, D. L. Cheney, R. M. Seyfarth, R. W. Wrangham & T. T. Struhsaker. Chicago, IL: University of Chicago Press, pp. 400–12.

Smuts, B. B. & Smuts, R. W. 1993. Male aggression and sexual coercion of females in nonhuman primates and other mammals: evidence and theoretical implications. *Advances in the Study of Behavior*, **22**, 1–63.

Soltis, J. 1999. Measuring male–female relationships during the mating season in wild Japanese macaques, *Macaca fuscata yakui*. *Primates*, **40**, 453–67.

Soltis, J., Thomsen, R., Matsubayashi, K. & Takenaka, O. 2000. Infanticide by resident males and female counter-strategies in wild Japanese macaques, *Macaca fuscata*. *Behavioral Ecology and Sociobiology*, **48**, 195–202.

Sommer, V. 1994. Infanticide among the langurs of Jodhpur: testing the sexual selection hypothesis with a long-term record. In *Infanticide and Parental Care*, ed. S. Parmigiani & F. S. vom Saal. London: Harwood Academic Publishers, pp. 155–98.

Sommer, V., Srivastava, A. & Borries, C. 1992. Cycles, sexuality, and conception in free-ranging langurs. *American Journal of Primatology*, **28**, 1–28.

Steenbeek, R. 1996. What a maleless group can tell us about the constraints on female transfer in Thomas's langurs, *Presbytis thomasi*. *Folia Primatologica*, **67**, 169–81.

Stein, D. M. 1984. *The Sociobiology of Infant and Adult Male Baboons*. Norwood, NJ: Ablex.

Stephens, D. W. & Krebs, J. R. 1986. *Foraging Theory*. Princeton, NJ: Princeton University Press.

Strier, K. B. 1994. Myth of the typical primate. *Yearbook of Physical Anthropology*, **37**, 233–71.

Struhsaker, T. T. & Leland, L. 1985. Infanticide in a patrilineal society of red colobus monkeys. *Zeitschrift für Tierpsychologie*, **69**, 89–132.

Sugiyama, Y. 1966. An artificial social change in a hanuman langur troop *Presbytis entellus*. *Primates*, **7**, 41–72.

Sussman, R. W., Cheverud, J. M. & Bartlett, T. Q. 1995. Infant kiling as an evolutionary strategy: reality or myth? *Evolutionary Anthropology*, **3**, 149–51.

Swedell, L. 2000. Two takeovers in wild hamadryas baboons. *Folia Primatologica*, **71**, 169–72.

Symons, D. 1982. Another woman that never existed. *Quarterly Review of Biology*, **57**, 297–300.

Takahata, Y., Sprague, D. S., Suzuki, S. & Okayasy, N. 1994. Female competition, co-existence, and the mating structure of wild Japanese macaques on Yakushima island, Japan. In *Animal Societies: Individuals, Interactions and Organisation*, ed. P. J. Jarman & A. Rossiter. Kyoto: Kyoto University Press, pp. 163–79.

Tutin, C. E. G. 1979. Responses of chimpanzees to copulation, with special reference to interference by immature individuals. *Animal Behaviour*, **27**, 845–54.

van Noordwijk, M. A. 1985. Sexual behaviour of Sumatran long-tailed macaques, *Macaca fascicularis*. *Zeitschrift für Tierpsychologie*, **70**, 277–96.

van Noordwijk, M. A. & van Schaik, C. P. 1985. Male migration and rank acquisition in wild long-tailed macaques (*Macaca fascicularis*). *Animal Behaviour*, **33**, 849–61.

  1988. Male careers in Sumatran long-tailed macaques, *Macaca fascicularis*. *Behaviour*, **107**, 24–43.

  2000. Reproductive patterns in eutherian mammals: adaptations against infanticide? In *Infanticide by Males and Its Implications*, ed. C. P. van Schaik & C. H. Janson. Cambridge: Cambridge University Press, pp. 322–60.

  2001. Career moves: transfer and rank challenge decisions by male long-tailed macaques. *Behaviour*, **138**, 359–95.

van Schaik, C. P. 1996. Social evolution in primates: the role of ecological factors and male behaviour. *Proceedings of the British Academy*, **88**, 9–31.

2000a. Infanticide by male primates: the sexual selection hypothesis revisited. In *Infanticide by Males and Its Implications*, ed. C. P. van Schaik & C. H. Janson. Cambridge: Cambridge University Press, pp. 27–60.

2000b. Vulnerability to infanticide: patterns among mammals. In *Infanticide by Males and Its Implications*, ed. C. P. van Schaik & C. H. Janson. Cambridge: Cambridge University Press, pp. 61–71.

van Schaik, C. P. & Dunbar, R. I. M. 1990. The evolution of monogamy in large primates: a new hypothesis and some crucial tests. *Behaviour*, **115**, 30–62.

van Schaik, C. P. & Janson, C. H. (eds.) 2000. *Infanticide by Males and Its Implications*. Cambridge: Cambridge University Press.

van Schaik, C. P. & Kappeler, P. M. 1997. Infanticide risk and the evolution of male–female association in primates. *Proceedings of the Royal Society of London, Series B*, **264**, 1687–94.

van Schaik, C. P., van Noordwijk, M. A. & Nunn, C. L. 1999. Sex and social evolution in primates. In *Comparative Primate Socioecology*, ed. P. C. Lee. Cambridge: Cambridge University Press, pp. 204–40.

van Schaik, C. P., Hodges, J. K. & Nunn, C. L. 2000. Paternity confusion and the ovarian cycles of female primates. In *Infanticide by Males and Its Implications*, ed. C. P. van Schaik & C. H. Janson. Cambridge: Cambridge University Press, pp. 361–87.

Virgin, C. E. & Sapolsky, R. M. 1997. Styles of male social behavior and their endocrine correlates among low-ranking baboons. *American Journal of Primatology*, **42**, 25–39.

vom Saal, F. S. 1984. Proximate and ultimate causes of infanticide and parental behavior in male house mice. In *Infanticide: Comparative and Evolutionary Perspectives*, ed. G. Hausfater & S. B. Hrdy. New York, NY: Aldine de Gruyter, pp. 401–24.

Wallis, J. & Bettinger, T. 1993. Sexual activity in wild chimpanzees: a comparison of cyclic, pregnant, and lactating females with full anogenital swellings. *American Journal of Primatology*, **30**, 353.

Watts, D. P. 1991. Mountain gorilla reproduction and sexual behavior. *American Journal of Primatology*, **24**, 211–26.

Wrangham, R. W. 1979. On the evolution of ape social systems. *Social Sciences Information*, **18**, 334–68.

Zhao, Q.-K. 1993. Sexual behavior of Tibetan macaques at Mt. Emei, China. *Primates*, **34**, 431–44.

Zinner, D. & Deschner, T. 2000. Sexual swellings in female hamadryas baboons after male take-over: 'deceptive' swellings as a possible female counter-strategy against infanticide. *American Journal of Primatology*, **52**, 157–68.

9 • Post-copulatory sexual selection in birds and primates

TIM R. BIRKHEAD
Department of Animal and Plant Sciences
University of Sheffield
Sheffield, UK

PETER M. KAPPELER
Department of Behaviour and Ecology
German Primate Centre
Göttingen, Germany

INTRODUCTION

Post-copulatory sexual selection comprises two processes: sperm competition and cryptic female choice (Birkhead & Møller, 1998). Sperm competition is the inevitable consequence of females copulating with and being inseminated by more than one male during a single reproductive cycle: the sperm of different males compete to fertilise the female's ova. Cryptic female choice can similarly occur only when females are inseminated by more than one male, but – so far, at least – it does not appear to be inevitable. The evolutionary significance of these two processes has become apparent only since the 1970s; before that it was assumed that females in many taxa typically copulated with only a single male during any reproductive cycle. Recognition of the ubiquity of these post-copulatory mechanisms of sexual selection has created new opportunities to study the dynamic interactions between the sexes, which are being fuelled by the fundamental conflict between males and females over the maximisation of their reproductive success at yet additional frontiers (e.g. Johnstone & Keller, 2000). Furthermore, consideration of post-copulatory mechanisms can inform analyses and interpretation of pre-copulatory reproductive strategies of both sexes (Parker, 1970, 1984). In this chapter, we review mechanisms and consequences of post-copulatory sexual selection in two groups of animals for which some of the most detailed behavioural data exist, and which differ fundamentally in reproductive physiology and their modal social and mating systems: birds and primates.

Birds have long since been considered as models of monogamy because the majority of species breed as pairs comprising one male and one female (Lack, 1968). Despite this apparent monogamy, the molecular revolution and discovery of DNA-fingerprinting as a method of unambiguously assigning paternity (Burke, 1989), showed that appearances can be deceptive and that extra-pair paternity is widespread in birds (Birkhead & Møller, 1992a). Indeed, in the majority of species designated as socially monogamous, some individual females copulate with more than one male and produce broods fathered by either the extra-pair male or by both the social partner and extra-pair male (Petrie & Kempenaers, 1998). Initially, it was assumed that females were reluctant or at best acquiescent participants in extra-pair copulations, but subsequent detailed behavioural observations revealed that in many species females actively seek extra-pair copulations (Smith, 1988; Kempenaers et al., 1992). Thus, sperm competition and opportunities for cryptic female choice are abundant among birds.

Primates, in contrast, are by and large more gregarious than birds, and habitually promiscuous. Even though they exhibit a great diversity of social systems, females in the vast majority of species typically mate with more than one male during a single reproductive cycle (Hrdy & Whitten, 1987; van Schaik et al., 1999). This is true for the archetypical primate living in permanent groups composed of several adult males and females (Strier, 1994), as well as for many small nocturnal prosimians with a solitary social organisation (Sterling, 1993; Kappeler, 1997a; Pullen et al., 2000; Eberle & Kappeler, 2002). Polyandrous matings occur in many callitrichids, who live in groups with only one reproductive female and several adult males (Dunbar, 1995; Garber, 1997; Heymann, 2000). Only a minority of about 10 per cent of primate species lives in pairs, but new studies are beginning to reveal previously unknown or ignored social flexibility (Palombit, 1994; Reichard, 1995; Fuentes, 1999; Sommer & Reichard, 2000), and the first genetic paternity studies have revealed high rates of extra-pair copulations, even in species where social pairs predominate (*Cheirogaleus*: Fietz et al., 2000; *Phaner*: Schülke et al., 2004; but see Oka & Takenaka, 2001 for *Hylobates*). Apart from some pair-living species, only group-living primate females that are permanently associated

Sexual Selection in Primates: New and Comparative Perspectives, ed. Peter M. Kappeler and Carel P. van Schaik. Published by Cambridge University Press. © Cambridge University Press 2004.

with a single male can a priori be expected to mate monan-drously, but reports of groups in these species with a second male or influxes of non-resident males during the mating season are increasing (e.g. Cords, 1987, 2000; Robbins, 1995, 1999). In addition, females of some group-living species leave their social group to copulate with extra-group males (summarised in Agoramoorthy & Hsu, 2000; see also van Noordwijk & van Schaik, this volume). Thus, the vast majority of primate females typically mate with more than one male during a given reproductive cycle, so that post-copulatory selection should be particularly intense in this order.

Why so many birds and primates engage in multiple matings has been discussed in detail elsewhere (e.g. Hunter et al., 1993; Reynolds, 1996; van Schaik et al., 1999; Hrdy, 2000a; Tregenza & Wedell, 2000; Zeh & Zeh, 2001; Birkhead & Pizzari, 2002), so that we will not dwell on the adaptive significance of multiple matings here. Briefly, the benefits of extra-pair copulations to male birds were never in doubt, at least not conceptually, since it was obvious that extra-pair copulations would yield extra-pair offspring. However, demonstrating that male fitness was increased through extra-pair copulations took some time (reviewed in Birkhead, 1998a). The benefits which female birds accrue through engaging in extra-pair copulations are less obvious, and remain obscure. A suite of benefits which females might gain from multiple matings has been proposed, including fertility insurance, sperm replenishment and genetic diversity of offspring (e.g. Hrdy, 1981; Keller & Reeve, 1995; Jennions & Petrie, 2000; Birkhead & Pizzari, 2002). While there are examples consistent with each of these, there is no overall consensus about the benefits female birds gain from having more than one copulation partner (Birkhead, 1998a).

Male primates should also not forgo mating opportunities, including those with already-mated females, whenever the associated benefits exceed the costs. Aggression from rivals with formidable weapons, investment in high dominance rank and its underlying physical features, and the risk of acquiring sexually transmitted diseases are among the potential proximate costs associated with a given copulation (Clutton-Brock, 1985; Cowlishaw & Dunbar, 1991; Fawcett & Muhumuza, 2000; Nunn & Altizer, this volume; van Noordwijk & van Schaik, this volume). However, the benefits of attempting to mate with as many fertile females as possible, i.e. the probability of achieving fertilisation, are relatively high, because ovulation in many primates is concealed and receptive periods typically extend over days or even weeks (Sillén-Tullberg & Møller, 1993; Hrdy, 2000b; Heistermann et al., 2001; van Schaik et al., this volume; Zinner et al., this volume), so that effective monopolisation of a receptive female by a single male is typically difficult. In addition, multiple matings are actively sought by proceptive and receptive females in many primate species (Hrdy & Whitten, 1987; Hrdy, 1995, 2000a, b). Female primates may seek to mate with several males primarily because they can effectively reduce their unusually high risk of infanticide (van Schaik & Kappeler, 1997) by confusing paternity (Hrdy, 2000a; van Schaik, 2000; van Schaik et al., this volume). Other female benefits include increased mate choice and the opportunity to bias paternity (Hrdy, 2000a; van Schaik et al., this volume), as well as a reduction of sexual harassment (Drukker et al., 1991; van Schaik et al., 1999), whereas fertilisation assurance in seasonally reproducing species, and genetic diversity benefits in species with a litter size of one are presumably of little importance (Harcourt et al., 1995; but see Sauermann et al., 2001). These benefits of multiple matings apparently outweigh the higher risks for females of contracting sexually transmitted diseases (Nunn & Altizer, this volume).

When Parker (1970) identified the evolutionary significance of sperm competition, he recognised that it created conflicting evolutionary pressures: on the one hand selection favoured males who could protect their paternity, but it simultaneously favoured males who could overcome the paternity guards of other males and fertilise the eggs of already-mated females. Such conflicting evolutionary pressures can result in the rapid evolution of a diverse suite of adaptations to sperm competition, spanning behavioural, anatomical and physiological traits. In addition, the evolutionary arms race between the sexes over the control of reproduction contributes yet another layer of complexity to the evolutionary dynamics of these traits (Rice, 1996; Johnstone & Keller, 2000; Moore et al., 2001; Pitnick et al., 2001). Building on early work (Harcourt, 1996, 1997; Kappeler, 1997b; Birkhead, 1998a, b; Birkhead & Møller, 1998; Dixson, 1998a; Gomendio et al., 1998), we review more recent studies of mechanisms of sperm competition and cryptic female choice in birds and primates in order to: (1) directly compare sex-specific post-copulatory reproductive strategies between birds and non-human primates, (2) identify topics for future comparative research, and (3) contribute to recent efforts to push back taxonomic blinkers in evolutionary biology (Birkhead, 2002; Maestripieri & Kappeler, 2002).

ADAPTATIONS TO SPERM COMPETITION

In this section, we discuss behavioural and anatomical adaptations to sperm competition, and review relevant mechanisms of competition between sperm of different males in birds and primates. A much more detailed account of many examples and adaptations discussed below can be found in the superb review by Dixson (1998a) and the contributions to Birkhead and Møller (1998).

BEHAVIOURAL ADAPTATIONS TO SPERM COMPETITION

Mate guarding

There are two main behavioural adaptations to sperm competition: mate guarding and frequent copulation. In birds, mate guarding involves the male following his female partner during the time she is fertile, in order to prevent her from engaging in copulations with other males. The period of time that female birds are fertile is protracted, compared with most mammals, for two reasons. First, in birds each egg of a clutch is fertilised separately 24 hours (or occasionally 48 hours or longer) before the egg is laid. Second, female birds store sperm, often for several days (values for different species range from 8 to 42 days; Birkhead & Møller, 1992b) prior to the start of egg-laying. Therefore a species that can store sperm for 6 days and lays a clutch of 12 eggs will be fertile for the 6 days before egg-laying starts and until the day the penultimate eggs are laid: 17 days in total. In principle, a copulation occurring at any time during this period could potentially fertilise one or more of the eggs, and so it would pay males to guard throughout this period. In reality, the most effective time for a copulation appears to be in the day or so before the start of egg-laying (Colegrave et al., 1995; see also Birkhead & Møller, 1993a); and males of most species guard their partner most intensively at this time. There have been several alternative explanations for the close proximity between males and females around the time of egg-laying, but experiments in which guarding males are temporarily removed provide convincing evidence that the primary function of male proximity is to reduce the likelihood that females will copulate with other males (e.g. Dickinson, 1997).

Mate guarding and consortships, defined as temporary associations between a male and a receptive female, are also widespread among polygynous primates, where they have been described in detail for about 20 species of New and Old World monkeys (Dixson, 1998a). However, mate guarding also occurs among many prosimians, even though most of them are receptive for only a few hours a year (Jolly, 1967; Brockman et al., 1998; M. Eberle, unpublished data). In other primates, consortships vary in duration between hours and weeks, and some females form consortships with several different males in succession (Hrdy & Whitten, 1987; Dixson, 1998a). A most effective form of mate guarding is, for example, exhibited by male hamadryas baboons (*Papio hamadryas*), who kidnap young females and aggressively herd them permanently, thereby completely monopolising them (Kummer, 1968). In some species, such as mountain baboons (*Papio cyncocephalus*), aggressive herding of females also occurs during between-group encounters, thereby curtailing opportunities for non-resident males to determine the number of females in a given group, as well as to inspect their reproductive condition (Henzi et al., 1998). A final twist is exhibited by male chimpanzees (*Pan troglodytes*), who have been observed to form, under rare demographic conditions, temporary mate-guarding coalitions that share matings among each other (Watts, 1998).

Whether male or female primates take the initiative to form a consortship varies both among and within species, but a high degree of reciprocal communication and coordination is required in any event (van Noordwijk, 1985; Bercovitch, 1995). Consortships can occur within a group, but a consorting pair may also leave their group for days or even weeks (e.g. Tutin, 1979). In species where consortships occur, most copulations take place during this time (summarised in Dixson, 1998a), thus increasing the consorting male's probability of mating, in addition to providing an opportunity for preventing rivals from doing so. However, consortships are energetically costly for males (e.g. Alberts et al., 1996), and they are often not very effective because females also engage in additional copulations (see Dixson, 1998a). A main benefit of effective mate guarding for females may be the reduction of harassment by other males (Smuts & Smuts, 1993). This idea is supported by the observation that high-ranking males are more often able to form successful consortships than lower-ranking males (Bercovitch, 1991, 1995; Cowlishaw & Dunbar, 1991).

Despite their widespread occurrence, the effectiveness of consortships in increasing a male's success in sperm competition is compromised by basic aspects of female primates' reproductive physiology. As in most other mammals, fertilisation in primates is limited to a narrow time-window following ovulation. The timing of copulations in relation to ovulation is therefore the most decisive determinant of male reproductive success (Gomendio *et al.*, 1998). Because ovulation in most anthropoid primates is concealed within a wider window of receptivity (as in humans: see Gangestad & Thornhill, this volume), the effectiveness of consortships and other forms of mate guarding by male primates is not absolute, but instead dependent on its timing and duration relative to an unknown: the moment of ovulation.

Frequent copulation

From the male's perspective, frequent copulation is likely to be another effective paternity guard because, all else being equal, more copulations mean more sperm inseminated, and a greater chance of fertilisation. Although it is difficult to measure accurately the frequency of copulation in many bird species, the available evidence indicates that copulation frequency varies dramatically between species. Some species copulate only a few times for each clutch, while others copulate hundreds of times per clutch (Birkhead *et al.*, 1987). In poultry, a single artificial insemination is sufficient to fertilise an entire clutch of eggs (Lake, 1975) and on the basis of this it was assumed that a single copulation would be sufficient for wild birds to fertilise an entire clutch (Birkhead & Møller, 1992a). However, it has since become clear that poultry biologists typically inseminate many more sperm than would occur in a natural ejaculate, and usually do so at the optimum time for fertilisation. It has been shown, for example, that feral fowl usually inseminate fewer than 5 million sperm per natural insemination (T. Pizzari & T. R. Birkhead, unpublished), whereas poultry biologists usually inseminate 100 million sperm (e.g. van Krey, 1990).

In addition, it is now clear that not all behaviourally successful copulations in wild or captive birds result in sperm transfer – in some copulations, ejaculation does not occur, and in others sperm fail to be transferred into the female's cloaca (Birkhead *et al.*, 1988; Adkins-Regan, 1995). What this means is that the minimum number of copulations to ensure that the female has an adequate supply of sperm is not one, but a few. However, it also means that there is still a lot of variation in copulation frequency between species to explain. Species that cannot guard their partners usually exhibit higher copulation rates than mate-guarding species (Birkhead *et al.*, 1987). Other factors that are likely to account for the variation in copulation frequency include the number of sperm transferred per ejaculate, the rate at which females utilise sperm, and the quality of sperm themselves.

Until recently the number of sperm which male birds transferred during natural matings was unknown. However, by using a false cloaca it was shown that male zebra finches (*Taeniopygia guttata*) transfer from fewer than 1 million to over 10 million sperm and that much of this variation was accounted for by the time since the last ejaculation or, to put it another way, by sperm depletion (Pellatt & Birkhead, 1994; Birkhead *et al.*, 1995a). In addition to there being interspecific differences in the number of sperm that males transfer in relation to the intensity of sperm competition, in some species individual males are able to facultatively adjust their sperm numbers in relation to the likelihood of sperm competition. For example, in the Adelie penguin (*Pygoscelis adeliae*), males are more likely to transfer sperm during an extra-pair copulation, than they are during a pair copulation (Hunter *et al.*, 1995). In the sand martin (*Riparia riparia*), males transfer more sperm to a model female in the presence of other males, than they do when copulating without other males present (Nicholls *et al.*, 2001). In feral fowl, males exhibit a marked 'Coolidge Effect', transferring more sperm to novel females (T. Pizzari, T. R. Birkhead & C. K. Cornwallis, unpublished). These studies reveal that not all copulations are equal in terms of the number of sperm they transfer.

After sperm are inseminated, a portion of them, probably only a few per cent, are stored in the female's sperm storage tubules, from which they are released at a constant rate over the next few days or weeks (Bakst *et al.*, 1994). The released sperm travel up the oviduct to the infundibulum, where fertilisation takes place, if an ovum is present. If no ovum is present these sperm are lost in the female's body cavity (Bakst *et al.*, 1994). The rate at which sperm are released from the sperm storage tubules varies markedly between species (Birkhead *et al.*, 1994; Sax *et al.*, 1998) and may account for the frequency with which females need to copulate in order to replenish their sperm supplies. High rates of sperm utilisation may also provide females with some control over the paternity of their offspring.

The fertilising efficiency of sperm from different males varies markedly (Froman & Feltmann, 1998; Birkhead *et al.*, 1999). Females copulating with males whose sperm is particularly efficient may not have to copulate as frequently as those copulating with other males. Part of the fertilising efficiency of sperm is determined by the number accepted by the female and stored in the sperm storage tubules, and by the rate at which they are released from them (Froman *et al.*, 2002), suggesting a complex interaction between these different aspects of reproduction (Pizzari & Birkhead, 2002).

Copulation frequency is also highly variable among and within different primate species. Detailed documentation of mating behaviour in wild populations is rare, however, and most quantitative data focus on the number of ejaculations a male can achieve and not on the number that individual females receive (see Dixson, 1998a). Several studies have pointed out, however, that females may copulate (repeatedly) with 90 to 100 per cent of all resident males on any given day (e.g. Taub, 1980; de Ruiter et al., 1992), and it has been reported, for example, that chimpanzee females can mate up to four times with 13 or more males within an hour, adding up to an estimated 6000 or more potentially fertile copulations in a lifetime (Wrangham, 1993) – to produce an average of four or so offspring! Despite these limitations of the data, much of the variation in male copulation frequency among species can be explained by the mating system (Dixson, 1995a). In promiscuous species, such as ring-tailed lemurs (Lemur catta), muriquis (Brachyteles arachnoides), stump-tail macaques (Macaca arctoides) and bonobos (Pan paniscus), individual males can ejaculate up to 30 times a day, which is in contrast to monogamous and polygynous species with much lower frequencies (one to three ejaculations per day; summarised in Dixson, 1998a). This difference in copulation frequency correlates well with that in the presumed intensity of sperm competition, so that frequent copulations may contribute to the reproductive success of individual males, even though the probability of a given ejaculation resulting in a conception is very low.

A related difference has been documented for copulatory patterns, which differ along several axes and which are fairly uniform within species (Dixson, 1998a). The main differences are related to the number (single versus multiple) and duration of intromissions. Again, the more complex patterns, characterised by multiple or prolonged intromissions, are primarily (but not exclusively) found in species where females mate with multiple males (Dixson, 1987a, 1995b; but see Dewsbury & Pierce, 1989). Some prolonged intromissions, lasting for an hour or even more, may functionally approach mate guarding. Multiple intromissions, in relation to pelvic thrusting, may serve to remove sperm plugs deposited by previous males or to facilitate sperm transport within the female tract, although there is no evidence for these speculative ideas (see Gomendio et al., 1998; but see below). Both of these copulatory patterns may therefore also enhance a male's success in sperm competition, but the classification and functional interpretation of different patterns remain both poorly documented and controversial (Shively et al., 1982; Dewsbury & Pierce, 1989; Gomendio

et al., 1998), providing an important area for future systematic comparisons.

The large variation in the number of sperm inseminated is not peculiar to birds; in mammals the variation in sperm numbers ejaculated is also considerable (Amann, 1981). Because theoretical considerations, as well as experimental studies with other mammals, indicated that sperm production is costly and that sperm delivery is compromised by successive ejaculations (Dewsbury, 1982; Preston et al., 2001), the ability of male primates to copulate repeatedly should also be constrained. This notion is supported by a demonstration of declining sperm counts obtained from masturbating captive chimpanzees (Marson et al., 1989). Regular electroejaculation of bonnet macaques (Macaca radiata) revealed considerable intra- and inter-individual variation in semen volume and sperm counts across an entire year (Kholkute et al., 2000), but nothing is known about potentially differential sperm allocation in a natural mating context, or differential fertilisation capability of sperm from different males in these or any other primate species. The only (indirect) evidence for a positive relation between sperm numbers and the intensity of sperm competition in primates comes from a comparative study that found a positive correlation between ejaculate volume, sperm numbers and relative testes size (Møller, 1988).

Although repeated copulations are very common among primates, it is not clear why many primate females copulate repeatedly with the same male. The millions of spermatozoa contained in a single primate ejaculate should be more than sufficient to fertilise their one or two eggs. Furthermore, sperm storage and genetic diversity benefits are also obviously unimportant for female primates. Only if females were aware of the exact time of ovulation would it be advantageous for them to mate repeatedly with their preferred partner at the critical time, but there are currently no data to evaluate this possibility. It may therefore well be that females do not benefit directly from these repeated copulations, but that they may incur high costs if they refuse to do so (see Gomendio et al., 1998), either in the form of harassment by the consorting male, or by harassment from other males that are being repelled by their consort (see van Schaik et al., this volume).

ANATOMICAL ADAPTATIONS TO SPERM COMPETITION

The most ubiquitous adaptation to sperm competition is the possession of relatively large testes: across a range of taxa, including birds, species that experience high levels of sperm competition have relatively large testes (Møller, 1991; Birkhead & Møller, 1998). The significance of this is that larger testes produce more sperm, allowing males to either inseminate more sperm per ejaculate or produce more ejaculates per unit time. Studies with poultry have shown that the relative number of sperm from two males is an important factor determining the pattern of paternity (Martin *et al.*, 1974). In species in which sperm competition is intense, males possess large testes and large male sperm stores (the seminal glomera in passerines). In the aquatic warbler (*Acrocephalus paludicola*), males appear to copulate infrequently but transfer very large numbers of sperm per ejaculate (Schulze-Hagen *et al.*, 1995); the same may be true in fairy wrens (*Malurus* sp.: Tuttle *et al.*, 1996). In the dunnock (*Prunella modularis*) copulation occurs much more frequently – over 100 times per clutch – suggesting that this species produces smaller ejaculates (Davies, 1992).

Most birds do not possess a penis, and sperm transfer is effected by the juxtapositioning of the male and female cloacae, often very rapidly. The lack of a penis is generally thought to be an adaptation to flight; a weight-saving device (Briskie & Montgomerie, 1997). A small proportion of bird species do possess an intromittent organ, notably the ducks, geese and swans, the ratites (ostrich, emus, rheas and cassowaries), kiwis, tinamous and a few others (Birkhead & Møller, 1992a). The record holder is the Argentinian lake duck (*Oxyura vittata*), whose 42-cm-long penis is longer than its owner's body (McCracken *et al.*, 2001; see also Coker *et al.*, 2002) – but the adaptive significance of which remains entirely unknown. In male passerine birds the seminal glomera protrude into the cloaca to form a cloacal protuberance, which superficially resembles a penis and which may facilitate copulation (Wolfson, 1954). A few passerine species have a highly modified cloaca, which almost certainly facilitates sperm transfer; for example, the bearded tit (*Panurus biarmicus*) – a species with intense sperm competition (Hoi & Hoi-Leitner, 1997) – has an extrusible phallus (Birkhead & Hoi, 1994). The greater vasa parrot (*Coracopsis niger*) possesses an extraordinary and very large cloacal protrusion, which is used to form a protracted copulatory tie; recent research shows that this parrot also has an unusual mating system in which sperm competition is intense (Wilkinson & Birkhead, 1995; J. Ekstrom & T. R. Birkhead, unpublished data).

Another example of bizarre reproductive anatomy being associated with sperm competition is the red-billed buffalo weaver (*Bubalornis niger*). The male of this species possesses a false penis that lies anterior to the cloaca. The bird is about

the size of a European starling (*Sturnus vulgaris*) and its permanently erect false penis is up to 2 cm in length. This species also has an unusual mating system, comprising coalitions of two unrelated males sharing a harem of females, but also sharing paternity both within and between coalitions (Winterbottom *et al.*, 1999, 2001). Copulations are protracted, with bouts lasting 25 minutes but, contrary to expectation, the false penis is not inserted into the female's cloaca; instead it is rubbed on the outside of the female's cloaca. The function of the male's stimulation of the female with his false penis is not known; the most likely explanation is that this stimulation is designed to persuade the female to retain more of a particular male's ejaculate, as has been elegantly demonstrated in a beetle (Edvardsson & Arnqvist, 2000). In other words, the buffalo weaver's false penis may provide a form of copulatory courtship (Eberhard, 1996; Winterbottom *et al.*, 2001).

In every case examined so far, unusual male reproductive anatomy appears to be associated with intense sperm competition. In contrast, the gross anatomy of the female reproductive tract appears to be much more conservative in birds. There are a few examples of coevolution between male and female reproductive traits, notably the positive correlation between sperm length and the length of the female's sperm storage tubules (Birkhead & Møller, 1992b; Briskie & Montgomerie, 1992). Other female coevolved responses may be physiological or behavioural (see below).

Primates also exhibit an array of anatomical adaptations to sperm competition. Short (1979) was the first to point out that testes size among the great apes varies in accordance with the intensity of sperm competition. Analyses of larger samples of anthropoids confirmed that promiscuous anthropoids have larger testes, in relation to body size, than monogamous and polygynous species (Harcourt *et al.*, 1981; Harvey & Harcourt, 1984). The same trend was later demonstrated for prosimians (Kappeler, 1993a, 1997b), where the large number of species with a solitary social organisation and an unknown mating system continue to hamper comparative analyses (Dixson, 1998a; Gomendio *et al.*, 1998). As already indicated by the much greater interspecific variation in relative testes size among solitary prosimians (Kappeler, 1997b), we now know that an unexpected diversity of mating systems is hidden within this category. Fat-tailed dwarf lemurs (*Cheirogaleus medius*) and some sportive lemurs (*Lepilemur* spp.), for example, have since been discovered to be pair-living (Fietz *et al.*, 2000; Zinner *et al.*, 2003), and they indeed have relatively small testes. Dwarf lemurs (*Mirza coquereli*) and mouse lemurs (*Microcebus* sp.), on the other hand, were found to engage in scramble-competition polygyny (Kappeler, 1997a; Fietz, 1999; Radespiel et al., 2001; Eberle & Kappeler, 2002), and thus in intense sperm competition (Schwagmeyer, 1988; Schwagmeyer & Parker, 1990), and they have some of the largest testes in relation to body size among primates (just one of their testes is larger and heavier than their brain; P. M. Kappeler, unpublished data). A more comprehensive comparative analysis of the evolution of primate testes size continues to be hampered by a lack of a unitary database (cf. Harcourt et al., 1995; Kappeler, 1997b) and many missing data points, especially from prosimians, but the major trends have clearly been identified.

Apart from body size and the mating system, part of the variation in primate testes size could also be owing to the requirements of seasonal reproduction. Reproductive activity of many primates is more or less seasonal, resulting in an increased frequency of matings during a relatively short time period, which may also select for increased testes size and sperm production, compared to year-round breeders (Short, 1977). While males of most seasonally breeding species exhibit marked annual fluctuations in testes size (e.g. Sade, 1964; Schmid & Kappeler, 1998; Pochron et al., 2002), a comparative study of a large representative sample of species did not detect the predicted systematic increase in relative testes size in seasonal breeders (Harcourt et al., 1995). Sperm characteristics, such as motility or morphology, can also vary seasonally (Brun & Rumpler, 1990; Hernandez-Lopez et al., 2002), but corresponding differences among species have not yet been examined systematically. More comprehensive comparative studies between seasonal and non-seasonal species are beginning to reveal how seasonal variation in testes size is closely integrated with other physiological adaptations to seasonality (Muehlenbein et al., 2002), and more such integrative studies are needed to illuminate further how patterns of pre-mating fat deposition and other adaptations correlate with different reproductive strategies in a wider range of species.

To what extent individual variation in primate testes size is positively correlated with competitive potential (rank, body size) on the one hand, and mating and reproductive success on the other hand, independent of potentially confounding co-variables, remains poorly studied and unresolved. For example, in promiscuous savannah baboons, neither body size nor testes size were related to inter-individual differences in male reproductive activity, measured as ejaculatory rate during consort with a fertile female (Bercovitch, 1989). In rhesus macaques (Macaca mulatta), in contrast, testes size was significantly larger in sires than in non-sires (as shown by genetic paternity testing), but testes size was also positively associated with body size, rank, and body condition (Bercovitch & Nürnberg, 1996). Again, additional studies from a wide range of primate taxa are needed to determine general interrelations among these traits and the potential advantage of large testes size at an individual level.

Other components of the male genital tract and characteristics of the spermatozoa itself have also been linked to sperm competition in primates. First, the seminal vesicles, which produce the bulk of the fluid proportion of the ejaculate, were found to be larger in species in which females copulate with more than one male (Dixson, 1998b), but little is known about variation in the composition of primate seminal fluid (Dixson, 1998a), which can have marked effects on female reproductive physiology in other taxa (see Johnstone & Keller, 2000). Second, because male primates store their sperm in the epididymis, its size may be even more crucial in sperm competition than testes size. In promiscuous rhesus macaques, epididymis size is indeed strongly correlated with testes size (Bercovitch & Rodriguez, 1993), but data from other taxa are not available for a comparative test of this prediction. Third, sperm anatomy may also reflect adaptations to sperm competition, and indeed their size and shape vary tremendously among primates and other mammals (Gage, 1998). Sperm length, in particular, has been linked to the intensity of sperm competition because longer sperm may have faster swimming speeds, and thus a competitive advantage (Gomendio & Roldan, 1991; Dixson, 1993). More recently, it has also been demonstrated that the volume of the midpiece of individual sperm, which is an indicator of mitochondrial loading and thus motility, is greater in primate species in which the females mate polyandrously (Anderson & Dixson, 2002).

Finally, primate males do possess a penis whose variation across species in size, shape and spinosity suggests important additional functions apart from simple intromission and sperm deposition. Existing studies indicate that much of the existing interspecific variation in penile morphology is functionally related to sperm competition. Primate males in species with polyandrous females tend to have a longer and morphologically more complex penis (Dixson, 1987a). There is also some indication that the degree of spinosity is positively associated with a promiscuous mating system (Dixson, 1987a; Verrell, 1992; Harcourt & Gardiner, 1994), but there is a lot of variation among higher taxa – prosimians, in general, have spines whereas anthropoids don't – that obscures functional relationships (Harcourt, 1996). In addition, penile morphologies vary considerably among species and genera, suggesting a potential function in species recognition, as among sympatric bushbabies (Galagoninae), for example (Anderson, 2000). Furthermore, the baculum, which is absent only in some New World primates, tarsiers and humans (see Hobday, 2000), is also highly variable in size and shape among the other primates. Here, much of this interspecific variation is explained by differences in copulatory pattern, with long-intromission species having relatively longer penis bones (Dixson, 1987b).

Thus, despite some unfounded scepticism (Brown *et al.*, 1995), virtually all aspects of primate genital anatomy and sperm morphology appear to be influenced by sperm competition in the predicted direction (Harcourt, 1997; Dixson, 1998a). However, many more studies of variation in these and other components of the male genital tract are needed to understand better how exactly they contribute to advantages in sperm competition.

Genitals of female primates generally show much less morphological variation than those of males, but there are a few striking exceptions. In female lemurs, the clitoris is hypertrophied (Petter-Rousseaux, 1964), albeit not as spectacularly as in spotted hyaenas (*Crocuta crocuta*; Frank, 1997). Furthermore, the vulva is sealed throughout the year in cheirogaleid lemurs (Cheirogaleidae), except for the few days around their brief annual oestrus and the subsequent birth (Foerg, 1982; see also Eberle & Kappeler, 2002). The morphology of female genitals of lemurs does not reflect obvious adaptations to sperm competition; rather, the prolonged clitoris may represent a by-product of the endocrinological correlates of female dominance (Kappeler, 1993b; Ostner *et al.*, 2003). A long, pendulous clitoris is also found in spider monkeys and their relatives (Atelinae), which also have relatively large testes, and mate promiscuously (Dixson, 1998a). The most striking feature of primate female genitals are the large perineal swellings of certain Old World monkeys and apes. They clearly have a function in sexual selection (Darwin, 1876), and their concentration in multi-male taxa suggests that sperm competition is involved in their origin and maintenance (Nunn, 1999). Their possible proximate and ultimate functions are discussed in detail in other contributions to this volume (Snowdon, this volume; Zinner *et al.*, this volume).

MECHANISMS OF SPERM COMPETITION

Studies of sperm competition in insects and birds share one feature in common: when two sequential inseminations (which are equal in all respects) are made, the second usually fertilises the majority of eggs. This phenomenon is referred to as 'last-male sperm precedence' (Birkhead & Møller, 1992a; Simmons, 2001). In reality, two inseminations made one after the other is a rather unnatural event in nature, but it has proved to be an extremely useful model for understanding the basic mechanisms of sperm competition in insects and birds alike.

Last-male sperm precedence had been known in poultry for a long time, and several different explanations were offered: (1) last in, first out – a kind of stratification of sperm within the female's sperm storage tubules; (2) sperm displacement: incoming sperm displaced, or otherwise disabled, previously inseminated sperm; or (3) passive sperm-loss: the longer the interval between two inseminations, the more likely it was that sperm from the earlier insemination had been used, and so more sperm from the second insemination were available for fertilisation. In fact, detailed experiments with domestic fowl (*Gallus gallus domesticus*), domestic turkey (*Meleagris gallopavo*) and the zebra finch, have demonstrated that passive sperm loss is the explanation for last-male sperm precedence (Colegrave *et al.*, 1995; Birkhead, 1996, 1998a; Birkhead & Biggins, 1998). Subsequent field studies of sperm competition in wild birds are consistent with this conclusion (Westneat, 1994; Lifjeld *et al.*, 1997).

Even though the sperm competition experiments involving two sequential artificial inseminations containing the same number of sperm gave results consistent with the passive sperm-loss model, they also always showed a lot of variation (Birkhead *et al.*, 1995b; Birkhead & Biggins, 1998), suggesting that other factors were also important in determining the outcome of sperm competition. One of these factors was the timing of insemination relative to when the female laid. Inseminations made close to the time of oviposition resulted in the uptake of relatively few sperm, biasing paternity in favour of the insemination made further from the time of oviposition (Birkhead *et al.*, 1995b).

Another factor that influences the outcome of sperm competition is differential fertilising capacity between males (Lanier *et al.*, 1979). If equal numbers of sperm from two or more males are mixed and inseminated into several different females, paternity is rarely shared equally among the males, and one male usually fathers a disproportionate number of offspring (Dziuk, 1996). Differential fertilising capacity has been known about for a long time but until recently the way it worked was not known. Froman and Feltmann (1998) used an ingenious but simple assay to measure the fertilising ability of different male domestic fowl: this is the net movement of a population of sperm in an inert medium, Accudenz®, over a standard time period, and referred to as sperm mobility (Froman & Feltman, 1998, 2000; Froman *et al.*, 1999). Sperm mobility is highly consistent within males and remains consistent over time. In a non-competitive situation, high-mobility sperm fertilise more eggs than low-mobility sperm (Froman & Feltman, 1998, 2000; Froman *et al.*, 1999), and the same is true in a competitive situation when equal numbers of high- and medium-mobility sperm were mixed and inseminated into females (Birkhead *et al.*, 1999). Remarkably, on a sperm-for-sperm basis, one sperm from a high-mobility male can be equivalent to ten of those from a medium-mobility male, so sperm mobility has a powerful effect on the outcome of sperm competition, and may help to explain some of the variation observed in previous experiments. High-mobility sperm out-compete low-mobility sperm for at least two reasons. First, a higher proportion of high-mobility sperm gain access to the female's sperm storage tubules; and second, it appears that more of the high-mobility sperm remain in the tubules for longer. That is, they are released from the sperm storage tubules more slowly, and hence retain their fertilising potential for longer (Froman *et al.*, 2002). Sperm mobility shows a normal distribution across a population of males, but it is not yet known whether similar variation and consistency within males occurs in wild birds. Nor is it known what maintains the variation in sperm mobility, although there are several indications (see below).

Mechanisms of sperm competition in primates remain virtually unstudied. Because of basic similarities in sperm longevity and egg lifespan with other mammals (see above), it is safe to assume that the timing of an ejaculation relative to ovulation is of greatest importance for its fertilisation success (see Huck *et al.*, 1985, 1989). However, what determines the success of ejaculations of different males, deposited during this critical time-window in primates, remains unknown. The situation in most primates is complicated by the fact that litter size is one and that most males ejaculate repeatedly, so that sequence or timing effects of particular ejaculations are impossible to determine, and to distinguish from a raffle because only one fertilisation is taking place (Parker, 1990). Only experimentally controlled matings will provide insights into mechanisms of sperm competition, but the required experimental control is difficult to achieve with most primates.

Grey mouse lemurs (*Microcebus murinus*) and other small prosimians are notable exceptions in this respect because they can be easily kept, bred and handled in captivity. In addition, they typically give birth to two infants (range 1–4), so that multiple paternity is also possible. Two recent studies set up small groups of two or three males and females, respectively, and measured the success of different males by determining paternity genetically. One study found that the dominant of three males sired all infants in 16 out of 17 litters in seven groups (Andrès *et al.*, 2001), whereas the other found that the dominant of two males in five groups fathered only six of 11 infants for which paternity and male dominance relations were determined (Radespiel *et al.*, 2002). Unfortunately, behavioural data were collected only for one hour per night in both studies so that these divergent results cannot be related to the number and timings of copulations by the competing males. Based on results of a detailed field study, where dominance among males appears to govern male mating success much less than in captivity (Eberle & Kappeler, 2002; P. M. Kappeler, M. Eberle & M. Perret, unpublished data) 12 oestrous females were presented with up to six males individually for one copulation each. In this way it was possible to control the number, sequence and intervals between copulations that each female received during continuous observations to test for effects of mating order and timing. The subsequent paternity analyses indicated that earlier-mating males sired more offspring.

Mouse lemurs, as well as several other lemurs, certain lorises, spider monkeys, macaques and chimpanzees are known to form copulatory plugs with parts of their ejaculate (Dixson, 1998a). These plugs harden shortly after ejaculation and completely block the female vaginal tract. It has been speculated that these plugs may act as physical barriers to subsequent matings and/or that they hamper sperm loss by the female (discussed in Dixson, 1998a), and thus may qualify as mechanisms of sperm competition that contribute to the reproductive success of males who deposit a plug. At the moment, we do not even know the distribution of copulatory plugs across primates, however, so that comparative tests of these hypotheses are impossible. Moreover, observations of active plug-ejection by a female ring-tailed lemur (*Lemur catta*) during the approach of another mate (P. M. Kappeler, unpublished observation), and the removal of plugs with the help of penile spines by male grey mouse lemurs (M. Eberle, unpublished observation) indicate that copulatory plugs may not be very effective barriers. On the other hand, our mating experiment with grey mouse lemurs revealed that a copulation of a few seconds can be sufficient to sire offspring (M. Eberle, personal communication), but copulating males in the wild were often interrupted by rivals before they could remove the plug (M. Eberle, unpublished data). Future studies should therefore compare latencies between subsequent matings as a function of plug presence or persistence, to begin illuminating sperm-plug function.

CRYPTIC FEMALE CHOICE

The outcome of sperm competition is also mediated by female factors, both before and after copulation has taken place. It has long been recognised that sexual selection operates more intensively on males than on females, in part because the reproductive potential of males is so much greater than that of females (Trivers, 1972; Parker, 1984; Clutton-Brock & Vincent, 1991). It was therefore assumed that females played a relatively passive or acquiescent role in sperm competition. In the early 1980s, however, Thornhill (1983) suggested that females might exhibit what he called 'cryptic female choice' – the differential utilisation of sperm from different males. The term 'cryptic' referred to the fact that this choice took place out of sight, inside the female reproductive tract. Thornhill's paper was more or less ignored, at least initially, presumably because at that time even pre-copulatory female choice – that is, choice of partner – was poorly documented. Subsequently, as the evidence for pre-copulatory female choice became more convincing (Andersson, 1994), researchers began to focus their attention on the role of females in sperm competition (Birkhead & Møller, 1993b; Birkhead, 2000a).

There is no hard and fast division between pre- and post-copulatory female choice, and the females of several bird species appear to choose particularly attractive males as their extra-pair copulation partners (reviewed in Birkhead, 1998a). The main impetus for exploring the idea of cryptic female choice came with the publication of Eberhard's book, *Female Control* (1996). Eberhard documented a large number of different processes by which females could potentially control events associated with insemination and fertilisation. These processes ranged from controlling the amount of sensory feedback they provided to males during copulation, which in turn might influence the number of sperm ejaculated, to being able to discriminate between the sperm of different males in the absence of any phenotypic cues. Cryptic female choice is difficult to demonstrate unequivocally because, like pre-copulatory female choice, it is hard to disentangle from male–male competition. In order to demonstrate cryptic female choice, one has to control for all male effects, such as sperm numbers or differential

fertilising ability (Birkhead, 1998c, 2000b; Pitnick & Brown, 2000).

There is now evidence for cryptic female choice in feral fowl, however. Female hens prefer to copulate with dominant males, but because males are substantially larger than females, subordinate males can coerce females into copulating. Coerced or forced copulations are precisely the type of situation where we might expect cryptic female choice to occur, because females are unable to operate any kind of pre-copulatory choice. In feral fowl, subordinate males frequently attempt coerced copulations, and females have three strategies for dealing with them. First, they attempt to run away. If this fails and the subordinate male grabs them (usually by their comb – which looks very painful!), the female can utter a specific distress call which attracts the dominant male who supplants the subordinate. If the dominant male fails to hear the female's distress call, and the subordinate male manages to inseminate the female, she can forcibly eject his ejaculate. Pizzari and Birkhead (2000) showed, in a combination of observations and experiments, that females were significantly less likely to eject the sperm of socially dominant males. In this instance, females reinforced their pre-copulatory choice through cryptic female choice.

It is also possible that females might be able to discriminate between the sperm of different males based solely on the attributes of the sperm themselves – sperm choice. We tested this hypothesis by artificially inseminating the same female domestic fowl over several successive clutches with sperm from the same pair of males, and predicted that any sperm choice would be reflected by some females having offspring fathered by one male, and other females having their offspring fathered by the other male of the pair. This is what we found in a small, but statistically significant, proportion of cases (T. R. Birkhead *et al.*, unpublished data). What this result shows is that, over and above all the sperm competition effects that mediate the outcome of sperm competition, some sperm–female compatibility also exists. Indeed, many of the cases of cryptic female choice appear to be instances of compatibility (Birkhead, 1998c; Tregenza & Wedell, 2000), i.e. females favour the sperm of males with compatible genotypes regardless of their phenotype, perhaps based on MHC genes expressed on the surface of spermatozoa (Martin-Villa *et al.*, 1999).

Cryptic female choice in primates is poorly documented, even though there are theoretical reasons to expect it to be common. There is an emerging consensus that female primates mate polyandrously to reduce the risk of infanticide for their offspring by confusing paternity (van Schaik, 2000; Paul, 2002). However, females may want to bias paternity in favour of a particular male with a preferred phenotype or genotype, or both (Smuts, 1987; Small, 1989; Keddy-Hector, 1992; Manson, 1995; van Schaik *et al.*, this volume). Cryptic female choice could be one mechanism contributing to the solution of the resulting female dilemma. Even though most primate females have concealed ovulations, there is evidence that they use various pre-copulatory mechanisms, such as friendships (Smuts, 1985; Pereira & McGlynn, 1997) or increased proximity (e.g. Soltis *et al.*, 1997; Matsumoto-Oda, 1999; Zehr *et al.*, 2000) with favoured males, copulation calls that are likely to attract particular males (O'Connell & Cowlishaw, 1994; Semple, 1998), active solicitation of copulations around the likely conception date (Janson, 1984; Zehr *et al.*, 2000; see also Gangestad & Thornhill, this volume), as well as changes in chemical signals (Epple, 1986; Converse *et al.*, 1995; Kappeler, 1998; Snowdon, this volume); unique vocalisations (Stanger *et al.*, 1995; Buesching *et al.*, 1998); sexual swellings (Snowdon, this volume; Zinner *et al.*, this volume) and increased frequencies of particular behaviour patterns during the peri-ovulatory phase (Carosi & Visalberghi, 2002) to signal impending ovulation and/or to increase the chances of fertilisation by favoured males. These mechanisms of facilitating copulations from favoured males at critical times could be interpreted as prerequisites for subsequent cryptic choice because they ascertain that sperm from preferred males is available at the right time.

Post-copulatory interactions among sperm of different males, potential differential interactions between sperm and the female reproductive tract, as well as details of the sperm–egg interactions are comparatively poorly documented among primates (see Primakoff & Myles, 2002), perhaps because of the heavy reliance on invasive methods for obtaining them. Proximate questions about potential mechanisms of cryptic female choice in primates were posed long before the topic became popular in other taxa (Quiatt & Everett, 1982), but little relevant information can be found on this subject. Indirect evidence from a recent study of rates of molecular evolution of genes, coding for sperm-associated proteins, clearly indicated that these genes exhibit much higher rates of non-synonymous substitution in promiscuous primates (Wyckoff *et al.*, 2000). This difference has been interpreted as indicating that 'potential competition among sperm from different males has contributed to the accelerated evolution of genes involved in sperm and seminal fluid production', but cryptic female choice provides an equally plausible and not incompatible mechanism. Because the present data are limited to the great apes, more data from additional taxa are required for a stronger test of this hypothesis.

The strongest indirect evidence for a mechanism of cryptic female choice in primates is provided by the observation that females of several species of anthropoids (mostly macaques, baboons and chimpanzees) exhibit orgasm (Allen & Lemmon, 1981; Dixson, 1998a). It should be noted that the taxonomic distribution of female orgasm remains poorly documented and that it may be limited to Old World primates (see Dixson, 1998a). Physiological measures during artificially induced orgasms demonstrated the occurrence of the same vaginal and uterine contractions that also characterise human orgasm (Burton, 1971; Goldfoot *et al.*, 1980; Allen & Lemmon, 1981) and are thought to accelerate and facilitate sperm transport towards the cervix and ovaries (Smith, 1984). Interestingly, the occurrence of female orgasm is highly variable, both among and within females, and the adaptive nature and underlying physiology of this variation remain poorly understood, even in humans (Mah & Binik, 2001).

Hrdy (1996) suggested that reaching orgasm requires cumulative stimulation from multiple sexual encounters, and, at least in Japanese macaques (*Macaca fuscata*), the frequency of orgasm was indeed positively related to the number of mounts and pelvic thrusts, and thus the duration of copulation (Troisi & Carosi, 1998). Importantly, when the level of physical stimulation was controlled statistically, female orgasm was observed more often in macaque pairs including high-ranking males (Troisi & Carosi, 1998). A comparable effect of male social status on female orgasm rates has also been reported for humans (Fisher, 1973; Thornhill *et al.*, 1995; Gangestad & Thornhill, this volume). Orgasm therefore has the potential to be used selectively by females to facilitate fertilisation of their eggs by particular males (Smith, 1984; Thornhill *et al.*, 1995). This hypothesis is indirectly supported by the observation that female orgasm apparently does not occur among prosimians, which have penises with extremely mechanically stimulating appendages (see above), but rather among Old World primates, where the potential for coercive matings by multiple males is highest (van Schaik *et al.*, this volume). Seen this way, female primate orgasm may therefore represent an evolutionary response to male sexual coercion that provided females with an edge in the dynamic competition over the control of fertilisation (see also Hrdy, 2000a; Gowaty, this volume; van Schaik *et al.*, this volume).

DISCUSSION

The study of post-copulatory sexual selection has come a long way since Parker (1970) and Trivers (1972) first introduced researchers to the evolutionary significance of multiple mating by females. The initial focus was on the adaptive significance of reproductive behaviours and morphological structures (Krebs & Davies, 1978). Subsequently, in order to understand the processes that determine the outcome of sperm competition and cryptic female choice, researchers also had to explore the underlying mechanisms. This combination of functional and mechanistic approaches to reproduction has provided a much more revealing and intellectually rewarding perspective (Krebs & Davies, 1997). Sperm competition and cryptic female choice are not peculiar to birds and primates, but birds, in particular, have provided useful opportunities for testing many of the basic ideas.

Our review indicates that post-copulatory selection in birds and primates is ubiquitous and a potentially powerful evolutionary mechanism (see also Birkhead & Pizzari, 2002) that has produced a number of convergences. First, mate guarding is a widespread behavioural mechanism employed by male birds and primates alike to defend their investment in a female that is willing to mate with them. Because most birds are pair-living and most primates live in larger groups with several rivals and additional potential mates, mate guarding may be easier, and hence more effective among birds. Second, males of both taxa can apparently always improve their fertilisation prospects by depositing large numbers of sperm. Repeated copulations between the same pair are found in many taxa, and they appear to be mainly motivated by male interests because female benefits are not obvious in many cases. Third, anatomical adaptations to sperm competition involving testes size and sperm morphology are strikingly similar between birds and primates. Presumably as a result of fundamental design constraints associated with flight, male and female genitals of birds are much less elaborated and used in sperm competition than in primates. Aquatic and terrestrial birds contribute most exceptions to this rule, supporting this interpretation. Fourth, mechanisms of sperm competition differ between birds and primates for similar reasons. Here, details of reproductive physiology associated with fertilisation have resulted in differences in sperm life and the ways female birds and primates use sperm. Finally,

there are mainly theoretical reasons, in addition to preliminary empirical evidence, to expect cryptic female choice in both taxa, but the mechanisms are still too poorly understood to identify potential similarities and differences, with the possible exception of orgasm in certain primates.

Studies of post-copulatory sexual selection also generate numerous general questions about relevant evolutionary mechanisms and their genetic consequences. For example, one of the key issues in evolutionary biology is concerned with variation in traits associated with fitness. In domestic fowl, males vary markedly but exhibit remarkable consistency in their sperm mobility scores, and sperm mobility is a key determinant of male reproductive success. What then maintains the variability in sperm mobility? One way in which variation in sperm mobility might be maintained is through antagonistic pleiotropy: that is, a negative association between two fitness-related traits. In domestic fowl and feral fowl, social dominance is a key determinant of male reproductive success. Socially dominant males acquire more copulations and more fertilisations than subordinate males (Cheng & Burns, 1988; Jones & Mench, 1991), partly because females appear to prefer socially dominant males and spend more time near them, and are more receptive to copulations from socially dominant males. We tested the hypothesis that sperm mobility would co-vary with social dominance and found that socially dominant males were more likely to have low-mobility sperm than subordinate males, suggesting that a trade-off might exist between these two life-history traits (Froman *et al.*, 2002). This idea could also be studied in primates in more detail because dominance relations among several males are pronounced and important for access to females in many species.

The relationship between different traits that contribute to male reproductive success is another area of evolutionary biology that has not received much attention. Researchers have often been satisfied with finding a single phenotypic or behavioural trait that correlates positively with male reproductive success (see Cowlishaw & Dunbar, 1991; Sheldon, 1994; Pizzari & Birkhead, 2000). But male reproductive success is determined by a suite of pre- and post-copulatory traits, which may co-vary either positively or negatively (Bercovitch, 1989; Bercovitch & Nürnberg, 1996; Birkhead & Pizzari, 2002; Pizzari & Birkhead, 2002). Only by measuring all traits is it possible to assess the fitness of males.

A second way in which variation in fitness-related traits might be maintained is if some traits, such as sperm quality, are inherited maternally, or if there is maternally biased transmission. This appears to be the case in the domestic fowl: sperm mobility shows additive variation and is highly heritable along the female line (Froman *et al.*, 2002). Sperm mobility is mediated by the ability of the sperm mitochondria to synthesise ATP, which in turn is partly controlled by mtDNA (Froman & Feltmann, 1998; Clayton, 2000), and hence may explain the maternal inheritance of sperm mobility. Similar patterns exist in mammalian sperm (Ruiz-Pesini *et al.*, 2000). Genes that are differentially expressed in males and females or have sex-limited expression may affect the fitness of males and females differently and their transmission will be limited (Pizzari & Birkhead, 2002).

SUMMARY AND CONCLUSIONS

In conclusion, the study of post-copulatory sexual selection covers a broad range of topics, spanning behaviour, morphology, physiology and genetics. Birds have proved to be ideal organisms in many respects for these studies, not least because researchers can often combine field and laboratory studies, and because both wild and domesticated birds are so well known. In primates, in contrast, controlled and experimental studies are much more difficult and, as a result, the study of many aspects of post-copulatory sexual selection is still in its infancy. However, careful descriptions of relevant traits in many more taxa could provide a basis for powerful comparative tests of many hypotheses (see Nunn & Barton, 2001).

Our review of the available evidence revealed that birds and primates share many behavioural adaptations to sperm competition, including mate guarding and frequent copulation. Testes size and sperm morphology also exhibit many similarities between the two groups, but only primates exhibit numerous adaptations to sperm competition in penile morphology. Mechanisms of sperm competition also differ in fundamental respects between these two taxa because of the requirements of their respective reproductive physiologies. Cryptic female choice is likely to be important in birds and primates, but the exact mechanisms remain obscure and await further study. Female orgasm in primates may represent a unique adaptation in this context.

ACKNOWLEDGEMENTS

We thank Patricia Gowaty, Tommaso Pizzari, Dan Rubenstein and Carel van Schaik for comments on the manuscript, and Manfred Eberle for numerous stimulating discussions.

REFERENCES

Adkins-Regan, E. 1995. Predictors of fertilization in the Japanese quail, *Coturnix japonica*. *Animal Behaviour*, **50**, 1404–15.

Agoramoorthy, G. & Hsu, M. J. 2000. Extragroup copulation among red howler monkeys in Venezuela. *Folia Primatologica*, **71**, 147–51.

Alberts, S. C., Altmann, J. & Wilson, M. L. 1996. Mate guarding constrains foraging activity of male baboons. *Animal Behaviour*, **51**, 1269–77.

Allen, M. L. & Lemmon, W. B. 1981. Orgasm in female primates. *American Journal of Primatology*, **1**, 15–34.

Amann, R. P. 1981. A critical review of methods for evaluation of spermatogenesis from seminal characteristics. *Journal of Andrology*, **2**, 37–58.

Anderson, M. J. 2000. Penile morphology and classification of bush babies (Subfamily *Galagoninae*). *International Journal of Primatology*, **21**, 815–36.

Anderson, M. J. & Dixson, A. F. 2002. Motility and the midpiece in primates. *Nature*, **416**, 496.

Andersson, M. 1994. *Sexual Selection*. Princeton, NJ: Princeton University Press.

Andrès, M., Gachot-Neveu, H. & Perret, M. 2001. Genetic determination of paternity in captive grey mouse lemurs: pre-copulatory sexual competition rather than sperm competition in a nocturnal prosimian? *Behaviour*, **138**, 1047–63.

Bakst, M. R., Wishart, G. J. & Brillard, J.-P. 1994. Oviductal sperm selection, transport and storage in poultry. *Poultry Science Reviews*, **5**, 117–43.

Bercovitch, F. B. 1989. Body size, sperm competition, and determinants of reproductive success in male savanna baboons. *Evolution*, **43**, 1507–21.

  1991. Social stratification, social strategies, and reproductive success in primates. *Ethology and Sociobiology*, **12**, 315–33.

  1995. Female cooperation, consortship maintenance, and male mating success in savanna baboons. *Animal Behaviour*, **50**, 137–49.

Bercovitch, F. B. & Nürnberg, P. 1996. Socioendocrine and morphological correlates of paternity in rhesus macaques (*Macaca mulatta*). *Journal of Reproduction and Fertility*, **107**, 59–68.

Bercovitch, F. B. & Rodriguez, J. F. 1993. Testis size, epididymis weight, and sperm competition in rhesus macaques. *American Journal of Primatology*, **30**, 163–8.

Birkhead, T. R. 1996. Mechanisms of sperm competition in birds. *American Scientist*, **84**, 254–62.

1998a. Sperm competition in birds: mechanisms and function. In *Sperm Competition and Sexual Selection*, ed. T. R. Birkhead & A. P. Møller. London: Academic Press, pp. 579–622.

1998b. Sperm competition in birds. *Reviews of Reproduction*, **3**, 123–9.

1998c. Cryptic female choice: criteria for establishing female sperm choice. *Evolution*, **52**, 1212–18.

2000a. *Promiscuity: An Evolutionary History of Sperm Competition and Sexual Conflict*. London: Faber & Faber.

2000b. Defining and demonstrating post-copulatory female choice – again. *Evolution*, **54**, 1057–60.

2002. Why don't primatologists come to ISBE meetings? *ISBE Newsletter*, **14**, 8–9.

Birkhead, T. R. & Biggins, J. D. 1998. Sperm competition mechanisms in birds: models and data. *Behavioral Ecology*, **9**, 253–60.

Birkhead, T. R. & Hoi, H. 1994. Reproductive organs and mating strategies of the bearded tit *Panurus biarmicus*. *Ibis*, **136**, 356–60.

Birkhead, T. R. & Møller, A. P. (eds.) 1992a. *Sperm Competition in Birds: Evolutionary Causes and Consequences*. London: Academic Press.

1992b. Numbers and size of sperm storage tubules and the duration of sperm storage in birds: a comparative study. *Biological Journal of the Linnean Society*, **45**, 363–72.

1993a. Why do male birds stop copulating while their partners are still fertile? *Animal Behaviour*, **45**, 105–18.

1993b. Female control of paternity. *Trends in Ecology and Evolution*, **8**, 100–4.

1998. (eds.) *Sperm Competition and Sexual Selection*. London: Academic Press.

Birkhead, T. R. & Pizzari, T. 2002. Postcopulatory sexual selection. *Nature Reviews Genetics*, **3**, 262–73.

Birkhead, T. R., Atkin, L. & Møller, A. P. 1987. Copulation behaviour in birds. *Behaviour*, **101**, 101–38.

Birkhead, T. R., Pellatt, J. E. & Hunter, F. M. 1988. Extra-pair copulation and sperm competition in the zebra finch. *Nature*, **334**, 60–2.

Birkhead, T. R., Sheldon, B. C. & Fletcher, F. 1994. A comparative study of sperm-egg interactions in birds. *Journal of Reproduction and Fertility*, **101**, 353–61.

Birkhead, T. R., Fletcher, F., Pellatt, E. J. & Staples, A. 1995a. Ejaculate quality and the success of extra-pair copulations in the zebra finch. *Nature*, **377**, 422–3.

Birkhead, T. R., Wishart, G. J. & Biggins, J. D. 1995b. Sperm precedence in the domestic fowl. *Proceedings of the Royal Society of London, Series B*, **261**, 285–92.

Birkhead, T. R., Martinez, J. G., Burke, T. & Froman, D. P. 1999. Sperm mobility determines the outcome of sperm competition in the domestic fowl. *Proceedings of the Royal Society of London, Series B*, **266**, 1759–64.

Briskie, J. V. & Montgomerie, R. 1992. Sperm size and sperm competition in birds. *Proceedings of the Royal Society of London, Series B*, **247**, 89–95.

1997. Sexual selection and the intromittent organ of birds. *Journal of Avian Biology*, **28**, 73–86.

Brockman, D. K., Whitten, P. L., Richard, A. E. & Schneider, A. 1998. Reproduction in free-ranging male *Propithecus verreauxi*: the hormonal correlates of mating and aggression. *American Journal of Physical Anthropology*, **105**, 137–51.

Brown, L., Shumaker, R. W. & Downhower, J. 1995. Do primates experience sperm competition? *American Naturalist*, **146**, 302–6.

Brun, B. & Rumpler, Y. 1990. Seasonal variation of sperm morphology in the mayotte brown lemur (*Eulemur fulvus mayottensis*). *Folia Primatologica*, **55**, 51–6.

Buesching, C. D., Heistermann, M., Hodges, J. K. & Zimmermann, E. 1998. Multimodal oestrus advertisement in a small nocturnal prosimian, *Microcebus murinus*. *Folia Primatologica*, **69:S1**, 295–308.

Burke, T. 1989. DNA fingerprinting and other methods for the study of mating success. *Trends in Ecology and Evolution*, **4**, 139–44.

Burton, F. 1971. Sexual climax in female *Macaca mulatta*. *Proceedings of the International Congress of Primatology*, **3**, 180–91.

Carosi, M. & Visalberghi, E. 2002. Analyses of tufted capuchin (*Cebus apella*) courtship and sexual behavior repertoire: changes throughout the female cycle and female interindividual differences. *American Journal of Physical Anthropology*, **118**, 11–24.

Cheng, K. M. & Burns, J. T. 1988. Dominance relationship and mating behavior of domestic cocks – a model to study mate-guarding and sperm competition in birds. *The Condor*, **90**, 697–704.

Clayton, D. A. 2000. Vertebrate mitochondrial DNA – a circle of surprises. *Experimental Cell Research*, **255**, 4–9.

Clutton-Brock, T. H. 1985. Size, sexual dimorphism, and polygyny in primates. In *Size and Scaling in Primate Biology*, ed. W. L. Jungers. New York, NY: Plenum Press, pp. 51–60.

Clutton-Brock, T. H. & Vincent, A. C. J. 1991. Sexual selection and the potential reproductive rates of males and females. *Nature*, **351**, 58–60.

Coker, C., McKinney, F., Hays, H., Briggs, S. & Cheng, K. M. 2002. Intromittent organ morphology and testes size in relation to mating system in waterfowl. *Auk*, **119**, 403–13.

Colegrave, N., Birkhead, T. R. & Lessells, C. M. 1995. Sperm precedence in zebra finches does not require special mechanisms of sperm competition. *Proceedings of the Royal Society of London, Series B*, **259**, 223–8.

Converse, L. J., Carlson, A. A., Ziegler, T. E. & Snowdon, C. T. 1995. Communication of ovulatory state to mates by female pygmy marmosets, *Cebuella pygmaea. Animal Behaviour*, **49**, 615–21.

Cords, M. 1987. Forest guenons and patas monkeys: male–male competition in one-male groups. In *Primate Societies*, ed. B. B. Smuts, D. L. Cheney, R. M. Seyfarth, R. W. Wrangham & T. T. Struhsaker. Chicago, IL: University of Chicago Press, pp. 98–111.

  2000. The number of males in guenon groups. In *Primate Males: Causes and Consequences of Variation in Group Composition*, ed. P. M. Kappeler. Cambridge: Cambridge University Press, pp. 84–96.

Cowlishaw, G. & Dunbar, R. I. M. 1991. Dominance-rank and mating success in male primates. *Animal Behaviour*, **41**, 1045–56.

Darwin, C. 1876. Sexual selection in relation to monkeys. *Nature*, **15**, 18–19.

Davies, N. B. 1992. *Dunnock Behaviour and Social Evolution*. Oxford: Oxford University Press.

de Ruiter, J., Scheffrahn, W., Trommelen, G. J. J. M., Uitterlinden, A. G., Martin, R. D. & van Hooff, J. A. R. A. M. 1992. Male social rank and reproductive success in wild long-tailed macaques. In *Paternity in Primates: Genetic Tests and Theory*, ed. R. D. Martin, A. F. Dixson & J. Wickings. Basel: Karger, pp. 175–90.

Dewsbury, D. A. 1982. Ejaculate cost and male choice. *American Naturalist*, **119**, 601–10.

Dewsbury, D. A. & Pierce, J. 1989. Copulatory patterns of primates as viewed in broad mammalian perspective. *American Journal of Primatology*, **17**, 51–72.

Dickinson, J. L. 1997. Male detention affects extra-pair copulation frequency and pair behavior in western blue birds. *Animal Behaviour*, **53**, 561–71.

Dixson, A. F. 1987a. Observations on the evolution of the genitalia and copulatory behaviour in male primates. *Journal of Zoology, London*, **213**, 423–43.

  1987b. Baculum length and copulatory behavior in primates. *American Journal of Primatology*, **13**, 51–60.

  1993. Sexual selection, sperm competition and the evolution of sperm length. *Folia Primatologica*, **61**, 221–7.

  1995a. Sexual selection and ejaculate frequencies in primates. *Folia Primatologica*, **64**, 146–52.

  1995b. Sexual selection and the evolution of copulatory behavior in nocturnal prosimians. In *Creatures of the Dark. The Nocturnal Prosimians*, ed. L. Alterman, G. Doyle & M. Izard. New York, NY: Plenum Press, pp. 93–118.

  1998a. *Primate Sexuality*. Oxford: Oxford University Press.

  1998b. Sexual selection and evolution of the seminal vesicles in primates. *Folia Primatologica*, **69**, 300–6.

Drukker, B., Nieuwenhuijsen, K, van der Werff-ten Bosch, J. J., van Hooff, J. A. R. A. M. & Slob, A. K. 1991. Harassment of sexual interactions among stumptail macaques, *Macaca arctoides. Animal Behaviour*, **42**, 171–82.

Dunbar, R. I. M. 1995. The mating system of callitrichid primates. I. Conditions for the coevolution of pair bonding and twinning. *Animal Behaviour*, **50**, 1057–70.

Dziuk, P. J. 1996. Factors that influence the proportion of offspring sired by a male following heterospermic insemination. *Animal Reproduction Science*, **43**, 65–88.

Eberhard, W. G. 1996. *Female Control: Sexual Selection by Cryptic Female Choice*. Princeton, NJ: Princeton University Press.

Eberle, M. & Kappeler, P. M. 2002. Mouse lemurs in space and time: a test of the socioecological model. *Behavioral Ecology and Sociobiology*, **51**, 131–9.

Edvardsson, M. & Arnqvist, G. 2000. Copulatory courtship and cryptic female choice in red flour beetles, *Tribolium castneum. Proceedings of the Royal Society of London, Series B*, **267**, 559–63.

Epple, G. 1986. Communication by chemical signals. In *Comparative Primate Biology 2A*, ed. G. Mitchell & J. Erwin. New York, NY: A. R. Liss, pp. 531–80.

Fawcett, K. & Muhumuza, G. 2000. Death of a wild chimpanzee community member: possible outcome of intense sexual competition. *American Journal of Primatology*, **51**, 243–7.

Fietz, J. 1999. Mating system of *Microcebus murinus. American Journal of Primatology*, **48**, 127–33.

Fietz, J., Zischler, H., Schwiegk, C. *et al.* 2000. High rates of extra-pair young in the pair-living fat-tailed dwarf lemur, *Cheirogaleus medius. Behavioral Ecology and Sociobiology*, **49**, 8–17.

Fisher, S. 1973. *The Female Orgasm: Psychology, Physiology, Fantasy*. New York, NY: Basic Books.

Foerg, R. 1982. Reproduction in *Cheirogaleus medius*. *Folia Primatologica*, **39**, 49–62.

Frank, L. 1997. Evolution of genital masculinization: why do female hyaenas have such a large "penis"? *Trends in Ecology and Evolution*, **12**, 58–62.

Froman, D. P. & Feltmann, A. J. 1998. Sperm mobility: a quantitative trait in the domestic fowl (*Gallus domesticus*). *Biology of Reproduction*, **58**, 379–84.

2000. Sperm mobility: phenotype in roosters (*Gallus domesticus*) determined by concentration of motile sperm and straight line velocity. *Biology of Reproduction*, **62**, 303–9.

Froman, D. P., Feltmann, A. J., Rhoades, M. L. & Kirby, J. D. 1999. Sperm mobility: a primary determinant of fertility in the domestic fowl (*Gallus domesticus*). *Biology of Reproduction*, **61**, 400–5.

Froman, D. P., Pizzari, T., Feltmann, A. J., Castillo-Juarez, H. & Birkhead, T. R. 2002. Sperm mobility: mechanisms of fertilising efficiency, genetic variation and phenotypic relationship with male status in the fowl, *Gallus g. domesticus*. *Proceedings of the Royal Society of London, Series B*, **269**, 607–12.

Fuentes, A. 1999. Re-evaluating primate monogamy. *American Anthropologist*, **100**, 890–907.

Gage, M. 1998. Mammalian sperm morphometry. *Proceedings of the Royal Society of London, Series B*, **265**, 97–103.

Garber. P.A. 1997. One for all and breeding for one: cooperation and competition as a tamarin reproductive strategy. *Evolutionary Anthropology*, **5**, 187–99.

Goldfoot, D. A., Westerborg-van Loon, H., Groeneveld, W. & Slob, A. K. 1980. Behavioral and physiological evidence of sexual climax in the female stump-tailed macaque (*Macaca arctoides*). *Science*, **208**, 1477–9.

Gomendio, M. & Roldan, E. R. S. 1991. Sperm competition influences sperm size in mammals. *Proceedings of the Royal Society of London, Series B*, **243**, 181–5.

Gomendio, M., Harcourt, A. H. & Roldan, E. R. S. 1998. Sperm competition in mammals. In *Sperm Competition and Sexual Selection*, ed. T. R. Birkhead & A. P. Møller. London: Academic Press, pp. 667–751.

Harcourt, A. H. 1996. Sexual selection and sperm competition in primates: what are male genitalia good for? *Evolutionary Anthropology*, **5**, 121–9.

1997. Sperm competition in primates. *American Naturalist*, **149**, 189–94.

Harcourt, A. H. & Gardiner, J. 1994. Sexual selection and genital anatomy of male primates. *Proceedings of the Royal Society of London, Series B*, **255**, 47–53.

Harcourt, A. H., Harvey, P. H., Larson, S. G. & Short, R. V. 1981. Testis weight, body weight, and breeding system in primates. *Nature*, **293**, 55–7.

Harcourt A. H. Purvis, A. & Liles. L. 1995. Sperm competition: mating system, not breeding season, affects testes size of primates. *Functional Ecology*, **9**, 468–76.

Harvey, P. H. & Harcourt, A. H. 1984. Sperm competition, testes size, and breeding system in primates. In *Sperm Competition and the Evolution of Animal Mating Systems*, ed. R. L. Smith. New York, NY: Academic Press, pp. 589–600.

Heistermann, M., Ziegler, T., van Schaik, C. P. *et al.* 2001. Loss of oestrus, concealed ovulation and paternity confusion in free-ranging Hanuman langurs. *Proceedings of the Royal Society of London, Series B*, **268**, 2445–51.

Henzi, S. P., Lycett, J. E. & Weingrill, T. 1998. Mate guarding and risk assessment by male mountain baboons during inter-troop encounters. *Animal Behaviour*, **55**, 1421–8.

Hernandez-Lopez, L., Cerezo Parra, G., Cerda-Molina, A. L. *et al.* 2002. Sperm quality differences between the rainy and dry seasons in captive black-handed spider monkeys (*Ateles geoffroyi*). *American Journal of Primatology*, **57**, 35–41.

Heymann, E. W. 2000. The number of males in callitrichine groups and its implications for callitrichine social evolution. In *Primate Males: Causes and Consequences of Variation in Group Composition*, ed. P. M. Kappeler. Cambridge: Cambridge University Press, pp. 64–71.

Hobday, A. 2000. Where is the human baculum? *The Mankind Quarterly*, **41**, 43–58.

Hoi, H. & Hoi-Leitner, M. 1997. An alternative route to coloniality in the bearded tit: females pursue extra-pair fertilizations. *Behavioral Ecology*, **8**, 113–9.

Hrdy, S. B. 1981. *The Woman that Never Evolved*. Cambridge, MA: Harvard University Press.

1995. The primate origins of female sexuality and their implications for the role of nonconceptive sex in the reproductive strategies of woman. *Human Evolution*, **10**, 131–44.

1996. The evolution of female orgasm: logic please but no atavism. *Animal Behaviour*, **52**, 851–2.

2000a. The optimal number of fathers. Evolution, demography, and history in the shaping of female mate preferences. In *Evolutionary Perspectives on Human Reproductive Behavior*, ed. D. LeCroy & P. Moller.

New York, NY: The New York Academy of Sciences, pp. 75–96.

2000b. Raising Darwin's consciousness: sexual selection and the prehominid origins of patriarchy. In *Gender and Society*, ed. C. Blakemore & S. Iverson. Oxford: Oxford University Press, pp. 143–199.

Hrdy, S. B. & Whitten, P. L. 1987. Patterning of sexual activity. In *Primate Societies*, ed. B. B. Smuts, D. L. Cheney, R. M. Seyfarth, R. W. Wrangham & T. T. Struhsaker. Chicago, IL: University of Chicago Press, pp. 370–84.

Huck, U. W., Quinn, R. P. & Lisk, R. D. 1985. Determinants of mating success in the golden hamster (*Mesocricetus auratus*). IV. Sperm competition. *Behavioral Ecology and Sociobiology*, **17**, 239–52.

Huck, U. W., Tonias, B. A. & Lisk, R. D. 1989. The effectiveness of competitive male inseminations in golden hamsters, *Mesocricetus auratus*, depends on an interaction of mating order, time delay between males, and the time of mating relative to ovulation. *Animal Behaviour*, **37**, 674–80.

Hunter, F. M., Petrie, M., Otronen, M., Birkhead, T. R. & Møller, A. P. 1993. Why do females copulate repeatedly with one male? *Trends in Ecology and Evolution*, **8**, 21–6.

Hunter, F. M., Miller, G. D. & Davis, L. S. 1995. Mate switching and copulation behaviour in the Adelie penguin. *Behaviour*, **132**, 691–707.

Janson, C. H. 1984. Female choice and mating system of the brown capuchin monkey *Cebus apella* (Primates: Cebidae). *Zeitschrift für Tierpsychologie*, **65**, 177–200.

Jennions, M. D. & Petrie, M. 2000. Why do females mate multiply? A review of the genetic benefits. *Biological Reviews*, **75**, 21–64.

Johnstone, R. A. & Keller, L. 2000. How males can gain by harming their mates: sexual conflict, seminal toxins, and the cost of mating. *American Naturalist*, **156**, 368–77.

Jolly, A. 1967. Breeding synchrony in wild *Lemur catta*. In *Social Communication among Primates*, ed. S. A. Altman. Chicago, IL: University of Chicago Press, pp. 3–14.

Jones, M. E. J. & Mench, J. A. 1991. Behavioral correlates of male mating success in multisire flocks as determined by DNA fingerprinting. *Poultry Science*, **70**, 1493–8.

Kappeler, P. M. 1993a. Sexual selection and lemur social systems. In *Lemur Social Systems and Their Ecological Basis*, ed. P. M. Kappeler & J. U. Ganzhorn. New York, NY: Plenum Press, pp. 223–40.

1993b. Female dominance in primates and other mammals. In *Perspectives in Ethology*. Vol. 10: *Behaviour and*

Evolution, ed. P. P. G. Bateson, P. H. Klopfer & N. S. Thompson. New York, NY: Plenum Press, pp. 143–58.

1997a. Intrasexual selection in *Mirza coquereli*: evidence for scramble competition polygyny in a solitary primate. *Behavioral Ecology and Sociobiology*, **41**, 115–28.

1997b. Intrasexual selection and testis size in strepsirrhine primates. *Behavioral Ecology*, **8**, 10–19.

1998. To whom it may concern: transmission and function of chemical signals in *Lemur catta*. *Behavioral Ecology and Sociobiology*, **42**, 411–21.

Keddy-Hector, A. C. 1992. Mate choice in non-human primates. *American Zoologist*, **32**, 62–70.

Kempenaers, B., Verheyen, G. R., van den Broeck, M. *et al.* 1992. Extra-pair paternity results from female preference for high quality males in the bue tit. *Nature*, **357**, 494–6.

Keller, L. & Reeve, H. K. 1995. Why do females mate with multiple males? The sexually selected sperm hypothesis. *Advances in the Study of Behavior*, **24**, 291–315.

Kholkute, S. D., Gopalkrishnana, K. & Puri, C. P. 2000. Variation in seminal parameters over a 12-month period in captive bonnet macaques. *Primates*, **41**, 393–405.

Krebs, J. R. & Davies, N.B. (eds.) 1978. *Behavioural Ecology: An Evolutionary Approach*. Oxford: Blackwell Scientific Publications.

Kummer, H. 1968. *Social Organization of Hamadryas Baboons*. Chicago, IL: University of Chicago Press.

Lack, D. 1968. *Ecological Adaptations for Breeding in Birds*. London: Chapman Hall.

Lake, P. E. 1975. Gamete production and the fertile period with particular reference to domesticated birds. *Symposia of the Zoological Society of London*, **35**, 225–44.

Lanier, D. L., Estep, D. G. & Dewsbury, D. A. 1979. Role of prolonged copulatory behaviour in facilitating reproductive success in a competitive mating situation in laboratory rats. *Journal of Comparative and Physiological Psychology*, **93**, 781–92.

Lifjeld, J. T., Slagsvold, T. & Ellegren, H. 1997. Experimental mate switching in pied flycatchers: male copulatory access and fertilization success. *Animal Behaviour*, **53**, 1225–32.

Maestripieri, D. & Kappeler, P. M. 2002. Evolutionary theory and primate behavior. *International Journal of Primatology*, **23**, 703–5.

Mah, K. & Binik, Y. 2001. The nature of human orgasm: a critical review of major trends. *Clinical Psychology Review*, **21**, 823–56.

Manson, J. 1995. Female mate choice in primates. *Evolutionary Anthropology*, **3**, 192–5.

Marson, J., Gervais, D., Cooper, R. W. & Jouannet, P. 1989. Influence of ejaculation frequency on semen characteristics in chimpanzees (*Pan troglodytes*). *Journal of Reproduction and Fertility*, **85**, 43–50.

Martin, P. A., Reimers, T. J., Lodge, J. R. & Dziuk, P. J. 1974. The effect of ratios and numbers of spermatozoa mixed from two males on proportions of offspring. *Journal of Reproduction and Fertility*, **39**, 251–8.

Martin-Villa, J. M., Longas, J. & Arnaiz-Villena, A. 1999. Cyclic expression of HLA class I and II molecules on the surface of purified human spermatozoa and their control by serum inhibin B levels. *Biology of Reproduction*, **61**, 1381–6.

Matsumoto-Oda, A. 1999. Female choice in the opportunistic mating of wild chimpanzees (*Pan troglodytes schweinfurthii*) at Mahale. *Behavioral Ecology and Sociobiology*, **46**, 258–66.

McCracken, K., Wilson, R. E., McCracken, P. J. & Johnson, K. P. 2001. Are ducks impressed by a drake's display? *Nature*, **413**, 128.

Møller, A. P. 1988. Ejaculate quality, testes size and sperm competition in primates. *Journal of Human Evolution*, **17**, 479–88.

1991. Sperm competition, sperm depletion, parental care and relative testis size in birds. *American Naturalist*, **137**, 882–906.

Moore, A. J., Gowaty, P. A., Wallin, W. G. & Moore, P. J. 2001. Sexual conflict and the evolution of female mate choice and male social dominance. *Proceedings of the Royal Society of London, Series B*, **268**, 517–23.

Muehlenbein, M. P., Campbell, B. C., Murchinson, M. A. & Phillippi, K. M. 2002. Morphological and hormonal parameters in two species of macaques: impact of seasonal breeding. *American Journal of Physical Anthropology*, **117**, 218–27.

Nicholls, E. H., Burke, T. & Birkhead, T. R. 2001. Ejaculate allocation by male sand martins *Riparia riparia*. *Proceedings of the Royal Society of London, Series B*, **268**, 1265–70.

Nunn, C. L. 1999. The evolution of exaggerated sexual swellings in primates and the graded-signal hypothesis. *Animal Behaviour*, **58**, 229–46.

O'Connell, S. M. & Cowlishaw, G. 1994. Infanticide avoidance, sperm competition and mate choice: the function of copulation calls in female baboons. *Animal Behaviour*, **48**, 687–94.

Oka, T. & Takenaka, O. 2001. Wild gibbons' parentage tested by non-invasive DNA sampling and PCR-amplified polymorphic microsatellites. *Primates*, **42**, 67–73.

Ostner, J., Heistermann, M. & Kappeler, P. M. 2003. Intrasexual dominance, masculinized genitals and prenatal steroids: comparative data from lemurid primates. *Naturwissenschaften*, **90**, 141–4.

Palombit, R. A. 1994. Extra-pair copulations in a monogamous ape. *Animal Behaviour*, **47**, 721–3.

Parker, G. A. 1970. Sperm competition and its evolutionary consequences in the insects. *Biological Reviews*, **45**, 525–67.

1984. Sperm competition and the evolution of animal mating strategies. In *Sperm Competition and the Evolution of Animal Mating Systems*, ed. R. L. Smith. London: Academic Press, pp. 1–60.

1990. Sperm competition games: raffles and roles. *Proceedings of the Royal Society of London, Series B*, **242**, 120–6.

Paul, A. 2002. Sexual selection and mate choice. *International Journal of Primatology*, **23**, 877–904.

Pellatt, E. J. & Birkhead, T. R. 1994. Ejaculate size in zebra finches *Taeniopygia guttata* and a method for obtaining ejaculates from passerine birds. *Ibis*, **136**, 97–101.

Pereira, M. E. & McGlynn, C. A. 1997. Special relationships instead of female dominance for red-fronted lemurs, *Eulemur fulvus rufus*. *American Journal of Primatology*, **43**, 239–58.

Petrie, M. & Kempenaers, B. 1998. Why does the proportion of extra-pair paternity vary within and between species? *Trends in Ecology and Evolution*, **13**, 52–7.

Petter-Rousseaux, A. 1964. Reproductive physiology and behavior of the Lemuroidea. In *Evolutionary and Genetic Biology of Primates*, ed. J. Buettner-Janusch. New York, NY: Academic Press, pp. 91–132.

Pitnick, S. & Brown, W. D. 2000. Criteria for demonstrating female sperm choice. *Evolution*, **54**, 1052–6.

Pitnick, S., Miller, G. T., Reagan, J. & Holland, B. 2001. Males' evolutionary responses to experimental removal of sexual selection. *Proceedings of the Royal Society of London, Series B*, **268**, 1071–80.

Pizzari, T. & Birkhead, T. R. 2000. Female fowl eject sperm of subdominant males. *Nature*, **405**, 787–9.

2002. The sexually selected sperm hypothesis: sex-biased inheritance and sexual antagonism. *Biological Reviews*, **77**, 183–209.

Pochron, S. T., Wright, P. C., Schaentzler, E. *et al.* 2002. Effects of season and age on the gonadosomatic index of Milne-Edwards' sifakas (*Propithecus diadema edwardsi*) in Ranomafana National Park, Madagascar. *International Journal of Primatology*, **23**, 355–64.

Preston, B. T., Stevenson, I. R., Pemberton, J. M. & Wilson, K. 2001. Dominant rams lose out by sperm depletion. A waning success in siring counters a ram's high score in competition for ewes. *Nature*, **409**, 681–2.

Primakoff, P. & Myles, D. G. 2002. Penetration, adhesion, and fusion in mammalian sperm-egg interaction. *Science*, **296**, 2183–5.

Pullen, S. L., Bearder, S. K. & Dixson, A. F. 2000. Preliminary observations on sexual behavior and the mating system in free-ranging lesser galagos (*Galago moholi*). *American Journal of Primatology*, **51**, 79–88.

Quiatt, D. & Everett, J. 1982. How can sperm competition work? *American Journal of Primatology*, **1**, 161–9.

Radespiel, U., Ehresmann, P. & Zimmermann, E. 2001. Contest versus scramble competition for mates: the composition and spatial structure of a population of gray mouse lemurs (*Microcebus murinus*) in north-west Madagascar. *Primates*, **42**, 207–20.

Radespiel, U., Dal Sekko, V., Drögenmüller, C. *et al.* 2002. Sexual selection, multiple mating and paternity in grey mouse lemurs, *Microcebus murinus*. *Animal Behaviour*, **63**, 259–68.

Reichard, U. 1995. Extra-pair copulations in a monogamous gibbon (*Hylobates lar*). *Ethology*, **100**, 99–112.

Reynolds, J. D. 1996. Animal breeding systems. *Trends in Ecology and Evolution*, **11**, 68–72.

Rice, W.R. 1996. Sexually antagonistic male adaptation triggered by experimental arrest of female evolution. *Nature*, **381**, 232–4.

Robbins, M. M. 1995. A demographic analysis of male life history and social structure of mountain gorillas. *Behaviour*, **132**, 21–47.

1999. Male mating patterns in wild multimale mountain gorilla groups. *Animal Behaviour*, **57**, 1013–20.

Ruiz-Pesini, E., Lapeña, A. C., Díez-Sánchez, C. *et al.* 2000. Human mtDNA haplogroups associated with high or reduced spermatozoa motility. *American Journal of Human Genetics*, **67**, 682–96.

Sade, D. S. 1964. Seasonal cycle in size of testes of free-ranging *Macaca mulatta*. *Folia Primatologica*, **2**, 171–80.

Sauermann, U., Nürnberg, P., Bercovitch, F. B. *et al.* 2001. Increased reproductive success of MHC class II heterozygous males among free-ranging rhesus macaques. *Human Genetics*, **108**, 249–54.

Sax, A., Hoi, H. & Birkhead, T. R. 1998. Copulation rate and sperm use by female bearded tits, *Panurus biarmicus*. *Animal Behaviour*, **56**, 1199–204.

Schmid, J. & Kappeler, P. M. 1998. Fluctuating sexual dimorphism and differential hibernation by sex in a primate, the gray mouse lemur (*Microcebus murinus*). *Behavioral Ecology and Sociobiology*, **43**, 125–32.

Schülke, O., Zischler, H. & Kappeler, P. M. 2004. Small testes size despite high extra-pair paternity in the pair-living nocturnal primate *Phaner furcifer*. *Behavioral Ecology and Sociobiology*, in press.

Schulze-Hagen, K., Leisler, B., Birkhead, T. R. & Dyrcz, A. 1995. Prolonged copulation, sperm reserves and sperm competition in the aquatic warbler *Acrocephalus paludicola*. *Ibis*, **137**, 85–91.

Schwagmeyer, P. L. 1988. Scramble-competition polygyny in an asocial mammal: male mobility and mating success. *American Naturalist*, **131**, 885–92.

Schwagmeyer, P. L. & Parker, G. A. 1990. Male mate choice as predicted by sperm competition in thirteen-lined ground squirrels. *Nature*, **348**, 62–4.

Semple, S. 1998. The function of Barbary macaque copulation calls. *Proceedings of the Royal Society of London, Series B*, **265**, 287–91.

Sheldon, B. C. 1994. Male phenotype, fertility and the pursuit of extra-pair copulations by female birds. *Proceeding of the Royal Society of London, Series B*, **257**, 25–30.

Shively, C., Clarke, S., King, N., Schapiro, S. & Mitchell, G. 1982. Patterns of sexual behavior in male macaques. *American Journal of Primatology*, **2**, 373–84.

Short, R. V. 1977. Sexual selection and the descent of man. In *Reproduction and Evolution*, ed. J. H. Calaby & C. H. Tyndale-Biscoe. Canberra: Australian Academy of Science, pp. 3–19.

1979. Sexual selection and its component parts, somatic and genital selection, as illustrated by man and the great apes. *Advances in the Study of Behavior*, **9**, 131–58.

Sillén-Tullberg, B. & Møller, A. P. 1993. The relationship between concealed ovulation and mating systems in anthropoid primates: a phylogenetic analysis. *American Naturalist*, **141**, 1–25.

Simmons, L.W. 2001. *Sperm Competition and Its Evolutionary Consequences in the Insects*. Princeton, NJ: Princeton University Press.

Small, M. F. 1989. Female choice in nonhuman primates. *Yearbook of Physical Anthropology*, **32**, 103–27.

Smith, R. 1984. Human sperm competition. In *Sperm Competition and the Evolution of Animal Mating Systems*, ed. R. Smith. New York, NY: Academic Press, pp. 601–59.

Smith, S. A. 1988. Extra-pair copulations in black-capped chickadees: the role of the female. *Behaviour*, **107**, 15–23.

Smuts, B. B. 1985. *Sex and Friendship in Baboons*. Hawthorne, NY: Aldine.

1987. Sexual competition and mate choice. In *Primate Societies*, ed. B. B. Smuts, D. L. Cheney, R. M. Seyfarth, R. W. Wrangham & T. T. Struhsaker. Chicago, IL: University of Chicago Press, pp. 385–99.

Smuts, B. B. & Smuts, R. 1993. Male aggression and sexual coercion of females in nonhuman primates and other mammals: evidence and theoretical implications. *Advances in the Study of Behavior*, **22**, 1–63.

Soltis, J., Mitsunaga, F., Shimizu, K. *et al.* 1997. Sexual selection in Japanese macaques. II. Female mate choice and male–male competition. *Animal Behaviour*, **54**, 737–46.

Sommer, V. & Reichard, U. 2000. Rethinking monogamy: the gibbon case. In *Primate Males: Causes and Consequences of Variation in Group Composition*, ed. P. M. Kappeler. Cambridge: Cambridge University Press, pp. 159–68.

Stanger, K. F., Coffman, B. S. & Izard, M. K. 1995. Reproduction in Coquerel's dwarf lemur (*Mirza coquereli*). *American Journal of Primatology*, **36**, 223–37.

Sterling, E. 1993. Patterns of range use and social organization in aye-ayes (*Daubentonia madagascariensis*) on Nosy Mangabe. In *Lemur Social Systems and Their Ecological Basis*, ed. P. M. Kappeler & J. U. Ganzhorn. New York, NY: Plenum Press, pp. 1–10.

Strier, K. 1994. Myth of the typical primate. *Yearbook of Physical Anthropology*, **37**, 233–71.

Taub, D. 1980. Female choice and mating strategies among wild Barbary macaques (*Macaca sylvanus*). In *The Macaques: Studies in Ecology, Behavior and Evolution*, ed. D. Lindburg. New York, NY: Van Nostrand Reinhold, pp. 287–344.

Thornhill, R. 1983. Cryptic female choice and its implications in the scorpionfly *Harpobittacus nigriceps*. *American Naturalist*, **122**, 765–88.

Thornhill, R., Gangestad, S. W. & Comer, R. 1995. Human female orgasm and mate fluctuating asymmetry. *Animal Behaviour*, **50**, 1601–15.

Tregenza, T. & Wedell, N. 2000. Genetic compatibility, mate choice and patterns of parentage: invited review. *Molecular Ecology*, **9**, 1013–27.

Trivers, R. L. 1972. Parental investment and sexual selection. In *Sexual Selection and the Descent of Man, 1871–1971*, ed. B. Campbell. Chicago, IL: Aldine-Atherton, pp. 136–79.

Troisi, A. & Carosi, M. 1998. Female orgasm rate increases with male dominance in Japanese macaques. *Animal Behaviour*, **56**, 1261–6.

Tutin, C. 1979. Mating patterns and reproductive strategies in a community of wild chimpanzees (*Pan troglodytes*

schweinfurthii*). *Behavioral Ecology and Sociobiology*, **6**, 29–38.

Tuttle, E. M., Pruett-Jones, S. & Webster, M. S. 1996. Cloacal protruberances and extreme sperm production in Australian fairy-wrens. *Proceedings of the Royal Society of London, Series B*, **263**, 1359–64.

van Krey, H. P. 1990. Reproductive biology in relation to breeding and genetics. In *Poultry Breeding and Genetics*, ed. R. D. Crawford. Amsterdam: Elsevier Science, pp. 61–90.

van Noordwijk, M. 1985. Sexual behaviour of Sumatran long-tailed macaques (*Macaca fascicularis*). *Zeitschrift für Tierpsychologie*, **70**, 277–96.

van Schaik, C. P. 2000. Infanticide by male primates: the sexual selection hypothesis revisited. In *Infanticide by Males and Its Implications*, ed. C. P. van Schaik & C. Janson. Cambridge: Cambridge University Press, pp. 27–60.

van Schaik, C. P. & Kappeler, P. M. 1997. Infanticide risk and the evolution of male–female association in primates. *Proceedings of the Royal Society of London, Series B*, **264**, 1687–94.

van Schaik, C. P., van Noordwijk, M. & Nunn, C. 1999. Sex and social evolution in primates. In *Comparative Primate Socioecology*, ed. P. C. Lee. Cambridge: Cambridge University Press, pp. 204–40.

Verrell, P. A. 1992. Primate penile morphologies and social systems: further evidence for an association. *Folia Primatologica*, **59**, 114–20.

Watts, D. P. 1998. Coalitionary mate guarding by male chimpanzees at Ngogo, Kibale National Park, Uganda. *Behavioral Ecology and Sociobiology*, **44**, 43–56.

Westneat, D. F. 1994. To guard or go forage: conflicting demands affect the paternity of male red-winged blackbirds. *American Naturalist*, **144**, 343–54.

Wilkinson, R. & Birkhead, T. R. 1995. Copulation behaviour in the vasa parrots *Coracopsis vasa* and *C. nigra*. *Ibis*, **137**, 117–9.

Winterbottom, M., Burke, T. & Birkhead, T. R. 1999. A stimulatory phalloid organ in a weaver bird. *Nature*, **399**, 28.

2001. The phalloid organ, orgasm and sperm competition in a polygynandrous bird: the red-billed buffalo weaver (*Bubalornis niger*). *Behavioral Ecology and Sociobiology*, **50**, 474–82.

Wolfson, A. 1954. Sperm storage at lower-than-body temperature outside the body cavity in some passerine birds. *Science*, **120**, 68–71.

Wrangham, R. W. 1993. The evolution of sexuality in chimpanzees and bonobos. *Human Nature*, **4**, 447–80.

Wyckoff, G., Wang, W. & Wu, C.-I. 2000. Rapid evolution of male reproductive genes in the descent of man. *Nature*, **403**, 304–9.

Zeh, J. A. & Zeh, D. W. 2001. Reproductive mode and the genetic benefits of polyandry. *Animal Behaviour*, **61**, 1051–63.

Zehr, J. L., Tannenbaum, P. L., Jones, B. & Wallen, K. 2000. Peak occurrence of female sexual initiation predicts day of conception in rhesus monkeys (*Macaca mulatta*). *Reproduction, Fertility and Development*, **12**, 397–404.

Zinner, D., Hilgartner, R., Kappeler, P. M., Pietsch, T. & Ganzhorn, J. U. 2003. Social organization of *Lepilemur ruficaudatus*. *International Journal of Primatology*, **24**, 869–88.

Part IV
Development and consequences

10 • Development and sexual selection in primates

JOANNA M. SETCHELL & PHYLLIS C. LEE

Department of Biological Anthropology
University of Cambridge
Cambridge, UK

INTRODUCTION

Studies of sexual selection in primates or other animals tend to focus on outcomes – sexual dimorphism, differential mating and reproductive success for adult males and females. However, adult sex differences represent the end-points of complex and interrelated developmental processes, and arise from differences in behaviour and physiology between males and females. In most vertebrates, including primates, the sexes are nearly identical in size and shape during early development, and adult differences are thus the product of divergent growth strategies (Badyaev, 2002). Sex differences in growth and development arise as a result of the different roles played by the two sexes in reproduction and the corresponding determinants of reproductive success for males and females, which are intricately linked to social organisation and mating system (Kappeler & van Schaik, 2002). Evolution shapes processes throughout the lifecycle, and the mechanisms for partitioning resources among growth, reproduction and survival are, to a large part, established during development, while consequences may not be observed until the end of the lifespan. A developmental perspective is therefore fundamental to studies of the action of sexual selection (see also Pereira & Leigh, 2003).

For mammals in general, and primates in particular, past work on sexual selection and development has concentrated on the influence of growth on sexual dimorphism (e.g. ungulates: Jarman, 1983; Georgiadis, 1985; Clutton-Brock *et al.*, 1992; seals: Trillmich, 1996; primates: Leigh, 1995; Pereira & Leigh, 2003) or growth and life-history traits such as rates of reproduction (e.g. Gordon, 1989; Pontier *et al.*, 1989; Lee & Kappeler, 2003). However, integration in the context of sexual selection of these elements – growth and development, attained adult sexual dimorphism and reproductive output – is generally lacking. In this chapter we explore interrelations between development and sexual selection, and, in so doing, highlight the paucity of available data for testing hypotheses concerning sexual selection and development in primates.

Primates are characterised by a lengthy pre-reproductive period relative to their body size (Brody, 1945; Schultz, 1956, 1969). This is a period of high risk from extrinsic sources of mortality (predation, infanticide, environmental stochasticity leading to starvation or catastrophic death), as well as death caused by growth faltering, disease or stress (Small & Smith, 1986; Janson & van Schaik, 1993; Lee, 1997; Altmann, 1998; Ross & Jones, 1999). Selection acting on the immature phase of the life history is therefore evolutionarily extremely important (Pereira & Fairbanks, 1993). Individuals who survive were either born in favourable times, or have specific genetic, maternal or learned characteristics advantageous for survival. Mortality during the pre-reproductive period is the pacemaker of life-history evolution (Promislow & Harvey, 1990), and life history underlies the maintenance of sexually dimorphic traits (Pereira & Leigh, 2003). The study of primate development from a perspective of sexual selection is particularly interesting because primate growth and development strategies differ from those of many other animals, and they attain sexual size dimorphism (where it occurs) differently. Mammals often show maximum growth rates shortly after birth, and in sexually dimorphic species dimorphism is achieved by faster post-natal growth rates from birth onwards (e.g. Jarman, 1983; Clutton-Brock *et al.*, 1992; Lee & Moss, 1995). In primates growth rates are rapid shortly after birth, then decline until puberty, at which point males (and, in some species, females) experience a peak in growth rate (the 'adolescent growth spurt', e.g. Watts, 1985, 1986; Leigh, 1992).

Investigations of sexual selection from a developmental perspective in primates have focused on sex differences in the duration of growth (bimaturism) and rates of development, and differences in costs and benefits of early or late

---

Sexual Selection in Primates: New and Comparative Perspectives, ed. Peter M. Kappeler and Carel P. van Schaik. Published by Cambridge University Press. © Cambridge University Press 2004.

maturation have been demonstrated by sex (e.g. Leigh, 1995; Leigh & Shea, 1995; Leigh & Terranova, 1998; Bercovitch, 2000, 2001). A further interesting question with regard to sexual selection is whether heterochrony explains some of the differences between species in the extent of sexual dimorphism, with selection acting to displace the timing of developmental events relative to an ancestral condition, and thus promoting or constraining sexual dimorphism as a function of patterns of development (Shea, 1983, 2000).

Development can be regarded as both a continuous process, from conception to old age, and as a series of events (discontinuities, e.g. Bateson, 1981) separating different stages, each with continuity. Underlying physiological systems can undergo abrupt changes: for example, lactase production terminates at nutritional weaning, and the sudden onset of the luteinising hormone-releasing hormone (LHRH) pulse generator determines puberty. By contrast, somatic growth processes are continuous: for example, long bone growth until epiphyseal closure (e.g. Hamada & Udono, 2002). The distinction between discontinuities and continuous processes is fundamental to assessing sex-specific vs. phase-specific selective pressures. The simultaneous occurrence of gradual and discontinuous processes during development suggests that an exploration of rates of processes and the timing of events will contribute towards understanding the mechanisms and outcomes of sexual selection (e.g. Watts, 1985; Leigh, 1992, 1995; Setchell & Dixson, 2002; Pereira & Leigh, 2003). Developmental phases are made up of a combination of physical traits, behavioural characteristics and mortality risks that can be compared between the sexes, and sexual selection may be a determinant of differences in either rate or timing or both.

MATURATIONAL 'STAGES'

We are not the first to highlight confusion and lack of consensus in the primatological literature concerning the definition of developmental phases (see also Altmann *et al.*, 1981; Pereira & Altmann, 1985; Altmann & Alberts, 1987; Caine, 1987; Bernstein *et al.*, 1991; Bercovitch, 2000). Lack of clarity or consistency in terminology means that concepts crucial to understanding development may reflect different stages or events to different authors, despite being particularly important for comparative studies of sexual selection. In this section we therefore examine developmental terms and definitions, associated maturational changes for the two sexes, and sex differences in mortality at each life-history stage. For each stage we attempt to assess the potential for the action of sexual selection on development, and show how sexual selection theory leads to different predictions for males and females growing up in specific social or reproductive contexts. The examples of sex differences in traits are selective rather than exhaustive. In many cases there are few data available for primates, and we thus use representative examples from the literature on other mammal species to make predictions for primates.

INFANCY

Definition

It is relatively easy to define a primate 'infant' (e.g. Altmann, 1980). Infancy begins at birth, continues while the individual is directly dependent on the mother for survival, and ends when the animal attains the capacity to provide its own nutrition and survive maternal death (Pereira & Altmann, 1985). Infants also rely on caretakers for transport and protection from elements, infanticide and predation (Pereira & Altmann, 1985). Weaning is a process of transition to general self-sufficiency, rather than a sharp cut-off, and includes the gradual nutritional shift from mother's milk to solid foods (Martin, 1984). Both suckling and time in contact with the mother (or caretaker) decline with age, as independence increases and infants explore their environment and learn to obtain solid food (reviewed in Pereira & Altmann, 1985; Janson & van Schaik, 1993; Lee, 1997).

Physical sex differences

Physical sex differences during infancy obviously include the primary sexual characteristics, but there can also be slight, but consistent, sex differences in body mass, with males being heavier than females in species with sexual dimorphism in adult mass (Smith & Leigh, 1998). Little is known about sex differences in growth and development during infancy (but see Leigh, 1992, 1995; Bowman & Lee, 1995; Setchell *et al.*, 2001). Rates of mass growth are highest soon after birth, and gains in body mass decelerate as infancy progresses (Leigh, 1992, 1995). The deciduous dentition erupts during infancy – effectively a discontinuity in development, as well as a process. The appearance of deciduous and then permanent teeth are often used as markers of life-history stages (e.g. Smith, 1992; Smith *et al.*, 1994; Godfrey *et al.*, 2001), and there appears to be less variation in the sequence of dental development than there is in infant mass growth (Lee, 1999), suggesting that the two phenomena may not be directly linked as life-history processes. Again, there are few data on sex differences in non-human primate dental development.

Sex differences in behaviour

In cercopithecid species where females remain in their natal group for life, associating in matrilines, but males disperse during adolescence, infant males and females both receive and exhibit differential interest towards adults of both sexes, and they differ in the patterns of interactions with peers (Berman, 1980; Brown & Dixson, 2000). Sex-differentiation of infantile behaviour or attractivity may be expected to be less marked or even non-existent in non-gregarious and monogamous taxa. However, adult behaviour is sexually differentiated in all primate species, and Nash (1993) has shown sex-differentiated sociosexual behaviour in pre-reproductive galagos, a non-gregarious species.

Sex differences in mortality

Infancy appears to include periods of sensitivity to environmental constraints, which can influence subsequent development (Worthman, 1993). Sex differences in body mass and size, and in the rate and duration of post-natal growth, although small, may be biologically significant if associated with either mortality risk or attained adult mass. If males in sexually dimorphic species grow faster, under conditions of limited nutrition they may suffer higher mortality than females. Support for this 'fragile-male' hypothesis (van Schaik & de Visser, 1990), that males are more susceptible to mortality during development, is found in some dimorphic mammal species (Lee & Moss, 1986; Clutton-Brock, 1991; Trillmich, 1996). As yet, very few data exist on primates to assess whether males are more vulnerable to early mortality, or whether differential early mortality is linked to the degree of sexual dimorphism or to a species' social system (which are, of course, all interrelated!). While male infants do have higher mortality among captive strepsirrhines (Debyser, 1995), Fedigan and Zohar (1997) found no sex difference in infant mortality in Japanese macaques. The lack of a sex difference in mortality for a relatively sexually dimorphic species may be explained by the fact that growth rates in primates diverge later in life than do those of other mammals (Leigh, 1995), or by hypotheses predicting excess female mortality (e.g. van Schaik & de Visser, 1990).

Sexual selection and infancy

The major influence on infant development is that of the mother or caretaker. Differences between individuals in maternal style (a function of dominance rank, parity, and experience) have significant consequences for the behavioural development of both male and female infant primates (e.g. Altmann, 1980), with impacts on growth processes. Maternal age and status (Setchell et al., 2001), mass (Bowman & Lee, 1995) and maternal condition (Johnson & Kapsalis, 1995) have been related to infant mass in cercopithecid primates. Sex and individual differences in neonate mass and growth rate relate to gestation costs, and raise interesting questions about maternal expenditure (see Brown, 2001; Bercovitch, 2002). It can be suggested that a skew in sex ratio at birth might represent either differential vulnerability to mortality in utero, or differential investment in the production of one sex, both of which are consequences of sexual selection (see Silk & Brown, this volume). Post-natal growth is clearly a function of maternal investment, which has the potential to vary by sex. If an infant must attain a certain threshold mass (weaning mass) before it is able to sustain itself nutritionally at weaning, then post-natal growth to weaning is under a metabolic constraint (Lee et al., 1991). The rate of attainment of weaning mass and the time taken to growing thus available for selection acting on the proximate control mechanisms of growth (hormones and growth factors). Unfortunately, almost no data exist to relate weaning mass (early growth) to weaning age for male and female infants (Lee, 1999).

While mothers respond protectively to infants as a function of mortality risks, which can potentially differ between the sexes, they do not show consistent sex biases in their treatment of infants or in care allocation (see Fairbanks, 1996). Weaning is later for sons in some primates (e.g. chimpanzees: Boesch, 1997), while it is later for daughters of subordinate mothers in rhesus macaques (Gomendio, 1990). Despite numerous analyses (e.g. van Schaik & Hrdy, 1991), no consistent sex differences in weaning age have been described. Factors such as maternal dominance, size and composition of matrilines, predation, and local food abundance interact to obscure any clear trends in sex-specific care allocation.

A further potential influence on the survival of primate infants is sexually selected infanticide by males (see van Schaik & Janson, 2000; van Schaik et al., this volume). This mortality risk can be expected to select for the evolution of traits in infants that counter the threat of infanticide. For example, Treves (1997) has shown that natal coat contrast is significantly associated with adult testes weight, suggesting a link between mating system and infant coloration, and that the natal coats found in many species of primate may serve as an infanticide-avoidance strategy.

THE JUVENILE PERIOD

Definition

A juvenile is defined as a weaned individual, capable of surviving the death of its mother (or other caretaker), but that has not yet entered puberty (Pereira & Altmann, 1985). Juveniles thus obtain their own food, travel independently, and avoid danger unaided, although case studies suggest that they may still be psychologically dependent on their mother (Pereira & Altmann, 1985). The infant–juvenile transition can be defined using various markers, including tooth eruption, lactational weaning, interbirth interval, and percentages of maternal mass. As these developmental markers do not necessarily co-occur, the 'boundaries' of this stage are difficult to determine, while the stage itself is relatively distinctive.

Physical sex differences

The rate of mass gain initially decreases during the juvenile period, then accelerates towards puberty. Although sex differences in mass remain slight, the velocity of mass growth may already differ between the sexes in sexually dimorphic species, with males growing faster than females (Leigh, 1992, 1995). This difference has implications for both foraging behaviour (i.e. energetics) and mortality risks. The deciduous dentition is completed during the juvenile period, and subsequently replaced by the eruption of the permanent dentition. A study of squirrel monkeys has shown that there are no sex differences in the eruption of the deciduous dentition, but that females are more precocious in the appearance of the permanent teeth (Galliari & Colillas, 1985).

Sex differences in behaviour

Juvenile physical and behavioural development underlies, at least in part, sex differences in adult reproductive strategies. Despite an edited volume synthesising knowledge of juvenile primate life history, development and behaviour (Pereira & Fairbanks, 1993), juveniles remain the least studied life-history stage in primates (Pereira & Leigh, 2003).

Behavioural sex differences occur in socialisation processes, interactions and competition with conspecifics, and foraging behaviour (Lee & Johnson, 1992; Nikolei & Borries, 1997), as a function of species social system (Pereira & Altmann, 1985). In sexually dimorphic species, sex differences in foraging behaviour are predicted: juvenile males should compete more for food because they are growing faster than their female counterparts and need additional energy (e.g. Pereira, 1988).

Sex differences in mortality

Juvenile mortality rates are lower than those during infancy, but are still approximately double those of adults. A recent summary of mortality for 20 strepsirrhine, New and Old World monkeys and apes found that, on average, 15 per cent of juveniles die annually (Ross & Jones, 1999). Both avoiding predation (e.g. Stanford, 1998) and maximising energy intake (e.g. Altmann, 1998) are specific problems faced by juveniles, where survival probability can be directly linked to behaviour in the juvenile period. As with infants, few data on sex-differentiated mortality exist for juvenile primates. The 'fragile-male' hypothesis (see above) predicts that juvenile mortality should be biased towards males in sexually dimorphic species because higher growth rates and consequently lower levels of fat reserves in males make them more vulnerable to nutritional stress than females. However, this hypothesis is not supported in analyses of the few data available for juvenile primates (van Schaik & de Visser, 1990; Hauser & Harcourt, 1992; van Schaik, 1992; Fedigan & Zohar, 1997).

Sexual selection and the juvenile period

Juveniles are, by definition, non-reproductive, and a juvenile's 'task' is to survive predation and starvation, in such a way as to maximise its chances of reaching the required age and/or size for reproductive maturation, and thus ensure its future reproductive success. If adult size is a consequence of either compromised or enhanced growth during the juvenile stage (owing to status and differential resource acquisition, e.g. Altmann, 1991; or environmental factors, e.g. Bercovitch & Strum, 1993), then this period underlies adult social status, morphology, reproductive tactics and reproductive success. In particular, negative energy balance during the juvenile stage may affect health and limit growth, delaying the age of reproductive maturation (Hamilton & Bulger, 1990), or altering the adult career (Pereira, 1995), and thus influencing lifetime reproductive success.

ADOLESCENCE

Definition

The transition between pre-reproductive juvenile to full adult has been termed 'puberty', 'subadulthood', 'young adulthood', and 'adolescence', depending to some extent on the definition of 'adult' used. The use of these terms is by no means standardised between species, between studies (Caine, 1987), or even between the two sexes in the same study (see references in Bernstein *et al.*, 1991).

Puberty is a suite of physiological changes that culminate in reproductive maturity (a process, not an event: Bernstein *et al.*, 1991), initiated by increases in gonadotrophic hormones and sex steroid production during sleep, and induced by the onset of activity in the hypothalamic-pituitary axis, and increased secretion of LHRH (Plant, 1994). The internal process can typically be assessed only in terms of visible outcomes (e.g. first swelling in females, first ejaculation of semen in males), which occur some time after the increase in sex steroid levels (faecal hormone analyses – e.g. Whitten *et al.* (1998) – may allow for more accurate determination of hormonal events during puberty).

Puberty ends when an individual attains reproductive competence – i.e. when a female can bear an infant to term, and a male can impregnate a female. Some authors (e.g. Pereira & Altmann, 1985) have used this criterion of reproductive maturity to define the beginning of adulthood. However, the end of puberty generally occurs prior to the attainment of adult size and the cessation of growth in primates (Watts, 1985; Bercovitch, 2000), and individuals therefore continue to invest in their own growth as well as in reproduction once they have attained reproductive maturity (see Table 10.1).

There is no doubt that reproductive maturity is an important milestone in an individual's life history, but should it be termed adulthood? A 4-year-old female mandrill suckling her first infant has reproduced successfully, but she is only two-thirds of her final adult body mass and has a further five years to grow in mass and size (Setchell *et al.*, 2001). She is unlikely to have her full permanent dentition (Setchell, unpublished observations). Similarly rhesus mothers first give birth at two-thirds of adult body mass, and many still have some deciduous teeth at this stage (Bernstein *et al.*, 1991). Are these mothers adult? A male human, aged 15 years, with large testes, a deep voice and a beard is reproductively capable, but will not attain his full adult height for another 5 years (Bogin, 1999). Male baboons of 6 years are reproductively competent, but may be too physically

and socially inferior to other males to copulate with a fertile female (Alberts & Altmann, 1985a). Unflanged male orangutans aged 23 years can (and do) reproduce, but have arrested secondary sexual development, and show none of the typical adult male secondary sexual traits of the species (Utami *et al.*, 2002; Utami & van Hooff, this volume). Can we term these males adult?

Caine (1987) defines reproductively capable, but non-reproducing individuals as adolescent, and we concur, defining 'adolescence' as the period from the onset of puberty to the attainment of full adult size (see also Watts, 1985). Puberty and adolescence thus have the same onset, but while puberty ends at reproductive maturity, adolescence continues until adult size and appearance are attained, and somatic growth is complete. If the term 'adolescence' is ambiguous, then 'subadult' is even more loosely applied in the literature (Caine, 1987), and is usually only used for males. We question the use of this term, finding it more useful to consider the chronological age and developmental stage of individuals studied.

Adolescence is thus a period of rapid morphological, physiological and behavioural development, leading to the cessation of physical growth, endocrine stability, and social competence. By the end of adolescence, female primates are sexually mature, fully grown, and have begun their reproductive career. Males have attained sexual maturity, adult dominance rank (in group-living species), and may have sired offspring. In males this last milestone is dependent on proximate, often stochastic variables; for example, the age at which a male baboon first consorts with a female depends on the number of females undergoing sexual cycles in the group and his relative rank (Alberts & Altmann, 1995a).

Physical sex differences

During puberty and adolescence physical sex differences become pronounced in sexually dimorphic primates. However, relatively little is known about physical development in pubertal and adolescent primates, particularly for strepsirrhines (reviewed in Caine, 1987). As well as initiating sexual maturation, pubertal hormones lead to events and new processes in physical growth and development. The skeletal system matures, with fusion of the epiphyses. Males, and females of some species, show a growth spurt in mass (Leigh, 1992, 1995). Complete adult dentition is attained with the eruption of the last permanent teeth (Watts, 1985). In strepsirrhines, puberty is indicated by the advent of particular kinds of scent-marking (e.g. genital scent-marking

Table 10.1 *Stages of development defined by physiological signals, with approximate age of occurrence for chimpanzees (*Pan troglodytes*), from Kraemer et al. (1982); and mandrills, from J. M. Setchell and E. J. Wickings (in preparation).*

Period	Development			Chimpanzees' age [range (mo)]		Mandrills' age [range (mo)]	
	Males	Males and Females	Females	Males	Females	Males	Females
Infant		Mass gain decelerates Deciduous dentition erupts		0–24		0–8	
Juvenile	Low T and LH levels Testes grow slowly	Mass gain decelerates early, accelerates towards puberty Deciduous dentition complete		24–60		8–46	8–43
Puberty	Testes growth accelerates Production of sperm SSCs begin to develop	Rise in LH results in increased gonadotrophic hormones and sex steroid production Loss of deciduous dentition and eruption of adult dentition Reproductive maturity attained	First, irregular sexual swellings Menarche First ovulation	60–84	60–96	46–?	43–48
Adolescence (includes puberty)	Mass growth spurt	All deciduous dentition lost	Continued investment in own growth as well as reproduction	84–132	96–144	46–120	43–84
Adult	T peaks SSCs continue to develop Adult levels of T Adult SSCs	Adult dentition completed Epiphyseal union Full size attained		132+	144+	120+	84+

LH: luteinising hormone

SSCs: secondary sex characters

T: testosterone

in ring-tailed lemurs: Pereira, 1995). The relationships between hormones and aspects of growth and development remain unclear, as do interrelationships between developmental markers (Watts, 1985).

In female primates, the release of luteinising hormone leads to the maturation of ovarian tissues during puberty. Females of many species show tumescence, with changes in facial or perineal coloration during this period (Caine, 1987). Pubertal sexual swellings may occur even in species where adults do not show sexual swellings (Dixson, 1998). However, although first sexual swelling, or first menstruation, are often used as an indicator of puberty, neither are signs of full sexual maturity in females. Both can occur prior to first ovulation, which is a late event in female puberty (Tanner, 1962). First ovulation itself can only be determined hormonally, and does not indicate capacity to carry a pregnancy to term. The first several menstrual cycles are often irregular and anovulatory, while complete maturation of luteal function, necessary to maintain embryo implantation, occurs last (Hobson et al., 1980). Female reproductive maturity is attained when the hormonal cycle is regular, and is not achieved until about 5 years after menarche in human females (Bogin, 1999). It is difficult to estimate the age at onset of puberty from external signals in females, and occurrence of the various proxies used is not necessarily simultaneous or inter-correlated (Bercovitch, 2000). Once physiologically mature, a female primate starts to reproduce, perhaps the most major change in a primate's life since her own birth. However, even after the first birth, adolescent females may not be as fertile as adult females, with longer interbirth intervals than fully grown females, most likely because they are still investing in their own growth or body condition to some extent (e.g. Paul & Thommen, 1984; Itoigawa et al., 1992; Setchell et al., 2002; see below).

In males, the first visible sign of puberty is descent or rapid enlargement of the testes, and this is often used as a marker for the onset of male puberty (Rowell & Dixson, 1975; Nigi et al., 1980; Altmann et al., 1981; Pereira & Altmann, 1985; Watts, 1985). This is more useful than the visible markers used for females, as testicular development occurs fairly early in puberty (e.g. Nieuwenhuijsen et al., 1987a). As the testes develop, the Leydig cells produce increasing quantities of testosterone, although there is a time lag of several months before circulating levels of testosterone begin to increase (Glick, 1979; Nigi et al., 1980; Kraemer et al., 1982). The tubules and cellular elements involved in sperm production proliferate, leading ultimately to the production of mature sperm (Dang & Meussy-Dessolle,

1984). Male primates are fertile before testes reach full 'adult' dimensions (e.g. Nielson et al., 1986; Wickings & Dixson, 1992), and the testes reach adult dimensions at approximately the same age as does the body (Kraemer et al., 1982; Nieuwenhuijsen et al., 1987a; Setchell & Dixson, 2002). Testosterone levels continue to increase, and secondary sex characters mature under the influence of the increased production of gonadal steroid hormones. Males of some species develop sexual ornamentation. Although the development (and individual variation in development) of these characteristics is rarely documented (but see Tanner, 1962; Watts, 1985; Liang et al., 2000; Setchell & Dixson, 2002), it is linked to dominance rank in some species (reviewed in Setchell, 2003). For example, only dominant male mandrills develop full adult male traits, and subordinate status has been linked to failure to develop secondary sexual traits in male orangutans (Utami & van Hooff, this volume). The maturational process takes time, and changes may occur over a period of years (e.g. 4 years in male rhesus: Bernstein et al., 1991; and mandrills: Setchell & Dixson, 2002).

Sex differences in behaviour

As hormonal and physical maturation proceed during adolescence, changes in behaviour also occur. This process of behavioural development is somewhat better understood than physical development and has been comprehensively reviewed elsewhere (Kraemer et al., 1982; Pereira & Altmann, 1985; Caine, 1987; Pusey, 1990), although adolescence remains under-represented in the primate literature. During this period, individuals continue to make the transition from maternal association to integration into the adult community, or dispersal, and adult behaviour develops. Natal dispersal is often sexually differentiated, with consequences for sexual selection (Pusey & Packer, 1987; van Noordwijk & van Schaik, 2001, this volume). The relationships between physiological maturation and behavioural development are not well understood (but see Kraemer et al., 1982; Pusey, 1990; Setchell, 1999; Table 10.1). Pusey (1990) has shown that stage of genital development is more closely correlated with a decline in association with the mother in male chimpanzees than is chronological age or body mass. In mandrills, male peripheralisation is more related to morphological development than to age (Setchell & Dixson, 2002). Behavioural changes, for example in male aggression and hierarchical behaviour (Kraemer et al., 1982; Nadler et al., 1987), appear to be associated with adult hormone production and function and adult size, rather than linked to the onset or event marking the increase in hormone levels.

Morphological maturation also elicits a change in the responses of other individuals. Adult male baboons (Scott, 1984) and mandrills (Setchell, 1999) show interest in adolescent female cycles, although this is less than that for full adults. Young cycling (tumescent) female baboons and mandrills often appear nervous (Scott, 1984; Setchell, unpublished observations), suggesting that proceptive and receptive behaviour have major learned components. Dominant males cease to tolerate sexual behaviour from what are now reproductively competent males, and therefore potential rivals. Juvenile male stump-tailed macaques copulate 'publicly', but post-pubertal males do so surreptitiously (Nieuwenhuijsen et al., 1987b), while 5–6-year-old sooty mangabeys receive more aggression and chases from dominants when they mount females than do younger males (Gust & Gordon, 1991). Males developing conspicuous secondary sex characters may also receive increased aggression (e.g. Rowell, 1977). Thus the frequency of copulation may decline in adolescent males, although adult behaviour – such as consortships – begin, and successful consortships may be achieved as males approach adult size (e.g. Pusey, 1990). For species where relationships and integration into a group of adult males are of paramount importance to male reproductive success, males cultivate relationships with other males (Pusey, 1990). Little is known about the reproductive success of adolescent male primates. Adolescent males are generally subordinate and physically inferior (less powerful or skilled in contests) to 'prime' males, and mate opportunistically with less attractive females (e.g. Kuester & Paul, 1999; Setchell, 1999). Ejaculate volume and sperm number have implications for fertility (Birkhead & Kappeler, this volume), but are unknown for most species.

Sex differences in mortality

Where variation in reproductive success is higher in males than in females, male reproductive strategies are expected to involve a higher mortality risk, but a higher potential reproductive gain than those of females (Trivers, 1985). In sexually active primates, males may range over larger areas than females in search of mates or while dispersing, and thus have a higher risk of death from predation or starvation (Isbell et al., 1990; Alberts & Altmann, 1995b). Males also risk serious injury or death along with energetic costs of male–male competition for access to receptive females. For example, Fedigan and Zohar (1997) found that adolescent and adult

Japanese macaque males have higher mortality than females and juvenile males, and males remain at greater risk than females until old age. In addition, high levels of circulating androgens can suppress immunity (Folstad & Karter, 1992); simply being male involves costs that are not incurred by females. The high-risk, high-benefit strategies employed by males are ultimately responsible for female-biased sex ratios in polygynous species (Clutton-Brock et al., 1977; Clutton-Brock, 1991; Kappeler, 1999).

The extent of sexual dimorphism is generally related to the extent of male–male competition (Plavcan, 1999, this volume) but also appears to be associated with which sex disperses from the natal unit. Using a measure of adult sexual dimorphism based on the residual of the regression of female mass against male mass (where the regression was calculated separately for strepsirrhines and anthropoids), species with male dispersal have significantly smaller females for male body size (ANOVA, hierarchical model, removing subfamily on step 1 to control for phylogenetic similarity, $F = 5.09$, $df = 2, 89, p = 0.009$, post hoc, all $p < 0.05$; methods and data set in Lee & Kappeler, 2003). It can thus be suggested that increased costs of ranging, exposure to predation, and risks of injury due to contest competition tend to be associated with male dispersal, and that attaining large male size relative to that of females may be advantageous in this context. As dispersal is often initiated in adolescence, a relationship could be predicted between the timing of the male growth spurt and which sex disperses, but such data are not currently available.

Sexual selection and adolescence

Individuals experience the same sequence of events during adolescence, but can vary considerably in the rate of changes and when these occur during the process (e.g. Tanner, 1962; Watts, 1985; Setchell & Dixson, 2002). Differences in the age at first reproduction, in attained mass or size, in the development of coloration or other signals, weaponry, and behaviour associated with reproduction, all influence reproductive success. Thus, all are the product of sexual selection, while the variation between individuals is the focus of selection.

The effect of nutrition or energy balance on the age at which female primates reach menarche suggests that a minimum investment in growth and body condition is required before reproduction occurs (e.g. Schwartz et al., 1988; Surbey, 1990; Bercovitch & Strum, 1993; Bercovitch, 2000; Setchell et al., 2002). Thus, early maternal influences on growth and individual success as a juvenile are factors that may pace processes in adolescence. Social factors are also important in the regulation of puberty (e.g. Vandenbergh *et al.*, 1972; Vandenbergh & Coppola, 1986; Surbey, 1990; Worthman, 1993), and intrasexual competition influences age at maturity in female strepsirrhines (Izard, 1990), Old World monkeys (Bercovitch & Goy, 1990), New World monkeys (Abbot *et al.*, 1990; Ziegler *et al.*, 1990) and apes (Graham & Nadler, 1990). In tamarins and common marmosets, males contribute to infant care, and reproductive skew is high among females, producing intense female–female competition. Reproductive function and puberty are physiologically suppressed in subordinate females in the presence of dominants, and high-ranking females interrupt the copulations of other females (Abbott *et al.*, 1990; Ziegler *et al.*, 1990; Dixson, 1998).

The timing of puberty in males is influenced by neonatal testosterone (Mann *et al.*, 1989, 1998; Eisler *et al.*, 1993; Lunn *et al.*, 1994), and conditions during the infant stage thus have consequences for pubertal development. The timing of sexual maturation may also be affected by the social environment. For example, the onset of puberty occurs earlier in higher-ranking male rhesus macaques (Bercovitch, 1993) and baboons (Alberts & Altmann, 1995a). High-ranking male mandrills have larger testes for their age than do lower-ranking males (Setchell & Dixson, 2002). There may be a mass threshold that is necessary, but not sufficient, for reproductive maturation in males (Bercovitch, 2000; Setchell *et al.*, 2001), although age at onset of puberty in male stump-tailed macaques is not significantly correlated with mass, rank or maternal rank (Nieuwenhuijsen *et al.*, 1987a). Maternal rank affects adolescent social and physical development (see also Colvin, 1983): male rhesus macaques born to high-ranking mothers have higher levels of circulating testosterone and larger testes during adolescence (Dixson & Nevison, 1997), and adolescent sons of high-ranking female mandrills are heavier for their age than are the sons of low-ranking females (Setchell *et al.*, 2001). Prepubertal differences in body mass are likely to determine the dominance rank of age-mates in adolescence, with heavier males continuing to dominate lighter age-mates (Lee & Johnson, 1992; Pereira, 1995). Dominant males have higher testosterone levels, and hence develop more conspicuous secondary sexual traits more quickly, although in captive primates, high levels of testosterone are a result of, rather than a predictor of, high dominance rank (Sapolsky, 1993). Low testosterone levels in subordinate males may be due to stress-induced suppression of testicular function (Sapolsky, 1985; Graham & Nadler, 1990), while differences in the hormone respon-

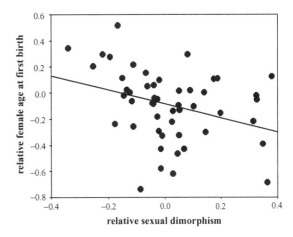

Fig. 10.1   The correlation between relative sexual dimorphism and relative female age at first birth for primate species. Sexual dimorphism is expressed as the unstandardised residual of male mass to female mass, and age at first birth has been expressed as the unstandardised residual of age for female body mass. Life-history data from Lee and Kappeler (2003); sexual dimorphism data from Smith and Jungers (1997).

siveness of tissues, feeding competition, and/or metabolic efficiency may play a part in constraining the rate and extent of development for subordinate males (Bercovitch, 2000).

The timing of full reproductive maturation represents an opportunity for the action of sexual selection, but this is not clear-cut. In general, primate species with relatively high male-biased sexual dimorphism tend to be those where females reproduce relatively early (Fig. 10.1), suggesting that dimorphism has costs for both sexes. Early maturation may represent a trade-off between decreased investment in growth against early reproduction, or it may be a consequence of higher risks taken early in life, thus attaining a threshold mass at a younger age. Individuals who delay reproduction face increased mortality risks as well as a loss of reproductive time. We need far more data on individual growth patterns up to sexual maturity to be able to relate sexual selection to strategies for growth and dispersal at adolescence.

ADULTHOOD

Definition

The problem of what constitutes an adult primate is probably the most important one for comparative studies of selection pressures (see Bernstein *et al.*, 1991; Bercovitch,

2000). As with all developmental stages, the characteristics used to define adults are neither uniform nor necessarily linked to one another. Comparative studies of 'adult' characteristics across mating systems, be they body size, mass, or differential reproductive success, are severely limited by differing usage of the term 'adult'. Furthermore, greater attention is needed regarding the several adult life-history stages that are usually classed together as one ('adulthood') or two ('adolescence' and 'adulthood') stages. For example, Pereira and Altmann (1985) have suggested that 'adulthood' be divided into 'young', 'mature', and 'aged' stages. From their perspective, 'adolescent' primates are gametogenic but not yet reproductive, young adult primates grow and reproduce (here termed 'adolescent'), mature adults reproduce but have ceased to grow, and senescent adults are past their prime, show signs of old age, and have reduced reproductive rates. We consider adults as socially integrated, reproductive individuals that have ceased physical growth and attained endocrine stability.

'Adulthood' itself does not represent an end-point. Rather, maturation is a life-long process that does not cease at the onset of adulthood. Age-graded changes in physical parameters, in fertility, in reproductive strategies, and in reproductive experience have all been widely discussed in relation to sexual selection. As such we will simply highlight some areas where patterns in early development lead to predictions about reproductive tactics used by the two sexes.

Physical sex differences

Although statural growth ceases at epiphyseal closure, growth in body mass does not stop abruptly, but declines gradually (Leigh, 1992), and gradual changes in fertility, bone mass and body mass occur throughout an individual's lifetime (e.g. Schwartz & Kemnitz, 1992; Johnson & Kapsalis, 1998). Captive male orangutans continue to increase in body mass throughout their adult life (Leigh & Shea, 1995). Dentition becomes worn further, affecting body mass (e.g. Phillips-Conroy et al., 2000). Secondary sexual characters may change during the adult phase. For example, male mandrill coloration can both increase and fade during adulthood (Setchell & Dixson, 2001b), and cheek flanges in orangutans develop relatively late in the lifespan of a male (see Utami & van Hooff, this volume).

Sex differences in behaviour

Sex differences in the behaviour of adult primates are well known and need not be reiterated here. Of interest here are those elements of behaviour that relate to sexual selection acting earlier in the developmental stages, and as there have been so few long-term studies on different primate species, much is speculation. For example, we know very little about the consequences of early reproduction in females and later mortality. Does early reproduction in fruit flies (Partridge, 1988) and captive elephants (Clubb & Mason, 2002) lead to earlier death, as has been suggested for humans (Lycett et al., 2000)? Does investing heavily in one offspring to ensure its growth and survival lead to depletion of maternal resources and thus lowered investment in subsequent offspring?

Primiparous females have longer interbirth intervals than do experienced females (e.g. Bercovitch et al., 1998; Setchell et al., 2002), suggesting that they have fewer resources to invest in offspring. Interbirth intervals increase again in old females (Strum & Western, 1982). As fertility declines, females may invest more in each individual offspring or litter, more in one sex of their offspring than in another, or employ more costly tactics to ensure fertility. However, convincing tests of hypotheses as to the extent and nature of sex biases in maternal investment are difficult to carry out (Brown & Silk, 2002; Silk & Brown, this volume).

As with females, male reproductive success is age-graded, related to experience, and associated with different behavioural tactics (e.g. Dunbar, 1984; Sommer & Rajpurohit, 1989; Alberts & Altmann, 1995b; van Noordwijk & van Schaik, 2001, this volume; Utami & van Hoof, this volume). Changes in reproductive potential and fighting ability over a lifetime should affect mate-choice criteria and competitive tactics. While prime males may rely on competition to gain access to fertile females and guard them from other males' mating attempts, younger males, or senescent males, may employ sneaky tactics to obtain matings. Other examples of tactical decisions in group-living species that depend on age and condition include: when to transfer from one group to another; whether to associate with a group of females, or with an all-male band, or to live alone; and whether to challenge higher-ranking or harem males for access to fertile females. In non-gregarious species, decisions also include dispersal and ranging tactics – such as whether to challenge for possession of a territory – and attempt to exclude other males from access to females, or to range as a 'floating' male and sneak copulations (e.g. galagos: Charles-Dominique, 1977; Bearder, 1987; orangutans: Utami & van Hoof, this volume). Again, more detailed information is required from long-term studies to determine the stability and reproductive pay-offs of alternative reproductive tactics over the lifespan in male primates, and specifically how these relate to early physical and social development.

CONSEQUENCES OF SEXUAL SELECTION FOR DEVELOPMENT

SEXUAL DIMORPHISM AND DEVELOPMENT

While it is relatively easy to measure sexual dimorphism, understanding the impact of sexual selection on dimorphism is more problematic (Plavcan, 1999, this volume). However, selection may be inferred from patterns and processes of growth and maturation. With the exception of growth in body mass (notably Leigh, 1992, 1995) and studies of humans (e.g. Tanner, 1962; Malina, 1978), detailed studies of growth to maturity are extremely rare for primates, particularly for wild populations (but see Altmann & Alberts, 1987; Strum, 1991; Cheverud *et al.*, 1992).

Leigh (1992, 1995) investigated how mass growth rates diverge between male and female primates, and sex differences in the length of the growth period. Using mixed longitudinal and cross-sectional body-mass data for 37 species of captive primate, he showed that, as predicted from patterns of adult size dimorphism, species with low levels of intermale competition do not show sex differences in development (monogamous/polyandrous mating systems). Where it occurs, adult sexual size dimorphism can arise via sex differences in the duration of growth (sexual bimaturism), by differential growth rates in the two sexes, or by a combination of the two (Gavan & Swindler, 1966; Shea, 1986). These different ontogenetic trajectories, bimaturism (male time hypermorphosis) and differential growth rates (male rate hypermorphosis) can, none the less, produce similar levels of adult dimorphism, as observed in relatively closely related species (Leigh, 1992). Species with a multi-male–multi-female mating system develop sexual size dimorphism through bimaturism, with minimal sex differences in growth rate, and males attain adult mass later than do females. By contrast, in species where a single male monopolises a group of females, dimorphism is attained through males growing faster than females, with less input from increased duration of growth. Leigh (1995) relates these ontogenetic differences to the distribution of risks during a male's development, suggesting that growth-rate differences occur in taxa where the lifetime distribution of risks changes rapidly (uni-male–multi-female groups), whereas bimaturism occurs in species that have a relatively uniformly changing or stable risk distribution (multi-male–multi-female groups).

In addition to linking mortality risks, patterns of development and sexual dimorphism, Leigh's data suggest independence between adult morphology and the specific pathway to attaining that morphology; developmental trajectories can vary in pattern but yield similar adult morphologies. Rather than selection targeting a particular degree of adult dimorphism, it may be targeting development, and adult dimorphism thus is not constrained by the pattern of development. Leigh (2001) has recently highlighted this flexibility of developmental stages in primates, demonstrating a lack of correlation in the relative duration of the infant, juvenile and adolescent growth periods. This has important implications for selection, which can alter these periods independently. The degree to which patterns of development and adult morphology are linked and constrained, however, remains an open question.

SEX DIFFERENCES IN TIMING OF REPRODUCTIVE MATURATION

Differential selection for enhanced male size and competitive ability, in combination with a divergence between the reproductive strategies of males and females (e.g. Wilner & Martin, 1985), has led to sex differences in reproductive development and maturity. Bercovitch (2000) collated data on age at puberty, first parturition, and full size for males and females of polygynous cercopithecid species, showing that females enter puberty significantly earlier, and reach adult body size significantly earlier than do males. He also makes the important distinction between the onset of reproductive capacity (which occurs around the same age in males and females), and the observed onset of reproduction (on average, later in males than females). As a result, the interval between potential and actual reproduction is shorter for females than for males. However, male reproduction can occur earlier in the final growth phase than it does for females. This is most likely because gametogenesis is relatively cheap for males and thus not in conflict with growth. By attaining physiological maturity early, the potential age at first reproduction can be reduced for males who employ surreptitious mating tactics.

INVESTMENT IN GROWTH VERSUS INVESTMENT IN REPRODUCTION

All else being equal, individuals should mature and reproduce as early as possible, since the risk of mortality increases with any delay to onset of reproduction. Only when the benefits of delayed reproduction outweigh the mortality costs associated with delay should reproduction be postponed. There is the potential for conflict between the allocation of resources to growth rather than reproduction and investment in progeny, if females are still growing when they reproduce for the first time. This conflict may be

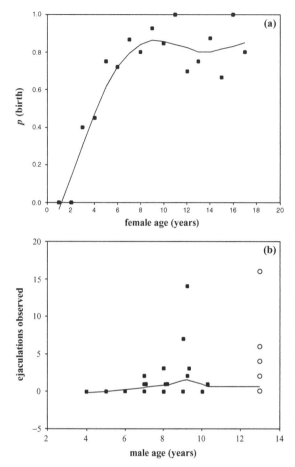

Fig. 10.2   Reproduction vs. age in mandrills: (a) probability that a female mandrill gives birth vs. age; (b) successful copulations (ejaculations) observed for males vs. age. Open circles are adults of unknown age.

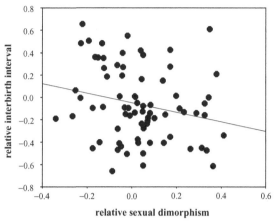

Fig. 10.3   The relationship between relative interbirth intervals and relative sexual dimorphism ($F = 4.54$, df $= 1, 81$, $p = 0.037$, controlling for phylogeny at the subfamily level) for primate species. Data as Fig. 10.1.

accentuated when there are energetic limitations (Strum, 1991; Martin et al., 1994). At the point where the advantages of extra growth (either for neonate or individual survival) are outweighed by the advantages of earlier reproduction, females allocate resources to maintenance, growth and reproduction, with consequences for reproductive outcomes in primiparous mothers (Wilner & Martin, 1985; Strum, 1991; Martin et al., 1994).

Male reproductive effort depends on gaining access to fertile females. Where male mating opportunities are related to contest ability, and males compete directly for access to females, they achieve only occasional successful matings before they attain full adult size (e.g. mandrills – Fig. 10.2), although they reach sexual maturity long before this point. Males of such species therefore invest in traits that will lead

to mating success: dominance rank, skills in assessment, large body size, weaponry and displays where contest competition is important (Wickings & Dixson, 1992; Setchell & Dixson, 2001a, b); and fat reserves where endurance rivalry is important (e.g. Bercovitch, 1983; Alberts et al., 1996). Sexual dimorphism has important costs for males in terms of time and energy. Males grow for longer (bimaturism) and/or invest more resources in growth during the growth period (rate dimorphism) than females, and the associated postponement of reproduction increases the risk of mortality before males begin their reproductive career. Thus the fitness benefits of large body size, in terms of high rank and access to females, must offset these costs. Large body size in males reduces vulnerability to predators, is energetically advantageous during dispersal, reduces susceptibility to disease (Scanlan et al., 1987; Raleigh & McGuire, 1990), and females may prefer to mate with larger males.

As we note above, early female reproduction (for body mass) is associated with higher sexual dimorphism, not just among the cercopithecids, suggesting some causal or selective relationship between the pace of growth and reproductive strategies of males and females. This observation lends additional support to the proposal that one mechanism for enhancing dimorphism is for females to start reproduction earlier and at a smaller size, and thus divert resources away from growth (Wilner & Martin, 1985). Interestingly, sexual dimorphism is also associated with relative reproductive rates – with shorter interbirth intervals (for female mass) found in species with higher dimorphism (Fig. 10.3).

More dimorphic species tend to lead more rapid reproductive lives, and we expect that early mortality should underlie at least some of these patterns. All in all, males of highly sexually dimorphic species appear to gain additional advantages for their reproductive potential – their females begin reproduction relatively early and they reproduce relatively frequently for their mass. This suggests to us that, over and above the energy, time and mortality cost inherent in growing rapidly and/or for longer, dimorphism contributes to the 'average' male's reproductive success. As dominant males in highly dimorphic species may also gain priority-of-access paternity (e.g. Altmann *et al.*, 1996), the advantages of enhanced male size appear to be significant.

CONSEQUENCES OF DEVELOPMENT FOR SEXUAL SELECTION

Developmental trajectories have potentially far-reaching consequences for the action of sexual selection (Leigh & Pereira, 2003). Ecological and social conditions during neonatal and pre-reproductive development influence adult reproductive potential and behavioural strategies (Harcourt & Stewart, 1981; Draper & Harpending, 1982; Altmann, 1998; Pereira & Leigh, 2003). Development is modulated by maternal factors, nutrition, disease and psychological stress, all of which impact on adult outcomes. Furthermore, the reproductive tactics employed by individuals to maximise fitness are influenced by environmental conditions, age, physical or reproductive condition, population density or structure (Brockman, 2001), which may act at any point from infancy onwards. Different ontogenetic pathways may be set early in life, as hypothesised for orangutans (Maggioncalda *et al.*, 1999). It is therefore important to include ontogeny in studies of alternative mating tactics (Caro & Bateson, 1986), and to relate differences in ontogeny to differences during adulthood. Maternal and allomaternal care have major consequences for infant survival and mortality in primates, which again interact with the potential for selection on female size, contest ability and reproductive strategies. Moreover, sexual selection may vary in action or intensity over the course of an individual's lifetime. For example, adolescent or subordinate adult males, faced with a competitive disadvantage and potentially increased mortality, will pursue alternative mating tactics, which should lead to a shifting pattern of sexual selection over time (e.g. Setchell & Dixson, 2001a, b; Utami *et al.*, 2002; Utami & van Hooff, this volume).

Furthermore, developmental rates and patterns of mortality may constrain the potential for the evolution of sexual dimorphism. In the interspecific comparisons attempted

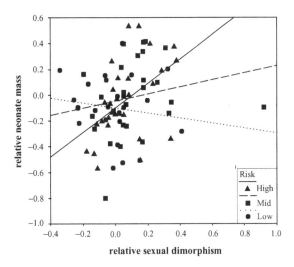

Fig. 10.4   The correlation between relative neonate mass and relative sexual dimorphism ($r = 0.256$, $n = 86$, $p = 0.017$ overall), shown separately for those primate species where predation risk categories were assigned ($n = 80$). High- and low-risk categories differ ($p < 0.05$ post hoc), controlling for phylogeny at the subfamily level. Data as Fig. 10.1.

here, relative birth mass was positively correlated with sexual dimorphism – the larger the male relative to the female, the greater the relative size of the neonate (Fig. 10.4). However, this relationship varies as a function of mortality risks due to predation. For species with little risk of predation, neonates are relatively smaller (and cheaper to produce) when dimorphism is greater, while the opposite is true for neonates born to species where predation risk is high. Predation may be driving both male mass and birth mass to increase, thus enhancing the ability of individuals to escape predation; but when predation risks are low, sexual dimorphism arises from male competition and females may be producing cheaper infants when infanticide risks due to this competition are high. A similar trend is found in age at weaning (Fig. 10.5). Relatively highly dimorphic species wean early when the risks of mortality due to predation are low (and possibly the risks of infanticide are high), but wean late when the predation risks are high. Strategies to ensure infant survival in risky environments (e.g. Janson & van Schaik, 1993) interact with sexual selection, and impact on growth patterns for different species. This is further emphasised when environmental risks are nutritional. The relatively low levels of mass dimorphism amongst neotropical primates may reflect the limited energy availability from patchy fruit supplies (e.g. Ford, 1994), which again constrains the potential for rapid or prolonged growth during the developmental period.

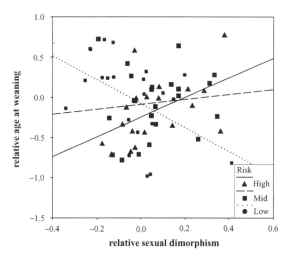

Fig. 10.5   The relationships between relative age at weaning and sexual dimorphism as a function of predation-risk category for 65 primate species. The correlations for high risk ($r = 0.493$, $n = 19$, $p = 0.032$) and low risk ($r = -0.490$, $n = 23$, $p = 0.025$) are significant. Data as Fig. 10.1.

Similarly, Leigh and Terranova (1998) suggest that extreme seasonality, reduced growth periods and relatively early male maturation in lemurs when compared with anthropoid primates of similar body sizes, preclude the evolution of sexual dimorphism through bimaturism, despite high levels of inter-male competition.

Examination of relationships among the timing of events during development, fecundity, mortality and reproductive tactics in primates requires long-term studies, following the fate of individuals and cohorts from birth to death. Clearly, studies linking experiences during development to reproductive tactics and fitness, such as those on red deer (Kruuk et al., 1999) and yellow baboons (Altmann, 1991), are required to identify ontogenetic strategies. Studies of sexual selection will benefit greatly from the application of a developmental and life-history perspective.

SUMMARY AND CONCLUSIONS

Our aim in this chapter has been to explore how sexual selection relates to development over the lifespan, and how patterns of sexual selection change over the course of an individual's lifetime. Sexual dimorphism in body size, behaviour and physiology results from selection pressures acting on male and female characters independently, and in different ways, during development. Within-species, sex-specific

growth rates and developmental trajectories evolve when age-dependent expenditure on reproductive effort differs by sex (Wiley, 1974). While we currently have insufficient longitudinal data on development and its consequences to specify mechanisms for either sex, some patterns are clear. Female primates are time-limited, needing to attain the minimum mass required to conceive and support the costs of pregnancy and lactation during maturation. Energetic and time constraints thus suppress female growth in favour of allocating resources to reproduction (Demment, 1983; Strum, 1991; Martin et al., 1994). Mortality risks also influence the life-history strategies of females, enhancing opportunities for sexual selection, and specifically acting on sexual size dimorphism. By contrast, where male reproductive success depends on fighting ability and body size, male maturation is delayed to achieve larger size and competitive ability. This delay, which involves greater expenditure and risk prior to first reproduction, may be profitable for male mating effort, especially as highly dimorphic species also appear to have rapid reproductive rates.

Before we can fully understand how sexual selection operates on development and how development constrains or facilitates sexual selection, we need more data on primate patterns of development (see also Pereira & Leigh, 2003) and, in particular, to be able to link discontinuities with the gradual processes of development. Individual variation in rates of growth, age at reproductive maturation, social contexts and paternity need to be brought together to gain a synthetic perspective on the reproductive outcomes of alternative ontogenetic pathways. Finally, sexual selection in the context of development may be represented most clearly by heterochronic changes in timing of developmental events and states between ancestors and descendents.

ACKNOWLEDGEMENTS

We are grateful to Peter Kappeler and Carel van Schaik for their invitation to contribute to this volume. Michael Pereira provided detailed criticism of an earlier draft of the manuscript; an anonymous reviewer and the editors also provided helpful comments and encouragement during the process of gestation.

REFERENCES

Abbott, D. H., George, L. M., Barrett, J. et al. 1990. Social control of ovulation in marmoset monkeys: a neuroendocrine basis for the study of infertility. In

Socioendocrinology of Primate Reproduction, ed. T. E. Ziegler & F. B. Bercovitch. New York, NY: Wiley Liss, pp. 135–58.

Alberts, S. C. & Altmann, J. 1995a. Preparation and activation – determinants of age at reproductive maturity in male baboons. *Behavioral Ecology and Sociobiology*, **36**, 397–406.

1995b. Balancing costs and opportunities – dispersal in male baboons. *American Naturalist*, **145**, 279–306.

Alberts, S. C., Altmann, J. & Wilson, M. 1996. Mate guarding constrains foraging activity of male baboons. *Animal Behaviour*, **51**, 1269–77.

Altmann, J. 1980. *Baboon Mothers and Infants*. Chicago, IL: University of Chicago Press.

Altmann, J. & Alberts, S. C. 1987. Body mass and growth rates in a wild primate population. *Oecologia*, **72**, 15–20.

Altmann, J., Alberts, S. C., Haines, S. A. *et al.* 1996. Social structure predicts genetic structure in a wild primate group. *Proceedings of the National Academy of Science, USA*, **93**, 5797–801.

Altmann, J., Altmann, S. A. & Hausfater, G. 1981. Physical maturation and age estimates of yellow baboons. *American Journal of Primatology*, **1**, 389–99.

Altmann, S. A. 1991. Diets of yearling female primates, *Papio cynocephalus*, predict lifetime fitness. *Proceedings of the National Academy of Science, USA*, **88**, 420–3.

1998. *Foraging for Survival*. Chicago, IL: University of Chicago Press.

Badyaev, A. V. 2002. Growing apart: an ontogenetic perspective on the evolution of sexual size dimorphism. *Trends of Ecology and Evolution*, **17**, 369–78.

Bateson, P. P. G. 1981. Ontogeny of behaviour. *British Medical Journal*, **37**, 159–64.

Bearder, S. K. 1987. Lorises, bushbabies and tarsiers: diverse societies in solitary foragers. In *Primate Societies*, ed. B. B. Smuts, D. L. Cheney, R. M. Seyfarth, R. W. Wrangham & T. T. Struhsaker. Chicago, IL: University of Chicago Press, pp. 11–24.

Bercovitch, F. B. 1983. Time budgets and consortships in olive baboons, *Papio anubis*. *Folia Primatologica*, **41**, 180–90.

1993. Dominance rank and reproductive maturation in male rhesus macaques, *Macaca mulatta*. *Journal of Reproduction and Fertility*, **99**, 113–20.

2000. Behavioral ecology and socioendocrinology of reproductive maturation in cercopithecine monkeys. In *Old World Monkeys*, ed. P. F. Whitehead & C. J. Jolly. Cambridge: Cambridge University Press, pp. 298–320.

2001. Reproductive ecology of Old World Monkeys. In *Reproductive Ecology and Human Evolution*, ed. P. T. Ellison. New York, NY: Aldine, pp. 369–96.

2002. Sex-biased parental investment in primates. *International Journal of Primatology*, **23**, 905–21.

Bercovitch, F. B. & Goy, R. 1990. The socioendocrinology of reproductive development and reproductive success in macaques. In *Socioendocrinology of Primate Reproduction*, ed. T. E. Ziegler & F. B. Bercovitch. New York, NY: Wiley-Liss, pp. 59–93.

Bercovitch, F. B. & Strum, S. 1993. Dominance rank, resource availability, and reproductive maturation in female savanna baboons. *Behavioral Ecology and Sociobiology*, **33**, 313–18.

Bercovitch, F. B., Lebron, M. R., Martinez, H. S. & Kessler, M. J. 1998. Primigravidity, body weight and the costs of rearing first offspring in rhesus macaques. *American Journal of Primatology*, **46**, 135–44.

Berman, C. M. 1980. Mother–infant relationships among free-ranging rhesus monkeys on Cayo Santiago: a comparison with captive pairs. *Animal Behaviour*, **28**, 860–73.

Bernstein, I. S., Ruehlmann, T. E., Judge, P. G., Lindquist, T. & Weed, J. L. 1991. Testosterone changes during the period of adolescence in male rhesus monkeys, *Macaca mulatta*. *American Journal of Primatology*, **24**, 29–38.

Boesch, C. 1997. Evidence for dominant wild chimpanzees investing more in sons. *Animal Behaviour*, **54**, 811–15.

Bogin, B. 1999. *Patterns of Human Growth*. Cambridge: Cambridge University Press.

Bowman, J. E. & Lee, P. C. 1995. Growth and threshold weaning weights among captive rhesus macaques. *American Journal of Physical Anthropology*, **96**, 159–75.

Brockmann, H. J. 2001. The evolution of alternative strategies and tactics. *Advances in the Study of Behavior*, **30**, 1–51.

Brody, S. 1945. *Bioenergetics and Growth*. New York, NY: Rheinhold.

Brown, G. R. 2001. Sex-biased investment in non-human primates: can Trivers & Willard's theory be tested? *Animal Behaviour*, **61**, 683–94.

Brown, G. R. & Dixson, A. F. 2000. The development of behavioural sex differences in infant rhesus macaques, *Macaca mulatta*. *Primates*, **41**, 63–77.

Brown G. R. & Silk, J. B. 2002. Reconsidering the null hypothesis: is maternal rank associated with birth sex ratios in primate groups? *Proceedings of the National Academy of Science, USA*, **99**, 11252–5.

Caine, N. G. 1987. Behavior during puberty and adolescence. In *Comparative Primate Biology*, Vol. 2A, ed. G. Mitchell & J. Erwin. New York, NY: Alan R. Liss, pp. 327–61.

Caro, T. M. & Bateson, P. P. G. 1986. Organisation and ontogeny of alternative tactics. *Animal Behaviour*, **34**, 1483–99.

Charles-Dominique, P. 1977. *Ecology and Behaviour of Nocturnal Primates*. London: Duckworth.

Cheverud, J. M., Wilson, P. & Dittus, W. P. J. 1992. Primate population studies at Polonnaruwa. 3. Somatometric growth in a natural population of toque macaques, *Macaca sinica*. *Journal of Human Evolution*, **23**, 51–77.

Clubb, R. & Mason, G. 2002. *A Review of the Welfare of Zoo Elephants in Europe*. Oxford: Department of Zoology, University of Oxford.

Clutton-Brock, T. H. 1991. The evolution of sex differences and the consequences of polygyny in mammals. In *The Development and Integration of Behaviour*, ed. P. P. G. Bateson. Cambridge: University of Cambridge Press, pp. 229–53.

Clutton-Brock, T. H., Harvey, P. H. & Rudder, B. 1977. Sexual dimorphism, socionomic sex ratio and body weight in primates. *Nature*, **269**, 797–800.

Clutton-Brock, T. H., Guinness, F. E. & Albon, S. D. 1992. *Red Deer: Behavior and Ecology of Two Sexes*. Edinburgh: Edinburgh University Press.

Colvin, J. 1983. Familiarity, rank and the structure of rhesus male peer networks. In *Primate Social Relationships*, ed. R. A. Hinde. Oxford: Blackwell, pp. 190–200.

Dang, D. C. & Meussy-Dessolle, N. 1984. Quantitative study of testis histology and plasma androgens at onset of spermatogenesis in the prepubertal laboratory-born macaque, *Macaca fascicularis*. *Archives of Andrology*, **12**, S43–51.

Debyser, I. 1995. Prosimian juvenile mortality in zoos and primate centers. *International Journal of Primatology*, **16**, 889–907.

Demment, M. W. 1983. Feeding ecology and the evolution of body size of baboons. *African Journal of Ecology*, **21**, 219–33.

Dixson, A. F. 1998. *Primate Sexuality. Comparative Studies of the Prosimians, Monkeys, Apes and Human Beings*. Oxford: Oxford University Press.

Dixson, A. F. & Nevison, C. 1997. The socioendocrinology of adolescent development in male rhesus monkeys, *Macaca mulatta*. *Hormones and Behavior*, **31**, 126–35.

Draper P. & Harpending, H. 1982. Father absence and reproductive strategy: an evolutionary perspective. *Journal of Anthropological Research*, **38**, 255–73.

Dunbar, R. I. M. 1984. *Reproductive Decisions: An Economic Analysis of Gelada Baboon Social Strategies*. Princeton, NJ: Princeton University Press.

Eisler, J. A., Tannenbaum, P. L., Mann, D. R. & Wallen, K. 1993. Neonatal testicular suppression with a GnRH agonist in rhesus monkeys: effect on adult endocrine function and behavior. *Hormones and Behavior*, **27**, 551–67.

Fairbanks, L. M. 1996. Individual differences in maternal style: causes and consequences for offspring. In *Parental Care: Evolutionary Mechanisms and Adaptive Significance*, ed. J. S. Rosenblatt & C. T. Snowdon. London: Academic Press, pp. 579–611.

Fedigan, L. M. & Zohar, S. 1997. Sex differences in mortality of Japanese macaques: twenty-one years of data from the Arashiyama West population. *American Journal of Physical Anthropology*, **102**, 161–72.

Folstad, I. & Karter, A. J. 1992. Parasites, bright males and the immunocompetence handicap. *American Naturalist*, **139**, 603–22.

Ford, S. M. 1994. Evolution of sexual dimorphism in body weight in platyrrhines. *American Journal of Primatology*, **34**, 221–44.

Galliari, C. A. & Colillas, O. J. 1985. Sequences and timing of dental eruption in Bolivian captive-born squirrel-monkeys, *Saimiri sciureus*. *American Journal of Primatology*, **8**, 195–204.

Gavan, J. A. & Swindler, D. R. 1966. Growth rates and phylogeny in primates. *American Journal of Physical Anthropology*, **38**, 69–82.

Georgiadis, N. 1985. Growth patterns, sexual dimorphism and reproduction in African ruminants. *African Journal of Ecology*, **23**, 75–87.

Glick, B. B. 1979. Testicular size, testosterone level and body weight in male *Macaca radiata*. *Folia Primatologica*, **32**, 268–89.

Godfrey, L., Samonds, K., Jungers, W. & Sutherland, M. 2001. Teeth, brains, and primate life histories. *American Journal of Physical Anthropology*, **114**, 192–214.

Gomendio, M. 1990. The influence of maternal rank and infant sex on maternal investment trends in macaques: birth sex ratios, inter-birth intervals and suckling patterns. *Behavioral Ecology and Sociobiology*, **27**, 365–75.

Gordon, I. J. 1989. The interspecific allometry of reproduction – do larger species invest relatively less in their offspring? *Functional Ecology*, **3**, 285–8.

Graham, C. & Nadler, R. 1990. Socioendocrine interactions in great ape reproduction. In *Socioendocrinology of Primate Reproduction*, ed. T. Ziegler & F. B. Bercovitch. New York, NY: Wiley Liss, pp. 33–58.

Gust, D. A. & Gordon, T. P. 1991. Male age and reproductive behavior in sooty mangabeys, *Cercocebus torquatus atys*. *Animal Behaviour*, **41**, 277–83.

Hamada, Y. & Udono, T. 2002. Longitudinal analysis of length-growth in the chimpanzee *Pan troglodytes*. *American Journal of Physical Anthropology*, **118**, 268–84.

Hamilton, W. J. III & Bulger, J. B. 1990. Natal male baboon rank rises and successful challenges to resident alpha males. *Behavioral Ecology and Sociobiology*, **26**, 357–62.

Harcourt, A. H. & Stewart, K. J. 1981. Gorilla male relationships – can differences during immaturity lead to contrasting reproductive tactics in adulthood? *Animal Behaviour*, **29**, 206–10.

Hauser, M. & Harcourt, A. H. 1992. Is there sex-biased mortality in primates? *Folia Primatologica*, **58**, 47–52.

Hobson, W. C., Winter, J. S. D., Reyes, F. I., Fuller, G. B. & Faiman, C. 1980. Nonhuman primates as models for studies on puberty. In *Animal Models in Human Reproduction*, ed. M. Serio & L. Martin. New York, NY: Raven Press, pp. 409–21.

Isbell, L. A., Cheney, D. L. & Seyfarth, R. M. 1990. Costs and benefits of home range shifts among vervet monkeys, *Cercopithecus aethiops*, in Amboseli National Park, Kenya. *Behavioral Ecology and Sociobiology*, **27**, 351–8.

Itoigawa, N., Tanaka, T., Ukai, N. *et al.* 1992. Demography and reproductive parameters of a free-ranging group of Japanese macaques, *Macaca fuscata*, at Katsuyama. *Primates*, **33**, 49–68.

Izard, M. 1990. Social influences on the reproductive success and reproductive endocrinology of prosimian primates. In *Socioendocrinology of Primate Reproduction*, ed. T. E. Ziegler & F. B. Bercovitch. New York, NY: Wiley-Liss, pp. 159–86.

Janson, C. H. & van Schaik, C. P. 1993. Ecological risk aversion in juvenile primates: slow and steady wins the race. In *Juvenile Primates: Life History, Development, and Behavior*, ed. M. E. Pereira & L. A. Fairbanks. New York, NY: Oxford University Press, pp. 57–74.

Jarman, P. J. 1983. Mating system and sexual dimorphism in large terrestrial mammalian herbivores. *Biological Reviews*, **58**, 485–520.

Johnson, R. L. & Kapsalis, E. 1995. Determinants of postnatal weight in infant rhesus monkeys: implications for the study of interindividual differences in neonatal growth. *American Journal of Physical Anthropology*, **98**, 343–53.

1998. Menopause in free-ranging rhesus macaques: estimated incidence, relation to body condition, and adaptive significance. *International Journal of Primatology*, **19**, 751–65.

Kappeler, P. M. 1999. Primate socioecology: new insights from males. *Naturwissenschaften*, **86**, 8–29.

Kappeler, P. M. & van Schaik, C. P. 2002. The evolution of primate social systems. *International Journal of Primatology*, **23**, 707–40.

Kraemer, H. C., Horvat, J. R., Doering, C. & McGinnis, P. R. 1982. Male chimpanzee development focusing on adolescence: integration of behavior with physiological changes. *Primates*, **23**, 393–405.

Kruuk, L. E. B., Clutton-Brock, T. H., Rose, K. E. & Guinness, F. E. 1999. Early determinants of lifetime reproductive success differ between the sexes in red deer. *Proceedings of the Royal Society of London, Series B*, **266**, 1655–61.

Kuester, J. & Paul, A. 1999. Male migration in Barbary macaques, *Macaca sylvanus*, at Affenberg Salem. *International Journal of Primatology*, **20**, 85–106.

Lee, P. C. 1997. The meanings of weaning: growth, lactation and life history. *Evolutionary Anthropology*, **5**, 87–96.

1999. Comparative ecology of post-natal growth and weaning among haplorhine primates. In *Comparative Primate Socioecology*, ed. P. C. Lee. Cambridge: Cambridge University Press, pp. 111–39.

Lee, P. C. & Johnson, J. 1992. Sex differences in the acquisition of dominance status among primates. In *Cooperation and Competition in Animals and Humans*, ed. A. H. Harcourt & F. B. de Waal. Oxford: Oxford University Press, pp. 391–414.

Lee, P. C. & Kappeler, P. M. 2003. Socio-ecological correlates of phenotypic plasticity in primate life history. In *Primate Life History and Socioecology*, ed. P. M. Kappeler & M. E. Pereira, in press. Chicago, IL: University of Chicago Press.

Lee, P. C. & Moss, C. J. 1986. Early maternal investment in male and female African elephant calves. *Behavioral Ecology and Sociobiology*, **18**, 353–61.

1995. Statural growth in known-age African elephants, *Loxodonta africana*. *Journal of Zoology, London*, **236**, 29–41.

Lee, P. C., Majluf, P. & Gordon, I. 1991. Growth, weaning and maternal investment from a comparative perspective. *Journal of Zoology, London*, **225**, 99–114.

Leigh, S. R. 1992. Patterns of variation in the ontogeny of primate body size dimorphism. *Journal of Human Evolution*, **23**, 27–50.

 1995. Socioecology and the ontogeny of sexual size dimorphism in anthropoid primates. *American Journal of Physical Anthropology*, **97**, 339–56.

 2001. Evolution of human growth. *Evolutionary Anthropology*, **10**, 223–36.

Leigh, S. R. & Shea, B. T. 1995. Ontogeny and the evolution of adult body size dimorphism in apes. *American Journal of Primatology*, **36**, 37–60.

Leigh, S. R. & Pereira, M. E. 2003. Modes of primate development. In *Primate Life Histories and Socioecology*, ed. P. M. Kappeler & M. E. Pereira. Chicago, IL: University of Chicago Press, pp. 149–76.

Leigh, S. R. & Terranova, C. J. 1998. Comparative perspectives on bimaturism, ontogeny, and dimorphism in lemurid primates. *International Journal of Primatology*, **19**, 723–49.

Liang, B., Zhang, S. Y. & Wang, L. X. 2000. Development of sexual morphology, physiology and behaviour in Sichuan golden monkeys, *Rhinopithecus roxellana*. *Folia Primatologica*, **71**, 413–16.

Lunn, S. F., Recio, R., Morris, K. & Fraser, H. M. 1994. Blockade of the neonatal rise in testosterone by a gonadotrophin-releasing hormone antagonist: effects on timing of puberty and sexual behaviour in the male marmoset monkey. *Journal of Endocrinology*, **141**, 439–47.

Lycett, J. E., Dunbar, R. I. M. & Volland, E. 2000. Longevity and the cost of reproduction in a historical human population. *Proceedings of the Royal Society of London, Series B*, **267**, 2355–8.

Maggioncalda, A. N., Sapolsky, R. M. & Czekala, N. M. 1999. Reproductive hormone profiles in captive male orangutans: implications for understanding developmental arrest. *American Journal of Physical Anthropology*, **109**, 19–32.

Malina, R. M. 1978. Adolescent growth and maturation: selected aspects of current research. *Yearbook of Physical Anthropology*, **21**, 63–94.

Mann, D. R., Gould, K. G., Collins, D. C. & Wallen, K. 1989. Blockade of neonatal activation of the pituitary testicular axis: effect on peripubertal luteinizing hormone and testicular secretion and on testicular development in male monkeys. *Journal of Clinical Endocrinology and Metabolism*, **68**, 600–7.

Mann, D. R., Akinbami, M. A., Gould, K. G., Paul, K. & Wallen, K. 1998. Sexual maturation in male rhesus monkeys: importance of neonatal testosterone exposure and social rank. *Journal of Endocrinology*, **159**, 493–501.

Martin, P. S. 1984. The meaning of weaning. *Animal Behaviour*, **32**, 1257–9.

Martin, R. D, Wilner, L. A. & Dettling, A. 1994. The evolution of sexual size dimorphism in primates. In *The Differences Between the Sexes*, ed. R. V. Short & E. Balaban. Cambridge: Cambridge University Press, pp. 159–200.

Nadler, R. D., Wallis, J., Rothmeyer, C., Cooper, R. W. & Baulieu, E. E. 1987. Hormones and behavior of prepubertal and peripubertal chimpanzees. *Hormones and Behaviour*, **21**, 118–31.

Nash, L. T. 1993. Juveniles in nongregarious primates. In *Juvenile Primates: Life History, Development, and Behavior*, ed. M. E. Pereira & L. A. Fairbanks. New York, NY: Oxford University Press, pp. 119–37.

Nielson, C. T., Skakkebaek, N. E., Richardson, D. W. *et al.* 1986. Onset of the release of spermatozoa spermarche in boys in relation to age, testicular growth, pubic hair and height. *Journal of Clinical Endocrinology and Metabolism*, **62**, 532–5.

Nieuwenhuijsen, K., Bonke-Jansen, M., de Neef, K. J., van der Werff ten Bosch, J. J. & Slob, A. K. 1987a. Physiological aspects of puberty in group-living stump-tail monkeys, *Macaca arctoides*. *Physiology and Behavior*, **41**, 37–45.

Nieuwenhuijsen, K., de Neef, K. J., van der Werff den Bosch, J. J. & Slob, A. K. 1987b. Testosterone, testes size, seasonality and behaviour in group living stump-tail macaques, *Macaca arctoides*. *Hormones and Behavior*, **21**, 153–67.

Nikolei, J. & Borries, C. 1997. Sex differential behaviour of immature hanuman langurs, *Presbytis entellus*, in Ramnagar, South Nepal. *International Journal of Primatology*, **18**, 415–37.

Nigi, H., Tiba, T., Yamamoto, S., Floescheim, Y. & Ohsawa, N. 1980. Sexual maturation and seasonal changes in the reproductive phenomena of female Japanese monkeys. *Primates*, **21**, 230–40.

Partridge, L. 1988. Lifetime reproductive success in *Drosophila*. In *Reproductive Success*, ed. T. H. Clutton-Brock. Chicago, IL: University of Chicago Press, pp. 11–23.

Paul, A. & Thommen, D. 1984. Timing of birth, female reproductive success and infant sex ratio in semifree-ranging Barbary macaques, *Macaca sylvanus*. *Folia Primatologica*, **42**, 2–16.

Pereira, M. E. 1988. Effects of age and sex on intra-group spacing behaviour in juvenile savannah baboons, *Papio cynocephalus cynocephalus*. *Animal Behavior*, **36**, 184–204.

1995. Development and social dominance among group-living primates. *American Journal of Primatology*, **37**, 143–75.

Pereira, M. E. & Altmann, J. 1985. Development of social behaviour in free-living non-human primates. In *Nonhuman Primate Models for Human Growth and Development*, ed. E. Watts. New York, NY: Alan R. Liss, pp. 217–309.

Pereira, M. E. & Fairbanks, L. A. (eds.) 1993. *Juvenile Primates: Life History, Development, and Behavior*. New York, NY: Oxford University Press.

Pereira, M. E. & Leigh, S. R. 2003. Modes of primate development. In *Primate Life History and Socioecology*, ed. P. M. Kappeler & M. E. Pereira, in press. Chicago, IL: University of Chicago Press.

Phillips-Conroy, J. E., Bergman, T. & Jolly, C. J. 2000. Quantitative assessment of occlusal wear and age estimation in Ethiopian and Tanzanian baboons. In *Old World Monkeys*, ed. P. F. Whitehead & C. J. Jolly. Cambridge: Cambridge University Press, pp. 321–40.

Plant, T. M. 1994. Puberty in primates. In *The Physiology of Reproduction*, Vol. 2, ed. E. Knobil & J. D. Neill. New York, NY: Raven Press, pp. 1763–88.

Plavcan, J. M. 1999. Mating systems, intrasexual competition and sexual dimorphism in primates. In *Comparative Primate Socioecology*, ed. P. C. Lee. Cambridge: Cambridge University Press, pp. 241–69.

Pontier, D., Gaillard, J. M., Allaine, D. *et al.* 1989. Postnatal-growth rate and adult body-weight in mammals – a new approach. *Oecologia*, **80**, 390–4.

Promislow, D. E. L. & Harvey, P. H. 1990. Living fast and dying young: a comparative analysis of life history variation among mammals. *Journal of Zoology, London*, **220**, 417–38.

Pusey, A. E. 1990. Behavioural changes at adolescence in chimpanzees. *Behaviour*, **115**, 203–46.

Pusey, A. E. & Packer, C. 1987. Dispersal and philopatry. In *Primate Societies*, ed. B. B. Smuts, D. L. Cheney, R. M. Seyfarth, R. W. Wrangham & T. T. Struhsaker. Chicago, IL: University of Chicago Press, pp. 250–66.

Raleigh, M. J. & MacGuire, M. T. 1990. Social influences on endocrine function in male vervet monkeys. In *Socioendocrinology of Primate Reproduction*, ed. T. E. Ziegler & F. B. Bercovitch. New York, NY: Wiley-Liss, pp. 95–111.

Ross, C. & Jones, K. E. 1999. Socioecology and the evolution of primate reproductive rates. In *Comparative Primate Socioecology*, ed. P. C. Lee. Cambridge: Cambridge University Press, pp. 73–110.

Rowell, T. E. 1977. Variation in age at puberty in monkeys. *Folia Primatologica*, **27**, 284–90.

Rowell, T. E. & Dixson, A. F. 1975. Changes in social organisation during the breeding season of wild talapoin monkeys. *Journal of Reproduction and Fertility*, **43**, 419–34.

Sapolsky, R. I. M. 1985. Stress-induced suppression of testicular function in the wild baboon – role of glucocorticoids. *Endocrinology*, **116**, 2273–8.

1993. The physiology of dominance in stable vs. unstable hierarchies. In *Primate Social Conflict*, ed. W. Mason & S. Mendoza. New York, NY: State University of New York Press, pp. 171–204.

Scanlan, J. M., Coe, C. L., Latts, A. & Suomi, S. J. 1987. Effects of age, rearing, and separation stress on immunoglobulin levels in rhesus monkeys. *American Journal of Primatology*, **13**, 11–22.

Schultz, A. H. 1956. Postembryonic changes. *Primatologia*, **1**, 887–964.

1969. *The Life of Primates*. New York, NY: Universe Books.

Schwartz, S. M. & Kemnitz, H. W. 1992. Age-related and gender-related changes in body size, adiposity, and endocrine and metabolic parameters in free-ranging rhesus macaques. *American Journal of Physical Anthropology*, **89**, 109–21.

Schwartz, S. M., Wilson, M. E., Walker, M. L. & Collins, D. C. 1988. Dietary influences on growth and sexual maturation in premenarchial rhesus monkeys. *Hormones and Behavior*, **22**, 231–51.

Scott, L. 1984. Reproductive behavior of adolescent female baboons, *Papio anubis*, in Kenya. In *Female Primates: Studies by Women Primatologists*, ed. M. F. Small. New York, NY: Alan R Liss, pp. 77–100.

Setchell, J. M. 1999. Socio-sexual development in the male mandrill *Mandrillus sphinx*. Ph.D. thesis, University of Cambridge.

2003. The evolution of alternative reproductive morphs in male primates. In *Sexual Selection and Reproductive Competition in Primates: New Perspectives and Directions*, ed. C. B. Jones, in press. American Society of Primatologists Special Topics in Primatology.

Setchell, J. M. & Dixson, A. F. 2001a. Arrested development of secondary sexual adornments in subordinate adult male mandrills, *Mandrillus sphinx*. *American Journal of Physical Anthropology*, **115**, 245–52.

2001b. Changes in the secondary sexual adornments of male mandrills *Mandrillus sphinx* are associated with gain and loss of alpha status. *Hormones and Behavior*, **39**, 177–84.

2002. Developmental variables and dominance rank in male mandrills *Mandrillus sphinx*. *American Journal of Primatology*, **56**, 9–25.

Setchell, J. M., Lee, P. C., Wickings, E. J. & Dixson, A. F. 2001. Growth and ontogeny of sexual size dimorphism in the mandrill *Mandrillus sphinx*. *American Journal of Physical Anthropology*, **115**, 349–60.

2002. Reproductive parameters and maternal investment in mandrills *Mandrillus sphinx*. *International Journal of Primatology*, **23**, 51–68.

Shea, B. T. 1983. Allometry and heterochrony in the African apes. *American Journal of Physical Anthropology*, **62**, 275–89.

1986. Ontogenetic approaches to sexual dimorphism in anthropoids. *Human Evolution*, **1**, 97–110.

2000. Current issues in the investigation of evolution by heterochrony, with emphasis on the debate over human neotony. In *Biology, Brains and Behaviour*, ed. S. Taylor Parker, J. Langer & M. L. McKinney. Santa Fe: School of American Research Press, pp. 181–214.

Small, M. F. & Smith, D. G. 1986. The influence of birth timing upon infant growth and survival in captive rhesus macaques, *Macaca mulatta*. *International Journal of Primatology*, **7**, 289–304.

Smith, B. H. 1992. Life history and the evolution of human maturation. *Evolutionary Anthropology*, **14**, 134–42.

Smith, B. H., Crummett, T. L. & Brandt, K. L. 1994. Ages of eruption of primate teeth: a compendium for ageing individuals and comparing life histories. *Yearbook of Physical Anthropology*, **37**, 177–231.

Smith, R. J. & Jungers, W. L. 1997. Body mass in comparative primatology. *Journal of Human Evolution*, **32**, 523–59.

Smith, R. J. & Leigh, S. R. 1998. Sexual dimorphism in primate neonatal body mass. *Journal of Human Evolution*, **34**, 173–201.

Sommer, V. & Rajpurohit, L. S. 1989. Male reproductive success in harem troops of Hanuman langurs, *Presbytis entellus*. *International Journal of Primatology*, **10**, 293–317.

Stanford, C. B. 1998. *Chimpanzee and Red Colobus: The Ecology of Predator and Prey*. Cambridge, MA: Harvard University Press.

Strum, S. C. 1991. Weight and age in wild olive baboons. *American Journal of Primatology*, **25**, 219–37.

Strum, S. C. & Western, J. D. 1982. Variations in fecundity with age and environment in olive baboons, *Papio anubis*. *American Journal of Primatology*, **3**, 61–76.

Surbey, M. K. 1990. Family composition, stress and the timing of human menarche. In *Socioendocrinology of Primate Reproduction*, ed. T. E. Ziegler & F. B. Bercovitch. New York, NY: Wiley Liss, pp. 11–32.

Tanner, J. M. 1962. *Growth at Adolescence*. Oxford: Blackwell.

Treves, A. 1997. Primate natal coats: a preliminary analysis of distribution and function. *American Journal of Physical Anthropology*, **104**, 47–70.

Trillmich, F. 1996. Parental investment in pinnipeds. In *Parental Care: Evolutionary Mechanisms and Adaptive Significance*, ed. J. S. Rosenblatt & C. T. Snowdon. London: Academic Press, pp. 533–77.

Trivers, R. M. 1985. *Social Evolution*. Menlo Park, CA: Benjamin Cummings.

Utami, S. S., Goossens, B., Bruford, M. W., de Ruiter, J. & van Hooff, J. A. R. A. M. 2002. Male bimaturism and reproductive success in Sumatran orangutans. *Behavioral Ecology*, **13**, 643–52.

Vandenbergh, J. G. & Coppola, D. M. 1986. The physiology and ecology of puberty modulation by primer pheromones. *Advances in the Study of Behavior*, **16**, 71–107.

Vandenbergh, J. G., Drickamer, L. C. & Colby, D. R. 1972. Social and dietary factors in the sexual maturation of female mice. *Journal of Reproduction and Fertility*, **28**, 515–23.

van Noordwijk, M. A. & van Schaik, C. P. 2001. Career moves: transfer and rank challenge decisions by male long-tailed macaques. *Behaviour*, **138**, 359–95.

van Schaik, C. P. 1992. Sex-biased juvenile mortality in primates: a reply to Hauser & Harcourt. *Folia Primatologica*, **58**, 53–5.

van Schaik, C. P. & de Visser, J. A. G. M. 1990. Fragile sons or harassed daughters? Sex differences in mortality among juvenile primates. *Folia Primatologica*, **55**, 10–23.

van Schaik, C. P. & Hrdy, S. B. 1991. Intensity of local resource competition shapes the relationship between maternal rank and sex ratios at birth in Cercopithecine primates. *American Naturalist*, **138**, 1555–61.

van Schaik, C. P. & Janson, C. H. (eds.) 2000. *Infanticide by Males and Its Implications*. Cambridge: Cambridge University Press.

Watts, E. S. 1985. Adolescent growth and development of monkeys, apes and humans. In *Nonhuman Primate Models for Human Growth and Development*, ed. E. S. Watts. New York, NY: Liss, pp. 41–65.

1986. Evolution of the human growth spurt. In *Human Growth*, ed. F. Falkner & J. M. Tanner. New York, NY: Plenum, pp. 153–66.

Whitten, P. L., Brockman, D. K. & Stavisky, R. C. 1998. Recent advances in noninvasive techniques to monitor hormone–behavior interactions. *Yearbook of Physical Anthropology*, **41**, 1–23.

Wickings, E. J. & Dixson, A. F. 1992. Development from birth to sexual maturity in a semi-free-ranging colony of mandrills, *Mandrillus sphinx* in Gabon. *Journal of Reproduction and Fertility*, **95**, 129–38.

Wiley, R. H. 1974. Effects of delayed reproduction on survival, fecundity, and the rate of population increase. *American Naturalist*, **108**, 705–9.

Wilner, L. A. & Martin, R. D. 1985. Some basic principles of mammalian sexual dimorphism. In *Human Sexual Dimorphism*, ed. J. Ghesquiere, R. D. Martin & F. Newcombe. London: Taylor & Francis, pp. 1–42.

Worthman, C. M. 1993. Biocultural interactions in human development. In *Juvenile Primates: Life History, Development, and Behavior*, ed. M. E. Pereira & L. A. Fairbanks. New York, NY: Oxford University Press, pp. 339–58.

Ziegler, T. E., Snowdon, C. T. & Uno, H. 1990. Social interactions and determinants of ovulation in tamarins, *Saguinus*. In *Monographs in Primatology*. Vol. 13: *Socioendocrinology of Primate Reproduction*, ed. T. E. Ziegler & F. B. Bercovitch. New York, NY: Wiley Liss, pp. 113–33.

11 • Alternative male reproductive strategies: male bimaturism in orangutans

SUCI UTAMI ATMOKO
Faculty of Biology
National University
Jakarta, Indonesia

JAN A. R. A. M. VAN HOOFF
Ethology and Socio-ecology
Utrecht University
Utrecht, The Netherlands

INTRODUCTION

An individual's lifetime reproductive success is determined by the degree to which it copes with the various challenges that it faces in the different stages of its life. These challenges change constantly from early juvenility until late adulthood (Setchell & Lee, this volume). They require corresponding adaptive variation in the tactics of survival and reproduction, as the animal passes through the successive phases of its life cycle. This dynamic change takes its most dramatic form in those species where an individual goes through one or several distinct larval stages before becoming a reproductively active adult. Apart from such successive changes, there are also variations in fitness-maximising tactics that coexist side by side. In this chapter we deal specifically with such parallel or alternative fitness trajectories in a primate, the orangutan, a species with two adult, sexually mature male morphs. A recent study by Utami *et al.* (2002) has shown that these two male morphs exist side by side in a natural population and that each morph can and does produce offspring, suggesting that they represent parallel alternative reproductive tactics. Here we review the pertinent evidence.

It has long been an established fact in ethology that interactions with social partners influence an individual's motivational state and vice versa, and, through interactions, its physiological development and condition. For example, the suppression of reproductive processes by the presence of a same-sex conspecific has been documented for many species, including primates. In some cases an individual adopts a 'waiting-room strategy', a subordinate manner of behaving that does not yield any direct reproductive success, but that allows it to bide its time for better social opportunities. However, there is another possibility often found among non-mammalian species that includes an option of alternative fitness-optimising trajectories.

In 1996, Gross proposed a categorisation of alternative reproductive strategies and tactics with regard to the underlying genetic diversity. He distinguished three categories: Alternative Strategies, a Mixed Strategy with Alternative Tactics and a Conditional Strategy with Alternative Tactics. Alternative Strategies are based on genetic polymorphism. These are genetically different strategies with equal average-fitness pay-offs, maintained by frequency-dependent selection. A Mixed Strategy with Alternative Tactics is monogenetic. One strategy can contain N tactics (i.e. more than one tactic) with equal average fitness pay-offs. Each individual plays the different tactics at random and in ESS proportions. Although this is a theoretical possibility, there is no convincing evidence that this strategy occurs in reality (Gross, 1996). The third category, however, is a real possibility: a Conditional Strategy with Alternative Tactics, which is genetically monomorphic. The one strategy contains N tactics that have equal fitness-returns at a point where an individual could switch between tactics in response to different conditions. The diversity of tactics is maintained by status- or condition-dependent selection.

So far there is evidence for genetically different alternative reproductive strategies within a sex from only a few species. A well-known example is the ruff (*Philomachus pugnax*) in which males congregate at leks where they court the attracted females. In a classic study, van Rhijn (1973) showed that there are two male morphs: dark-collared dominant males who defend a courting site on the lek, and white-collared satellite males that are tolerated because their presence increases the attractiveness of the lek. While helping to attract females, satellite males may sneak matings. Lank

Sexual Selection in Primates: New and Comparative Perspectives, ed. Peter M. Kappeler and Carel P. van Schaik. Published by Cambridge University Press. © Cambridge University Press 2004.

et al. (1995) demonstrated that the different morphs are determined by two alleles at a single autosomal locus. Similarly, Shuster and Wade (1991) could show that the presence of three male morphs with equal reproductive success in the isopod *Paracerceis* (namely Large Fighters, Intermediate Female Mimics and Small Sneakers) is determined by three alleles at a single autosomal locus. Ryan and Wagner (1987) and Ryan *et al.* (1992) demonstrated for the swordfish *Xiphiphorus* that three alternative male strategies exist, determined by three alleles at a single Y-locus. Likewise, there is female polymorphism in many odonate insects. Andrés *et al.* (2002) present evidence that a genetically determined colour polymorphism in female damselflies is indeed a balanced polymorphism maintained by a combination of density- and frequency-dependent selection. Similar evidence for genetically determined balanced diversity in morphological forms in birds and mammals is lacking, although there is evidence for balanced and genetically determined *behavioural* polymorphisms, namely of a bimodality in coping styles that coexist in equilibrium in a population. These polymorphisms have been studied particularly in rodents, but they continue to be discovered in other species as well (e.g. Koolhaas *et al.*, 2001).

The existence of a conditional strategy with different tactics has been demonstrated in several species of mammals. To mention but one clear example: in savannah baboons a male may decide what tactic to follow in its relationships with females after assessing what others do. Smuts (1985) has shown that dominant males follow a sexual tactic in which they monopolise access to fertile females by contest competition. A subordinate male may use another tactic. He may persuade a female to choose him for mating by rendering services to the female (e.g. protecting her in between-female competition) and thus forming a 'friendship' with the female. Similar variation in tactics has been found in other primates (e.g. in rhesus macaques, Berard *et al.*, 1994).

Whether the polymorphisms are an aspect of phenotypic plasticity or genetically predisposed, a condition-dependent choice or selection mechanism must play a role, acting either at the level of ontogenetic development or natural selection. Ontogenetic developments may be reversible, and thus appear to be flexible 'conditional' tactics, but sometimes they are not, in particular when they are associated with morphological variation. Instances of such different morphs, as coexisting alternative fitness-maximising trajectories, as in the examples given above, exist in invertebrates and fishes. It has been doubtful whether they exist in mammals. How-

ever, there is one exceptional case in a primate: bimaturism in male orangutans (*Pongo pygmaeus*).

ORANGUTANS: TWO KINDS OF MALES

The orangutan is the only representative of the three great-ape genera that lives in Asia. It is distinct because of two remarkable characteristics. First, orangutans are almost exclusively arboreal – in fact they are the largest and heaviest arboreal mammals on earth – and, second, they are comparatively solitary: the adult individuals spend most of their time on their own (Horr, 1975, 1977; Galdikas, 1979; Rijksen, 1978).

The arboreal nature explains the orangutan's comparatively solitary nature (Horr, 1975, 1977; Mitani *et al.*, 1991; van Hooff, 1995). As long as orangutans stay high up in the trees, adults as well as youngsters associated with their mother are virtually immune from predation. Their large size and heavy weight put considerable constraints on their ability to cover large distances in their arboreal habitat, but, at the same time, require them to consume large amounts of food. As a result, moving in social groups is largely inhibited by the attendant intense resource competition (Rodman, 1977; Mitani *et al.*, 1991). Indeed, orangutans aggregate only occasionally, namely in large fruit patches (MacKinnon, 1974; Rijksen, 1978; Schürmann & van Hooff, 1986), and, sometimes, if there is a rich fruiting season, these aggregations may develop into small social groups of animals travelling and foraging together temporarily (Sugardjito *et al.*, 1987; Utami *et al.*, 1997). Although they are largely solitary, the members of a population clearly know one another and maintain differentiated relationships (van Schaik & van Hooff, 1996; Delgado & van Schaik, 2000). The most common associations occur when a male and a female join one another in a consortship characterised by regular sexual contact. Such a consort pair may range together for several days or even weeks (Rijksen, 1978; Schürmann, 1982).

The orangutan is also one of the most sexually dimorphic species of mammals with dimorphism in size and in adornments. Fully grown adult males are on average 2.0 to 2.3 times as heavy as adult females (Leigh & Shea, 1995). In addition, previous studies of the social organisation of orangutans have reported the existence of two forms of sexually mature males: fully developed large adult males and others lacking secondary sexual characteristics. In the orangutan, male

Table 11.1 *Bimaturism in male orangutans.*

Unflanged males	Flanged males
• Sexually mature around 8 years of age and sexually active	• Sexually active
• Secondary sexual characteristics (SSCs) undeveloped	• Fully developed SSCs: hair coat, cheek flanges, throat pouch, 'long-call'
• Comparatively 'social' and tolerant toward other males	• Highly intolerant toward other flanged males
	• Relatively tolerant toward unflanged males

growth is associated with fattening (Leigh, 1995), and, on average, males keep getting heavier the older they get. In this respect there is what Leigh and Shea (1995) called 'indeterminate male growth'. Such indeterminate increase in weight is uncommon in mammals, and has so far only been found in elephant seals (Jolicœur, 1985), grey kangaroos (Jarman & Southwell, 1986) and African elephants (Jarman, 1983), all species with strong between-male contest competition for access to fertile females. The fully developed orangutan males also possess a number of secondary sexual characteristics, such as cheek flanges, a throat pouch and a long coat of hairs. Furthermore, they produce 'a long-call' by which they make their presence known far beyond the reach of visual contact (Table 11.1). Sexual-size dimorphism is indicative of a regime of strong between-male contest competition for receptive females (Darwin, 1871; Alexander *et al.*, 1979; Plavcan & van Schaik, 1992; Weckerly, 1998; Plavcan, 1999, this volume). The male adornments suggest an additional influence of female choice.

In addition to these flanged males, there are males that have often been referred to as 'subadult'. They may be as big or bigger than fully adult females and they resemble them because they lack the male secondary sexual characteristics (SSC), such as the long-call. Since the pioneering study of MacKinnon (1974), it has been known that these so-called subadults are sexually active and can sire offspring (e.g. Nadler, 1977, 1981). The term subadult reflects the idea that the unflanged males are in transition to full adulthood. This is not devoid of truth, because unflanged males do change into flanged males. And indeed, a few well-documented observations on captive orangutans (Nadler, 1977, 1981; Kingsley,

1982) and some field observations (Galdikas, 1985a, b; Utami Atmoko, 2000) show that the development of SSCs can occur quite suddenly (in a matter of months). There is evidence that this change is influenced by social factors, i.e. it can be retarded by the presence of a fully flanged male and, therefore, indicates some kind of psychoneuroendocrinological inhibition (Graham, 1988; Kingsley, 1988; Maggioncalda, 1995; Maggioncalda *et al.*, 1999).

Still, the term subadult is misleading because, as we shall argue, it de-emphasises the fact that this maturational arrest may last for a major part, if not all, of an individual's adult lifetime (e.g. te Boekhorst *et al.*, 1990) and, therefore, the undeveloped stage is better regarded as a separate morph representing a parallel, alternative reproductive strategy. This idea is supported by the long-term behavioural studies of Sumatran orangutans carried out in the Gunung Leuser National Park in north-west Sumatra during more than 25 years, combined with molecular-genetic data on 39 animals, showing that these males are reproductively successful (Utami *et al.*, 2002).

DIFFERENT ULTIMATE EXPLANATIONS

To explain orangutan bimaturism, several hypotheses have been proposed in the past. MacKinnon (1974, 1979) was the first to note that unflanged adults share home-ranges with a flanged adult male. During his field study he never saw flanged males mate. However, he regularly observed unflanged males mating, often in the form of forced copulations, resisted by the females, which he called 'rapes'. He therefore postulated that the adult males are postreproductive and territorial, defending a range for their subadult sons. Only then would the adult's tolerance be explainable through inclusive fitness benefits. Utami *et al.* (2002) called this idea the 'range-guardian' hypothesis.

Subsequent field studies (Schürmann, 1982; Galdikas, 1985a, b; Mitani, 1985a, b; Schürmann & van Hooff, 1986; Rodman & Mitani, 1987; MacKinnon, 1989) revealed that flanged males mate as often as unflanged males. This does refute the post-reproductive status of flanged males, but it does not refute McKinnon's original suggestion that the adult males tolerate subadults and their sexual activity because of an inclusive fitness benefit. This hypothesis would be refuted if resident flanged males were related not more than average to the unflanged males sharing their range.

If these males can be shown to be unrelated, then there is a theoretical problem with respect to the mechanisms of sexual selection underlying bimaturism. Flanged males are very intolerant toward one another (e.g. Horr, 1972, 1977; Rodman, 1973, 1977; Galdikas, 1979, 1985b; Knott, 1998) and there can be little doubt that the extreme sexual size dimorphism is the result of male–male competition. Rodman and Mitani (1987) argued that male competition is the sole factor explaining sexual dimorphism. These authors felt that it was premature to accept female choice as an important selective factor. However, Galdikas (1985a, b) and Schürmann and van Hooff (1986) argued that female choice is an essential element for an explanation: if dimorphism was based exclusively on male competition it would leave unanswered the question of why flanged males should be more tolerant (or less intolerant) toward unflanged than flanged males. They could be tolerant because the unflanged males pose no serious reproductive competition as a result of potential female preferences for flanged males during periods of female fertility. That females do approach males selectively with regard to the phase of their menstrual cycle has been demonstrated in experiments with captive orangutans (see also Gangestad & Thornhill, this volume). When a female cannot escape, a male will force her into copulation, irrespective of her reproductive condition. If a female controls male access, males tend to become more courteous and she will approach then and initiate copulations when she is in the ovulatory phase (Nadler, 1995; cf. Maple *et al.*, 1979).

Thus, female choice could be the crucial factor that promotes the stable coexistence of flanged and unflanged males. This reasoning also implies that the unflanged males are in a 'waiting-room' situation, making the best of a bad job. They are biding their time until a vacancy for a flanged male position occurs. In other words, the unflanged stage is not an alternative tactic but a transitional stage. A prediction that follows from this hypothesis is that unflanged males do not (or only exceptionally) reproduce successfully.

Both the data for Bornean (e.g. Galdikas, 1979) and Sumatran orangutans (e.g. Schürmann & van Hooff, 1986) seem to support this latter hypothesis. In Borneo, females were shown to have a preference for fully adult males (Galdikas, 1995). Matings with flanged males took place primarily in the context of sexual consorts in which the female willingly engaged in sexual interactions, whereas matings with unflanged males took place mostly outside sexual consorts, with the female resisting the copulation (Galdikas, 1979, 1995; Fig. 11.1a).

The data for the Sumatran orangutans (Fig. 11.1b) point in the same direction; however, the difference in the types of mating between flanged and unflanged males was not as great

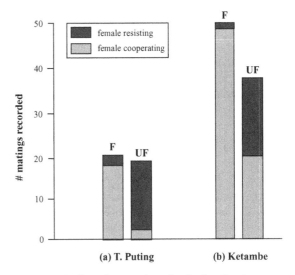

Fig. 11.1 Numbers of cooperative and resisted matings by flanged (F) and unflanged (UF) male orangutans; recorded in Tanjung Puting (data from Galdikas, 1979) and Ketambe (data after Schürmann & van Hooff, 1986).

because females cooperated in an appreciable number of matings with unflanged males. This observation hints at another possibility, namely that there might be two, alternative parallel tactics or strategies coexisting in some frequency- or density-dependent ratio with equal reproductive pay-offs. On the one hand, there are flanged males who rely on a female preference and advertise their presence with long-calls, and wait for females to join them ('long-call-and-wait'). On the other hand, there are unflanged males who look for females and try to seduce them into mating, hoping for the possible chance that a female agrees to a consortship ('go-and-find'). In this case, the prediction is that both male morphs are equally successful in reproduction in the equilibrium situation.

TESTING THE HYPOTHESES WITH PATERNITY DATA

Since 1972, the Ethology and Socio-ecology Group of the Universiteit Utrecht, the Netherlands, in collaboration with the Universitas Nasional, Jakarta, Indonesia, has conducted almost continuous socioecological studies of an orangutan population living in the Ketambe research area, a stretch of tropical lowland rain forest situated in the Gunung Leuser National Park in north-west Sumatra, Indonesia (e.g. Rijksen, 1978; Schürmann, 1982; Sugardjito, 1986;

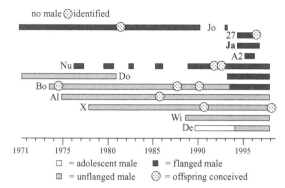

Fig. 11.2   Presence of 12 males (11 individuals identified by their name codes; one unknown individual) in the research area between 1972 and 1999, their status as adolescent, unflanged and flanged males, and the moment an infant was born that they had fathered. The males Do (Doba) and Bo (Boris) changed from unflanged to flanged during this period. After Utami *et al.* (2002).

Sugardjito *et al.*, 1987; te Boekhorst *et al.*, 1990; Utami Atmoko, 2000).

To test predictions of the above hypotheses, we determined paternities and relatedness between adult males by means of microsatellite polymorphism analyses in a cooperative project with Dr Michael Brufford, from Cardiff University, UK, and colleagues. The methods and specific results of these genetic analyses have been presented in detail by Utami *et al.* (2002) and Goossens *et al.* (2000).

Briefly, between 1993 and 1998, faecal samples of 28 individuals were obtained, including those of 11 offspring born during the 25-year period of field studies on the Ketambe population, their mothers ($n = 6$) and 11 males known to be residing in the area, representing 69 per cent of males observed during this period. Because the genetic method used for the present paternity analyses had become available only at the end of the 1990s, not all offspring born since the early 1970s could be sampled before their disappearance. Moreover, the resident adult male that had been present during most of this period had disappeared lately. So the analysis was necessarily restricted to the individuals present at the time when the molecular-genetic techniques could be applied.

We set out to test three hypotheses. The first is the 'tolerance-of-sons' hypothesis, which predicts that the unflanged males enjoy substantial reproductive success and that they are related to the resident flanged male. If there is no such father–son relationship, two possibilities remain.

The 'waiting-room' hypothesis predicts that flanged males monopolise reproduction and father all or a substantial proportion of the offspring. The Alternative Tactics hypothesis predicts that both male morphs are successful and will have a roughly equal number of offspring.

The results of the paternity analysis are summarised in Fig. 11.2. For ten of the 11 offspring, one of the males still present could not be excluded as the father. For four offspring this was a flanged male, for the other six it was an unflanged male. Clearly unflanged males are equally well represented as flanged males, and the hypothesis that unflanged males are in a waiting-room position is not supported. This conclusion is reinforced when we consider the moment during development that the unflanged males sired offspring. Figure 11.2 shows that some offspring were fathered by unflanged males (e.g. Boris), who continued to spend many years in the unflanged condition.

Boris, in particular, must have been more than 8 years old in 1974 when he fathered an offspring. This means that Boris must have been more than 30 years old in late 1994, when he finally changed into a flanged male. In other words: Boris has been a reproductively successful unflanged male for more than 23 years! In addition, a relatedness analysis showed convincingly that the 'tolerance-of-sons' hypothesis can be refuted. Relatedness between the two dominant/resident flanged males, Jon and Nur, and the unflanged males found in the study area were significantly lower than 0.5; and for some unflanged/flanged male pairs, relatedness values were even negative (Utami *et al.*, 2002).

Thus, in the Ketambe population there are two coexisting male morphs, each of which is reproductively successful. Moreover, the relative tolerance of the flanged males toward the unflanged males is not facilitated by an inclusive fitness advantage by the flanged males. The co-occurrence of flanged and unflanged males must, therefore, be explained as a result of the practical impossibility for flanged males to keep unflanged males out of their home-ranges.

The results of these genetic analyses suggest that we must understand and explain orangutan bimaturism in terms of two coexisting alternative tactics. The flanged-male condition is associated with the tactic of advertising one's presence with long-calls and waiting for females, which are attracted when sexually motivated. The unflanged condition is associated with a tactic of keeping a low profile (no long-calls), thus avoiding the provocation of flanged males, and at the same time searching for females and trying to seduce them, or occasionally even to force them into matings.

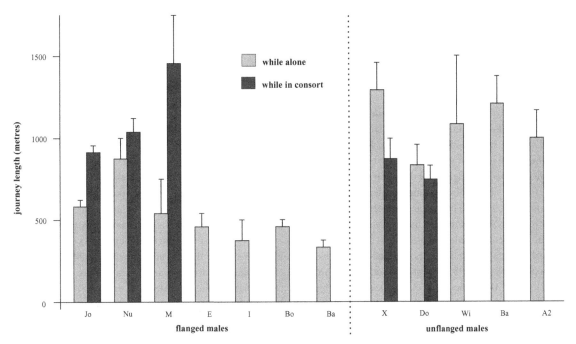

Fig. 11.3 Day journey lengths of Ketambe males. (1) When alone, unflanged males travel significantly farther than flanged males ($p < 0.05$); (2) when in association with a female, flanged males travel significantly farther than alone ($p < 0.05$); (3) when in association with a female, unflanged males tend to travel less far than alone. (After Utami Atmoko, 2000.)

DIFFERENCES IN THE RELATIONSHIPS OF FEMALES WITH FLANGED AND NON-FLANGED MALES

There is behavioural evidence from the Ketambe population for the existence of different male tactics. Utami Atmoko (2000) analysed male–female contacts during a 6-year period and she distinguished between periods of high and low stability in male hierarchical relationships. Moreover, she distinguished between non-reproductive and potentially reproductive females. The latter had been defined as non-pregnant and non-lactating females without an infant or with an infant older than 8 years. The stability factor was related to male–female relationships and their mating patterns. During unstable periods, mating promiscuity was more pronounced for both sexes, including both male classes and both female reproductive stages, suggesting that sexual interactions may be used to regulate relation-

ships (e.g. van Schaik & van Hooff, 1996; Delgado & van Schaik, 2000). In stable periods, only females in the potentially reproductive stage often took the initiative to mate. They initiated copulations more often in relationships with flanged males than with unflanged males. The flanged males copulated more with potentially reproductive females, and only occasionally with non-reproductive females. Unflanged males, on the other hand, took the initiative equally often toward non-reproductive females and potentially reproductive females (S. Utami Atmoko *et al.*, submitted for publication).

Furthermore, there was a difference in the day-journey lengths travelled by Ketambe males that reflected the two tactics: namely, the flanged male tactic of advertising one's presence and waiting for females, and the unflanged tactic of silently searching for females (Fig. 11.3). When alone, flanged males travelled significantly less distance than unflanged males. When in consort, flanged males travelled significantly farther than when alone. However, there was a trend in the opposite direction for unflanged males; they tended to travel less in consort than when alone. In other words, the two morphs adjusted when travelling with a female, but in different directions (Utami Atmoko, 2000; Utami Atmoko *et al.*, submitted). There is an interesting difference here between Sumatran and Bornean orangutans. In Borneo, adult males travel large distances on the ground.

They adjust by ranging shorter distances when in consort with adult females (Galdikas, 1995).

PHENOTYPIC DIVERSITY IN REPRODUCTIVE TACTICS

The preliminary data from this small sample suggest that orangutans follow a conditional strategy with different tactics. Evidence from studies in captivity leaves no doubt that the switch between male tactics may be triggered by a change in the social situation (the disappearance of a flanged male from the vicinity of an unflanged male; Kingsley, 1982; Maggioncalda, 1995). The switch is associated with different levels of certain hormones in certain stages of development. Adolescent males have considerably higher levels of growth hormone than juveniles, arrested adolescents and developed adults, which is therefore associated with their growth spurt (Maggioncalda et al., 2000), as well as higher levels of sex steroids and luteinising hormone (Maggioncalda et al., 1999). Nevertheless, arrested males produce sufficient levels of sex steroids for their primary sexual development and fertility.

Our field data also suggest that the development of secondary sexual characteristics may be associated with changes in the social environment. For example, the transition from the unflanged to flanged state in Boris coincided with the disappearance of the resident flanged male Jon and a subsequent situation of social instability, with many strange flanged males passing through (see also Utami & Mitra Setia, 1995). Maggioncalda et al. (1999) have pointed out that in Ketambe, where orangutan density is comparatively high, the ratio of flanged males to adult females is 1:3, and the ratio of flanged males to unflanged males is 1:2, whereas in Tanjung Puting in Borneo, where orangutan density is low, the ratio of flanged males to females is 1:1, and unflanged males are rare. In other words, there seems to be a density-dependent effect on male sexual development.

The case of orangutan bimaturism is rather unique in two respects. First, males undergo not only a behavioural change but also a corresponding change in physical appearance. Second, for lack of contrary evidence, this change must be regarded as irreversible. In other words, changing from the unflanged to flanged state is a once-in-a-lifetime decision. The remarkable result of our study is that this change does not necessarily occur during the transition from adolescence to adulthood, but that it can be postponed for many years or even decades. With an estimated maximum longevity of about 45 years (Leighton et al., 1995), this implies that certain males may never reach this developmental stage while nevertheless reproducing successfully. This conclusion is also supported by a study of museum skeletons by Uchida (1996) who found that there were males who must have died at an old age, but who had maintained a 'female appearance'.

'FIXED' OR 'PLASTIC' ALTERNATIVE PHENOTYPES?

Alternative phenotypes may be realised as either developmentally fixed or plastic alternatives (Moore, 1991). In the first case the relevant hormonal influences are of the organisational type (the hormonal mechanisms that turn animals into males or females are an example); in the second case the influences are of the activational or motivational type (the effects fluctuate with circumstances and are, in principle, reversible). A study by Moore et al. (1998) is exemplary in pointing out another relevant distinction. They studied dewlap polymorphism in male tree lizards (Urosaurus ornatus). This flap beneath the chin may turn orange when a male grows adult. In some males the lower part turns bright blue. These orange–blue males are sedentary and highly territorial. The presence of an orange–blue dewlap is associated with a high level of both progesterone and testosterone. The males with an orange dewlap are roaming in dry years, but they may be either roaming or sedentary in wet years. These orange animals have comparatively low progesterone and testosterone levels. In the dry years corticosterone levels are higher, both in the orange–blue and in the orange males. However, whereas this high level is associated with a decrease of testosterone levels in the orange males, turning sedentary orange males into roamers, there is no reduction in testosterone level in the orange–blue males. Thus orange–blue and orange males represent 'fixed' alternative phenotypes, reflecting an organisational endocrine influence, whereas the orange males represent two 'plastic' alternative phenotypes (i.e. high-testosterone residents and low-testosterone roamers), the switch depending on environmentally determined differences in activational endocrine variation.

With the evidence available at the moment we have to conclude that the switch in orangutan males is of the 'fixed alternatives' type. Whatever the endocrinological background of the switch may be, it is obviously an organisational influence: once the switch occurred, there is no return. Is the orangutan an exceptional case in this respect or do we have other similar examples among primates?

MANDRILLS AND ORANGUTANS: SPECIES WITH DIFFERENT MALE MORPHS

The best candidate within the primate taxon for such alternative phenotypes of adult male strategies or tactics is the mandrill (*Mandrillus sphinx*). In this species two morphs of adult males can be distinguished: so-called fatted males with bright red–blue coloration of nose, cheeks, buttocks and scrotum, and non-fatted males with pale colours. Dixson *et al.* (1993) studied a captive population living in very spacious, naturally forested enclosures in Gabon. The two types of males did not differ greatly in body weight. However, the fatted males had considerably larger testes and higher plasma testosterone levels. Moreover, whereas the fatted males spent nearly 100 per cent of their time in a bisexual group, the non-fatted males did not. As males matured they became peripheral from their group and, at about 6 years of age, their secondary sexual characteristics began to develop. This development varied enormously among males. It was arrested in subordinate males, which remained peripheral (Setchell & Dixson, 2002).

Thus, there seems to be a difference between the mandrill and the orangutan. Whereas, in orangutans, fully developed and arrested adult males were both reproductively successful, this was not the case in the mandrills. Paternity studies revealed that all of 12 offspring born over a period of 5 years were fathered by the two most dominant of three fatted males and none by the three non-fatted males present (Dixson *et al.*, 1993; Wickings *et al.*, 1993). It remains uncertain whether this is an effect of male exclusion competition alone or whether the outcome also reflects female choice. Given the large size of free-ranging mandrill groups, it remains to be seen whether the extremely asymmetrical outcome was due to unnatural conditions of dominant male control in this captive situation.

There is another notable difference between the orangutan and the mandrill. The secondary sexual adornments of the adult male mandrills gain and lose in intensity as they gain and lose alpha status (Setchell & Dixson, 2001a, b). This reversibility is particularly evident for the red colour and the fattening. Here, a plastic polymorphism is manifested, which is due to the activational effect of steroid hormones. Such an effect on the red skin coloration has also been noted in other primate species (e.g. rhesus monkeys: Vandenburgh, 1965; Rhodes *et al.*, 1998). The blue colour, however, appears to be a fixed trait, as it is in most other investigated species (e.g. the patas monkey: Bercovitch, 1996) and the talapoin monkey

(Dixson & Herbert, 1974). The vervet monkey, where the intensity of scrotal blue is correlated with dominance rank, appears to be an exception in this respect (e.g. Isbell, 1995).

All these observations suggest that orangutans and mandrills differ in the nature of their male bimaturism. Whereas in mandrills the non-fatted, pale morph seems to represent a 'waiting-room' tactic and is largely of the plastic type, in the orangutan bimaturism may reflect alternative successful tactics that are developmentally fixed in one direction (*sensu* Moore, 1991). However, we have to realise that the data sets for both species are still very limited, and the mandrill data come from a captive population with unknown potential differences in wild groups.

An intriguing question about orangutan male bimaturism remains: how did it evolve? According to one scenario, sexual dimorphism evolved while ancestral orangutans were still living a more terrestrial life, permitting the formation of small cohesive groups. The fact that supposed close relatives of the orangutan, the Pleistocene *Gigantopithecus* and the Miocene *Sivapithecus*, led a terrestrial way of life supports this assumption (Fleagle, 1988). Another candidate relative, either as a sister group or as an ancestor, is the Miocene *Sivapithecus* (Andrews & Cronin, 1982; Schwartz, 1984; Andrews & Martin, 1987). The skeletal characteristics of this species indicate that it must have been more quadrupedal and terrestrial (Pilbeam *et al.*, 1990). The size of groups of these comparatively large individuals would have been constrained by the availability and concentration of food resources and, at the same time, the formation of small cohesive groups might have been promoted by the threat of large terrestrial predators, such as tigers. This constellation allowed for the monopolisation of access to fertile females. It would have resulted in intense male contest competition (e.g. van Hooff & van Schaik, 1994; van Hooff, 2000) and, consequently, in selecting for male contest potential, i.e. for sexual dimorphism.

This situation could have persisted after orangutans became more and more arboreal and solitary. However, this shift would have opened up the opportunity for a sneaker tactic. A pre-existing sensitivity to maturation-inhibiting influences of social stress, which is an adaptive phenomenon in many species, could have become emancipated into an opportunity for a long-lasting, parallel alternative reproductive option. According to this speculation, the possibility for an opportunistically used, extremely retarded maturation is the outcome of a growing lack of control from which the dimorphic males in an originally contest-driven system of male relationships suffered, once their rivals and partners became scattered in an environment where home-ranges were large and the possibility of visual detection became small.

SUMMARY AND CONCLUSIONS

The orangutan is exceptional among primates in that it is a comparatively solitary diurnal species with an almost exclusively arboreal life style. In addition, it is exceptional among mammals in that there is pronounced male bimaturism. Within their populations, two coexisting adult and sexually mature male morphs occur: flanged males with fully developed secondary sexual characteristics and unflanged males. Flanged males are very intolerant of other flanged males, but often tolerate the presence of unflanged males in their home-range. This behavioural difference could be explained as an evolutionarily stable situation if the unflanged males are reproductively unsuccessful, i.e. when the development of maturity is suppressed by the presence of a flanged male, when the females prefer mating with flanged males, at least when they are fertile, and when the unflanged males bide their time till a 'vacancy' occurs (the 'waiting-room' hypothesis). If, however, they are reproductively successful, two explanations offer themselves. The relative tolerance could be understood if flanged males defend a range in which they tolerate reproductively successful unflanged males that are relatives, thus obtaining inclusive fitness benefits (the 'range-guardian' hypothesis). A second possibility is that the flanged and unflanged condition represent two parallel alternative tactics with roughly equal fitness returns in the equilibrium situation, a 'sit-call-and-wait' tactic of the flanged males, relying much on a female preference for flanged males, and a 'search-and-find' tactic of the unflanged males. By means of paternity analysis Utami et al. (2002) provided evidence supporting the latter Alternative Tactics hypothesis. They have shown that unflanged males may stay in an 'arrested" unflanged condition for up to 20 years after reaching sexual maturity, i.e. for a major period of their life, if not their entire life. When the socio-demographic situation is favourable, an unflanged male may take a once-in-a-lifetime decision to switch to the other tactic. It is plausible that this remarkable constellation has evolved after the ancestral orangutan changed from a more terrestrial life in small groups to an arboreal and more solitary life style with reduced control by the contest-oriented flanged males after their rivals and partners had become scattered in large home-ranges.

ACKNOWLEDGEMENTS

We are much indebted to Michael Brufford, Benoit Goossens and Jan de Ruiter for their most important and appreciated contribution to the research on which this chapter is based, namely the molecular genetic analyses of the genealogical relationships and the paternities in the Ketambe population of orangutans. We are grateful to Peter Kappeler and Carel van Schaik for organising the Göttinger Freilandtage (conference) at which this chapter was presented, and for the constructive comments of referees.

REFERENCES

Alexander, R. D., Hoogland, J. L., Howard, R. D., Noonan, K. M. & Sherman, P. W. 1979. Sexual dimorphisms and breeding systems in pinnipeds, ungulates, primates and humans. In *Evolutionary Biology and Human Social Behaviour: An Anthropological Perspective*, ed. N. A. Chagnon & W. A. Irons. Belmont: Wadsworth, pp. 402–603.

Andrés, J. A., Sánchez-Guillén, R. A. & Cordero Rivera, A. 2002. Evolution of female colour polymorphism in damselflies: testing the hypotheses. *Animal Behaviour*, **63**, 677–85.

Andrews, P. & Cronin, J. E. 1982. The relationship of *Sivapithecus* and *Ramapithecus* and the evolution of the orang-utan. *Nature*, **297**, 541–6.

Andrews, P. & Martin, L. 1987. Cladistic relationships of extant and fossil hominoids. *Journal of Human Evolution*, **16**, 101–18.

Berard, J. D., Nurnberg, P., Epplen, J. T. & Schmidtke, J. 1994. Alternative reproductive tactics and reproductive success in male rhesus macaques. *Behaviour*, **129**, 177–201.

Bercovitch, F. B. 1996. Testicular function and scrotal coloration in patas monkeys. *Journal of Zoology, London*, **107**, 93–100.

Darwin, C. 1871. *The Descent of Man and Selection in Relation to Sex*. Princeton, NJ: Princeton University Press.

Delgado, R. A. & van Schaik, C. P. 2000. The behavioral ecology and conservation of the orangutan (*Pongo pygmæus*): a tale of two islands. *Evolutionary Anthropology*, **9**, 201–18.

Dixson, A. F. & Herbert, J. 1974. The effects of testosterone on the sexual skin and genitalia of the male talapoin monkey. *Journal of Reproduction and Fertility*, **38**, 217–19.

Dixson, A. F., Bossi, T. & Wickings, E. J. 1993. Male dominance and genetically determined reproductive success in mandrills (*Mandrillus sphinx*). *Primates*, **34**, 525–32.

Fleagle, J.G. 1988. *Primate Adaptation and Evolution.* New York, NY: Academic Press.

Galdikas, B. M. F. 1979. Orangutan adaptation at Tanjung Puting Reserve: mating and ecology. In *The Great Apes*, ed. D. A. Hamburg & E. R. McCown. New York, NY: Academic Press, pp. 195–233.

1985a. Subadult male orangutan sociality and reproductive behavior at Tanjung Puting. *International Journal of Primatology*, **8**, 87–99.

1985b. Adult male sociality and reproductive tactics among orangutans at Tanjung Puting Reserve. *Folia Primatologica*, **45**, 9–24.

1995. Social and reproductive behavior of wild adolescent female orangutans. In *The Neglected Ape*, ed. R. D. Nadler, B. M. F. Galdikas, L. K. Sheeran & N. Rosen. New York, NY: Plenum Press, pp. 163–82.

Goossens, B., Chikhi, L., Utami, S. S., de Ruiter, J. & Bruford, M. W. 2000. A multi-samples, multi-extracts approach for microsatellite analysis of faecal samples in an arboreal ape. *Conservation Genetics*, **1**, 157–62.

Graham, C. E. 1988. Reproductive physiology. In *Orang-Utan Biology*, ed. J. H. Schwartz. New York, NY: Oxford University Press, pp. 91–103.

Gross, M. R. 1996. Alternative reproductive strategies and tactics: diversity within sexes. *Trends in Ecology and Evolution*, **11**, 92–8.

Horr, D. A. 1975. The Borneo orang utan: population structure and dynamics in relationship to ecology and reproductive strategy. In *Primate Behavior*, ed. L. A. Rosenblum, vol. 4. New York, NY: Academic Press, 307–23.

1977. Orang utan maturation: growing up in a female world. In *Primate Biosocial Development: Biological, Social and Ecological Determinants*, ed. S. Chevalier-Skolnikoff & F. E. Poirer. New York, NY: Garland, pp. 289–322.

Isbell, L. A. 1995. Seasonal and social correlates of changes in hair, skin and scrotal condition in vervet monkeys (*Cercopithecus æthiops*) of Amboseli National Park, Kenya. *American Journal of Primatology*, **36**, 61–70.

Jarman, P. J. 1983. Mating system and sexual dimorphism in large terrestrial mammalian herbivores. *Biological Reviews*, **58**, 485–520.

Jarman, P. J. & Southwell, C. J. 1986. Grouping, associations, and reproductive strategies in eastern grey kangaroos. In *Ecological Aspects of Social Evolution*, ed. D. I. Rubenstein & R. W. Wrangham. Princeton, NJ: Princeton University Press, pp. 399–428.

Jolicœur, P. 1985. A flexible 3-parameter curve for limited or unlimited somatic growth. *Growth*, **49**, 271–81.

Kingsley, S. R. 1982. Causes of non-breeding and the development of secondary sexual characteristics in the male orang-utan: a hormonal study. In *The Orang-Utan, Its Biology and Conservation*, ed. L. E. M. de Boer. Den Haag: Junk, pp. 215–29.

1988. Physiological development of male orang-utans and gorillas. In *Orang-Utan Biology*, ed. J. H. Schwartz. New York, NY: Oxford University Press, pp. 123–31.

Knott, C. 1998. Orangutans in the wild. *National Geographic Society*, **194**, 30–57.

Koolhaas, J. M., de Boer, S. F., Buwalda, B. *et al.* 2001. How and why coping systems vary among individuals. In *Coping with Challenge: Welfare in Animals Including Humans*, ed. D. Broom. Dahlem: Dahlem University Press, pp. 199–211.

Lank, D. B., Smith, C. M., Hanotte, O., Burke, T. & Cooke, F. 1995. Genetic polymorphism for alternative mating behavior in lekking male ruff *Philomachus pugnax*. *Nature*, **378**, 59–62.

Leigh, S. R. 1995. Socioecology and the ontogeny of sexual dimorphism in anthropoid primates. *American Journal of Physical Anthropology*, **97**, 339–56.

Leigh, S. R. & Shea, B. T. 1995. Ontogeny and the evolution of adult body size dimorphism in apes. *American Journal of Primatology*, **36**, 37–60.

Leighton, M., Seal, U. S., Soemarma, K. *et al.* 1995. Orangutan life history and vortex analysis. In *The Neglected Ape*, ed. R. D. Nadler, B. M. F. Galdikas, L. K. Sheeran & N. Rosen. New York, NY: Plenum Press, pp. 97–107.

MacKinnon, J. R. 1974. The behaviour and ecology of wild orang-utans (*Pongo pygmaeus*). *Animal Behaviour*, **22**, 3–74.

1979. Reproductive behaviour in wild orangutan populations. In *The Great Apes*, ed. D. A. Hamburg & E. R. McCown. Menlo Park, CA: Benjamin/Cummings, pp. 256–73.

1989. Field studies of wild orang-utans: current state of knowledge. In *Perspectives in Primate Biology*, Vol. 3, ed. P. K. Seth & S. Seth. New Dehli: Today's and Tomorrow's Publishers, pp. 173–86.

Maggioncalda, A. N. 1995. The socioendocrinology of orangutan growth, development, and reproduction – an analysis of endocrine profiles of juvenile, developing adolescent, developmentally arrested adolescent, adult, and aged captive male orangutans. Ph.D. dissertation, Duke University, Durham.

Maggioncalda, A. N., Sapolsky, R. M. & Czekala, N. M. 1999. Reproductive hormone profiles in captive male orangutans: implications for understanding developmental arrest. *American Journal of Physical Anthropology*, **109**, 19–32.

Maggioncalda A. N., Czekala, N. M. & Sapolsky, R. M. 2000. Growth hormone and thyroid stimulating hormone concentrations in captive male orangutans: implications for understanding developmental arrest. *American Journal of Primatology*, **50**, 67–76.

Maple, T. L., Zucker, E. L. & Dennon, M. B. 1979. Cyclic proceptive behavior in a captive female orang-utan (*Pongo pygmæus abelii*). *Behavioural Processes*, **4**, 53–9.

Mitani, J. C. 1985a. Sexual selection and adult male orang utan long calls. *Animal Behaviour*, **33**, 272–83.

1985b. Mating behaviour of male orang utans in the Kutai Game Reserve, Indonesia. *Animal Behaviour*, **33**, 392–402.

Mitani, J. C., Grether, G. F., Rodman, P. S. & Priatna, D. 1991. Associations among wild orang-utans: sociality, passive aggregations or chance? *Animal Behaviour*, **42**, 33–46.

Moore, M. C. 1991. Application of organization-activation theory to alternative male reproductive strategies: a review. *Hormones and Behavior*, **28**, 96–115.

Moore, M. C., Hews, D. K. & Knapp, R. 1998. Evolution and hormonal control of alternative male phenotypes. *American Zoologist*, **38**, 133–51.

Nadler, R. D. 1977. Sexual behavior of captive orangutans. *Archives of Sexual Behavior*, **6**, 457–75.

1981. Laboratory research on sexual behavior of the great apes. In *Reproductive Biology of the Great Apes: Biomedical and Comparative Perspectives*, ed. C. E. Graham. New York, NY: Academic Press, pp. 191–238.

1995. Sexual behavior of orangutans (*Pongo pygmæus*). In *The Neglected Ape*, ed. R. D. Nadler, B. M. F. Galdikas, L. K. Sheeran & N. Rosen. New York, NY: Plenum Press, pp. 223–37.

Pilbeam, D., Rose, M. D., Barry, J. C. & Shah, S. M. I. 1990. New *Sivapithecus* humeri from Pakistan and the relationship of *Sivapithecus* and *Pongo*. *Nature*, **284**, 447–8.

Plavcan, J. M. 1999. Mating systems, intrasexual competition and sexual dimorphism in primates. In *Comparative Primate Socioecology*, ed. P. C. Lee. Cambridge: Cambridge University Press, pp. 241–26.

Plavcan, J. M. & van Schaik, C. P. 1992. Intrasexual competition and canine dimorphism in primates. *American Journal of Physical Anthropology*, **87**, 461–77.

Rhodes, L., Argersinger, M. E., Gantert, L. T. *et al.* 1998. Effects of administration of testosterone, dihydrotestosterone, oestrogen and fadrazole, an aromatase inhibitor, on sex skin colour in intact male rhesus macaques. *Journal of Reproduction and Fertility*, **111**, 51–7.

Rijksen, H. D. 1978. *A Fieldstudy on Sumatran Orang Utans* (Pongo pygmaeus abelii *Lesson 1827*). Wageningen: H. Veenman & B. V. Zonen.

Rodman, P. S. 1973. Population composition and adaptive organisation among orang-utans of the Kutai Reserve. In *Comparative Ecology and Behaviour of Primates*, ed. H. D. Crook. London: Academic Press, pp. 171–209.

1977. Feeding behaviour of orang-utans in the Kutai Nature Reserve, East Kalimantan. In *Primate Ecology*, ed. T. H. Clutton Brock. London: Academic Press, pp. 383–413.

Rodman, P. S. & Mitani, J. C. 1987. Orangutans: sexual dimorphism in a solitary species. In *Primate Societies*, ed. B. B. Smuts, D. L. Cheney, R. M. Seyfarth, R. W. Wrangham & T. T. Struhsaker. Chicago, IL: University of Chicago Press, pp. 146–54.

Ryan, M. J. & Wagner, B. A. 1987. 'Alternative' mating behavior in the swordtails *Xiphophorus nigrensis* and *Xiphophorus pygmaeus* (Pisces: Poeciliidae). *Behavioral Ecology and Sociobiology*, **24**, 341–8.

Ryan, M. J., Pease, C. M. & Morris, R. M. 1992. A genetic polymorphism is the swordtail *Xiphophorus nigrensis*: testing the prediction of equal fitness, *American Naturalist*, **139**, 21–31.

Schürmann, C. L. 1982. Courtship and mating behavior of wild orangutans in Sumatra. In *Primate Behavior and Sociobiology*, ed. A. B. Chiarelli & R. S. Corrucini. Berlin: Springer Verlag, pp. 129–35.

Schürmann, C. L. & van Hooff, J. A. R. A. M. 1986. Reproductive strategies of the orang-utan: new data and a reconsideration of existing sociosexual models. *International Journal of Primatology*, **7**, 265–87.

Schwartz, J. H. 1984. The evolutionary relationships of man and orang-utans. *Nature*, **308**, 501–5.

Setchell, J. M. & Dixson, A. F. 2001a. Changes in the secondary sexual adornments of male mandrills (*Mandrillus sphinx*) are associated with gain and loss of alpha status. *Hormones and Behavior*, **39**, 177–84.

2001b. Circannual changes in the secondary sexual adornments of semifree-ranging male and female mandrills (*Mandrillus sphinx*). *American Journal of Primatology*, **53**, 109–21.

2002. Developmental variables and dominance rank in adolescent male mandrills (*Mandrillus sphinx*). *American Journal of Primatology*, **56**, 9–25.

Shuster, S. M. & Wade, M. J. 1991. Equal mating success among male reproductive strategies in a marine isopod. *Nature*, **350**, 608–10.

Smuts, B. B. 1985. *Sex and Friendship in Baboons*. New York, NY: Aldine.

Sugardjito, J. 1986. Ecological constraints on the behaviour of Sumatran orang-utans (*Pongo pygmæus abelii*) in the Gunung Leuser National Park, Indonesia. Ph.D. thesis, University of Utrecht.

Sugardjito, J., te Boekhorst, I. J. A. & van Hooff, J. A. R. A. M. 1987. Ecological constraints on the grouping of wild orang-utans (*Pongo pygmaeus*) in the Gunung Leuser National Park, Sumatra, Indonesia. *International Journal of Primatology*, **8**, 17–41.

te Boekhorst, I. J. A., Schürmann, C. L. & Sugardjito, J. 1990. Residential status and seasonal movements of wild orang-utans in the Gunung Leuser Reserve (Sumatra, Indonesia). *Animal Behaviour*, **39**, 1098–109.

Uchida, A. 1996. Craniodental variation among the great apes. *Peabody Museum Bulletin*, **4**, 1–186.

Utami Atmoko, S. 2000. Bimaturism in orang-utan males: reproductive and ecological strategies. Ph.D. thesis, University of Utrecht.

Utami, S. & Mitra Setia, T. 1995. Behavioral changes in wild male and female Sumatran orangutans (*Pongo pygmaeus abelii*) during and following a resident male take-over. In *The Neglected Ape*, ed. R. D. Nadler, B. M. F. Galdikas, L. K. Sheeran & N. Rosen. New York, NY: Plenum Press, pp. 183–90.

Utami, S. S., Wich, S. A., Sterck, E. H. M. & van Hooff, J. A. R. A. M. 1997. Food competition between wild orangutans in large fig trees. *International Journal of Primatology*, **18**, 909–27.

Utami, S. S., Goossens, B., Bruford, M. W., de Ruiter, J. R. & van Hooff, J. A. R. A. M. 2002. Male bimaturism and reproductive success in Sumatran orang-utans. *Behavioral Ecology*, **13**, 643–52.

Vandenburgh, J. G. 1965. Hormonal basis of the sex skin in male rhesus monkeys. *General and Comparative Endocrinology*, **5**, 31–4.

van Hooff, J. A. R. A. M. 1995. The orangutan: a social outsider and a social-ecology test case. In *The Neglected Ape*, ed. R. D. Nadler, B. M. F. Galdikas, L. K. Sheeran & N. Rosen. New York, NY: Plenum Press, pp. 153–62.

van Hooff, J. A. R. A. M. 2000. Relationships among non-human primate males: a deductive framework. In *Primate Males*, ed. P. M. Kappeler. Cambridge University Press: Cambridge, pp. 183–91.

van Hooff, J. A. R. A. M. & van Schaik, C. P. 1994. Male bonds: affiliative relationships among nonhuman primate males. *Behaviour*, **130**, 143–51.

van Rhijn, J. G. 1973. Behavioural dimorphism in male ruffs, *Philomachus pugnax* (L.). *Behaviour*, **47**, 153–229.

van Schaik, C. P. & van Hooff, J. A. R. A. M. 1996. Toward an understanding of the orangutan's social system. In *Great Ape Societies*, ed. W. C. McGrew, L. F. Marchant & T. Nishida. Chicago, IL: University of Chicago Press, pp. 3–15.

Weckerly, F. 1998. Sexual-size dimorphism: influence of mass and mating systems in the most dimorphic mammals. *Journal of Mammalogy*, **79**, 33–52.

Wickings, E. J., Bossi, T. & Dixson, A. F. 1993. Reproductive success in the mandrill, *Mandrillus sphinx*: correlations of male dominance and mating success with paternity, as determined by DNA fingerprinting. *Journal of Zoology, London*, **231**, 563–74.

12 • Sexual selection and the careers of primate males: paternity concentration, dominance-acquisition tactics and transfer decisions

MARIA A. VAN NOORDWIJK
and CAREL P. VAN SCHAIK
*Department of Biological
Anthropology and Anatomy
Duke University
Durham, NC, USA*

INTRODUCTION

Life-history theory suggests that natural selection has shaped an organism's development so that it optimally positions the young adult for the challenges of reproductive life. Sex differences in development are usually the product of sexual selection (Setchell & Lee, this volume). Thus, in polygamous organisms, sons are generally dependent on their mother for longer and make greater demands on maternal resources, so that adult males tend to be larger and stronger, allowing them to seek out and compete for mates efficiently (Clutton-Brock *et al.*, 1985; Trivers, 1985; Clutton-Brock, 1991). Sexual selection should also have had a profound effect on the behavioural decisions made by males, but studies so far have focused mainly on broad sex differences in migration and risk taking, and thus mortality rates (Clutton-Brock *et al.* 1985; Trivers, 1985; Clutton-Brock, 1991).

Especially in long-lived organisms such as primates, a male's success in competing for mates and protecting his offspring should be affected by the nature of major social decisions, such as whether and when to transfer to other groups or to challenge dominants. Several studies indicate dependence of male decisions about transfer and acquisition of rank on age and local demography (e.g. Phillips-Conroy *et al.*, 1992; Sprague, 1992; Watts, 2000). Likewise, our work on male long-tailed macaques (scientific names are listed in Table 12.1) indicated a remarkably tight fit between the behavioural decisions of males and expectations based on known determinants of success (van Noordwijk & van Schaik, 2001), suggesting that natural selection has endowed males with decision rules that, on average, produce optimal life-history trajectories (or careers) for a given set of conditions. Because

conditions vary among species, we expect a diverse array of career profiles. In this chapter, we explore the impact of variation in the intensity of male–male mating competition on the careers of individual males among a number of different anthropoid primate species and populations.

Most non-human primates live in groups with continuous male–female association, in which group membership of reproductively active (usually non-natal) males can last many years. For a male living in such a mixed-sex group, dominance rank reflects his relative power in excluding others from resources. However, the impact of dominance on mating success is variable (de Ruiter & Inoue, 1993; de Ruiter & van Hooff, 1993; Paul, 1997), and this variation may have produced variation in developmental profiles of male primates.

Although rank acquisition is usually considered separately from transfer behaviour and mating success, the hypothesis examined here is that they are interdependent (e.g. van Noordwijk & van Schaik, 2001). We predict that the degree of paternity concentration in the dominant male, determined by his ability to exclude other males from mating, determines the relative benefits of various modes of acquisition of top rank (challenge, succession, and presence or absence of the use of coalitions), and that these together determine patterns of male transfer (duration of stay in subsequent groups, conditions precipitating emigration, and features of immigration groups). Paternity concentration should also affect the number of males in a group: additional males can coexist in a group when dominants cannot monopolise all potentially fertile matings (Nunn, 1999). The main driving variable, paternity concentration, in turn, depends on the interaction between potential for monopolisation of fertile matings, which is affected by the number of females

---

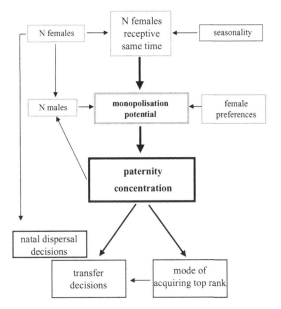

Fig. 12.1   Predicted relationships among the main variables. Arrows indicate that the variable at its origin is considered to have favoured selective changes in ('selected for') the variable at the end of the arrow. Bold arrows indicate the relationships examined in this chapter.

and their synchrony in sexual attractivity and by female mating preferences. Figure 12.1 illustrates the hypothesised relationships between these variables.

We begin by reviewing the variation in, and determinants of, paternity concentration, dominance acquisition and between-group transfer among male primates, and develop predictions for their interrelations. After that, we present the results of a comparative analysis, based on a search of the primate literature for data on groups and populations with known paternity distribution, in combination with data on social relationships, group size, breeding seasonality, mode of acquisition of the top-rank position, and transfer behaviour.

MALE CAREERS IN PRIMATES: PATTERNS AND PREDICTIONS

In this section, we briefly review the literature on male mating competition, dominance and transfer behaviour in primates, and develop predictions based on our working hypothesis. These predictions are placed in italics, and will be tested in the next section.

PATERNITY CONCENTRATION

Ever since Altmann (1962), the expectation for paternity distribution has been based on the priority-of-access model (PoA), which focuses on the degree to which the dominant male can monopolise mating access. It predicts a decreasing probability of paternity according to dominance rank, with all paternity concentrated in the dominant male if there is never more than one sexually attractive female at the same time. This model has rarely been tested with the appropriate data on female synchrony, except for baboons (Noë & Sluijter, 1990; Bulger, 1993; Alberts et al., 2003). Most studies report correlations between male dominance rank and siring success, but correlations are not the ideal measure for the relative proportion of paternities of a single male, i.e. the dominant (cf. Bulger, 1993). First, their values are affected by cohort sizes (see Barton & Simpson, 1992; Cowlishaw & Dunbar, 1992). Second, they are not suitable for characterising the degree of expected skew among the males, where many may have zero paternity. Third, correlations cannot take into account that some infants may be sired by extra-group males (Sprague et al., 1996, 1998; Keane et al., 1997; Borries, 2000; Soltis et al., 2001). We will therefore define 'paternity concentration' as the proportion of infants sired by the dominant male in a social group.

Empirical studies have noted variation in the relationship between dominance rank and paternity among species, populations within a species, groups within a population, and years within a group (Inoue et al., 1992; Paul et al., 1993; Paul, 1997; Berard, 1999). Some of this variation is consistent with the PoA model – paternity concentration is negatively affected by the number of females in the group (Altmann, 1962) and breeding seasonality (Cowlishaw & Dunbar, 1991; Paul, 1997): both affect the likelihood that females will be sexually active at the same time. However, additional sources of variation also exist.

The PoA model assumes that females are indifferent to male monopolisation. However, female behaviour may affect the monopolisability of matings by the dominant male. First, the cost of inbreeding may cause females to avoid mating with male relatives (Alberts & Altmann, 1995; Takahata et al., 1999; Constable et al., 2001). This tendency has been invoked to explain an apparent female preference for novel (recently immigrated) males (Takahata, 1982; Huffman, 1991b, 1992; Berard, 1999; but see Manson, 1995). Second, females may have mating preference for special male friends (Smuts, 1983, 1985). However, the existence of such friendships may not be widespread; more often female–male friendships were shown to result in a higher chance of survival of offspring already sired (Takahata, 1982; Smuts, 1983, 1985; Bercovitch, 1991; Huffman, 1991b; Manson, 1994; Palombit *et al.*, 1997, 2001). Third, variation and flexibility in the duration of receptivity and the degree to which females actively seek matings with multiple males may affect monopolisation potential (e.g. van Schaik *et al.*, 1999; Hrdy, 2000; see van Schaik *et al.*, this volume).

Male behavioural tactics may affect the dominant male's monopolisation as well. First, in some populations males are known to form coalitions that attack a higher-ranking male guarding a sexually attractive female (Noë & Sluijter, 1990). By forming coalitions, the participating males may increase their chances of siring offspring, above expectations based on PoA. These coalitions are more likely when more and older males are present in the group, although this does not fully explain observed variation among baboons (see Bulger, 1993; Weingrill *et al.*, 2000; Alberts *et al.*, 2003). Second, the dominant male's mate guarding may become less effective when too many males are present; in chimpanzees (*Pan troglodytes*) this leads to coalitionary mate guarding and sharing of matings among the allies (Watts, 1998). These phenomena may underlie the observed effect of the number of males on the dominant's (estimated) paternity concentration (Cowlishaw & Dunbar, 1991, 1992).

The interaction between male and female behaviour may also reduce the dominant's potential to monopolise females. In several situations, males visit groups in which they do not become a resident (for more than a few weeks), either alone (Mehlman, 1986; Huffman, 1991a; Sprague, 1991; Sprague *et al.*, 1998) or as part of an 'influx' (Takahata *et al.*, 1994; Borries, 2000; Cords, 2000). Such males may be members of a nearby group or not reside in a mixed-sex group at all, and their stay is often too short to establish a clear dominance relationship with the resident male(s). In general, females mate cooperatively with these visitors. Paternity studies in natural populations show that matings by extra-group males actually resulted in offspring (positively identified: Berard *et al.*, 1993; Keane *et al.*, 1997; inferred: Launhardt *et al.*, 2001; Soltis *et al.*, 2001). Extra-group males are likely to gain access to the group's females when they enter in large numbers, a high number of females are simultaneously in oestrus relative to the number of males in the group (Takahashi, 2001), or the resident male's or males' strength is diminished (Borries, 2000). Information on fluctuating numbers of oestrous females relative to the number of resident males

or the number or density of extra-group males needed to test this idea is often not available (Cords, 2000; Carlson & Isbell, 2001).

This overview shows that variations on the PoA model can be formulated that modify the assumptions of female indifference and invariant monopolisation by the dominant male (see, for example, the 'mating-conflict' model: van Schaik *et al.*, this volume). These models all predict that *male paternity concentration decreases with increasing group size and seasonality*. In practice, the data of existing studies are not reported in a way that allow us to distinguish between the models, even though future studies should be able to do so. We can also predict that *extra-group males sire more offspring in more seasonal populations*.

In sum, the main factors in monopolisation potential and consequent paternity concentration are thought to be the number of (unrelated) females and seasonality (or more precisely female synchrony), but additional female and male behavioural tactics may reduce the expected degree of paternity concentration.

ACQUISITION OF TOP-DOMINANCE RANK

Observations have shown that a male can attain top rank in a mixed-sex primate group in three different ways. First, he can defeat the current dominant male during an aggressive challenge, either coinciding with immigration or after having been resident in the group for some time (Rudran, 1973; Hrdy, 1977; Henzi & Lucas, 1980; Sigg *et al.*, 1982; Dunbar, 1984; van Noordwijk & van Schaik, 1985, 1988, 2001; Robinson, 1988a; Hamilton & Bulger, 1990; Rajpurohit, 1991; Kummer, 1995; Robbins, 1995; Sprague *et al.*, 1996; Perry, 1998a, b; Borries, 2000; Watts, 2000). Second, he can attain top rank during the formation of a new group, either by acquiring females from other groups (Harcourt, 1978; Sigg *et al.*, 1982; Watts, 1990; Kummer, 1995; Steenbeek *et al.*, 2000) or as the result of fission of an existing group (Koyama, 1970; Chepko-Sade & Sade, 1979; Dunbar, 1984; Yamagiwa, 1985; Kuester & Paul, 1997). In general, in these cases, the male attaining a group's top-dominance rank acts alone (van Noordwijk & van Schaik, 1985, 2001; Hamilton & Bulger, 1990; Sprague, 1992; Suzuki *et al.*, 1998a; Borries, 2000) with, at most, 'passive' help from others. However, there are exceptions to this pattern (see also Dittus *et al.*, 2001). Male red howler monkeys present a well-documented example of challenge take-overs by coalitions of (usually related) males who oust the previous male(s). It is almost impossible for a single male to defeat two cooperating resident males (Crockett & Sekulic, 1984; Pope, 1990). The allies tend to achieve reproductive success consecutively in the same group.

A third way to achieve top rank is by default, or through 'succession', after the departure or death of the previous top-ranking male, not preceded by challenges from other males (Drickamer & Vessey, 1973; Sugiyama, 1976; Itoigawa, 1993; Nakamichi et al., 1995; Sprague et al., 1996; Berard, 1999; Watanabe, 2001). In this condition, a male's dominance rank increases with his duration of residence and age. Some coalitionary support from both males and females enables a male's rise to top rank after the incumbent is gone, but severe fighting is rare (Itoigawa, 1993; Nakamichi et al., 1995). Although males may challenge others to rise in rank, the dominant is not challenged (Huffman, 1991a; Watanabe, 2001).

High potential for monopolisation of sexual access to females, if common over time, would favour the evolution of high-risk tactics of rank acquisition, such as challenging of the dominant through individual action and escalated fights. At intermediate levels of monopolisation potential, coalitionary challenges for mating access could be expected. With declining reproductive benefits of dominance rank, i.e. low paternity concentration, we expect lower-risk tactics of rank acquisition, such as a queuing system, or dominance acquisition through succession. In the extreme of mating scrambles, the only benefits of high rank, if any, may be increased longevity (Watanabe, 2001). Thus, at high paternity concentration we expect that *top rank is mainly acquired through challenge, whereas at low paternity concentration top rank is achieved more often through succession.*

Sprague et al. (1996) noted that succession was common in large (provisioned) groups of Japanese macaques, whereas take-overs were more common in the small (natural) groups on Yakushima. This pattern is expected if paternity concentration is reduced in larger groups. Therefore, we predict that *group size is correlated with the way in which top rank is achieved.*

To succeed in attaining top rank through challenge, a male should be in prime physical condition, whereas top rank through succession (and perhaps through coalitions) may be achieved by males with lower individual fighting ability. Because in most primate species young adult males are the most powerful fighters, we expect that *there is a difference in the age distribution of the dominant males under the two different kinds of top-rank acquisition.*

TRANSFER

In many group-living non-human primate species, males leave their natal group whereas females tend to stay (Pusey & Packer, 1987; Pope, 2000). The benefit of male transfer is generally thought to be improved mating access to females. Whether a male stays and breeds in his natal group is expected to depend on the cost of inbreeding, weighted for its probability, and the magnitude of the expected net benefit of transfer (access to unrelated mating partners minus costs).

The cost of inbreeding varies between species and populations, but in most known wild populations mortality is higher among inbred immatures (Crnokrak & Roff, 1999). Hence, we expect inbreeding avoidance where feasible. If a male has mating access to females, the probability of inbreeding depends on the number of closely related females relative to the number of unrelated females in the group. Indeed, in the large free-ranging groups of Barbary macaques at Affenberg, males with more maternal relatives were found to be significantly more likely to leave the natal group than those with few, and all males born in small groups left the group as subadults (Kuester & Paul, 1999). Similarly, the only male among 52 natal male long-tailed macaques in the Ketambe population known to have stayed in its natal group throughout subadulthood was an orphan who lived in the largest study group and had only one maternal sister (van Noordwijk & van Schaik, 2001). In the absence of genealogical data, *we expect that the average male born in a group with fewer females is more likely to leave before reproduction than one in a larger group.*

The benefit of transfer is generally expected to be improved mating access. Unfortunately, it is difficult to develop predictions for the frequency of secondary transfer and its relation to natal transfer, because too many factors affect the costs of transfer. The costs of transfer are composed of the cost of the transition period, e.g. predation and starvation risk (Alberts & Altmann, 1995), and the cost of immigration, e.g. risk of injury by members of the target group (Cheney & Seyfarth, 1983; Zhao, 1996). The transition period may be non-existent, as when males transfer during between-group encounters (Melnick et al., 1984); brief, as when transfer is into adjacent groups (Packer, 1979a; Henzi & Lucas, 1980; Cheney & Seyfarth, 1983; Melnick et al., 1984; van Noordwijk & van Schaik, 1985, 2001; Zhao, 1994; Pope, 2000); or long. When it is long, males may be alone (Alberts & Altmann, 1995; Mehlman, 1986; Muroyama et al., 2000) or in all-male bands (Rajpurohit et al., 1995;

Sprague et al., 1998; Yamagiwa & Hill, 1998; Steenbeek et al., 2000). Costs of immigration may be reduced by transferring together with peers or into adjacent groups with familiar or related males (Meikle & Vessey, 1981; Cheney & Seyfarth, 1983; van Noordwijk & van Schaik, 1985; Jack & Fedigan, 2001). For reasons that are poorly understood, males in some species delay natal dispersal until adult size is attained (Alberts & Altmann, 1995; Zhao, 1996).

Despite this complexity, we can none the less predict how paternity concentration would affect transfer decisions, all other things being equal. If paternity concentration is high, a male's best chance to achieve high reproductive success is to attain top rank and strive for long tenure. Thus, we should see transfer decisions that reflect the (possibly long-term) prospects of acquiring and maintaining top rank since a low- or even medium-ranking male has little chance of siring offspring. Therefore, a young male's best group of residence is the one in which he can expect to defeat the resident male cohort at some point in the future (van Noordwijk & van Schaik, 2001).

If potential paternity concentration is intermediate, the distribution of paternities among the non-dominant males is correlated with rank. Hence, a male is predicted to select the group in which he can reach the highest possible rank for his physical condition. Here, the number of females directly affects paternity concentration and thus the probability for lower-ranking males to sire offspring. Relative and absolute number of females are thus expected to affect a male's choice, as well as opportunities to attain high rank. Older males are unlikely to be able to challenge an incumbent dominant successfully, so we expect that in populations with high potential paternity concentration older males switch to this transfer tactic (including the option of staying in their current group).

If potential paternity concentration is low, i.e. mating competition is largely by scramble, males of all ranks have roughly equal chances of siring offspring. Their siring success will largely depend on the number of available females relative to the number of male competitors, i.e. the operational sex ratio. Therefore, a male is expected to reside in, or transfer into, the group that has the most favourable operational sex ratio, from among the groups that are available as immigration targets, including the current one. Whether or not a male transfers should depend to a large extent on the availability of accurate information on other groups.

Thus, we predict that *with increasing potential for paternity concentration, the prospects of attaining dominance in the destination group will become a more important criterion for the preferred group of residence than the number of females relative* *to number of males. We also predict that with age the balance of these criteria may shift, especially at higher potential paternity concentration.*

These predictions assume that males can make optimal decisions based on accurate assessment of their options. Indeed, individual primates can have considerable 'knowledge' about the identity of members of often-encountered adjacent groups (Cheney & Seyfarth, 1990), and many studies report that males quietly observe a group before deciding to immigrate, move on to another group, or return to their group of residence (e.g. Henzi & Lucas, 1980; Cords, 1984; Hamilton & Bulger, 1990; Takahata et al., 1994; Henzi et al., 1998; Muroyama et al., 2000; Oluput & Waser, 2001; van Noordwijk & van Schaik, 2001). Thus, we expect higher transfer rates, on average, in populations with higher encounter frequency and more home-range overlap. Unfortunately, we were unable to test this prediction owing to a lack of detailed published data on characteristics of all nearby possible destination groups and encounter frequencies.

TESTS OF THE PREDICTIONS
PROCEDURES AND DEFINITIONS

We searched the primate behaviour literature for data on group size, breeding seasonality, male dominance relationships and mode of acquisition of the top-rank position, and male transfer behaviour in populations with known paternity distribution. We limited our search to species with male transfer. Most data on captive groups could not be used because of their artificial group compositions and lack of transfer options. Most of the data found were on cercopithecines. The data on populations used for most analyses and references for each species are listed in Table 12.1.

Dominance ranks are usually based on priority of access to resources or (ritualised) unidirectional submission signals. We assume that males generally maximise their paternity opportunities, hence that the observed paternity concentration (in the dominant male) is close to the potential paternity concentration achievable under the given ecological and demographic conditions.

Analyses were done on two data sets: one including only populations with DNA-based paternity assessments, from which we derived the 'percentage paternity concentration'; the other combining DNA-based and behaviour-based paternity estimates (using the males' mating activity around

Table 12.1 *Correlates of male group transfer among primates.*

Species		Location	No. of females	No. of males	% births 3 months	% pat[a] topdom[b]	All patconc[c]	Reference
Wild								
Alouatta seniculus	red howler	Hato Masaguaral	3	2	30	95	high	Pope, 1990; Crockett & Janson, 2000
Cebus olivaceus	wedge-capped capuchin	Hato Masaguaral	5.5	2	68		high	Robinson, 1988a, b
Cercopithecus aethiops	vervet monkey	Amboseli	8	3	86		high	Cheney, 1983; Cheney & Seyfarth, 1983; Andelman, 1986; Cheney et al., 1988
Cercopithecus aethiops		Burman Bush	7	2.5	95		high	Henzi & Lucas, 1980
Erythrocebus patas	patas monkey	Kala Maloue		3	>90	65	high	Ohsawa et al., 1993
Gorilla g. beringei	mountain gorilla	Virunga		1.5			high	Watts, 2000
Macaca fascicularis	long-tailed macaque	Ketambe	11	7	49	75	high	van Noordwijk & van Schaik, 1985, 2001; de Ruiter et al., 1992
Macaca fuscata fuscata	Japanese macaque	Kinkazan	20	14	100		low	Sprague et al., 1998; Suzuki et al., 1998a; Yamagiwa & Hill, 1998
Macaca fuscata yakui		Yakushima	10	7	100	33	medium	Sprague et al., 1998; Suzuki et al., 1998a, b; Yamagiwa & Hill, 1998; Soltis, 1999; Soltis et al., 2000, 2001
Macaca mulatta	rhesus macaque	Dunga Gali			100		high	Melnick et al., 1984; Melnick & Hoelzer, 1996
Macaca radiata	bonnet macaque	Dharwar	9	8	95			Sugiyama, 1971
Macaca silenus	lion-tailed macaque	South India	8	1.1	60		high	Kumar & Kurup, 1985
Macaca sinica	toque macaque	Polonnaruwa	7	3	>95	43	medium	Dittus, 1975, 1988; Keane et al., 1997; Dittus et al., 2001
Macaca sylvanus	Barbary macaque	Akfadou	11	10	100		?	Ménard & Vallet, 1996
Macaca sylvanus		Tigounatine	13	13	100		high	Ménard & Vallet, 1996

(cont.)

Table 12.1 (*cont.*)

Species		Location	No. of females	No. of males	% births 3 months	% pat[a] topdom[b]	All patconc[c]	Reference
Macaca sylvanus		Ghomaran	9	5	100			Mehlman, 1986
Papio c. anubis	olive baboon	Gombe	13	8	25		high	Packer, 1979a, b
Papio c. ursinus	chacma baboon	Drakensberg	10.5	5.5			high	Henzi *et al.*, 1998; Weingrill *et al.*, 2000
Papio c. ursinus		Moremi	14	7			high	Bulger & Hamilton, 1987, 1988; Hamilton & Bulger, 1990; Bulger, 1993
Papio cynocephalus		Mikumi	37	14	30			Rasmussen, 1981, 1986
Semnopithecus entellus	hanuman langur	Ramnagar	11	4	86	57*	high	Koenig *et al.*, 1997; Borries, 2000; Launhardt *et al.*, 2001
Semnopithecus entellus		Ramnagar	6	1	86	100*	high	Koenig *et al.*, 1997; Borries, 2000; Launhardt *et al.*, 2001
Presbytis thomasi	Thomas' langur	Ketambe	3.2	1	27		high	Steenbeek *et al.*, 2000; R. Steenbeek, personal communication
Wild: food-enhanced								
Macaca fuscata fuscata	Japanese macaque	Arashiyama	80	31	90		low	Huffman, 1991a, b; Koyama *et al.*, 1992; Sprague *et al.*, 1996; Takahata, *et al.*, 1999
Macaca fuscata fuscata		Katsuyama	58	22	98			Itoigawa *et al.*, 1992; Itoigawa, 1993; Nakamichi, 1995
Macaca fuscata fuscata		Ryozenyama	17	5	98			Sugiyama & Ohsawa, 1982
Macaca fuscata fuscata		Koshima	40	17	100		low	Watanabe *et al.*, 1992; Watanabe, 2001
Macaca maurus	Moor macaque	Karaente	10	5	56			Okamoto *et al.*, 2000
Macaca thibetana	Tibetan macaque	Mt Emei	12	9	68		medium	Zhao, 1994, 1996

Species	Common name	Location						Reference
Papio cynocephalus	yellow baboon	Amboseli – Lodge	19	8	40	71	high	Altmann *et al.*, 1985; Alberts & Altmann, 1995; Altmann *et al.*, 1996; S.C. Alberts, pers. com.
Semnopithecus entellus	hanuman langur	Jodhpur	17.4	1	33		high	Sommer & Rajporohit, 1989
Free-ranging: food-enhanced								
Macaca mulatta	rhesus macaque	Sabana Seca	47	21	100			Bercovitch, 1993; Bercovitch & Nürnberg, 1996, 1997
Macaca mulatta		Cayo Santiago S	24	17	100	18	medium	Berard *et al.*, 1994; Berard, 1999
Macaca sylvanus	Barbary macaque	Affenberg	47	27	100	16	medium	Paul, 1989; Kuester & Paul, 1992; Paul *et al.*, 1993; Kuester *et al.*, 1995; Kuester & Paul, 1999
Mandrillus sphinx	mandrill	Gabon CIRMF	11	6	75	76	high	Dixson *et al.*, 1993; Wickings *et al.*, 1993; Setchell & Dixson, 2001
Captive								
Macaca arctoides	stump-tailed macaque	Madison	14	2	25	95	high	Bauers & Hearn, 1994
Macaca fascicularis	long-tailed macaque	CPRC Davis	26	7	25	15	low	Shively & Smith, 1985
Macaca fuscata fuscata	Japanese macaque	Kyoto PRI	18	11	100	12.5	low	Inoue *et al.*, 1991, 1993
Macaca mulatta	rhesus macaque	CPRC Davis	35	10	100	18	low	Smith, 1993, 1994
Macaca radiata	bonnet macaque	CPRC Davis	29	6	76		high	Samuels *et al.*, 1984; Silk, 1989, 1993

Note: For some analyses the population average of 69 per cent paternity concentration for *Semnopithecus entellus*, Ramnagar, was used; for others paternity concentrations for single-male and multi-male groups were both used.

[a]pat = paternity; [b]topdom = top-dominant; [c]patconc = paternity concentration.

the estimated time of conception). In this second set, we used three paternity concentration classes: high = dominant estimated to have majority (>50 per cent) of paternities; medium = dominant has more paternities than any other male but less than 50 per cent; low = dominant is not the male with highest percentage of estimated paternities. At least in (semi-) natural populations, behavioural assessments of paternity concentration were shown to agree with genetic assessments (Paul *et al.*, 1993; Wickings *et al.*, 1993; Altmann *et al.*, 1996; Bercovitch & Nürnberg, 1997; Soltis *et al.*, 2000).

Because we postulate that paternity concentration affects a male's transfer and rank-acquisition tactics, we expect that at similar levels of paternity concentration the same patterns should hold at the level of groups, populations and species, despite species differences in sexual dimorphism and male harassment potential. We also expect variation within an individual over its lifetime. Because variation at all these levels was indeed found, there was no indication for species-specific characteristics that would necessitate phylogenetically controlled analyses.

We employed the following definitions. The number of females or males refers to the (average) number of sexually mature females or males reported in the original studies. In species where subadult males are able to fertilise females, they are included among the males. We used birth seasonality to index synchrony in female sexual activity, because it is more readily available in the literature. We adopted Oi's (1996) quantitative measure of seasonality: the average percentage of births during the 3-month period with the highest number of births (preferably over multiple years and groups). We use the term 'residence' for the duration of a male's membership of a group and 'tenure' for the duration of a male's occupation of top rank (highest rank).

PATERNITY CONCENTRATION

As predicted, an increase in the number of adult males or females in a group was significantly correlated with a decrease in the percentage (DNA-based) paternity concentration, both among and within populations (Fig. 12.2). The effect of the number of males seems stronger than that of the number of females, but the two were so tightly correlated, both in this sample ($r = +0.940$, $n = 9$, $P < 0.001$) and in much larger comparisons (Nunn, 1999), that their effects cannot be reliably untangled without experiments. The correlation between paternity concentration and group size was supported by various additional analyses. First,

we saw the same relationship within three wild populations (Fig. 12.2). Second, the relationship is retained in the second data set containing both behavioural and genetic estimates of paternity concentration (Spearman rank correlations: for males $r_s = -0.719$, $n = 26$, $P < 0.001$; for females $r_s = -0.567$, $n = 24$, $P = 0.008$). Third, paternity concentration also decreased with increasing numbers of males in captive populations of Japanese and rhesus macaques (data sources: Inoue *et al.*, 1992; Smith, 1993, 1994).

For ten populations with known birth seasonality and DNA-based paternity concentration, a significant correlation between the two was found (Fig. 12.3). A multiple regression confirmed that both number of males and seasonality have independent effects on percentage paternity concentration ($r = 0.914$, $n = 10$, $P = 0.002$; for males $P = 0.04$; for seasonality $P = 0.01$). For 13 populations we found data on the estimated proportion of paternities by extra-group males, based on either DNA analysis or an indication of absence of matings by extra-group males in spite of intensive observations (Fig. 12.4). The relationship between seasonality and the percentage of paternities by extra-group males was significant. In our sample, 6 of 8 seasonal populations (i.e. having more than 75 per cent of births in a 3-month period) had more than 5 per cent extra-group paternities, whereas none of the 5 less seasonal populations did ($G_{adj} = 7.95$, $P < 0.01$). Group size, expressed either as number of males or as number of females, was not significantly correlated with extra-group paternities. In addition, the African guenons (not included in our data set owing to a lack of information on paternity concentration), for which occasional influxes of males are reported, are also highly seasonal breeders (Cords, 1984, 2000; Harding & Olson, 1986; Henzi & Lawes, 1988; Ohsawa *et al.*, 1993; Carlson & Isbell, 2001). The data support the prediction that mating with extra-group males (individually or as 'influxes') is more common in seasonally reproducing populations. Mating by extra-group males is often made possible by active female cooperation (e.g. Takahata *et al.*, 1994; Agoramoorthy & Hsu, 2000; Soltis *et al.*, 2001).

ACQUISITION OF TOP RANK

If the benefits of top rank are marginal, a male cannot be expected to take high risks to achieve this position. Indeed, in natural and provisioned populations with very low paternity concentration, the dominant position was acquired through succession rather than through challenge, whereas at high paternity concentration challenges were the norm

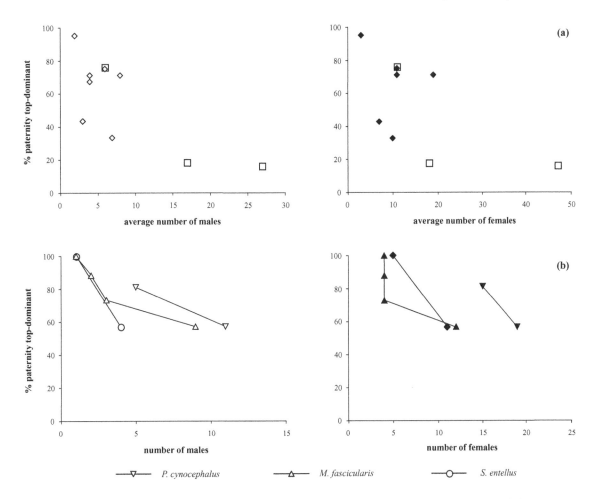

Fig. 12.2   The relationship between the (average) number of sexually mature males and adult females in a group, and the percentage of paternities by the top-dominant male: (a) comparisons between populations; squares represent free-ranging provisioned populations; (b) comparisons within populations (same group at different times or different groups). Correlations are significant for males ($r = -0.753$, $n = 10$, $P = 0.01$ for all, and $r = -0.319$, $n = 7$, n.s. for wild only), but not for females ($r = -0.613$, $n = 9$, $P = 0.08$ for all, and $r = -0.126$, $n = 6$, n.s. for wild only). Note that the comparisons within populations follow the overall trend.

(Fig. 12.5). The relationship was significant and adding data from three captive populations strengthened this relationship even further (Fig. 12.5).

Because paternity concentration is smaller in larger groups, we expect that mode of dominance acquisition is correlated with group size. Table 12.2 shows that in small groups (and thus high paternity concentration), acquisition

of top rank is always by challenge, whereas in large groups it is always by succession. Within species, we see the same effect of group size in Japanese macaques (Sprague et al., 1996) and in Barbary macaques (compare small groups with take-over (Witt et al., 1981) vs. large groups with succession, see Table 12.1).

As expected from the strong relationship between group size and seasonality, all populations with succession of top rank were strongly seasonal (>90 per cent births in 3 months). However, the few highly seasonal populations with small group sizes (≤ 10 males) all had medium to high paternity concentrations and top-dominance acquisition through challenge. Thus, paternity concentration, not seasonality, is the causal factor.

As expected, males attaining top rank through challenge are described as young or prime adult, i.e. at peak physical strength at the moment of their challenge (Table 12.3). In contrast, in populations with succession to top rank, mostly

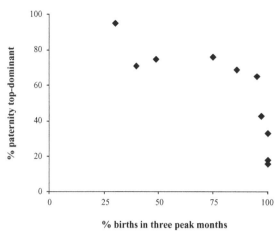

Fig. 12.3   The relationship between reproductive seasonality and paternity concentration in the top-dominant male ($r = -0.793$, $n = 10$, $P = 0.006$) in wild and free-ranging populations.

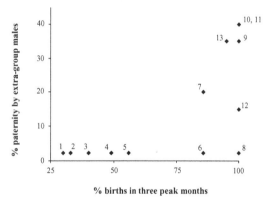

Fig. 12.4   The percentage of births in the three peak months in relation to the reported percentage of paternities by males not residing in the group. (To allow for the chance of conception after occasionally observed extra-group matings, studies reporting zero extra-group paternities were conservatively classified as having < 5 per cent.) The relationship is significant (Spearman $r_s = 0.704$, $n = 13$, $P < 0.05$). 1. *Alouatta seniculus*, Hato Masaguaral; 2. *Semnopithecus entellus*, Jodhpur; 3. *Papio cynocephalus*, Amboseli; 4. *Macaca fascicularis*, Ketambe; 5. *M. maurus*, Karaente; 6. *S. entellus* (single male), Ramnagar; 7. *S. entellus* (multi-male), Ramnagar; 8. *M. sylvanus*, Affenberg; 9. *M. mulatta*, Cayo Santiago; 10. *M. fuscata yakui*, Yakushima; 11. *M. f. fuscata*, Kinkazan; 12. *M. sinica*, Polonnaruwa; 13. *Erythrocebus patas*, Kala Maloue.

middle-aged or even old males have been reported to become the dominant, even when younger and supposedly stronger males are present (Itoigawa, 1993; Nakamichi *et al.*, 1995; Watanabe, 2001). Thus, the age profiles of top-dominants

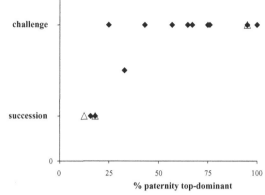

Fig. 12.5   The relationship between paternity concentration and the way top rank is acquired. In Japanese macaques from Yakushima top rank can be acquired either through succession or through challenge. (Spearman rank correlation corrected for ties, $r_s = 0.696$, $n = 10$, $P < 0.05$; including the Yakushima macaques as an intermediate case improves the fit.) Open triangles represent captive groups.

of the two modes of rank acquisition differ. Indeed, mode of dominance acquisition has profound implications for the trajectories of a male's dominance. Figure 12.6 illustrates this with several empirical examples, in order of decreasing paternity concentration. Where males challenge, individuals may reach a rather brief and sharp peak in dominance when in their prime (Fig. 12.6a, b). As the dominant's ability to monopolise matings declines, male dominance rank initially continues to peak in the prime years (Fig. 12.6c), but at the near-scramble extreme, male dominance rank increases with age, with the oldest males ranking highest (Fig. 12.6d). Siring success closely follows the dominance-rank trajectory in the first three cases, but in the last example siring success still tends to be higher during the prime years than in old age, when rank is highest (Bercovitch & Nürnberg, 1997; Takahata *et al.*, 1998).

Although we did not develop a prediction for it, we also noted a possible pattern among challengers. Challengers are either recent immigrants or more long-term residents. Obviously, in a group with a single male, challengers must be immigrants. In larger groups both resident and immigrating males could mount the challenge, but there may be an effect of group size: in groups with very few males, top rank was more often attained by immigrants, whereas in groups with more males (non-natal) resident males were more often successful in attaining top rank (wedge-capped capuchins: Robinson, 1988a; long-tailed macaques: van Noordwijk

Table 12.2 *Paternity concentration (in three classes) and mode of top-dominance rank acquisition: through succession or through challenge (including populations with both modes: $G_{adj} = 12.94$, $P < 0.005$; excluding these populations: $G_{adj} = 14.26$, $P < 0.001$; for references see Table 12.1).*

Paternity concentration	Top rank acquisition		No. males average	Species	Site
	[succession	challenge]			
low	✓		31	*Macaca f. fuscata*	Arashiyama
	✓		14	*Macaca f. fuscata*	Kinkazan
	✓		16?	*Macaca f. fuscata*	Koshima
	✓		27	*Macaca sylvanus*	Affenberg
medium	✓	(✓)	17	*Macaca mulatta*	Cayo Santiago S
	(✓)	✓	7	*Macaca f. yakui*	Yakushima
		✓	3	*Macaca sinica*	Polonnaruwa
		✓	9	*Macaca thibetana*	Mt Emei
		✓	5–8	*Papio cynocephalus*	Amboseli
		✓	8	*Papio c. anubis*	Gombe
		✓	7	*Papio c. ursinus*	Moremi
high		✓	2	*Alouatta seniculus*	Hato Masaguaral
		✓	2	*Cebus olivaceus*	Hato Masaguaral
		✓	3	*Cercopithecus aethiops*	Amboseli
		✓	2.5	*Cercopithecus aethiops*	Burman Bush
		✓	7	*Macaca fascicularis*	Ketambe
		✓	1–4+	*Erythrocebus patas*	Kala Maloue
		✓	1–4	*Semnopithecus entellus*	Ramnagar
		✓	1	*Semnopithecus entellus*	Jodhpur

& van Schaik, 1985, 2001). One possible reason for this group-size effect is that mounting a challenge as a resident is only a viable option in larger groups, but the preferred one because a resident challenger has an information advantage over immigrant challengers. Thus, long-tailed macaque males were more often successful when they tried to defeat the dominant who had already lost some of his strength after he had held tenure for at least one year. However, the group-size effect does not always hold. For example, the dominant male in a larger unit of gelada baboons (*Theropithecus gelada*) was more likely to be defeated by an immigrant challenger than the dominant in a small unit, probably due to weaker female loyalty to the dominant in a larger group (Dunbar, 1984).

NATAL TRANSFER

As expected, a strong negative correlation was found between the percentage of natal males emigrating before breeding, often as subadults, and the number of females in the group (Fig. 12.7). Intraspecific comparisons across a wide enough range of female group sizes in Japanese monkeys showed the same trend (Yamagiwa & Hill, 1998). Although baboon males sometimes leave the natal group only after the onset of breeding, actual inbreeding still appears to be rare: males who become sexually active in their natal group are reported to have very low reproductive success in that group (Amboseli: Alberts & Altmann, 1995), or have such short tenures that they are unlikely to breed with maternal relatives (Moremi: Hamilton & Bulger, 1990).

Low natal emigration rates in some populations of Barbary macaques and Japanese macaques may be due to the absence of nearby groups to migrate into (cf. Mehlman, 1986; Muroyama et al., 2000).

SECONDARY TRANSFER

We predicted that dominance acquisition tactics affect the criteria used to select their target groups by males who

Table 12.3 *Age of top-dominant at start of tenure, way in which top-dominance rank is achieved and paternity concentration (for references see Table 12.1).*

Top-dominance acquisition	Age-top-dominant	Paternity concentration	Species	Site
succession	adult – old		*Macaca f. fuscata*	Katsuyama
	adult – old	low	*Macaca f. fuscata*	Koshima
	adult – post-prime	low	*Macaca f. fuscata*	Arashiyama
	adult – post-prime	low	*Macaca f. fuscata*	Kinkazan
	adult – post-prime	medium	*Macaca mulatta*	Cayo Santiago S
	adult – post-prime	medium	*Macaca sylvanus*	Affenberg
challenge	young adult	medium	*Macaca f. yakui*	Yakushima
	young adult	medium	*Macaca thibetana*	Mt Emei
	young adult	medium–high	*Papio cynocephalus*	Amboseli
	young adult	high	*Alouatta seniculus*	Hato Masguaral
	prime	high	*Cebus olivaceus*	Hato Masguaral
	young adult?	high	*Cercopithecus aethiops*[a]	Amboseli
	young adult	high	*Gorilla g. beringei*	Virunga
	young adult	high	*Macaca fascicularis*	Ketambe
	young adult	high	*Papio c. ursinus*	Moremi
	young adult	high	*Semnopithecus entellus*	Jodhpur
	young adult	high	*Presbytis thomasi*	Ketambe

[a] *Cercopithecus aethiops* males are probably top-dominant only once in their lives (Cheney & Seyfarth, 1983).

transfer. Indeed, most males in populations with high paternity concentration did not selectively immigrate into groups with more females per male, but apparently based their choice on characteristics of the male cohort, e.g. the tenure of the resident top-dominant or the identity of the resident males (Table 12.4; see also Borries, 2000; van Noordwijk & van Schaik, 2001), which should affect their prospects for mating access. In contrast, in populations where at least half the paternities go to non-dominant males, most males tended to immigrate into a local group with a more favourable sex ratio, fewer males or more sexually active females. Thus, the basic prediction is upheld.

We also predicted that in challenge situations, the criteria for transfer and destination would change with age, more than in succession situations. The limited available data are consistent with this suggestion. In Japanese macaques at Yakushima, young subadult males made different choices from (adult) males ready to challenge (Suzuki *et al.*, 1998a). In olive baboons, some adult males with high consort success secondarily transferred to an adjacent group with a high(er) number of sexually active females, but natal transfer in this population was into an adjacent group irrespective of the number of females or sex ratio, presumably into one in which the male could soon attain high rank (Packer, 1979a).

In some populations, males may live outside mixed-sex groups. Here, prime adult males achieve much higher reproductive success as the residential male in a mixed-sex group, especially if top ranking (Ohsawa *et al.*, 1993; Borries, 2000; Takahashi, 2001). Young or post-prime males with low siring chances were found to achieve slightly better siring success by temporarily visiting the local group with the best instantaneous access to females, than by being residential in a mixed-sex group (Henzi & Lawes, 1988; Rajpurohit & Mohnot, 1988; Borries, 2000).

We noted one more pattern in the data concerning emigration. In general, non-natal males often emigrate after experiencing a drop in rank, but they are rarely evicted from multi-male groups unless they are challenging the dominant (Pusey & Packer, 1987; Pusey, 1992). Emigration patterns differ, however, depending on the mode of dominance acquisition. Where the dominant position is acquired through challenge, we found no evidence that males ever transfer without having been deposed from this position, although they may emigrate while still high ranking. Where top rank is acquired through succession, dominants have been known to emigrate (Huffman, 1991a; Berard, 1999), although their departure may have been preceded by a decrease in support by group members (Huffman, 1991a).

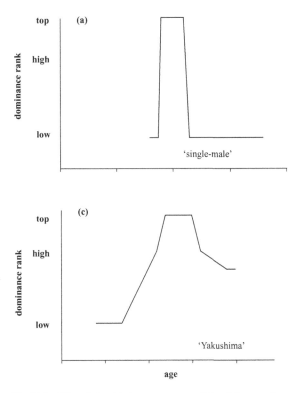

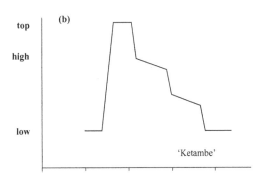

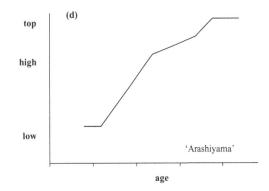

Fig. 12.6   The relationship between a successful male's age and his dominance rank in a mixed-sex group, in situations with high (a) and (b), intermediate (c) and low (d) paternity concentration. (a) Typical single-male mixed-sex group: a male generally only succeeds in obtaining a breeding position in the few years of his prime (Rajpurohit *et al.*, 1995; Borries, 2000; Steenbeek *et al.*, 2000); (b) age-rank profile for long-tailed macaques at Ketambe; note that males reside in several groups during their lifetime (van Noordwijk & van Schaik, 2001); (c) age-rank profile for Japanese macaques at Yakushima (Sprague, 1992); (d) age-rank profile for male in group with succession at Arashiyama (based on Japanese macaques: Takahata *et al.*, 1998; and rhesus macaques: Berard, 1999).

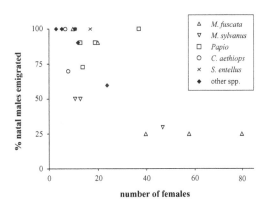

Fig. 12.7   The relationship between the number of females in the group and the percentage of natal males emigrating before breeding ($r = -0.755$, $n = 23$, $P < 0.0001$). Other species: *A. seniculus; C. olivaceus; M. mulatta; M. fascicularis; M. thibetana; P. thomasi.*

Overall, the limited data on secondary transfer by cercopithecines support our hypothesis that transfer decisions are strongly affected by the degree of paternity concentration in the local groups, and vary predictably with age.

DISCUSSION

We have adopted a career approach to connect seemingly disparate aspects of male life histories, such as rank acquisition tactics and transfer decisions, to the impact of male dominance rank on mating success. This approach is based

on the idea that optimum male decisions should depend on the nature of mating competition. We found that the nature of male competition for access to mates, with the contest component varying from weak (i.e. mainly scramble) to strong, predicted various aspects of their careers. Potential

Table 12.4 *Paternity concentration (PC) and whether the choice of immigration group is based on the (relative) number of females or high-rank potential (for references see Table 12.1).*

% PC	PC class	Choice immigration group		Species	Site
		more females better SR	rank increase or potential		
16	medium	√		*Macaca sylvanus*	Affenberg
33		√ secondary	√	*Macaca fuscata yakui*	Yakushima
		√	(√)	*Macaca thibetana*	Mt Emei
45		√	(√)[a]	*Papio cynocephalus*	Amboseli
	high	√	first[b]	*Papio c. anubis*	Gombe
		post prime	first?	*Papio c. ursinus*	Moremi
57		no	√	*Semnopithecus entellus*	Ramnagar
75		no	√	*Macaca fascicularis*	Ketambe
		no[c]	?	*Cercopithecus aethiops*	Burman Bush
		no	√	*Cercopithecus aethiops*	Amboseli

[a] Personal communication, S. C. Alberts.

[b] *Papio c. anubis* secondary transfer only by males with consort success (but not top-dominant).

[c] *Cercopithecus aethiops* at Burman Bush: males do not prefer groups with better sex ratio.

paternity concentration in the top-ranking male in a group showed strong correlations with the mode of acquisition of the top position, with the age at which it is attained, and with the features of the target group selected by immigrant males at different stages in their career. At this stage, most data come from catarrhines, and information on other primates or non-primate mammals is either not available or not summarised in ways that allow for easy comparisons with our findings. Moreover, tests could not pit the predictions developed here against those derived from alternative hypotheses, because we are not aware of formally developed alternatives.

This study focused on species in which males transfer. In species with male philopatry, we can only examine how paternity concentration affects rank acquisition strategies. The presence of close male relatives may affect career paths. Thus, the net benefit of escalated challenge fights may be less if the incumbent dominant is a close relative. Moreover, group-level male alliances against neighbouring groups may be more common, as in chimpanzees, complicating the within-group acquisition and maintenance of dominance, although top rank is still acquired through challenge (Nishida, 1983; Goodall, 1986; Takahata, 1990). In species with female transfer in addition to male transfer, other factors may come to the fore as well: for example, a possibly stronger role for female preferences (e.g. Steenbeek *et al.*, 2000).

We found not only variation between species but also remarkable variation within species, or even populations, in the effect of group size on paternity concentration and thus transfer decisions, as well as mode of rank acquisition and likelihood of natal transfer. This variability suggests that a primate male's behaviour is guided by a set of conditional rules that allow him to respond to a variety of local situations. Species may none the less differ in whether increasingly large group size is tolerated or leads to fissioning into smaller units. Extremely large groups with more than 100 members are mainly found among the very seasonal populations, where monopolisation potential is low anyway. Careful intraspecific comparisons are needed to show the full extent of behavioural flexibility. For example, in high-contest situations the closer a young male approaches his prime, the better his near future take-over chances should be in order to stay in a group. Thus, dispersal of mountain gorilla males is strongly affected by the age and tenure of the incumbent dominant male and the number of other young males in the group. A young male transfers when the expected 'waiting' time to achieve a breeding position in his natal group is too long (Watts, 2000). Male long-tailed macaques time their challenges to top dominants when their chances of success are highest, e.g. after the incumbent has been dominant for more than a year (van Noordwijk & van Schaik, 2001).

Similarly, natal chacma baboon males can be mid-ranking for a variable period before quickly rising in rank and taking over top rank, apparently waiting for the 'right' time (Hamilton & Bulger, 1990). These examples show that the conditional rules may reach a surprising level of sophistication.

A recent model of levelling coalitions among primate males (Pandit & van Schaik, in press) shows that low-contest situations favour levelling coalitions by lower-ranking males that can effectively level ranks to the point that the situation is turned into a scramble. This explains much of what we found at the low end of the paternity concentration spectrum. Higher rank for young males may mean earlier ascension to mating status (Bercovitch, 1993; Paul *et al.*, 1993), but the older males in these conditions still maintain high formal dominance-ranks. In the most extreme form, the succession hierarchies are accompanied by very low rates of transfer, giving males ample opportunity to form long-lasting alliances with each other and with high-ranking females to form a central 'clique' (Takahata, 1982; Huffman, 1991a; Itoigawa, 1993; Nakamichi *et al.*, 1995; Watanabe, 2001). Central clique members maintain their ranks over younger males.

A near-scramble among males is also the situation in which females are most often credited with sanctioning, or actively defending, the top-dominant male, although the latter is neither the most successful in obtaining matings and siring offspring nor the greatest provider of physical protection to offspring (Chapais, 1986; Huffman, 1991a; Itoigawa, 1993; Nakamichi *et al.*, 1995; Berard, 1999; Watanabe, 2001).

These observations raise the question of benefits accruing from high rank to these ageing males or their supporters. It is not simply an artefact of low transfer rates, because a predictable increase in rank with residence combined with the lack of challenges is also found in large groups with frequent transfers (Yamagiwa & Hill, 1998; Berard, 1999). In one such population, male siring success was found to increase over the first few years of a male's residence, followed by a steady decrease while rank increased (Berard, 1999). Unfortunately, no information is published on the ages of these males and whether this cycle was repeated in their subsequent group. Thus, we need new ideas to explain rank profiles and the role of alliances among males in the near-scramble situation of large, seasonally breeding groups.

SUMMARY AND CONCLUSIONS

The way a male can achieve high reproductive success guides all aspects of his career choices, with respect to the risk taken in rank acquisition, transfer between groups, and even his tendency to form coalitions. Thus, various phenomena that have so far been considered mainly in isolation can be linked together using an integrated, career-level approach. Primate males appear to have a set of conditional rules that allow them to respond flexibly to variation in the potential for paternity concentration. Before mounting a challenge, they assess the situation in their current group, and before making their transfer decisions they monitor the situation in multiple potential-target groups, where this is possible. Further improvements in our understanding of male careers and the decision rules males use for transfer and rank-acquisition tactics are likely to come from long-term studies of known individuals followed through all groups in which they reside.

ACKNOWLEDGEMENTS

We thank Susan Alberts and Sagar Pandit for discussion, and Peter Kappeler and two anonymous reviewers for comments on an earlier version of this chapter.

REFERENCES

Agoramoorthy, G. & Hsu, M. J. 2000. Extragroup copulation among wild red howler monkeys in Venezuela. *Folia Primatologica*, 71, 147–51.

Alberts, S. C. & Altmann, J. 1995. Balancing costs and opportunities: dispersal in male baboons. *American Naturalist*, 145, 279–306.

Alberts, S. C., Watts, H. & Altmann, J. 2003. Queuing and queue jumping: long term patterns of dominance rank and mating success in male savannah baboons. *Animal Behaviour*, in press.

Altmann, J., Hausfater, G. & Altmann, S. A. 1985. Demography of Amboseli baboons, 1963–83. *American Journal of Primatology*, 8, 113–25.

Altmann, J., Alberts, S. C., Haines, S. A. *et al.* 1996. Behavior predicts genetic structure in a wild primate group. *Proceedings of the National Academy of Science*, 93, 5797–801.

Altmann, S. A. 1962. A field study of the sociobiology of the rhesus monkey, *Macaca mulatta*. *Annual Proceedings of the New York Academy of Science*, 102, 338–435.

Andelman, S. J. 1986. Ecological and social determinants of cercopithecine mating patterns. Princeton, NJ: Princeton University Press.

Barton, R. A. & Simpson, A. J. 1992. Does the number of males influence the relationship between dominance and mating success in primates? *Animal Behaviour*, **44**, 1159–61.

Bauers, K. A. & Hearn, J. P. 1994. Patterns of paternity in relation to male social rank in the stumptailed macaque, *Macaca arctoides*. *Behaviour*, **129**, 149–76.

Berard, J. D. 1999. A four-year study of the association between male dominance rank, residence status, and reproductive activity in rhesus macaques, *Macaca mulatta*. *Primates*, **40**, 159–75.

Berard, J. D., Nürnberg, P., Epplen, J. T. & Schmidtke, J. 1993. Male rank, reproductive behavior, and reproductive success in free-ranging rhesus macaques. *Primates*, **34**, 481–9.

1994. Alternative reproductive tactics and reproductive success in male rhesus macaques. *Behaviour*, **127**, 177–201.

Bercovitch, F. B. 1991. Mate selection, consortship formation, and reproductive tactics in adult female savanna baboons. *Primates*, **32**, 437–52.

1993. Dominance rank and reproductive maturation in male rhesus macaques, *Macaca mulatta*. *Journal of Reproduction and Fertility*, **99**, 113–20.

Bercovitch, F. B. & Nürnberg, P. 1996. Socioendocrine and morphological correlates of paternity in rhesus macaques, *Macaca mulatta*. *Journal of Reproduction and Fertility*, **107**, 59–68.

1997. Genetic determination of paternity and variation in male reproductive success in two populations of rhesus macaques. *Electrophoresis*, **18**, 1701–5.

Borries, C. 2000. Male dispersal and mating season influxes in Hanuman langurs living in multi-male groups. In *Primate Males: Causes and Consequences of Variation in Group Composition*, ed. P. M. Kappeler. Cambridge: Cambridge University Press, pp. 146–58.

Bulger, J. B. 1993. Dominance rank and access to estrous females in male savanna baboons. *Behaviour*, **127**, 67–103.

Bulger, J. B. & Hamilton, W. J. III. 1987. Rank and density correlates of inclusive fitness measures in a natural chacma baboon *Papio ursinus* troop. *International Journal of Primatology*, **8**, 635–50.

1988. Inbreeding and reproductive success in a natural chacma baboon *Papio ursinus* population. *Animal Behaviour*, **36**, 574–8.

Carlson, A. A. & Isbell, L. A. 2001. Causes and consequences of single-male and multimale mating in free-ranging patas monkeys, *Erythrocebus patas*. *Animal Behaviour*, **62**, 1042–58.

Chapais, B. 1986. Why do adult male and female rhesus monkeys affiliate during the birth season? In *The Cayo Santiago Macaques. History, Behavior and Biology*, ed. R. G. Rawlins & M. J. Kessler. Albay, NY: State University of New York Press, pp. 173–200.

Cheney, D. L. 1983. Proximate and ultimate factors related to the distribution of male migration. In *Primate Social Relationships. An Integrated Approach*, ed. R. A. Hinde. Oxford: Blackwell, pp. 241–9.

Cheney, D. L. & Seyfarth, R. M. 1983. Nonrandom dispersal in free-ranging vervet monkeys: social and genetic consequences. *American Naturalist*, **122**, 392–412.

1990. *How Monkeys See the World*. Chicago, IL: Chicago University Press.

Cheney, D. L., Seyfarth, R. M., Andelman, S. J. & Lee, P. C. 1988. Reproductive success in vervet monkeys. In *Reproductive Success*, ed. T. H. Clutton-Brock. Chicago, IL: University of Chicago Press, pp. 384–402.

Chepko-Sade, B. D. & Sade, D. S. 1979. Patterns of group splitting within matrilineal groups. *Behavioral Ecology and Sociobiology*, **5**, 67–86.

Clutton-Brock, T. H. 1991. The evolution of sex differences and the consequences of polygyny in mammals. In *The Development and Integration of Behaviour*, ed. P. Bateson. Cambridge: University of Cambridge Press, pp. 229–53.

Clutton-Brock, T. H., Albon, S. D. & Guinness, F. E. 1985. Parental investment and sex differences in juvenile mortality in birds and mammals. *Nature*, **313**, 131–3.

Constable, J. L., Ashley, M. V., Goodall, J. & Pusey, A. E. 2001. Noninvasive paternity assignment in Gombe chimpanzees. *Molecular Ecology*, **10**, 1279–300.

Cords, M. 1984. Mating patterns and social structure in redtail monkeys, *Cercopithecus ascanius*. *Zeitschrift für Tierpsychologie*, **64**, 313–29.

2000. The number of males in guenon groups. In *Primate Males: Causes and Consequences of Variation in Group Composition*, ed. P. M. Kappeler. Cambridge: Cambridge University Press, pp. 84–96.

Cowlishaw, G. & Dunbar, R. I. M. 1991. Dominance rank and mating success in male primates. *Animal Behaviour*, **41**, 1045–56.

1992. Dominance and mating success: a reply to Barton & Simpson. *Animal Behaviour*, **44**, 1162–3.

Crnokrak, P. & Roff, D. A. 1999. Inbreeding depression in the wild. *Heredity*, **83**, 260–70.

Crockett, C. M. & Janson, C. H. 2000. Infanticide in red howlers: female group size, male membership, and a possible link to folivory. In *Infanticide by Males and Its Implications*, ed. C. P. van Schaik & C. H. Janson. Cambridge: Cambridge University Press, pp. 75–98.

Crockett, C. M. & Sekulic, R. 1984. Infanticide in red howler monkeys, *Alouatta seniculus*. In *Infanticide: Comparative and Evolutionary Perspectives*, ed. G. Hausfater & S. B. Hrdy. New York, NY: Aldine, pp. 173–91.

de Ruiter, J. R. & Inoue, M. 1993. Paternity, male social rank, and sexual behaviour: general discussion. *Primates*, **34**, 553–5.

de Ruiter, J. R. & van Hooff, J. A. R. A. M. 1993. Male dominance rank and reproductive success in primate groups. *Primates*, **34**, 513–23.

de Ruiter, J. R., Scheffrahn, W., Trommelen, G. J. J. M. *et al.* 1992. Male social rank and reproductive success in wild long-tailed macaques. In *Paternity in Primates: Genetic Tests and Theories*, ed. R. D. Martin, A. F. Dixson & E. J. Wickings. Basel: Karger, pp. 175–91.

Dittus, W. P. J. 1975. Population dynamics of the toque monkey, *Macaca sinica*. In *Socioecology and Psychology of Primates*, ed. R. H. Tuttle. Den Haag: Mouton, pp. 125–51.

1988. Group fission among wild toque macaques as a consequence of female resource competition and environmental stress. *Animal Behaviour*, **36**, 1626–45.

Dittus, W. P. J., Keane, B. & Melnick, D. 2001. The effects of age and rank on the reproductive success of wild male toque macaques, *Macaca sinica*. In *XVIIIth Congress of the International Primatological Society*. Adelaide, Australia.

Dixson, A. F., Bossi, T. & Wickings, E. J. 1993. Male dominance and genetically determined reproductive success in the mandrill *Mandrillus sphinx*. *Primates*, **34**, 525–32.

Drickamer, L. C. & Vessey, S. 1973. Group changing in free-ranging male rhesus monkeys. *Primates*, **14**, 359–68.

Dunbar, R. I. M. 1984. *Reproductive Decisions: An Economic Analysis of Gelada Baboon Social Strategies*. Princeton, NJ: Princeton University Press.

Goodall, J. 1986. *The Chimpanzees of Gombe*. Cambridge, MA: Harvard University Press.

Hamilton, W. J. & Bulger, J. B. 1990. Natal male baboon rank rises and successful challenges to resident alpha males. *Behavioral Ecology and Sociobiology*, **26**, 357–62.

Harcourt, A. H. 1978. Strategies of emigration and transfer by primates, with particular reference to gorillas. *Zeitschrift für Tierpsychologie*, **48**, 401–20.

Harding, R. S. O. & Olson, D. K. 1986. Patterns of mating among male patas monkeys, *Erythrocebus patas*, in Kenya. *American Journal of Primatology*, **11**, 343–58.

Henzi, S. P. & Lawes, M. 1988. Strategic responses of male samango monkeys, *Cercopithecus mitis*, to a reduction in the availability of receptive females. *International Journal of Primatology*, **9**, 479–95.

Henzi, S. P. & Lucas, J. W. 1980. Observations on the inter-troop movement of adult vervet monkeys, *Cercopithecus aethiops*. *Folia Primatologica*, **33**, 220–35.

Henzi, S. P., Lycett, J. E. & Weingrill, T. 1998. Mate guarding and risk assessment by male mountain baboons during inter-troop encounters. *Animal Behaviour*, **55**, 1421–8.

Hrdy, S. B. 1977. *The Langurs of Abu. Female and Male Strategies of Reproduction*. Cambridge, MA: Harvard University Press.

2000. The optimal number of fathers: evolution, demography, and history in the shaping of female mate preferences. *Annual Proceedings of the New York Academy of Sciences*, **907**, 75–96.

Huffman, M. A. 1991a. History of the Arashiyama Japanese macaques in Kyoto, Japan. In *The Monkeys of Arashiyama: Thirty-five Years of Research in Japan and the West*, ed. L. M. Fedigan & P. J. Asquith. Albany, NY: State University of New York Press, pp. 21–53.

1991b. Mate selection and partner preferences in female Japanese macaques. In *The Monkeys of Arashiyama: Thirty-five Years of Research in Japan and the West*, ed. L. M. Fedigan & P. J. Asquith. Albany, NY: State University of New York Press, pp. 101–22.

1992. Influences of female partner preference on potential reproductive outcome in Japanese macaques. *Folia Primatologica*, **59**, 77–88.

Inoue, M., Mitsunaga, F., Ohsawa, H. *et al.* 1991. Male mating behaviour and paternity discrimination by DNA fingerprinting in a Japanese macaque group. *Folia Primatologica*, **56**, 202–10.

Inoue, M., Mitsunaga, F., Ohsawa, H. *et al.* 1992. Paternity testing in captive Japanese macaques *Macaca fuscata* using DNA fingerprinting. In *Paternity in Primates: Genetic Tests and Theories*, ed. R. D. Martin, A. F. Dixson & E. J. Wickings. Basel: Karger, pp. 131–40.

Inoue, M., Mitsunaga, F., Nozaki, M. *et al.* 1993. Male dominance rank and reproductive success in an enclosed group of Japanese macaques: with special reference to post-conception mating. *Primates*, **34**, 503–11.

Itoigawa, N. 1993. Social conflict in adult male relationships in a free-ranging group of Japanese monkeys. In *Primate Social Conflict*, ed. W. A. Mason & S. P. Mendoza. Albany, NY: State University of New York Press, pp. 145–69.

Itoigawa, N., Tanaka, T., Ukai, N. *et al.* 1992. Demography and reproductive parameters of a free-ranging group of Japanese macaques *Macaca fuscata* at Katsuyama. *Primates*, **33**, 49–68.

Jack, K. & Fedigan, L. 2001. Life history of male white-faced capuchins *Cebus capucinus*, Santa Rosa National Park, Costa Rica. *American Journal of Primatology, Supplement 1*, **54**, 50.

Keane, B., Dittus, W. P. J. & Melnick, D. J. 1997. Paternity assessment in wild groups of toque macaques *Macaca sinica* at Polonnaruwa, Sri Lanka using molecular markers. *Molecular Ecology*, **6**, 267–82.

Koenig, A., Borries, C., Chalise, M. K. & Winkler, P. 1997. Ecology, nutrition, and timing of reproductive events in an Asian primate, the hanuman langur *Presbytis entellus*. *Journal of Zoology, London*, **243**, 215–35.

Koyama, N. 1970. Changes in dominance rank and kinship of a wild Japanese monkey troop in Arashiyama. *Primates*, **11**, 335–90.

Koyama, N., Takahata, Y., Huffman, M. A., Norikoshi, K. & Suzuki, H. 1992. Reproductive parameters of female Japanese macaques: thirty years from the Arashiyama troops, Japan. *Primates*, **33**, 33–47.

Kuester, J. & Paul, A. 1992. Influence of male competition and female mate choice on male mating success in Barbary macaques, *Macaca sylvanus*. *Behaviour*, **120**, 192–217.

　1997. Group fission in Barbary macaques *Macaca sylvanus* at Affenberg Salem. *International Journal of Primatology*, **18**, 941–66.

　1999. Male migration in Barbary Macaques *Macaca sylvanus* at Affenberg Salem. *International Journal of Primatology*, **20**, 85–106.

Kuester, J., Paul, A. & Arnemann, J. 1995. Age-related and individual differences of reproductive success in male and female Barbary macaques, *Macaca sylvanus*. *Primates*, **36**, 461–76.

Kumar, A. & Kurup, G. U. 1985. Sexual behavior of the lion-tailed macaque, *Macaca silenus*. In *The Lion-tailed Macaque: Status and Conservation*, ed. P. G. Heltne. New York, NY: Alan R. Liss, pp. 109–30.

Kummer, H. 1995. *In Quest of the Sacred Baboon. A Scientist's Journey*, translated by M. A. Biederman-Thorson. Princeton, NJ: Princeton University Press.

Launhardt, K., Borries, C., Hardt, C., Epplen, J. T. & Winkler, P. 2001. Paternity analysis of alternative male reproductive routes among langurs *Semnopithecus entellus* of Ramnagar. *Animal Behaviour*, **61**, 53–64.

Manson, J. H. 1994. Mating patterns, mate choice, and birth season heterosexual relationships in free-ranging rhesus macaques. *Primates*, **35**, 417–33.

　1995. Do female rhesus macaques choose novel males? *American Journal of Primatology*, **37**, 285–96.

Mehlman, P. 1986. Male intergroup mobility in a wild population of the Barbary macaque *Macaca sylvanus*, Ghomaran Rif Mountains, Morocco. *American Journal of Primatology*, **10**, 67–81.

Meikle, D. B. & Vessey, S. H. 1981. Nepotism among rhesus monkey brothers. *Nature*, **294**, 160–1.

Melnick, D. J. & Hoelzer, G. A. 1996. The population genetic consequences of macaque social organization and behaviour. In *Evolution and Ecology of Macaque Societies*, ed. J. E. Fa & D. G. Lindburg. Cambridge: Cambridge University Press, pp. 413–43.

Melnick, D. J., Pearl, M. C. & Richard, A. F. 1984. Male migration and inbreeding avoidance in wild rhesus monkeys. *American Journal of Primatology*, **7**, 229–43.

Ménard, N. & Vallet, D. 1996. Demography and ecology of Barbary macaques *Macaca sylvanus* in two different habitats. In *Evolution and Ecology of Macaque Societies*, ed. J. E. Fa & D. G. Lindburg. Cambridge: Cambridge University Press, pp. 106–31.

Muroyama, Y., Imae, H. & Okuda, K. 2000. Radio tracking of a male Japanese macaque emigrated from its group. *Primates*, **41**, 351–6.

Nakamichi, M., Kojima, Y., Itoigawa, N., Imakawa, S. & Machida, S. 1995. Interactions among adult males and females before and after the death of the alpha male in a free-ranging troop of Japanese macaques. *Primates*, **36**, 185–96.

Nishida, T. 1983. Alpha status and agonistic alliance in wild chimpanzees. *Primates*, **24**, 318–36.

Noë, R. & Sluijter, A. A. 1990. Reproductive tactics of male savanna baboons. *Behaviour*, **113**, 117–70.

Nunn, C. L. 1999. The number of males in primate social groups: a comparative test of the socioecological model. *Behavioral Ecology and Sociobiology*, **46**, 1–13.

Ohsawa, H., Inoue, M. & Takenaka, O. 1993. Mating strategy and reproductive success of male patas monkeys *Erythrocebus patas*. *Primates*, **34**, 533–44.

Oi, T. 1996. Sexual behaviour and mating system of the wild pig-tailed macaque in West Sumatra. In *Evolution and Ecology of Macaque Societies*, ed. J. E. Fa & D. G. Lindburg. Cambridge: Cambridge University Press, pp. 342–68.

Okamoto, K., Matsumura, S. & Watanabe, K. 2000. Life history and demography of wild moor macaques *Macaca maurus*: summary of ten years of observations. *American Journal of Primatology*, **52**, 1–11.

Olupot, W. & Waser, P. M. 2001. Correlates of intergroup transfer in male grey-cheeked mangabeys. *International Journal of Primatology*, **22**, 169–87.

Packer, C. 1979a. Inter-troop transfer and inbreeding avoidance in *Papio anubis*. *Animal Behaviour*, **27**, 1–36.
1979b. Male dominance and reproductive activity in *Papio anubis*. *Animal Behaviour*, **27**, 37–45.

Palombit, R. A., Seyfarth, R. M. & Cheney, D. L. 1997. The adaptive value of 'friendships' to female baboons: experimental and observational evidence. *Animal Behaviour*, **54**, 599–614.

Palombit, R. A., Cheney, D. L. & Seyfarth, R. M. 2001. Female–female competition for male 'friends' in wild chacma baboons, *Papio cynocephalus ursinus*. *Animal Behaviour*, **61**, 1159–71.

Pandit, S. & van Schaik, C. P. 2004. A model for leveling coalitions among primate males: toward a theory of egalitarianism. *Behavioural Ecology and Sociobiology*, in press.

Paul, A. 1989. Determinants of male mating success in a large group of Barbary macaques *Macaca sylvanus* at Affenberg Salem. *Primates*, **30**, 461–76.
1997. Breeding seasonality affects the association between dominance and reproductive success in non-human male primates. *Folia Primatologica*, **68**, 344–9.

Paul, A., Kuester, J., Timme, A. & Arnemann, J. 1993. The association between rank, mating effort, and reproductive success in male Barbary macaques *Macaca sylvanus*. *Primates*, **34**, 491–502.

Perry, S. 1998a. A case report of a male rank reversal in a group of wild white-faced capuchins *Cebus capucinus*. *Primates*, **39**, 51–70.
1998b. Male–male social relationships in wild white-faced capuchins, *Cebus capucinus*. *Behaviour*, **135**, 139–72.

Phillips-Conroy, J. E., Jolly, C. J., Nystrom, P. & Hemmalin, H. A. 1992. Migration of male Hamadryas baboons into anubis groups in the Awash National Park, Ethiopia. *International Journal of Primatology*, **13**, 455–76.

Pope, T. R. 1990. The reproductive consequences of male cooperation in the red howler monkey: paternity exclusion in multi-male and single-male troops using genetic markers. *Behavioral Ecology and Sociobiology*, **27**, 439–46.
2000. The evolution of male philopatry in neotropical monkeys. In *Primate Males: Causes and Consequences of Variation in Group Composition*, ed. P. M. Kappeler. Cambridge: Cambridge University Press, pp. 219–35.

Pusey, A. E. 1992. The primate perspective on dispersal. In *Dispersal: Small Mammals as a Model*, ed. N. C. Stenseth & W. Z. J. Lidicker. London: Chapman and Hall, pp. 243–59.

Pusey, A. E. & Packer, C. 1987. Dispersal and philopatry. In *Primate Societies*, ed. B. B. Smuts, D. L. Cheney, R. M. Seyfarth, R. W. Wrangham & T. T. Struhsaker. Chicago, IL: Chicago University Press, pp. 250–66.

Rajpurohit, L. S. 1991. Resident male replacement, formation of a new male band and paternal behaviour in *Presbytis entellus*. *Folia Primatologica*, **57**, 159–64.

Rajpurohit, L. S. & Mohnot, S. M. 1988. Fate of ousted male residents of one-male bisexual troops of Hanuman langurs *Presbytis entellus* at Jodhpur, Rajasthan, India. *Human Evolution*, **3**, 309–18.

Rajpurohit, L. S., Sommer, V. & Mohnot, S. M. 1995. Wanderers between harems and bachelor bands: male hanuman langurs *Presbytis entellus* at Jodhpur in Rajasthan. *Behaviour*, **132**, 255–99.

Rasmussen, D. R. 1981. Communities of baboon troops *Papio cynocephalus* in Mikumi National Park, Tanzania. *Folia Primatologica*, **36**, 232–42.

Rasmussen, K. L. R. 1986. Spatial patterns and peripheralisation of yellow baboons *Papio cynocephalus* during sexual consortships. *Behaviour*, **97**, 161–80.

Robbins, M. M. 1995. A demographic analysis of male life history and social structure of mountain gorillas. *Behaviour*, **132**, 21–47.

Robinson, J. G. 1988a. Group size in wedge-capped capuchin monkeys, *Cebus olivaceus*, and the reproductive success of males and females. *Behavioral Ecology and Sociobiology*, **23**, 187–97.
1988b. Demography and group structure in wedge-capped capuchin monkeys, *Cebus olivaceus*. *Behaviour*, **104**, 202–32.

Rudran, R. 1973. Adult male replacement in one-male troops of purple-faced langurs *Presbytis senex senex* and its effect on population structure. *Folia Primatologica*, **19**, 166–92.

Samuels, A., Silk, J. B. & Rodman, P. S. 1984. Changes in the dominance rank and reproductive behaviour of male bonnet macaques *Macaca radiata*. *Animal Behaviour*, **32**, 994–1003.

Setchell, J. M. & Dixson, A. F. 2001. Arrested development of secondary sexual adornments in subordinate adult male mandrills *Mandrillus sphinx*. *American Journal of Physical Anthropology*, **115**, 245–52.

Shively, C. & Smith, D. G. 1985. Social status and reproductive success of male *Macaca fascicularis*. *American Journal of Primatology*, **9**, 129–35.

Sigg, H., Stolba, A., Abeggglen, J.-J. & Dasser, V. 1982. Life history of hamadryas baboons: physical development, infant mortality, reproductive parameters and family relationships. *Primates*, **23**, 473–87.

Silk, J. B. 1989. Reproductive synchrony in captive macaques. *American Journal of Primatology*, **19**, 137–46.

1993. Does participation in coalitions influence dominance relationships among male bonnet macaques? *Behaviour*, **126**, 171–89.

Smith, D. G. 1993. A 15-year study of the association between dominance rank and reproductive success of male rhesus monkeys. *Primates*, **34**, 471–80.

1994. Male dominance and reproductive success in a captive group of rhesus macaques *Macaca mulatta*. *Behaviour*, **129**, 225–42.

Smuts, B. B. 1983. Special relationships between adult male and female olive baboons: selective advantages. In *Primate Social Relationships. An Integrated Approach*, ed. R. A. Hinde. Oxford: Blackwell, pp. 262–6.

1985. *Sex and Friendship in Baboons*. Hawthorne, NY: Aldine.

Soltis, J. 1999. Measuring male–female relationships during the mating season in wild Japanese macaques *Macaca fuscata yakui*. *Primates*, **40**, 453–67.

Soltis, J., Thomsen, R., Matsubayashi, K. & Takenaka, O. 2000. Infanticide by resident males and female counter-strategies in wild Japanese macaques *Macaca fuscata*. *Behavioral Ecology and Sociobiology*, **48**, 195–202.

Soltis, J., Thomsen, R. & Takenaka, O. 2001. The interaction of male and female reproductive strategies and paternity in wild Japanese macaques, *Macaca fuscata*. *Animal Behaviour*, **62**, 485–94.

Sommer, V. & Rajpurohit, L. S. 1989. Male reproductive success in harem troops of hanuman langurs *Presbytis entellus*. *International Journal of Primatology*, **10**, 293–317.

Sprague, D. S. 1991. Mating by nontroop males among the Japanese macaques of Yakushima Island. *Folia Primatologica*, **57**, 156–8.

1992. Life history and male intertroop mobility among Japanese macaques *Macaca fuscata*. *International Journal of Primatology*, **13**, 437–54.

Sprague, D. S., Suzuki, S. & Tsukahara, T. 1996. Variation in social mechanisms by which males attained the alpha rank among Japanese macaques. In *Evolution and Ecology of Macaque Societies*, ed. J. E. Fa & D. G. Lindburg. Cambridge: Cambridge University Press, pp. 444–58.

Sprague, D. S., Suzuki, S., Takahashi, H. & Sato, S. 1998. Male life history in natural populations of Japanese macaques: migration, dominance rank, and troop participation of males in two habitats. *Primates*, **39**, 351–63.

Steenbeek, R., Sterck, E. H. M., de Vries, H. & van Hooff, J. A. R. A. M. 2000. Costs and benefits of the one-male, age-graded and all-male phase in wild Thomas's langur groups. In *Primate Males: Causes and Consequences of Variation in Group Composition*, ed. P. M. Kappeler. Cambridge: Cambridge University Press, pp. 130–45.

Sugiyama, Y. 1971. Characteristics of the social life of bonnet macaques *Macaca radiata*. *Primates*, **12**, 247–66.

1976. Life history of male Japanese monkeys. *Advances in the Study of Behavior*, **7**, 255–84.

Sugiyama, Y. & Ohsawa, H. 1982. Population dynamics of Japanese monkeys with special reference to the effect of artificial feeding. *Folia Primatologica*, **39**, 238–63.

Suzuki, S., Hill, D. A. & Sprague, D. S. 1998a. Intertroop transfer and dominance structure of nonnatal male Japanese macaques in Yakushima, Japan. *Primates*, **39**, 703–22.

Suzuki, S., Noma, N. & Izawa, K. 1998b. Inter-annual variation of reproductive parameters and fruit availability in two populations of Japanese macaques. *Primates*, **39**, 313–24.

Takahashi, H. 2001. Influence of fluctuation in the operational sex ratio to mating of troop and non-troop male Japanese macaques for four years on Kinkazan island, Japan. *Primates*, **42**, 183–91.

Takahata, Y. 1982. The socio-sexual behavior of Japanese monkeys. *Zeitschrift für Tierpsychologie*, **59**, 89–108.

1990. Adult males' social relationships with adult females. In *The Chimpanzees of the Mahale Mountains: Sexual and Life History Strategies*, ed. T. Nishida. Tokyo: University of Tokyo Press, pp. 133–48.

Takahata, Y., Sprague, D. S., Suzuki, S. & Okayasu, N. 1994. Female competition, co-existence, and the mating structure of wild Japanese macaques on Yakushima island, Japan. In *Animal Societies: Individuals, Interactions and Organisation*, ed. P. J. Jarman & A. Rossiter. Kyoto: Kyoto University Press, pp. 163–79.

Takahata, Y., Suzuki, S., Agetsuma, N. *et al.* 1998. Reproduction of wild Japanese macaque females of Yakushima and Kinkazan islands: a preliminary report. *Primates*, **39**, 339–49

Takahata, Y., Huffman, M. A., Suzuki, S., Koyama, N. & Yamagiwa, J. 1999. Why dominants do not consistently attain high mating and reproductive success: a review of longitudinal Japanese macaques studies. *Primates*, **40**, 143–58.

Trivers, R. 1985. *Social Evolution*. Menlo Park, CA: Benjamin Cummings.

van Noordwijk, M. A. & van Schaik, C. P. 1985. Male migration and rank acquisition in wild long-tailed macaques *Macaca fascicularis*. *Animal Behaviour*, **33**, 849–61.

1988. Male careers in Sumatran long-tailed macaques *Macaca fascicularis*. *Behaviour*, **107**, 24–43.

2001. Career moves: transfer and rank challenge decisions by male long-tailed macaques. *Behaviour*, **138**, 359–95.

van Schaik, C. P., van Noordwijk, M. A. & Nunn, C. L. 1999. Sex and social evolution in primates. In *Comparative Primate Socioecology*, ed. P. C. Lee. Cambridge: Cambridge University Press, pp. 204–40.

Watanabe, K. 2001. A review of 50 years of research on the Japanese monkeys of Koshima: status and dominance. In *Primate Origins of Human Cognition and Behavior*, ed. T. Matsuzawa. Tokyo: Springer, pp. 405–17.

Watanabe, K., Mori, A. & Kawai, M. 1992. Characteristic features of the reproduction of Koshima monkeys, *Macaca fuscata fuscata*: a summary of thirty-four years of observation. *Primates*, **33**, 1–32.

Watts, D. P. 1990. Ecology of gorillas and its relation to female transfer in mountain gorillas. *International Journal of Primatology*, **11**, 21–45.

1998. Coalitionary mate guarding by male chimpanzees at Ngogo, Kibale National Park, Uganda. *Behavioral Ecology and Sociobiology*, **44**, 43–56.

2000. Causes and consequences of variation in male mountain gorillas' life histories and group membership. In *Primate Males: Causes and Consequences of Variation in Group Composition*, ed. P. M. Kappeler. Cambridge: Cambridge University Press, pp. 169–79.

Weingrill, T., Lycett, J. E. & Henzi, S. P. 2000. Consortship and mating success in chacma baboons *Papio cynocephalus ursinus*. *Ethology*, **106**, 1033–44.

Wickings, E. J., Bossi, T. & Dixson, A. F. 1993. Reproductive success in the mandrill, *Mandrillus sphinx*: correlations of male dominance and mating success with paternity, as determined by DNA fingerprinting. *Journal of Zoology, London*, **231**, 563–74.

Witt, R., Schmidt, C. & Schmitt, J. 1981. Social rank and Darwinian fitness in a multimale group of Barbary macaques *Macaca sylvana* Linnaeus, 1758. *Folia Primatologica*, **36**, 201–11.

Yamagiwa, J. 1985. Socio-sexual factors of troop fission in wild Japanese monkeys *Macaca fuscata yakui* on Yakushima Island, Japan. *Primates*, **26**, 105–20.

Yamagiwa, J. & Hill, D. A. 1998. Intraspecific variation in the social organization of Japanese macaques: past and present scope of field studies in natural habitats. *Primates*, **39**, 257–73.

Zhao, Q.-K. 1994. Mating competition and intergroup transfer of males in Tibetan macaques *Macaca thibetana* at Mt. Emei, China. *Primates*, **35**, 57–61.

1996. Etho-ecology of Tibetan macaques at Mount Emei, China. In *Evolution and Ecology of Macaque Societies*, ed. J. E. Fa & D. G. Lindburg. Cambridge: Cambridge University Press, pp. 263–89.

13 • Sexual selection, measures of sexual selection, and sexual dimorphism in primates

J. MICHAEL PLAVCAN
Department of Anthropology
University of Arkansas
Fayetteville, AR, USA

INTRODUCTION

Sexual dimorphism in anthropoid primates is commonly viewed as a product of sexual selection (Clutton-Brock *et al.*, 1977; Harvey *et al.*, 1978; Gaulin & Sailer, 1984; Clutton-Brock, 1985; Milton, 1985; Rodman & Mitani, 1987; Kay *et al.*, 1988; Ely & Kurland, 1989; Greenfield, 1992a, b; Plavcan & van Schaik, 1992, 1994, 1997; Ford, 1994; Martin *et al.*, 1994; Mitani *et al.*, 1996b; Lindenfors & Tullberg, 1998; Plavcan, 1999, 2001; Barton, 2000; Lindenfors, 2002a; Mitani *et al.*, 2002). Yet dimorphism in anthropoids is highly variable, and is expressed not as a single character, but rather to different degrees in different traits. This naturally raises the question of whether this variation is owing to variation in the strength of sexual selection, phylogenetic effects or the action of other selective factors on the dimorphic characters. While numerous papers have examined the causes and correlates of dimorphism in anthropoids, the relative contribution of sexual selection and other factors to variation in dimorphism remains unclear.

Part of this problem lies in the way that both dimorphism and sexual selection are measured. Both of these variables are estimated with error, not only in a simple statistical sense, but also in the assumptions that are used to justify measures as appropriate for analysis. Some of these biases are obvious, while others are not. Thus, if dimorphism is poorly correlated with an estimate of sexual selection, we can legitimately ask whether dimorphism is affected by factors other than sexual selection, or whether our measures fail to capture variation in either sexual selection or the targets of sexual selection. A careful examination of these biases should lead to a better understanding of not only how dimorphism is related to sexual selection, but also the relationship between sexual selection, behaviour and mating systems.

The first problem is that dimorphism is, by definition, a proportional difference between two sexes. The most com-mon measure – the ratio of male and female trait values – is an intuitively sound representation of dimorphism. However, a ratio may change value by altering either the numerator or the denominator. Recent studies emphasise that dimorphism in animals can be a function of independent variation in either or both sexes, or that the expression of dimorphism can be constrained by genetic correlated response (Greenfield, 1992a; Leigh, 1992, 1995; Martin *et al.*, 1994; Leigh & Shea, 1995; Plavcan *et al.*, 1995; Lindenfors, 2002b). Thus the product we are interested in – the proportional difference in trait values between the sexes – is clearly not the sole consequence of sexual selection acting on male characters. This mandates that in order to understand the influence of a single factor such as sexual selection on dimorphism, we must investigate variation in both male and female traits separately, and how that variation relates back to dimorphism.

More problematic with studies of dimorphism in anthropoids is the estimate of sexual selection. Sexual selection ideally should be measured as the reproductive skew generated as a function of either mate choice or mate competition. Such information is not available for most anthropoids (van Noordwijk & van Schaik, this volume). Consequently, comparative studies of dimorphism in anthropoids (and most other animals) use surrogate measures of sexual selection. However, different measures are based on different assumptions about the relationship between sexual selection and behaviour, and may not be directly comparable. Therefore, a careful evaluation of the a priori justification for using each measure, and the a-posteriori differences between the measures, is necessary if we are to make any progress in understanding why animals differ in magnitude and patterns of dimorphism.

The rest of this chapter first reviews various measures of sexual selection in anthropoid primates, exploring how these might be updated with more recent information on behavioural ecology. The chapter then turns to how an

Sexual Selection in Primates: New and Comparative Perspectives, ed. Peter M. Kappeler and Carel P. van Schaik. Published by Cambridge University Press. © Cambridge University Press 2004.

understanding of dimorphism as a function of male and female trait variation might help clarify the role of sexual selection in the evolution of dimorphism. The analysis focuses primarily on canine-tooth-size dimorphism because the relative contributions of male and female canine size to dimorphism are easily quantified. However, many published analyses have exclusively focused on either body-mass dimorphism, or canine-tooth-size dimorphism, while the methods used in different analyses have not necessarily been comparable. This creates some problems in comparing the results and conclusions of different analyses. Therefore, in addition to presenting some new analyses, the relationship between both canine-tooth-size dimorphism and body-mass dimorphism to various estimates of sexual selection will be reviewed using comparable methods.

ESTIMATES OF MALE–MALE COMPETITION

There are two basic components of sexual selection theory – mate competition and mate choice. Both factors have been demonstrated to contribute to the evolution of dimorphism in size, weaponry and ornamentation in a wide variety of animals (e.g. Andersson, 1994; Weckerly, 1998; Davies, 2000; Jarman, 2000; Lindenfors, 2002a; Perez-Barberia et al., 2002). In anthropoid primates, mate competition is thought to be the primary factor leading to the evolution of sexual size dimorphism and canine-tooth-size dimorphism (Plavcan, 2001). Male anthropoids compete overtly for access to females, and such competition has been demonstrated to lead to reproductive skew in male primates in several well-studied species (van Hooff, 2000; van Noordwijk & van Schaik, this volume). It is presumed that large male body size and large canine teeth assist males in winning fights, even though this supposition has not actually been formally tested in the wild for any primate species.

Mate choice is also thought to have a significant impact on dimorphism in at least some anthropoids (Small, 1989; van Hooff, 2000). Because of the difficulty in quantifying female choice, no study has evaluated the impact of female choice on the evolution of dimorphism across species. However, as evidence of female mate choice in primates has grown, it is becoming more feasible to consider its impact. The following analysis will therefore present a limited test of the hypothesis that female choice affects dimorphism, in addition to considering the role of male–male competition.

Measuring sexual selection due to male–male competition in primates is not straightforward. The long lifespans, slow reproductive rates, relatively limited group sizes, and practical difficulty in studying primates in the wild precludes a direct measure of male reproductive skew associated with mate competition in most species.

The alternative is to estimate male–male competition using a surrogate measure – either of the degree to which males fight, or the degree to which males are excluded from access to females. Emlen and Oring (1977) clearly laid down the principle that male–male competition should be proportional to the number of females available for mating, and the degree to which such females can be monopolised. This is referred to as the operational sex ratio (OSR), and should be proportional to the strength of sexual selection.

Emlen and Oring pointed out that the OSR should be strongly contingent on the clumping of the limiting sex, the degree to which males are sexually active, and the degree of asynchrony in female receptivity. Given that most anthropoid females live in groups, males are continuously capable of breeding, and females of most species tend to show at least some degree of oestrus asynchrony, the majority of anthropoid primates follow a pattern of female defence polygyny. Where females are dispersed, either monogamy develops, or a form of female defence polygyny develops in which males defend extensive territories overlapping those of several females (Rodman & Mitani, 1987; Dunbar, 2000; Sommer & Reichard, 2000; van Hooff, 2000).

The basic distinction between monogamy (and polyandry) versus polygyny has been widely used in studies of dimorphism in anthropoids. The fact that monogamous and polyandrous species tend to be monomorphic, while polygynous species tend to be dimorphic, has been repeatedly demonstrated (e.g. Clutton-Brock et al., 1977; Leutenegger & Kelley, 1977; Lindenfors & Tullberg, 1998; Plavcan, 1999). However, this dichotomous measure of mating system provides little resolution in measuring the relative intensity of sexual selection – there is a wide range of dimorphism among polygynous species (Clutton-Brock et al., 1977; Cheverud et al., 1986; Plavcan & van Schaik, 1997; Plavcan, 2001). Furthermore, there is significant heterogeneity in mating systems among polygynous anthropoids. Most obviously, some species form multi-male, multi-female groups, while others form single-male, multi-female groups. This distinction is easily justified as a measure of sexual selection, assuming that reproductive skew is lower in multi-male groups because mating is more promiscuous than in single-male groups (Harvey et al., 1978; Lindenfors, 2002a, b). In this sense the distinction between single-male and multi-male groups is just a categorical approximation of the OSR. Unfortunately, no study has clearly demonstrated a significant difference in dimorphism between single-male and multi-male species (Harvey *et al.*, 1978; Barton, 2000; Plavcan, 2001).

An alternative is to focus on observed behavioural differences in male–male competition. The degree of male–male competition for access to mates should be proportional to the potential monopolisability of mates, and hence the strength of sexual selection (Emlen & Oring, 1977). Ideally, male–male competition should be measured in terms of its frequency and intensity (Clutton-Brock, 1985). Unfortunately, comparable direct measures of male–male competition are lacking for most species. However, male primates do show fairly obvious differences in the intensity and frequency of male–male competition, and these cut across divisions of mating systems. For example, both *Brachyteles arachnoides* and *Papio cynocephalus* live in multi-male, multi-female groups. However, the former are clearly characterised by little if any fighting among males (Milton, 1985), while the latter are characterised by overt, frequent, agonistic male–male competition (Barton, 2000).

Competition levels estimate male–male competition using categorical definitions of the intensity and potential frequency of male–male competition (Kay *et al.*, 1988; Plavcan & van Schaik, 1992, 1997). Competition intensity is divided into 'high' and 'low' classes. High-intensity species are those where males are reported to be intolerant of one another, or maintain stable dominance hierarchies through agonistic encounters. Low-intensity species are those where males are reported to be relatively tolerant of one another. The potential frequency of competition is likewise dichotomised into 'high' and 'low' classes. High-frequency species are classified as those where more than one breeding male typically occurs in a group such that competition can occur on a daily basis at any time of the year. Low-frequency species are those where groups typically contain either a single adult male, or male–male competition is clearly limited to a short breeding season. These classifications are combined into four competition levels, with the intensity category identified as the dominant signal. Finally, because monogamous and polyandrous species are predicted to show little differential reproductive success from competition over mates, they are placed within competition level 1. Competition levels are associated with dimorphism in the predicted direction (Plavcan & van Schaik, 1992, 1997).

Lindenfors (2002a, b) has critiqued competition levels as poor estimates of sexual selection on three bases:

(1) that they conflate sexual and natural selection,
(2) that the definitions of competition frequency and intensity are not adhered to in classifying species, and
(3) that 'promiscuous' multi-male species classified into competition level 4 are predicted to show greater dimorphism than single-male species classified into competition level 3, in total contrast to the normal expectation of sexual selection theory' (Lindenfors, 2002b, p. 598).

These criticisms serve as an excellent basis for discussing the relationship between primate mating systems, sexual selection, and male–male competition. To begin, though Lindenfors (2002a) never specifies how competition levels conflate sexual and natural selection, this criticism probably arises from the hypothesis that multi-male dominance hierarchies which are maintained year-round, even when females are not available for mating, might reflect male contest competition for resources as well as mates (Kay *et al.*, 1988; Plavcan & van Schaik, 1992, 1997). These mechanisms are not mutually exclusive. Many year-round dominance hierarchies in primates are associated with aseasonal breeding, suggesting that they are maintained by mate competition. For those species that show more seasonal breeding, maintenance of male dominance hierarchies could either reflect male tactics for ensuring priority of access to females when breeding begins, or competition for access to resources. Male resource defence forms the basis for intense polygyny, and hence sexual selection, in a number of mammalian (and non-mammalian) species (Emlen & Oring, 1977). There is little evidence that this mechanism operates in primates (Kappeler & van Schaik, 2002). However, the hypothesis should not be summarily dismissed without careful documentation, and does not conflate sexual and natural selection.

Next, Lindenfors (2002a, b) notes that some multi-male species are placed in the low-frequency category, while some single-male species are classified as high-frequency. This is true, but hardly represents an a-posteriori shifting of species to make the competition levels fit the data better. Rather it reflects a focus on behaviour, and not just group composition. Multi-male species (specifically *Saimiri*) in which male–male competition is limited to a short breeding season are placed in the low-frequency category to distinguish them from species that compete year-round. In contrast, species in which males maintain single-male breeding units within multi-male bands (*Papio hamadryas*, *Theropithecus gelada*) are placed in the high-frequency category because males come into proximity daily, allowing for year-round potential male–male competition on a daily basis.

Most importantly, Lindenfors notes that competition levels predict that some multi-male species should show greater dimorphism than single-male species. That this is in 'total contradiction to sexual selection theory' is highly debatable. An understanding of why this is so is critical for making progress in understanding the relationship between variation in mating systems, behaviour and sexual selection. It has long been recognised that multi-male primate species show great variation in degrees and types of male–male competition (Crook, 1972; Clutton-Brock *et al.*, 1977; Harvey *et al.*, 1978; Leutenegger & Kelley, 1977). Kappeler and van Schaik (2002) note that demographic data on male and female group composition is not a reliable indicator of the mating system, and that care needs to be exercised in distinguishing between social organisation and mating system. As already noted, taxa such as *Brachyteles* and *Papio* show profoundly different patterns of mating and male–male competition, though both are multi-male. Males of *Papio* compete intensely and often, while males of *Brachyteles* show little male agonistic competition by comparison to *Papio*. Not surprisingly, dominant males in *Papio* show a reproductive advantage over lower-ranking males, while males of *Brachyteles* mate promiscuously and show far less reproductive skew (Strier, 1992, 2000; Altmann, 2000). Competition levels were not defined on the observed promiscuity or reproductive skew, but post hoc interpretation of the results suggests an agreement between degrees of male–male competition and reproductive skew that is consistent with sexual selection theory.

Lindenfors points out that sexual selection should be more intense in single-male species because multi-male species are more promiscuous, and hence should show less reproductive skew. In theory, this would be true if there were a simple correspondence between the number of males in a group and reproductive skew. This has long been questioned (Clutton-Brock *et al.*, 1977). Altmann (2000, p. 247) succinctly points out that

> [w]hereas single-male groups are ubiquitously assumed
> to be reproductively advantageous to males, this
> assumption is based in short term perspectives and one
> that focuses solely on mating success . . . Because male
> tenure is usually short in single-male primate
> populations and because tenure changes often entail the
> risk of infanticide, reproductive skew in single-male

populations will not necessarily be greater than in multi-male populations.

Hence, the theoretical basis for assuming that reproductive skew in multi-male species is typically less than that of single-males species, at least in primates, is weak.

All of the above categorical estimates of sexual selection are limited in how much variation in dimorphism among species can be explained. The socionomic sex ratio (SSR) is a simple, continuous measure of sexual selection (Clutton-Brock *et al.*, 1977) that should be correlated with the OSR if there is no overlap in female receptivity (Emlen & Oring, 1977). SSR is correlated with dimorphism in body mass and canine tooth size, but only if monogamous and polyandrous species are included in the analysis. There is no reported correlation between dimorphism and the SSR in polygynous species (Clutton-Brock *et al.*, 1977; Plavcan, 1999). Problems with the SSR have been noted for years, including the diversity of types of male–male competition in multi-male mating systems, and the fact that oestrus overlap can effectively lower the OSR in multi-male groups (Clutton-Brock *et al.*, 1977; Emlen & Oring, 1977; Mitani *et al.*, 1996b).

To overcome this, Mitani *et al.* (1996b) offer a calculation of the OSR that weights the SSR by the ratio of interbirth interval to the average number of days that a female is typically in oestrus (the product of the number of cycles to conception and the number of receptive days), and the duration of the breeding season. This measure of the OSR is correlated with body-mass dimorphism in a sample of 18 primates. It is not correlated with canine-tooth-size dimorphism, however (Plavcan, 1999). This latter observation suggests either that canine- and body-size dimorphism respond differently to sexual selection, that each character is under both sexual selection and a series of factors not necessarily held in common, or that the OSR calculations of Mitani *et al.* do not capture variation in sexual selection.

While the first two options seem most likely, the last should not be summarily dismissed. Two problems appear with the Mitani *et al.* formulation of the OSR. First, reliable field data necessary to calculate the OSR are available for a limited number of species, making it difficult to include in a broader analysis attempting to quantify the effects of factors other than sexual selection on dimorphism. Second, the formula presented by Mitani *et al.* cannot be applied to monogamous species. This is because the value of the OSR in the formula is contingent on the sex ratio – the closer the ratio is to one, the more skewed the OSR. This said, the calculations and analysis of Mitani *et al.* are internally consistent, and are the first to attempt explicitly to control for variation in the OSR: the importance of this cannot be understated.

Using the data of Mitani *et al.*, SSR is correlated with both body-mass and canine dimorphism employing the same methods used to demonstrate a relationship between dimorphism and OSR (Plavcan, 1999). This observation is intriguing, because one of the fundamental reasons for calculating the OSR is to control for oestrus overlap and variation in interbirth intervals, which presumably renders SSR an inaccurate measure of male–male competition. Within the Mitani *et al.* data set, SSR and OSR are significantly correlated ($n = 18$, $r = 0.650$, $p = 0.005$). This raises the question of how important oestrus overlap and long interbirth intervals actually are in constraining reproductive skew among many male primates. That oestrus overlap occurs in many species is not in doubt (Altmann, 2000). However, males living with large groups of females may still realise a significant reproductive advantage if dominant males sire more offspring than subordinates, while oestrus overlap does not necessarily preclude the ability of a male to monopolise access to more than one female as long as there is asynchrony in fertile periods (Altmann, 2000; Pereira *et al.*, 2000). The issue here is not whether the calculations of Mitani *et al.* are correct. Rather, the question is whether the SSR also picks up variation in the OSR among species.

Since the analysis of Harvey *et al.* (1977), there has been no analysis of SSR and dimorphism in primates using phylogenetic comparative methods, and using updated information on primate sex ratios. Furthermore, there has been considerable work done on understanding the relationship between female group size and male distributions in primates.

It is now well established that the number of males in primate groups is facultatively dependent on the number of females (Mitani *et al.*, 1996b; Nunn, 1999; Altmann *et al.*, 2000; Barton, 2000). Importantly, in multi-male groups this relationship is negatively allometric, meaning that SSR is more skewed in larger groups (Altmann, 2000). Given that male dominance is associated with reproductive success in wild primates (Altmann, 2000; van Hooff, 2000), these observations suggest that in multi-male groups dimorphism should be correlated with SSR.

In contrast, within-group SSRs for single-male groups vary as a simple function of the number of females in a group. However, reproductive skew should actually be a function of male tenure, infanticide, and extra-group copulations by females (Altmann, 2000). Unfortunately, comparative data on these variables for single-male groups are unavailable. If reproductive skew in single-male groups is a simple function of female group size, then we should expect a positive correlation between SSR and dimorphism. Otherwise, the effect of these other factors might render a more complex relationship between dimorphism and SSRs.

FEMALE CHOICE

So far, no study has successfully quantified female choice in primates in such a way that a comparative study of its influence on dimorphism can be carried out. However, there is increasing information on female choice in a few species (Paul, 2002). Generally, we can hypothesise that female choice should either reinforce male reproductive skew, or dampen it. For example, females of *Cebus apella*, *Mandrillus sphinx* and *Pongo pygmeaus* prefer to mate with local dominant males (Janson, 1984; Setchell & Dixson, 2001a, b; Setchell *et al.*, 2001). Recent work on *Mandrillus* and *Pongo* suggests that this female choice has produced dual male mating strategies – one where dominant males acquire large size and ornamentation that is attractive to females and intimidating to other males, and the other where subordinate males attempt to mate opportunistically with unguarded females (Rogers *et al.*, 1996; van Schaik & van Hooff, 1996; Maggioncalda *et al.*, 1999; Setchell & Dixson, 2001a, b; Setchell *et al.*, 2001; Utami & van Hooff, this volume). In a similar vein, females of *Theropithecus gelada*, *Papio hamadryas* and *Gorilla gorilla* form strong bonds with a single dominant male with which they prefer to mate (Watts, 1996; Barton, 2000). Males of all these species compete agonistically. Females may prefer to mate with these males because of male quality, or to ensure a male of paternity and thereby gain protection from infanticide. Regardless, female choice in all of these species should reinforce reproductive skew, leading to the evolution of strong dimorphism.

In contrast, female promiscuity in other species is widely interpreted as a counter-strategy to male coercion (Smuts & Smuts, 1993; van Schaik *et al.*, 1999, 2000). Female matings with subordinate or extra-group males are commonly reported in a wide variety of primates. Females are active players in selecting mates, so this promiscuity should be viewed as a form of female choice. Compared to the above situation, such female choice should counter male attempts to monopolise matings, and hence should reduce male reproductive skew and result in comparatively less dimorphism than that seen in the above species.

These hypotheses will be evaluated below, but first we need to consider the other variable: dimorphism itself.

VARIATION IN DIMORPHISM AS A FUNCTION OF MALE AND FEMALE TRAIT VALUES

In order to understand the relationship between sexual selection and dimorphism, it is important to consider the forces that affect both male and female traits (Leigh, 1992, 1995; Martin *et al.*, 1994; Leigh & Shea, 1995; Plavcan *et al.*, 1995; Lindenfors, 2002b). Few comparative analyses have been carried out on this topic so far, but those that have underscore the importance of this point.

For this purpose, canine dimorphism is an easy system to work with for two reasons. First, large canine teeth in either males or females are most obviously hypothesised to function as weapons. There is little evidence of sexual divergence in canine function in most primates (Greenfield, 1992a; Plavcan, 2001). Consequently, males and females might compete for different things with different relative fitness consequences (mates versus resources), but as long as there is a fitness consequence to winning and losing fights, and canines help to determine the outcome of fights, then canine size should vary as a function of competition in either sex (Plavcan *et al.*, 1995). Second, because the effect of body size on canine size is easily calculated, allometric effects are easily controlled.

Plavcan *et al.* (1995) apply classifications of potential frequency and intensity of intrasexual competition to male and female primates independently. Additionally, they hypothesise that where fighting regularly occurs between groups, as opposed to dyadic fighting, selection for the development of weaponry should be less intense because individual weaponry will be less important than the number of coalition partners in determining the outcome of contests. Hence, they classify males and females as showing coalitionary or non-coalitionary competition. Notably, species classified as coalitionary by definition can only be compared to high-intensity, high-frequency species. Using species values, both male and female primates classified as high-intensity have larger canines than those classified as low-intensity. The frequency effect is more ambiguous. High-frequency males on average have larger canines than low-frequency, but the effect is not significant. The opposite result is attained in females. The effect of coalitions is in the predicted direction in both sexes. Independent contrasts paralleled the analysis of species values for the intensity and frequency

effects in both sexes. However, there were too few contrasts to obtain a significant result for the effect of coalitions in either sex.

Plavcan *et al.* (1995) noted that the potential-frequency classification produces a reverse effect in females from that predicted. Among species classified as showing 'high-intensity' female competition, low-frequency females have smaller canines than high-frequency females. They speculated that this may reflect higher-intensity competition in solitary females for resources, but provided no independent test.

Importantly, Janson and Goldsmith (1995) demonstrate that feeding competition is negatively associated with group size in female primates. This model posits that for smaller groups of primates, the costs of including additional group members are proportionally higher, leading to higher competition to limit group size. Janson and Goldsmith (1995) specifically find that the correlation between competition and group size is significant in frugivorous, but not folivorous species. These findings can be applied to the analysis of female canine size in primates. As noted by Plavcan *et al.* (1995) the classifications of high- and low-intensity female intrasexual competition correspond roughly to distinctions between scramble and contest competition, which in turn are broadly associated with diet (Plavcan & van Schaik, 1994). Consequently, if both models are true, then we should see a strong correlation between relative female canine tooth size and female group size in those species classified as 'high-intensity' female competition, and a weak or non-significant correlation in the species classified as showing 'low-intensity' female competition.

MATERIALS AND METHODS

There has been little uniformity over the years in the methods used for analysing the relationship between dimorphism and different measures of sexual selection. Furthermore, many analyses focus on either body-mass dimorphism or canine-tooth-size dimorphism, raising the question of whether results differ for different anatomical systems. The following presents a brief re-analysis of the relationship between canine-tooth-size dimorphism and body-mass dimorphism (Table 13.1). Following that are new analyses of the relationship between dimorphism and updated SSR data, a test of the hypothesis that female choice influences dimorphism, and an analysis of the relationship between relative male and female canine size, the SSR, and female group size. Partitioning of body-mass dimorphism into separate male

Table 13.1 *Data used in the present analyses of sexual dimorphism.*

Species	MN	FN	lmwgt	lfwgt	lwdim	lcdim	mchgt	fchgt	freshgt	CL	MS	FI	FF	References
Alouatta belzebul	4	11.5	1.971	1.645	0.326	0.593	−0.193	−0.379	−0.142	4	m			Mitani et al. (1996a)
Alouatta caraya	1.9	2.6	1.859	1.466	0.394	0.399	−0.092	−0.05	0.118	4	m(v)			Crockett & Eisenberg (1987)
Alouatta fusca			1.907	1.47	0.436	0.412	−0.075	−0.081	0.075	4	m			Crockett & Eisenberg (1987)
Alouatta palliata	2.7	6.3	1.950	1.735	0.215	0.412	−0.076	−0.106	0.051	4	m	2	2	Crockett & Eisenberg (1987)
Alouatta pigra	1.3	1.8	2.434	1.861	0.573	0.378	−0.226	−0.035	0.225	4	m(v)			Crockett & Eisenberg (1987)
Alouatta seniculus	1.6	2.6	1.901	1.651	0.250	0.405	−0.124	−0.15	0.040	4	m(v)	1	2	Crockett & Eisenberg (1987)
Aotus trivirgatus	1	1	−0.086	−0.135	0.049	0.077	−0.279	−0.149	0.147	1	mp	2	1	Robinson et al. (1987)
Ateles belzebuth	4	11.5	2.115	2.061	0.055	0.419				1	m	2	1	Robinson & Janson (1987)
Ateles geoffroyi			2.052	1.987	0.065	0.451	−0.316	−0.363	−0.042	2	m	1	2	
Ateles paniscus	5	15.5	2.209	2.133	0.076	0.285	−0.298	−0.416	−0.107	2	m	1	2	Mitani et al. (1996a)
Brachyteles arachnoides	7	9	2.263	2.126	0.137		−0.550	−0.493	−0.011	2	m	1	2	Mitani et al. (1996a)
Cebus albifrons	4.8	7.2	1.157	0.829	0.328					4	m			Robinson & Janson (1987)
Cebus apella	3.4	4.4	1.166	0.728	0.439	0.344	0.166	0.229	0.220	4	s	c	2	Robinson & Janson (1987)
Cebus capucinus	5.5	4	1.303	0.932	0.371					3				Mitani et al. (1996a)
Cebus nigrivittatus	1.4	6.1	1.191	0.924	0.267					4	m(v)			Robinson & Janson (1987)
Chiropotes albinasus	9	8	1.147	0.912	0.235					2	m			Robinson et al. (1987)
Chiropotes satanas			1.065	0.948	0.117	0.179	0.017	0.165	0.258	2	s			
Cacajao calvus			1.238	1.058	0.181	0.227	0.152	0.284	0.285	3				
Callicebus moloch	1	1	0.020	−0.045	0.065	0.077	−0.598	−0.345	0.170	1	mp	1	1	Robinson et al. (1987)
Callicebus personatus	1	1	0.239	0.322	−0.083					1	mp			Robinson et al. (1987)
Callicebus torquatus	1	1	0.247	0.191	0.056	0.039	−0.622	−0.424	0.107	1	mp	1	1	Robinson et al. (1987)
Callithrix argentata			−1.109	−1.022	−0.087					1	mp			
Callithrix humeralifer	1	1	−0.744	−0.751	0.006	0.030				1	mp			Goldizen (1987); Dunbar (2000)
Callithrix jacchus	2	1	−1.016	−0.965	−0.051					1	mp			Dunbar (2000)
Lagothrix lagthricha	3.25	9.75	1.898	1.712	0.186	0.571	−0.022	−0.239	−0.119	2	m			DiFiore & Rodman (2001)
Leontopithecus rosalia			−0.478	−0.514	0.036	0.166				2	mp			
Pithecia pithecia	2.7	1	0.588	0.412	0.176	0.277	−0.051	0.030	0.170	1	mp			Robinson et al. (1987)
Saimiri oerstedi	10	16	−0.288	−0.511	0.223	0.519	0.077	−0.071	−0.019	3	m	1	2	Mitani et al. (1996a)
Saimiri sciureus	7	23	−0.250	−0.412	0.163	0.344	−0.071	−0.053	0.101	3	m	c	2	Mitani et al. (1996a)
Saguinus fuscicollis	1.88	1	−1.070	−1.027	−0.043	0.030	−0.014	0.251	0.366	1	mp	2	2	Dunbar (2000)
Saguinus midas			−0.664	−0.553	−0.110	−0.010	−0.124	0.163	0.353	1	mp	2	2	
Saguinus mystax	1.88	1	−0.673	−0.618	−0.055					1	mp			Dunbar (2000)
Saguinus nigricollis	1.3	1.3	−0.759	−0.726	−0.034					1	m			Heyman (2000)
Saguinus oedipus	2	1	−0.734	−0.693	−0.041	0.000				1	mp			Dunbar (2000)
Cercopithecus aethiops	3.8	5.8	1.522	1.182	0.340	0.588	0.320	0.140	0.026	4	m	c	2	Melnick & Pearl (1987)

(cont.)

Species														Reference
Cercopithecus ascanius	1	8.2	1.308	1.072	0.237	0.445	0.257	0.190	0.119	3	s(i)	2	2	Cords (2000)
Cercopithecus cephus	1	3.4	1.456	1.058	0.398	0.571	0.279	0.123	0.037	3	s			Cords (2000)
Cercopithecus diana	1	7	1.649	1.361	0.288	0.476				3	s			Cords (2000)
Cercopithecus hamlyni			1.703	1.212	0.491	0.610	0.235	0.121	0.065	3	s			
Cercopithecus lhoesti	1	7.8	1.787	1.238	0.548	0.678	0.248	0.054	−0.011	3	s(i)	2	2	Cords (2000)
Cercopithecus mitis			1.953	1.399	0.554	0.642	0.339	0.182	0.055	3	s	2	2	Cords (2000)
Cercopithecus mona	1	2	1.482	0.916	0.566					3	s			
Cercopithecus neglectus	1	3.6	1.844	1.376	0.467	0.536	0.338	0.252	0.125	3	s			Cords (2000)
Cercopithecus nictitans	1	5	1.907	1.454	0.453	0.476	0.223	0.193	0.145	3	s			Cords (2000)
Cercopithecus petaurista	1	4.6	1.500	1.065	0.435					3	s			Cords (2000)
Cercopithecus pogonias	1		1.449	1.065	0.385	0.438	0.210	0.199	0.16	3	s			Cords (2000)
Cercopithecus preussi						0.571				3	s			
Cercopithecus wolfi			1.364	1.054	0.309	0.631	0.071	−0.153	−0.097	3	s			
Colobus angolensis			2.264	2.050	0.214	0.610	0.223	0.198	0.150	4	m			
Colobus guereza	1	3	2.233	2.058	0.175	0.378	0.143	−0.069	−0.062	3	s(v)	1	2	Mitani et al. (1996a)
Colobus polykomos			2.293	2.116	0.176	0.565	−0.069	−0.28	−0.110	4	m(v)			
Colobus satanas			2.342	2.004	0.338	0.560	−0.095	−0.15	−0.099	4	m(v)			
Procolobus badius	7.75	16.5	2.123	2.105	0.018	0.631	0.079	0.149	−0.099	4	m	1	2	Struhsaker (2000)
Procolobus verus	1.25	2	1.548	1.435	0.112	0.693	0.149	−0.209	−0.206	4	m			Mitani et al. (1996a)
Cercocebus agilis			2.251	1.733	0.518	0.673				4	m			
Cercocebus galeritus	2	6	2.320	1.699	0.621					4	m			Mitani et al. (1996a)
Cercocebus torquatus			2.398	1.825	0.573	0.900	0.303	−0.105	−0.208	4	m	c	2	Cords (2000)
Erythrocebus patas	1	9.6	2.518	1.872	0.646	0.775	0.431	0.163	−0.027	3	s	2		Sterck & van Hooff (2000)
Kasi johnii	1	4.5	2.485	2.416	0.069	0.489	0.208	−0.069	−0.062	3	s			Sterck & van Hooff (2000)
Kasi vetulus	1	4	1.735	1.631	0.104	0.593	0.104	−0.054	−0.091	3	s			
Lophocebus aterrimus	3.8	7	2.059	1.751	0.308	0.637	0.041	−0.186	−0.109	4	m	c	2	Melnick & Pearl (1987)
Macaca fascicularis	5.7	9.9	1.571	1.115	0.456	0.811	0.633	0.243	−0.086	4	m	c	2	Melnick & Pearl (1987)
Macaca fuscata	7	24	2.398	2.083	0.315	0.708	0.110	−0.219	−0.189	4	m	c	2	Melnick & Pearl (1987)
Macaca mulatta	4.6	14.3	2.042	1.681	0.362	0.728	0.148	−0.198	−0.194	4	m	c	2	Melnick & Pearl (1987)
Macaca nemestrina	2	18	2.416	1.872	0.544	0.859	0.545	0.144	−0.124	4	m	c	2	Melnick & Pearl (1987)
Macaca nigra			2.292	1.699	0.592	0.959	0.655	0.172	−0.172	4	m			
Macaca simica	2.7	6.2	1.737	1.163	0.574	0.678	0.312	0.146	0.037	4	m	c	2	Melnick & Pearl (1987)
Macaca tonkeana			2.701	2.197	0.504	0.658				4	m			
Mandrillus leucophaeus			2.833	2.303	0.531	1.411	0.883	−0.052	−0.552	4	m			
Mandrillus sphinx	96	272	3.453	2.557	0.896	0.525	0.277	0.119	0.034	4	m			Rogers et al. (1996)
Miopithecus talapoin	13	27	0.322	0.113	0.209	0.277	0.072	0.119	0.034	4	m			Melnick & Pearl (1987)
Nasalis larvatus	1	5	3.016	2.284	0.731	0.788	0.072	−0.168	−0.112	4	m(s)			Mitani et al. (1996a)

Table 13.1 (cont.)

Species	MN	FN	lmwgt	lfwgt	lwdim	lcdim	mchgt	fchgt	freshgt	CL	MS	FI	FF	References
Papio anubis	6.8	11.4	3.186	2.761	0.425	0.798				4	m			Melnick & Pearl (1987)
Papio cynocephalus	8.4	12.6	2.847	2.277	0.570	1.030	0.249	−0.292	−0.358	4	m	c	2	Melnick & Pearl (1987)
Papio hamadryas	15.7	22.4	3.045	2.434	0.611	1.008	0.387	−0.097	−0.257	4	m(s)	2	2	Dunbar (2000)
Papio ursinus	9.5	14	3.312	2.676	0.637	1.343				4	m			Melnick & Pearl (1987)
Presbytis comata	1	2.1	1.899	1.904	−0.004	0.405	−0.043	−0.154	−0.020	3	s			Sterck & van Hooff (2000)
Presbytis melalophos			1.869	1.844	0.025	0.536	−0.031	−0.261	−0.135	3	s(v)	1	2	Sterck & van Hooff (2000)
Presbytis potenziani	1	1	1.82	1.856	−0.037	0.344	−0.060	−0.106	0.040	1	mp			Watanabi (1981)
Presbytis rubicunda	1	2.5	1.813	1.822	−0.009	0.405	−0.108	−0.187	−0.008	3	s(v)	1	2	Sterck & van Hooff (2000)
Semnopithecus entellus	2.4	6.3	2.365	1.902	0.463	0.525	0.117	0.003	0.028	4	m(v)	1	2	Borries (2000)
Simias concolor			2.214	1.917	0.297	0.542		−0.311		3	s			
Trachypithecus cristata	1	9.3	1.889	1.751	0.138	0.529	0.249	0.068	0.002	3	s			Sterck & van Hooff (2000)
Trachypithecus obscurus	1	5	1.992	1.899	0.092	0.571	0.004	−0.226	−0.124	3	s(v)	1	2	Sterck & van Hooff (2000)
Trachypithecus pileatus	1	3.6	2.485	2.288	0.196	0.631	0.360	0.063	−0.079	3	s			Sterck & van Hooff (2000)
Pygathrix nemaeus			2.398	2.133	0.265	0.663				3	s			
Theropithecus gelada	34	86	2.944	2.460	0.485	1.169	0.654	−0.057	−0.400	4	m(s)	c	2	Dunbar (2000)
Hylobates concolor	1	1	2.053	2.031	0.022	0.140	0.332	0.472	0.350	1	mp			
Hylobates hoolock	1	1	1.927	1.929	−0.001	0.030	0.189	0.470	0.446	1	mp	2	1	
Hylobates klossi	1	1	1.735	1.778	−0.043	−0.062	0.131	0.475	0.490	1	mp	2	1	
Hylobates lar	1	1	1.742	1.668	0.075	0.148	0.283	0.459	0.370	1	mp	2	1	
Hylobates syndactylus	1	1	2.384	2.361	0.023	0.166	0.203	0.337	0.303	1	mp	2	1	
Gorilla gorilla	1	3	5.138	4.270	0.868	0.548	−0.323	−0.414	−0.088	3	s	1	2	Mitani *et al.* (1996a)
Pan paniscus	8	8	3.807	3.503	0.304	0.322	−0.567	−0.492	0.001	2	m	1	2	Mitani *et al.* (1996a)
Pan troglodytes	6.3	10.7	4.089	3.824	0.265	0.351	−0.219	−0.186	0.069	2	m	1	2	Wrangham (2000)
Pongo pygmaeus	1	1	4.363	3.578	0.785	0.525	−0.233	−0.193	0.072	3	s	1	1	Mitani *et al.* (1996b)

Abbreviations: MN = male group size; FN = female group size; lmwgt = ln(male body mass); lfwgt = ln(female body mass); lwdim = ln(body-mass dimorphism); lcdim = ln(maxillary canine crown height dimorphism); mchgt = residual male maxillary canine height; fchgt = residual female maxillary canine height; freshgt = female adjusted maxillary canine height; CL = competition levels (from Plavcan & van Schaik, 1997); MS = mating system (mp = monogamous or polyandrous, m = multi-male, m(v) = primarily multi-male but variable, m(s) = single-male with periodic multi-male influxes; s = single-male, s(v) = primarily single-male but variable, s(i) = single-male with multi-male bands, c = coalitionary); FI = female competition intensity (1 = low, 2 = high, c = coalitionary); FF = female competition frequency (1 = low, 2 = high). References for group-size data only. See text for other sources. Note that male competition classifications are not listed, but generally follow the competition levels (1 = low frequency, low intensity; 2 = high frequency, low-intensity; 3 = low frequency, high intensity; 4 = high frequency, high intensity) with the exception of some monogamous and polyandrous species, which are classified as low frequency, high intensity (see Plavcan *et al.*, for explanation and details).

and female components is not attempted in this analysis because of the difficulty in controlling for allometric effects of body size itself (but see Lindenfors, 2002b, for an approach using phylogenetic analysis).

Strepsirrhine primates are not evaluated as they show little dimorphism (the maximum is males about 20 per cent larger than females in some lorisids – Kappeler, 1990; Smith & Jungers, 1997). The reasons for the lack of dimorphism in strepsirrhine primates remain unclear (Kappeler, 1990; Godfrey *et al.*, 1993; Plavcan *et al.*, 1995; van Schaik & Kappeler, 1996; Leigh & Terranova, 1998; Plavcan, 1998, 2001; Lindenfors, 2002a), though most agree that anthropoids and strepsirrhines differ fundamentally in the relationship between mating systems, competition and dimorphism.

Data on mating systems, group size and sex ratios, OSRs, SSRs and competition levels were gathered from the literature (Plavcan & van Schaik, 1992, 1997; Plavcan *et al.*, 1995; Mitani *et al.*, 1996a, b; Dunbar, 2000; Lindenfors, 2002a). Data on male and female body size were gathered from the primary literature or from museum records (Plavcan & van Schaik, 1997; Smith & Jungers, 1997; Plavcan, 1999; Delson *et al.*, 2000). Body-mass data were selected from subspecies to match the canine data wherever possible. Data on maxillary canine crown height were gathered from museum specimens of known sex, representing restricted geographic distributions of species (Plavcan, 1990).

Because information on body mass, canine tooth size, and various behavioural variables is not available for all species uniformly, sample sizes differ substantially between analyses.

Mating and breeding system classifications differ slightly among analyses. Several species show single-male breeding units within multi-male bands, leading to different classifications (e.g. Harvey & Harcourt, 1984; Plavcan & van Schaik, 1992, 1997; Rogers *et al.*, 1996; Lindenfors & Tullberg, 1998; Barton, 2000; Dunbar, 2000). Analyses are repeated using both classifications. *Mandrillus*, often classified as single-male, is here classified only as multi-male on the basis of recent field work (Rogers *et al.*, 1996). *Simias* and *Cercopithecus neglectus* sometimes show pairs and sometimes single-male groups; however, field studies suggest that these species are fundamentally single-male (Watanabe, 1981; Brennan, 1985; Rowell, 1988). Several presbytine and cercopithecine species show variation between single-male and multi-male mating systems (Borries, 2000; Steenbeck, 2000; Steenbeck *et al.*, 2000; Sterck & van Hooff, 2000), or multi-male influxes during the breeding season (Cords, 1988, 2000). Because male reproductive skew in these species could

be considered intermediate between consistently single-male and multi-male species, they are evaluated as potentially showing intermediate levels of sexual dimorphism.

All analyses are carried out on ln-transformed data. Sexual dimorphism is estimated as the ratio of the male trait value divided by that of the female. The natural logarithm of this ratio is equivalent to the formula ln(male) − ln(female). Dimorphism is not corrected for body size because current models suggest that factors tied to sexual selection are causally related to variation in body size, while models suggesting a direct effect of size on dimorphism receive little support (Mitani *et al.*, 1996b; Plavcan & van Schaik, 1997; Plavcan, 2001). Hence, allometric adjustment would actually remove variation in the hypothesised causal variable.

Estimates of relative male and female canine size are generated as the least-squares residual from an isometric line passed through the relationship between male and female canine size and male and female body mass respectively (details are presented in Plavcan *et al.*, 1995; Plavcan, 1998). Both sexes are combined in a single analysis, so residuals between male and female canine size may be directly compared.

Analysis of relative female canine size is complicated by correlated response of female canine size to male canine size (Lande, 1980). Relative male and female canine size are strongly correlated across primates. Plavcan (1998) demonstrated that this correlation is only significant in species where females are classified as showing 'low-intensity' intrasexual competition. Consequently, relative female canine size is adjusted for correlated response to male canine size by passing a reduced major-axis regression line through the relationship between relative male and female canine size for the 'low-intensity' female competition species, then calculating least-squares residuals for all species from this line. The resulting values are referred to as 'adjusted relative female canine size' in subsequent analyses.

Analyses are carried out on both raw species data and using independent contrasts (Felsenstein, 1985). The phylogeny reported in Smith and Cheverud (2002) is used. Species not included by Smith and Cheverud are interpolated based on information from Purvis and Webster (1999). Species of *Presbytis* (*sensu lato*) are grouped into separate genera of *Presbytis*, *Trachypithecus*, *Kasi*, and *Semnopithecus* (Napier, 1985). Polytomies in these groups, and *Alouatta*, are resolved arbitrarily for the analysis of continuous data. The effect of branching sequence is checked by swapping branches. In no case are the results significantly altered.

For comparisons of continuous characters, independent contrasts were calculated using the 'PDTREE' program of Garland (Garland *et al.*, 1993, 1999). Contrasts were initially calculated by setting branch lengths to the divergence times in Smith and Cheverud (2002). Unknown divergence times were assigned values intermediate between known nodes, while arbitrarily resolved polytomies were assigned branch lengths of 0.01. Subsequently, correlations between absolute values of contrasts and their standard deviations were evaluated. Where correlations were significant (most comparisons), branch lengths were adjusted using the procedures recommended by Pagel, Grafen and Nee (provided in the program), as well as setting all branch lengths to a length of 1. The method of Pagel provided the least-biased estimates. Hence, all analyses use this method of branch transformation.

Two basic methods are used for evaluating the effect of categorical variables on continuous data – simple contrasts between species and clades differing in a categorical variable, and the 'matched-pairs' test (Nunn & Barton, 2001). Close scrutiny of the data indicates that results of comparisons can change depending on the representation of species, the specific phylogeny adopted, and the method used, especially given the small number of contrasts involved in this analysis. Consequently, results from both methods are presented. Most comparisons follow the matched-pairs method. Other independent contrasts are calculated between pairs of species, or between groups that differed consistently in a categorical variable. An exception was made for groups wherein the large majority of species shared a common categorical variable. Thus, African and Asian colobines were compared because most of the former are classified as multi-male, and most of the latter as single-male species, leading to the common supposition that these are primitive mating systems within the groups, with exceptions being derived. One-tailed sign tests are reported for results using the matched-pairs comparisons only.

RESULTS

RE-ANALYSES OF DIMORPHISM

As is well known, monogamous and polyandrous primates all show little or no sexual dimorphism, while dimorphism is limited to polygynous species (Fig. 13.1). This result is generally upheld in phylogenetic analyses (see also Lindenfors & Tullberg, 1998). Using species values, monogamous species show less dimorphism than either single-male or

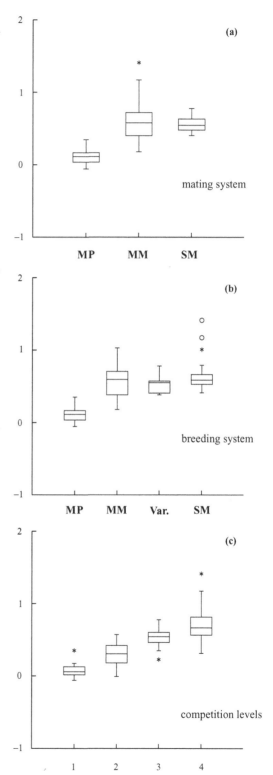

multi-male species, while single-male and multi-male species almost entirely overlap in their range (Fig. 13.1), with no significant difference between the two. These results do not change if multi-level species are classified as single-male or multi-male. Results are similar for canine and body-mass dimorphism.

Analysis of independent contrasts (Table 13.2) parallels the analysis of species values (see also Barton, 2000). Overall, matched-pairs comparisons are significant for body-mass dimorphism (15/6, $p = 0.05$), but not canine dimorphism (12/7, $p > 0.05$). Most contrasts comparing monogamous or polyandrous species to either single-male or multi-male species corroborate the sexual selection hypothesis (Lindenfors & Tullberg, 1998). The only negative contrast for canines involves pithecines, in which the canines are specialised for hard-object feeding in both sexes, lending doubt to the validity of the contrast. Overall, contrasts between single-male and multi-male anthropoids are about evenly split. Those contrasts involving 'pure' multi-male and single-male species are all positive for body-mass dimorphism. However, there are only five of these, with two involving multi-level species that might be considered multi-male (Dunbar, 2000). Those species showing variable mating systems exhibit neither significantly more nor less dimorphism than either single-male or multi-male species (Fig. 13.1, Table 13.2). This result holds for species values, and contrasts, and for both canine and mass dimorphism. Dimorphism in both body mass and canine size is strongly correlated with competition levels using species values (canine crown height; $r_s = 0.742$, $n = 84$, $p < 0.001$: body mass; $r_s = 0.741$, $n = 85$, $p < 0.001$). Phylogenetic contrasts (Table 13.3) corroborate this for body-mass dimorphism (12/3, $p < 0.05$), but not canine dimorphism (9/5, $p > 0.05$), though the latter is biased in the predicted direction.

Fig. 13.1   Box-and-whisker plots showing the relationship between maxillary canine crown height dimorphism and mating system (a), breeding system (b) and competition levels (c) among anthropoid primates. Mating system is divided into monogamous and polyandrous species (MP), multi-male, multi-female species (MM), and single-male, multi-female species (SM). *Theropithecus gelada*, *Papio hamadryas* and *Nasalis larvatus* are classified as multi-male species in this graph. Breeding system is divided into the same classes, except that these species are classed as single-male, while species showing variation among populations between multi-male and single-male units are separated into their own group (Var.). Competition levels are labelled 1 through 4.

Contrasts limited to only competition levels 2, 3 and 4 are biased in the predicted direction for mass dimorphism, but not statistically significant. Contrasts opposite to the prediction direction are limited to comparisons between competition levels 3 and 4, which involve the frequency effect (Plavcan *et al.*, 1995; Plavcan & van Schaik, 1997). All contrasts involving competition intensity are positive.

The OSR is correlated with body-size dimorphism for both species values and phylogenetic contrasts, but is not correlated with canine dimorphism in either analysis (Table 13.4).

Using the data provided in Table 13.1, SSR is significantly correlated with canine dimorphism for both species values and independent contrasts in anthropoids (Table 13.4). Excluding monogamous and polyandrous species, the relationship is no longer significant for species values for either canine or mass dimorphism. Using phylogenetic contrasts, however, there is a significant relationship for canine dimorphism, but not for body-mass dimorphism.

Restricting the analysis to multi-male and single-male species, there is a significant relationship between canine dimorphism and SSR in multi-male species, but not single-male species using phylogenetic contrast data (Table 13.4, Fig. 13.2). This result is partly contingent on the classification of *Theropithecus* and *Papio hamadryas*. If these species are classified as single-male, there is no significant relationship between SSR and canine dimorphism in multi-male species. However, this lack of significance is driven by a single clear statistical outlier, removal of which results in a very strong relationship ($n = 22$, $r = 0.57$, $p = 0.007$).

FEMALE CHOICE

Analyses of species values and contrasts (Table 13.5) support the hypothesis that where female choice reinforces male–male competition, dimorphism in body mass and canine size is exaggerated. In fact, *Mandrillus*, *Pongo*, *Gorilla* and *Theropithecus* are among the most dimorphic primates, while *C. apella* body-mass dimorphism is the second highest among platyrrhines. *C. apella* canine dimorphism is not greater than that of *Saimiri oerstedi*. However, this reflects large female canines in *C. apella*. Male relative male canine size is the greatest of all platyrrhines. A similar observation holds for the contrast of *P. hamadryas* canine dimorphism to other *Papio*. Finally, though not included formally because of uncertainty in female–male relationships, *Nasalis* is reported by Yeager (1990) to show a mating system similar to that

Table 13.2 *Results of independent contrast analyses for mating system in anthropoid primates.*

Contrast	m/p vs. s or m[a]		s vs. m		Notes
	Mass	Canine	Mass	Canine	
Matched-pairs comparisons					
Saguinus nigricollis (m)[2] vs. *S. fuscicollis* (m/p)	+				No canine data for *S. nigricollis*. Females larger than males in both species.
Callitrichids (m/p) vs. *Cebus/Saimiri* (m(v): m)	+	+			
Pithecia (m/p) vs. *Cacajao/Chiropotes* (m)	+	−			Canines of all species specialised for diet.
Callicebus (m/p) vs. Atelines (m)	+	+			
P. potenziani (m/p) vs. *Presbytis* (s: s(v))	+	+			
Hylobatids (m/p) vs. Great Apes (s: m)	+	+			
C. capucinus (s) vs. *C. albifrons* (m)			+		No canine data for *C. albifrons*.
P. hamadryas (m(s))[2] vs. *Papio* (m)			+	−	*P. hamadryas* mating system is single-male groups in multi-male bands, considered multi-male by Dunbar (2000).
T. gelada (m(s)) vs. *Lophocebus* (m)			+	+	*T. gelada* mating system is single-male groups in multi-male bands, considered multi-male by Dunbar (2000).
Miopithecus (m) vs. *Cercopithecus* (s: s(i))			+	+	
Gorilla (s) vs. *Pan* (m)			+	+	
Alouatta (m(v)) vs. *Alouatta* (m)			+	−	*Alouatta* is an unresolved polytomy.
C. nigrivittatus (m(v)) vs. *C. apella* (m)			−	+	
Nasalis (m(s))[2] vs. *Simias* (s)[2]			−	−	Mating system of *Nasalis* not well known. Comparable to *T. gelada* (Yeager, 1991).
P. melalophos/rubicunda (s(v))[2] vs. *P. comata* (s)			−	−	
S. entellus (m(v)) vs. *Kasi* (s)			−	+	
T. obscurus (s(v)) vs. *T. cristata/pileatus* (s)			+	+	Based on Smith & Cheverud (2002) phylogeny.
C. cephus (s) vs. *C. ascanius* (s(i))			+	+	Classification of *C. ascanius* reflects reports of multi-male influxes.
C. nictitans (s) vs. *C. mitis* (s(i))			−	−	Classification of *C. mitis* reflects reports of multi-male influxes.
C. aethiops (m) vs. *Erythrocebus* (s(i))			+	+	Based on Smith & Cheverud (2002) phylogeny. Classification of *Erythrocebus* reflects reports of multi-male influxes.
C. angolensis (m) vs. *C. guereza/polykomos* (m(v))			−	−	

(*cont.*)

Table 13.2 (*cont.*)

Contrast	m/p vs. s or m[a]		s vs. m		Notes
	Mass	Canine	Mass	Canine	
Other contrasts					
Theropithecus (m(s)) vs. *Papio* (m)			+	−	Contrast of *T. gelada* with *Papio* (not including *P. hamadryas*) as sister group.
Presbytis/Trachypithecus (s(v)) vs. *Presbytis/Trachypithecus* (s)			−	−	Based on uncertainty of relationships within presbytins.
S. entellus (m(v)) vs. *Presbytis* (s)			−	−	Based on uncertainty of relationships among presbytins.
Colobus/Procolobus (m: m(v)) vs. Presbytines (s: s(v))			+	−	Based on majority of *Colobus/Procolobus* being multi-male, majority of presbytines being single-male.
Cercopithecines (s: s(i)) vs. Papionines (m)			−	−	Excludes *Miopithecus* and *C. aethiops*.
Totals (+/−)					
Matched pairs	6/0	4/1	9/6	8/6	
All comparisons	6/0	4/1	11/9	8/11	

[a] Abbreviations as in Table 13.1.

of *Theropithecus*. If true, then the observation that *Nasalis* shows extreme canine dimorphism, male canine size, and body-mass dimorphism by comparison to *Simias* and all other colobines, would support the hypothesis.

MALE AND FEMALE CANINE SIZE

Males

Repeating the analyses of the previous section, relative male canine size co-varies with mating system, competition levels, SSR and OSR in much the same way that dimorphism does, though the relationships tend to be weaker (e.g. Table 13.4). Relative male canine size is less strongly associated with variation in competition levels – only the distinction between competition intensity is maintained for both species values and independent contrasts (Plavcan *et al.*, 1995). An important consideration in the analysis is the fact that male relative canine size is calculated with reference to male body mass. If male body mass and male canine size are both under sexual selection, then the correction for body mass may remove co-variation between male canine size and an estimate of sexual selection. Unfortunately, there seems to be no obvious way around this problem.

Females

Adjusted relative female canine size is strongly correlated with female group size using species values ($n = 57$, $r = 0.732$, $p < 0.001$). However, independent contrasts fail to confirm these results using all species. Limiting the analysis to high-intensity or low-intensity female competition species (consistent with the Janson & Goldsmith (1995) model), there is a strong correlation between female group size and relative female canine size ($n = 22$, $r = 0.675$, $p < 0.001$; Fig. 13.3), but only a weak, non-significant correlation in the low-intensity species ($n = 15$, $r = 0.434$, $p = 0.093$). These results are obtained for both species values and independent contrasts. This parallels the Janson and Goldsmith (1995) results.

These analyses raise the question of whether we can apply the results of the above analyses to observed levels of dimorphism using simple species values. Across anthropoids, relative female canine size varies almost as much as that of males (standard deviations are 0.244 and 0.309, respectively). However, as previously noted, relative female canine size is strongly correlated with that of males. Canine dimorphism is strongly correlated with relative male canine size ($n = 69$, $r = 0.615$, $p < 0.001$), but not relative female canine size

Table 13.3   *Results of independent contrast analyses for competition levels.*

Contrast	Mass	Canine	Notes
Matched pairs			
Leontopithecus (2) vs. *Callithrix* (1)	+	+	
Pithecia (1) vs. *Cacajao* (3)/*Chiropotes* (2)	+	−	Canines specialised in males and females.
Callicebus (1) vs. Atelines (2/3/4)	+	+	
P. potenziani (1) vs. *Presbytis* (3)	+	+	
Hylobates (1) vs. Great Apes (2/3)	+	+	
Callitrichids (1/2) vs. *Cebus*/*Saimiri* (3/4)	+	+	
Alouatta (4) vs. other Atelines (2)	+	+	
Pan (2) vs. *Gorilla* (3)	+	+	
Saimiri (3) vs. *Cebus apella* (4)	+	−	
Cebus capucinus (3) vs. *Cebus albifrons* (4)	−		No canine data.
Colobus guereza (3) vs. *Colobus polykomos* (4)	+	+	
Trachypithecus (3) vs. *Semnopithecus* (4)	+	−	
Simias (3) vs. *Nasalis* (4)	+	+	
Erythrocebus patas (3) vs. *Cercopithecus aethiops* (4)	−	−	
Cercopithecus (3) vs. *Miopithecus* (4)	−	−	
Other contrasts			
C. capucinus/*nigrivittatus* (3) vs. *C. apella*/*albifrons* (4)	+		No canine data.
Colobus guereza (3) vs. other *Colobus* (4)	+	+	
Semnopithecus (4) vs. other Presbytins (3)	+	+	
Presbytins (3) vs. *Colobus*/*Procolobus* (4)	−	+	
Cercopithecines (3) vs. Papionins (4)	+	+	Excludes *Miopithecus* and *C. aethiops*.
Totals			
Matched pairs	12/3	9/5	
All contrasts	16/4	13/5	

($n = 70$, $r = 0.201$, $p = 0.095$). Thus, canine dimorphism does not vary as a simple function of female canine size. But, restricting the analysis to the 'high-intensity' female competition species, canine dimorphism has a strong negative correlation with relative female canine size ($n = 26$, $r = 0.624$, $p < 0.001$). No such effect is apparent for the 'low-intensity' female competition species. This fits well with the above analyses of female canine size and group size. Finally, it is interesting to note that male relative canine size is positively correlated with female group size both across all anthropoids ($n = 54$, $r = 0.315$, $p = 0.02$, and within the high-intensity female competition classification ($n = 22$, $r = 0.440$, $p = 0.04$). Combing these results, in the high-intensity female competition group, canine dimorphism is reduced with decreasing female group size as a function of both a decrease in SSR (and probably sexual selection) and an increase in female canine size.

DISCUSSION

Comparative analyses since the 1970s have more than adequately demonstrated that dimorphism in primates, as in many other animals, is basically a consequence of sexual selection. All of the above analyses for both canine and body-mass dimorphism confirm this to some degree at least. Yet our understanding of dimorphism is far from complete. As noted by Pereira *et al.* (2000, p. 273), '[u]nderstanding of the relationships among mating system, dimorphism, and qualities of male–female relationship remains an area requiring

Table 13.4 *Analysis of species values and independent contrasts for the relationships between canine dimorphism, body-mass dimorphism, socionomic sex ratio, and operational sex ratio.*

Measurement	Species values			Contrasts		
	n	r	p	n	r	p
Socionomic sex ratio						
All anthropoids						
canine dimorphism	55	0.524	<0.001	54	0.368	0.006
mass dimorphism	73	0.374	0.001	72	0.216	0.068
male canine size[a]	54	0.371	0.006	54	0.092	0.506
Polygynous species						
canine dimorphism	44	0.160	0.298	43	0.378	0.012
mass dimorphism	56	0.042	0.760	55	0.140	0.308
male canine size[a]	43	0.384	0.011	43	0.142	0.358
Multi-male species[b]						
canine dimorphism	28	0.290	0.134	27	0.536	0.003
mass dimorphism	36	0.159	0.354	35	0.178	0.305
male canine size[a]	28	0.306	0.114	27	0.226	0.247
Single-male species[b]						
canine dimorphism	16	0.361	0.169	15	0.200	0.457
mass dimorphism	20	0.149	0.531	19	0.174	0.462
male canine size[a]	15	0.581	0.023	15	0.157	0.562
Operational sex ratio						
canine dimorphism	15	0.041	0.898	14	0.433	0.107
mass dimorphism	18	0.614	0.007	16	0.635	0.006
male canine size[a]	14	0.454	0.103	13	0.320	0.265

[a] Relative male maxillary canine crown height, estimated as a least-squares residual from an isometric line passed through the relationship with male body mass.
[b] *Theropithecus*, *Papio hamadryas* and *Nasalis larvatus* classified as multi-male breeding units for reported numbers. Results change for phylogenetic analysis of canine dimorphism in multi-male species – see text.

Table 13.5 *Independent contrast analyses evaluating female choice and dimorphism.*

Contrast	Body mass	Canines	Notes
Cebus apella vs. *Saimiri*	+	−	*C. apella* canine dimorphism reduced by large female canines. Male canine size is the largest of all platyrrhines.
Mandrillus vs. *Cercocebus*	+	+	
Papio hamadryas vs. *Papio*	+	−	*P. hamadryas* male canine size is larger than that of *P. cynocephalus*.
Theropithecus vs. *Lophocebus*	+	+	
Gorilla vs. *Pan*	+	+	
Pongo vs. *Pan*	+	+	Not a matched-pair contrast, but included for comparison.

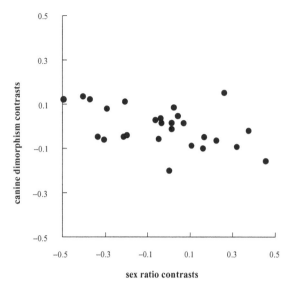

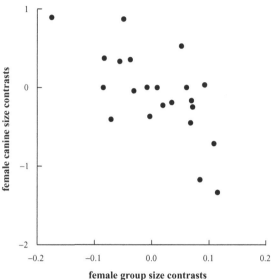

Fig. 13.2 Bivariate plot of independent contrasts between maxillary canine crown height dimorphism and sex ratio in multi-male anthropoid primates.

Fig. 13.3 Bivariate plot of independent contrasts in relative female maxillary canine crown height, adjusted for co-variation with relative male maxillary canine crown height, and female group size in species classified as 'high-intensity' female intrasexual competition (Plavcan *et al.*, 1995).

more research . . .'. From the viewpoint of comparative analysis, our understanding of the relationship between sexual selection and sexual dimorphism is constrained by our ability to measure sexual selection and our understanding of dimorphism as a joint function of variation in male and female traits.

The largest problem that we face is measuring the strength of sexual selection. Even though dimorphism is measured with error, it is trivial by comparison to the problems faced in deriving a comparable measure of sexual selection that can be applied to a wide variety of species. Certainly, male reproductive skew is related to patterns of male and female dispersal, and to observed patterns of male–male competition. But primates are complex social animals with long life histories. Hence, we should expect the intensity of male reproductive skew to be highly variable, even among species sharing broadly similar mating systems (van Noordwijk & van Schaik, this volume). In addition to male–male competition, reproductive skew is likely to be a function of various sex-specific life-history variables and reproductive tactics, including male coercion, female counter-tactics to such coercion, and female choice (Smuts & Smuts, 1993; van Schaik *et al.*, 1999; van Schaik *et al.*, 2000; van Schaik *et al.*, this volume).

The most interesting aspect of the previous analyses is not whether one measure is theoretically better than another,

or even better correlated with dimorphism. Rather, the interactions among the variables suggest something of the complexity in relationships between sexual selection, mating systems, male–male competition, female choice, and operational sex ratios. For example, comparisons of mating-system classifications and competition levels point to the variety in degrees of male–male competition among multi-male species, which in turn correlates with observed degrees of promiscuity, patterns of male scramble and contest competition for females, and ultimately male reproductive skew. The correlation between the OSR calculations and mass dimorphism clearly suggests again that male–male competition is tied to the ability of males to monopolise access to females on the basis of female life-history traits. But the broader correlation between the simple SSR and canine dimorphism in multi-male species also suggests that the negative allometric relationship between male and female group sizes is directly tied to male competition and male reproductive skew. Finally, the lack of correlation between canine dimorphism and SSR in single-male groups is consistent with the view that male reproductive skew is tied to male tenure and other factors in these species, and suggests that future studies of sexual selection and dimorphism should focus on explaining patterns of male lifetime reproductive skew as they relate to different mating systems. While these analyses hardly constitute a complete synthesis of the relationship between dimorphism and sexual selection, they nevertheless suggest that comparisons among various measures of sexual selection are far more productive than reliance upon one or the other alone in an analysis.

Quantifying female choice remains a major problem for understanding the evolution of dimorphism. Female choice is viewed as either a female counter-tactic to male coercion (usually through paternity confusion), or a phenomenon whereby females attempt to improve offspring survival by choosing males that confer good genes or some degree of male parental care (see Gangestad & Thornhill, this volume). In terms of dimorphism, however, female choice needs to be viewed in terms of its effect on male reproductive skew. In this sense, we can dichotomise female choice as either reinforcing male reproductive skew, or tempering it. The analysis presented above is only preliminary, but tentatively supports the hypothesis that female choice can have an important, if not major, impact on the evolution of sexual dimorphism, at least in some species. Again, this suggests that understanding dimorphism as a function of sexual selection is a matter of integrating information not just on male–male competition and the pattern of male demographics over a male's lifetime, but also female mating tactics. Female primates are not simply passive commodities fought over by males, and considerably more consideration needs to be given to the effect of female mating tactics on male reproductive skew.

The flip side of the coin is that sexual dimorphism cannot be understood except as a joint function of male and female trait values. This does not mean that dimorphism as a proportional difference between males and females is uninteresting. On the contrary, it is exactly what we are interested in explaining. However, if we are to understand the contribution of a single factor such as male–male competition to the evolution of dimorphism, then we need to break down the trait and study its component parts. The findings that female relative canine size co-varies with male canine size, and that female canine size co-varies with patterns of female contest competition and female group size, are both interesting in themselves, and essential in understanding how variation in dimorphism is a product of sexual selection. Studies of body-mass dimorphism have made the same point (Shea, 1986; Leigh, 1992, 1995; Leigh & Shea, 1995; Leigh & Terranova, 1998; Lindenfors, 2002b), leading to considerable advances in our understanding of this trait.

Importantly, canine dimorphism and body-mass dimorphism are not the same trait. Dimorphism is often spoken of as a single phenomenon, with interchangeable findings. But the above analyses underscore that results from one analysis do not necessarily apply to another. Canine dimorphism is more strongly associated with competition levels and SSR data, while body-mass dimorphism is more strongly associated with the OSR estimates. These findings might reflect different patterns of measurement error, and/or different constellations of factors that influence the evolution and expression of canine and body-mass dimorphism.

Measurement error in dimorphism cannot be ignored as an explanation for this difference. The most comprehensive listing of body-mass data is that of Smith and Jungers (1997). There is a temptation to use these data as absolute species-specific values for body mass. However, body-mass data are notoriously subject to uncontrollable errors, and many reports mix data from subspecies that themselves vary. For example, what is the true body-mass dimorphism of *Colobus badius*? Smith and Jungers list data for six subspecies that range in dimorphism between 2 per cent and 49 per cent (males larger than females). While one might simply average out these data, other studies suggest that population differences in body mass and dimorphism are biologically real (Albrecht & Miller, 1993; Turner *et al.*, 1994).

The canine data used here control for subspecific variation, and all were collected under the same protocol controlling for tooth wear. Furthermore, subspecific variation in canine dimorphism tends to be less than that of body-mass dimorphism. However, this raises the question of whether canine data are as responsive to immediate selective pressures as body-mass dimorphism might be. Unfortunately, this question cannot be answered except with long-term field data.

Regardless of measurement error, canine and body-mass dimorphism are clearly under the influence of different selective pressures and constraints, even though there is overwhelming evidence that both are a product of sexual selection (Leutenegger & Kelley, 1977; Plavcan & van Schaik, 1992, 1997; Ford, 1994; Leigh, 1995; Leigh & Shea, 1995; Plavcan, 1998; Lindenfors, 2002b). This point is underscored by simple observation that canine and body-mass data are only modestly correlated across species ($n = 79$, $r = 0.626$, $p < 0.001$). Rather than confounding analyses, this observation should actually provide further material for exploring hypotheses of the relationship between sexual selection and dimorphism in either trait. For example, why is the OSR calculation strongly correlated with body-mass dimorphism, but not canine dimorphism, while the opposite is true for the SSR data? Arguing that either canine or body-mass data are too error-prone seems unsupportable – such error is likely to yield a lack of correlations with the sample sizes used in these analyses. It seems likely that an understanding of why these traits differ in the pattern of correlations to different variables will provide further insight into the relationship between dimorphism in general, and sexual selection in particular.

SUMMARY AND CONCLUSIONS

That anthropoid sexual dimorphism is a function of sexual selection at some level seems not to be an issue. However, comparative analyses attempting to evaluate whether the magnitude of dimorphism corresponds to the intensity of sexual selection pressure are hindered by limitations in estimates of sexual selection, and the fact that dimorphism must be evaluated as both a relative difference between male and female traits and a consequence of separate factors affecting the expression of male and female traits. Future analyses should strive to understand the relationships between sexual selection, patterns of male and female life history and behaviour, mating systems and dimorphism. Lacking a direct measure of male reproductive skew, we cannot progress further than we have by relying on a single surrogate measure of sexual selection. Instead, we need to tease apart carefully relationships among different measures of sexual selection and dimorphism as a function of male and female traits.

ACKNOWLEDGEMENTS

I am sincerely grateful to Peter Kappeler for inviting me to present my views at the Göttinger Freilandtage (conference). Carel van Schaik, Robert Trivers, Timothy Clutton-Brock, Patrick Lindenfors, Charlie Nunn, Joseph Manson, Dan Rubenstein and Patty Gowaty provided stimulating conversation and penetrating questions. I thank the editors and two anonymous reviewers for comments. This work was supported by National Science Foundation (NSF) grants BNS 8814060 and SBR 9616671.

REFERENCES

Albrecht, G. H. & Miller, J. M. A. 1993. Geographic variation in primates: a review with implications for interpreting fossils. In *Species, Species Concepts, and Primate Evolution*, ed. W. H. Kimbel & L. B. Martin. New York, NY: Plenum Press, pp. 211–37.

Altmann, J. 2000. Models of outcome and process: predicting the number of males in primate groups. In: *Primate Males: Causes and Consequences of Variation in Group Composition*, ed. P. M. Kappeler. Cambridge: Cambridge University Press, pp. 236–47.

Andersson, M. 1994. *Sexual Selection*. Princeton, NJ: Princeton University Press.

Barton, R. A. 2000. Socioecology of baboons: the interaction of male and female strategies. In *Primate Males: Causes and Consequences of Variation in Group Composition*, ed. P. M. Kappeler. Cambridge: Cambridge University Press, pp. 97–107.

Borries, C. 2000. Male dispersal and mating season influxes in hanuman langurs living in multi-male groups. In *Primate Males: Causes and Consequences of Variation in Group Composition*, ed. P. M. Kappeler. Cambridge: Cambridge University Press, pp. 146–58.

Brennan, E. J. 1985. De Brazza's monkeys (*Cercopithecus neglectus*) in Kenya: census, distribution and conservation. *American Journal of Primatology*, **8**, 269–77.

Cheverud, J. M., Dow, M. & Leutenegger, W. 1986. The quantitative assessment of phylogenetic constraints in comparative analysis: sexual dimorphism in body weight among primates. *Evolution*, **38**, 1335–51.

Clutton-Brock, T. H. 1985. Size, sexual dimorphism, and polygyny in primates. In *Size and Scaling in Primate Biology*, ed. W. L. Jungers. New York, NY: Plenum Press, pp. 51–60.

Clutton-Brock, T. H., Harvey, P. H. & Rudder, B. 1977. Sexual dimorphism, socionomic sex ratio and body weight in primates. *Nature*, **269**, 797–800.

Cords, M. 1988. Mating systems of forest guenons: a preliminary review. In *A Primate Radiation: Evolutionary Biology of the African Guenons*, ed. A. Gautier-Hion, F. Bourliere, J.-P. Gautier & J. Kingdon. Cambridge: Cambridge University Press, pp. 323–39.

2000. The number of males in guenon groups. In *Primate Males: Causes and Consequences of Variation in Group Composition*, ed. P. M. Kappeler. Cambridge: Cambridge University Press, pp. 84–96.

Crockett, C. M. & Eisenberg, J. F. 1987. Howlers: variations in group size and demography. In *Primate Societies*, ed. B. B. Smuts, D. L. Cheney, R. M. Seyfarth, R. W. Wrangham & T. T. Struhsaker. Chicago, IL: University of Chicago Press, pp. 54–68.

Crook, J. H. 1972. Sexual selection, dimorphism, and social organization in primates. In *Sexual Selection and the Descent of Man 1871–1971*, ed. B. Campbell. Chicago, IL: Aldine, pp. 231–81.

Davies, N. B. 2000. Multi-male breeding groups in birds: ecological causes and social conflicts. In *Primate Males: Causes and Consequences of Variation in Group Composition*, ed. P. M. Kappeler. Cambridge: Cambridge University Press, pp. 11–20.

Delson, E., Terranova, C. J., Jungers, W. L. *et al.* 2000. Body mass in cercopithecidae (Primates, Mammalia): estimation and scaling in extinct and extant taxa. *American Museum of Natural History Anthropological Papers*, # 83.

DiFiore, A. & Rodman, P. S. 2001. Time allocation patterns of lowland woolly monkeys (*Lagothrix lagothricha poeppigii*) in a Neotropical Terra Firma forest. *International Journal of Primatology*, 22, 449–80.

Dunbar, R. I. M. 2000. Male mating strategies: a modeling approach. In *Primate Males: Causes and Consequences of Variation in Group Composition*, ed. P. M. Kappeler. Cambridge: Cambridge University Press, pp. 259–68.

Ely, J. & Kurland, J. A. 1989. Spatial autocorrelation, phylogenetic constraints, and the causes of sexual dimorphism in primates. *International Journal of Primatology*, 10, 151–71.

Emlen, S. T. & Oring, T. 1977. Ecology, sexual selection, and the evolution of mating systems. *Science*, 191, 215–33.

Felsenstein, J. 1985. Phylogenies and the comparative method. *American Naturalist*, 125, 1–15.

Ford, S. M. 1994. Evolution of sexual dimorphism in body weight in platyrrhines. *American Journal of Primatology*, 34, 221–4.

Garland, T. Jr., Dickerman, A. W., Janis, C. M. & Jones, J. A. 1993. Phylogenetic analysis of covariance by computer simulation. *Systematic Biology*, 42, 265–92.

Garland, T. Jr., Midford, P. E. & Ives, A. R. 1999. An introduction to phylogenetically based statistical methods, with a new method for confidence intervals on ancestral states. *American Zoologist*, 39, 374–88.

Gaulin, S. J. C. & Sailer, L. D. 1984. Sexual dimorphism in weight among primates: the relative impact of allometry and sexual selection. *International Journal of Primatology*, 5, 515–35.

Godfrey, L. R., Lyon, S. K. & Sutherland, M. R. 1993. Sexual dimorphism in large-bodied primates: the case of the subfossil lemurs. *American Journal of Physical Anthropology*, 90, 315–34.

Goldizen, A. W. 1987. Tamarins and marmosets: communal care of offspring. In *Primate Societies*, ed. B. B. Smuts, D. L. Cheney, R. M. Seyfarth, R. W. Wrangham & T. T. Struhsaker. Chicago, IL: University of Chicago Press, pp. 34–43.

Greenfield, L. O. 1992a. Origin of the human canine: a new solution to an old enigma. *Yearbook of Physical Anthropology*, 35, 153–85.

1992b. Relative canine size, behaviour, and diet in male ceboids. *Journal of Human Evolution*, 23, 469–80.

Harvey, P. H. & Harcourt, A. H. 1984. Sperm competition, testes size, and breeding system in primates. In *Sperm Competition and the Evolution of Animal Mating Systems*, ed. R. L. Smith. New York, NY: Academic Press, pp. 589–600.

Harvey, P. H., Kavanagh, M. & Clutton-Brock, T. H. 1978. Sexual dimorphism in primate teeth. *Journal of Zoology, London*, 186, 474–85.

Heymann, E. W. 2000. The number of adult males in callitrichine groups and its implications for callitrichine social evolution. In *Primate Males: Causes and Consequences of Variation in Group Composition*, ed. P. M. Kappeler. Cambridge: Cambridge University Press, pp. 64–71.

Janson, C. H. 1984. Female choice and mating system of the brown capuchin monkey *Cebus apella* (Primates: Cebidae). *Zeitschrift für Tierpsychologie*, 65, 177–200.

Janson, C. H. & Goldsmith, M. L. 1995. Predicting group size in primates: foraging costs and predation risks. *Behavioural Ecology*, 6, 326–36.

Jarman, P. J. 2000. Males in macropod society. In *Primate Males: Causes and Consequences of Variation in Group Composition*, ed. P. M. Kappeler. Cambridge: Cambridge University Press, pp. 21–33.

Kappeler, P. M. 1990. The evolution of sexual size dimorphism in prosimian primates. *American Journal of Primatology*, 21, 201–14.

Kappeler, P. M. & van Schaik, C. P. 2002. Evolution of primate social systems. *International Journal of Primatology*, 23, 707–40.

Kay, R. F., Plavcan, J. M., Glander, K. E. & Wright, P. C. 1988. Sexual selection and canine dimorphism in New World monkeys. *American Journal of Physical Anthropology*, 77, 385–97.

Lande, R. 1980. Sexual dimorphism, sexual selection, and adaptation in polygenic characters. *Evolution*, 33, 292–305.

Leigh, S. R. 1992. Patterns of variation in the ontogeny of primate body size dimorphism. *Journal of Human Evolution*, **23**, 27–50.

1995. Socioecology and the ontogeny of sexual size dimorphism in anthropoid primates. *American Journal of Physical Anthropology*, **97**, 339–56.

Leigh, S. R. & Shea, B. T. 1995. Ontogeny and the evolution of adult body size dimorphism in apes. *American Journal of Primatology*, **36**, 37–60.

Leigh, S. R. & Terranova, C. J. 1998. Comparative perspectives on bimaturism, ontogeny, and dimorphism in lemurid primates. *International Journal of Primatology*, **19**, 723–49.

Leutenegger, W. & Kelly, J. T. 1977. Relationship of sexual dimorphism in canine size and body size to social, behavioural and ecological correlates in anthropoid primates. *Primates*, **18**, 117–36.

Lindenfors, P. 2002a. Phylogenetic analysis of sexual size dimorphism. Doctoral thesis, University of Stockholm.

2002b. Sexually antagonistic selection on primate size. *Journal of Evolutionary Biology*, **15**, 595–607.

Lindenfors, P. & Tullberg, B. S. 1998. Phylogenetic analysis of primate size evolution: the consequences of sexual selection. *Biological Journal of the Linnean Society*, **64**, 413–47.

Maggioncalda, A. N., Sapolsky, R. M. & Czekala, N. M. 1999. Reproductive hormone profiles in captive male orangutans: implications for understanding developmental arrest. *American Journal of Physical Anthropology*, **109**, 19–32.

Martin, R. D., Willner, L. A. & Dettling, A. 1994. The evolution of sexual size dimorphism in primates. In *The Differences Between the Sexes*, ed. R. V. Short & E. Balaban. Cambridge: Cambridge University Press, pp. 159–200.

Melnick, D. J. & Pearl, M. C. 1987. Cercopithecines in multimale groups: genetic diversity and population structure. In *Primate Societies*, ed. B. B. Smuts, D. L. Cheney, R. M. Seyfarth, R. W. Wrangham & T. T. Struhsaker. Chicago, IL: University of Chicago Press, pp. 121–34.

Milton, K. 1985. Multimale mating and absence of canine tooth dimorphism in woolly spider monkeys (*Brachyteles arachnoides*). *American Journal of Physical Anthropology*, **68**, 519–23.

Mitani, J. C., Gros-Louis, J. & Richards, A. F. 1996a. Number of males in primate groups: comparative tests of competing hypotheses. *American Journal of Primatology*, **38**, 315–32.

1996b. Sexual dimorphism, the operational sex ratio, and the intensity of male competition in polygynous primates. *American Naturalist*, **147**, 966–80.

Mitani, J. C., Watts, D. P. & Muller, M. N. 2002. Recent developments in the study of wild chimpanzee behaviour. *Evolutionary Anthropology*, **11**, 9–25.

Napier, P. H. 1985. *Catalogue of Primates in the British Museum (Natural History) and Elsewhere in the British Isles. Part III: Family Cercopithecidae, Subfamily Colobinae*. London: British Museum (Natural History).

Nunn, C. L. 1999. The number of males in primates social groups: a comparative test of the socioecological model. *Behavioural Ecology and Sociobiology*, **46**, 1–13.

Nunn, C. L. & Barton, R. A. 2001. Comparative methods for studying primate adaptation and allometry. *Evolutionary Anthropology*, **10**, 81–98.

Paul, A. 2002. Sexual selection and mate choice. *International Journal of Primatology*, **23**, 877–903.

Pereira, M. A., Clutton-Brock, T. H. & Kappeler, P. M. 2000. Understanding male primates. In *Primate Males: Causes and Consequences of Variation in Group Composition*, ed. P. M. Kappeler. Cambridge: Cambridge University Press, pp. 271–7.

Perez-Barberia, F. J., Gordon, I. J. & Pagel, M. 2002. The origins of sexual dimorphism in body size in ungulates. *Evolution*, **56**, 1276–85.

Plavcan, J.M. 1990. Sexual dimorphism in the dentition of extant anthropoid primates. Ph.D. thesis, University Microfilms, Ann Arbor.

1998. Correlated response, competition, and female canine size in primates. *American Journal of Physical Anthropology*, **107**, 401–16.

1999. Mating systems, intrasexual competition and sexual dimorphism in primates. In *Comparative Primate Socioecology*, ed. P. C. Lee. Cambridge: Cambridge University Press, pp. 241–69.

2001. Sexual dimorphism in primate evolution. *Yearbook of Physical Anthropology*, **44**, 25–53.

Plavcan, J. M. & van Schaik, C. P. 1992. Intrasexual competition and canine dimorphism in anthropoid primates. *American Journal of Physical Anthropology*, **87**, 461–77.

1994. Canine dimorphism. *Evolutionary Anthropology*, **2**, 208–14.

1997. Intrasexual competition and body weight dimorphism in anthropoid primates. *American Journal of Physical Anthropology*, **103**, 37–68.

Plavcan, J. M., van Schaik, C. P & Kappeler, P. M. 1995. Competition, coalitions and canine size in primates. *Journal of Human Evolution*, **28**, 245–76.

Purvis, A. & Webster, A. J. 1999. Phylogenetically independent comparisons and primate phylogeny. In *Comparative Primate Socioecology*, ed. P. C. Lee. Cambridge: Cambridge University Press, pp. 44–70.

Robinson, J. G. & Janson, C. H. 1987. Capuchins, squirrel monkeys, and atelines: socioecological convergence with Old World primates. In *Primate Societies*, ed. B. B. Smuts, D. L. Cheney, R. M. Seyfarth, R. W. Wrangham & T. T. Struhsaker. Chicago, IL: University of Chicago Press, pp. 69–82.

Robinson, J. G., Wright, P. C. & Kinzey, W. G. 1987. Monogamous cebids and their relatives: intergroup calls and spacing. In *Primate Societies*, ed. B. B. Smuts, D. L. Cheney, R. M. Seyfarth, R. W. Wrangham & T. T. Struhsaker. Chicago, IL: University of Chicago Press, pp. 44–53.

Rodman, P. S. & Mitani, J. C. 1987. Orangutans: sexual dimorphism in a solitary species. In *Primate Societies*, ed. B. B. Smuts, D. L. Cheney, R. M. Seyfarth, R. W. Wrangham & T. T. Struhsaker. Chicago, IL: University of Chicago Press, pp. 146–54.

Rogers, M. E., Abernethy, K. A., Fontaine, B. *et al.* 1996. Ten days in the life of a mandrill horde in the Lope Reserve, Gabon. *American Journal of Primatology*, **40**, 297–313.

Rowell, T. 1988. The social system of guenons, compared with baboons, macaques and mangabeys. In *A Primate Radiation: Evolutionary Biology of the African Guenons*, ed. A. Gautier-Hion, F. Bourliere, J.-P. Gautier & J. Kingdon. Cambridge: Cambridge University Press, pp. 439–51.

Setchell, J. M. & Dixson, A. F. 2001a. Arrested development of secondary sexual adornments in subordinate adult male mandrills (*Mandrillus sphinx*). *American Journal of Physical Anthropology*, **115**, 245–52.

2001b. Changes in the secondary sexual adornments of male mandrills (*Mandrillus sphinx*) are associated with gain and loss of alpha status. *Hormones and Behaviour*, **39**, 177–84.

Setchell, J. M., Lee, P. C., Wickings, E. J. & Dixson, A. F. 2001. Growth and ontogeny of sexual size dimorphism in the mandrill (*Mandrillus sphinx*). *American Journal of Physical Anthropology*, **115**, 349–60.

Small, M. F. 1989. Female choice in nonhuman primates. *Yearbook of Physical Anthropology*, **32**, 103–27.

Smith, R. J. & Cheverud, J. M. 2002. Scaling of sexual dimorphism in body mass: a phylogenetic analysis of Rensch's Rule in primates. *International Journal of Primatology*, **23**, 1095–135.

Smith, R. J. & Jungers, W. L. 1997. Body mass in comparative primatology. *Journal of Human Evolution*, **32**, 523–59.

Smuts, B. B. & Smuts, R. W. 1993. Male aggression and sexual coercion of females in nonhuman primates and other mammals: evidence and theoretical implications. *Advances in the Study of Behavior*, **22**, 1–63.

Sommer, V. & Reichard, U. 2000. Rethinking monogamy: the gibbon case. In *Primate Males: Causes and Consequences of Variation in Group Composition*, ed. P. M. Kappeler. Cambridge: Cambridge University Press, pp. 159–68.

Steenbeck, R. 2000. Infanticide by males and female choice in wild Thomas's langurs. In *Infanticide by Males and Its Implications*, ed. C. P. van Schaik & C. H. Janson. Cambridge: Cambridge University Press, pp. 153–77.

Steenbeck, R., Sterck, E. H. M., de Vries, H. & van Hooff, J. A. R. A. M. 2000. Costs and benefits of the one-male, age-graded, and all-male phases in wild Thomas's langur groups. In *Primate Males: Causes and Consequences of Variation in Group Composition*, ed. P. M. Kappeler. Cambridge: Cambridge University Press, pp. 130–45.

Sterck, E. H. M. & van Hooff, J. A. R. A. M. 2000. The number of males in langur groups: monopolizability of females or demographic processes? In *Primate Males: Causes and Consequences of Variation in Group Composition*, ed. P. M. Kappeler. Cambridge: Cambridge University Press, pp. 120–9.

Strier, K. B. 1992. Atelinae adaptations: behavioral strategies and ecological constraints. *American Journal of Physical Anthropology*, **88**, 515–24.

Strier, K. B. 2000. From binding brotherhoods to short-term sovereignty: the dilemma of male Cebidae. In *Primate Males: Causes and Consequences of Variation in Group Composition*, ed. P. M. Kappeler. Cambridge: Cambridge University Press, pp. 72–83.

Struhsaker, T. T. 2000. Variation in adult sex ratios of red colobus monkey social groups: implications for interspecific comparisons. In *Primate Males: Causes and Consequences of Variation in Group Composition*, ed. P. M. Kappeler. Cambridge: Cambridge University Press, pp. 108–19.

Turner, T. R., Anapol, F. & Jolly, C. J. 1994. Body weights of adult vervet monkeys (*Cercopithecus aethiops*) at four sites in Kenya. *Folia Primatologica*, **63**, 177–9.

van Hooff, J. A. R. A. M. 2000. Relationships among non-human primate males: causes and consequences of variation in group composition: a deductive framework. In *Primate Males: Causes and Consequences of Variation in Group Composition*, ed. P. M. Kappeler. Cambridge: Cambridge University Press, pp. 183–91.

van Schaik, C. P. & van Hooff, J. A. R. A. M. 1996. Toward an understanding of the orangutan's social system. In *Great Ape Societies*, ed. W. C. McGrew, L. F. Marchant & T. Nishida. Cambridge: Cambridge University Press, pp. 3–15.

van Schaik, C. P. & Kappeler, P. M. 1996. The social systems of gregarious lemurs: lack of convergence with anthropoids due to evolutionary disequilibrium? *Ethology*, 102, 915–41.

van Schaik, C. P., Hodges J. K. & Nunn, C. L. 2000. Paternity confusion and the ovarian cycles of female primates. In *Infanticide in Males and Its Implications*, ed. C. P. van Schaik & C. H. Janson. Cambridge: Cambridge University Press, pp. 361–87.

van Schaik, C. P, van Noordwijk, M. A. & Nunn, C. L. 1999. Sex and social evolution in primates. In *Comparative Primate Socioecology*, ed. P. C. Lee. Cambridge: Cambridge University Press, pp. 204–31.

Watanabe, K. 1981. Variation in group composition and population density of the two sympatric Mentawian leaf-monkeys. *Primates*, 22, 145–60.

Watts, D. P. 1996. Comparative socio-ecology of gorillas. In *Great Ape Societies*, ed. W. C. McGrew, L. F. Marchant & T. Nishida. Cambridge: Cambridge University Press, pp. 16–28.

Weckerly, F. W. 1998. Sexual size dimorphism: influences of mass and mating systems in the most dimorphic mammals. *Journal of Mammology*, 79, 33–52.

Wrangham, R. W. 2000. Why are male chimpanzees more gregarious than mothers? A scramble competition hypothesis. In *Primate Males: Causes and Consequences of Variation in Group Composition*, ed. P. M. Kappeler. Cambridge: Cambridge University Press, pp. 248–58.

Yeager, C. P. 1990. Proboscis monkey (*Nasalis larvatus*) social organization: group structure. *American Journal of Primatology*, 20, 95–106.

14 • Sex ratios in primate groups

JOAN B. SILK
Department of Anthropology
University of California
Los Angeles, CA, USA

GILLIAN R. BROWN
Department of Zoology
University of Cambridge
Cambridge, UK

*I formerly thought that when a tendency to produce the two
sexes in equal numbers was advantageous to the species, it
would follow from natural selection, but I now see the whole
problem is so intricate that it is safer to leave its solution for
the future.*

Charles Darwin, 1871

INTRODUCTION

Sexual selection is an important evolutionary force in mammalian species because of one simple fact – males are a glut on the market. From the females' point of view, there are more males than needed to meet their reproductive requirements. And from the males' point of view, there are not nearly enough females to go around to satisfy their reproductive potential. The relative abundance of males generates strong intrasexual competition among them.

The goals of this chapter are to explore the selective factors that influence the evolution of birth sex ratios, and to weigh the empirical evidence that primate females facultatively manipulate birth sex ratios to enhance their own fitness. We will begin by briefly enumerating some of the ways in which adult sex ratios influence the evolution of male and female life histories, morphology and reproductive strategies in primate groups. Then, we will explain how natural selection shapes the evolution of birth sex ratios, and consider the empirical evidence for adaptive manipulation of birth sex ratios in primate groups.

OPERATIONAL SEX RATIOS AND SELECTION IN RELATION TO SEX

The nature and intensity of selection pressures on males and females are related to the 'operational sex ratio': that is, the number of adult males and adult females in the social group that are potentially able to reproduce (Emlen & Oring,

1977). Because female mammals bear the costs of gestation and lactation, the number of females that are able to conceive at any given time is generally lower than the number of males that are available to inseminate them. This can lead to intense competition among males over access to fertile females. Below, we will very briefly consider some of the consequences that the operational sex ratio can have for the life-history strategies, morphology and reproductive tactics of primates. Nearly all of these topics are considered at greater length by other authors in this volume. They are briefly discussed here as a prologue to the main subject of this chapter: that is, how evolution shapes primate birth sex ratios.

In primates, age-specific mortality is generally higher for males than females, and males have shorter life expectancies than females (Fedigan & Zohar, 1997). Sex differences in mortality are likely to be the product of sexual selection that favours high-risk, high-gain strategies among males (Trivers, 1985). Primatologists have debated the pattern and source of sex differences in mortality among immature primates. Selective pressures favouring rapid growth and large body size may make immature males more vulnerable to nutritional stress and more susceptible to disease than immature females. Thus, young males may be more 'fragile' than females. Van Schaik and de Visser (1990) questioned this idea, citing evidence that young females are particularly vulnerable to harassment in social groups, especially in matrilineal species (Silk, 1983). However, sex differences in mortality become male-biased by adolescence (reviewed by Debyser, 1995a–c). This may be related to the fact that adolescent males in many primate species engage in a variety of high-risk behaviours, including emigration to new groups (Pusey & Packer, 1987). Dispersing males run greater risks of being attacked by predators or harassed by rivals when they attempt to join or take over new groups, and males in many species compete for high-ranking positions in multi-male social groups (Alberts & Altmann, 1995). Males sometimes

Sexual Selection in Primates: New and Comparative Perspectives, ed. Peter M. Kappeler and Carel P. van Schaik. Published by Cambridge University Press. © Cambridge University Press 2004.

suffer serious, even lethal, injuries in agonistic contests with other males (Smuts, 1987; Drews, 1996).

Competition for access to mates may also lead to selection for physical characteristics that enhance success in competitive encounters with other males and increase males' attractiveness to females. The extent of sexual dimorphism in body size and canine length, and the relative size of male testes, are generally related to adult sex ratios and social organisation in anthropoid primates (Clutton Brock *et al.*, 1977; Harcourt *et al.*, 1981; Harvey & Harcourt, 1984; Mitani *et al.*, 1996b; Plavcan, this volume). Sexual dimorphism in body size and canine size is most pronounced in species in which adult sex ratios are most skewed toward females. Relative testes size is influenced, on the other hand, by the type of competitive regime that males face. In species that form one-male groups, males have relatively small testes, relying on their physical ability to defend access to groups of females. In many species that form polyandrous and multi-male groups, males have relatively large testes for their body size. In these cases, sperm competition plays a more prominent role in gaining fertilisations.

Finally, operational sex ratios are linked to male reproductive tactics. When females mainly mate with only one partner, males are likely to profit more from investing in their mates' offspring than from trying to obtain additional matings. In some sexually monogamous species, this leads males to become helpful fathers and reliable defenders of their territories, even though it does not necessarily guarantee fidelity (Palombit, 1994; Sommer & Reichard, 2000). In species that live in uni-male–multi-female groups, males compete vigorously for residence in female groups. However, non-resident males sometimes mate with receptive females, and males attempt to form new groups by luring females away from other males. In multi-male groups, males compete for positions of power within their groups. Genetic data from several free-ranging groups indicate that high-ranking males dominate paternity in these sorts of groups (e.g. de Ruiter *et al.*, 1992; Altmann *et al.*, 1996; Launhardt *et al.*, 2001; van Noordwijk & van Schaik, this volume).

Intense competition among males over access to females shapes male reproductive tactics in several other important ways. In many primate species, males commit infanticide when they enter new groups and attain dominant positions (van Schaik & Janson, 2000). Infanticide is adaptive because it directly reduces the fitness of rival males, and hastens resumption of female cycling, thus increasing the infanticidal males' likelihood of inseminating females. Although controversy has dogged this hypothesis (Sommer, 2000), there is now substantial evidence that male infanticide is a sexually selected strategy and occurs in all major taxa in the primate order (van Schaik & Janson, 2000). The threat of infanticide may have shaped female reproductive tactics as well. Females may solicit matings from multiple males to confuse males about paternity and thereby reduce males' likelihood of harming their infants (Hrdy & Whitten, 1987; Soltis & McElreath, 2001; van Schaik *et al.*, this volume). Moreover, in some species, the threat of infanticide seems to have favoured the evolution of close ties between males and mothers of newborn infants (Palombit *et al.*, 1997; Palombit, 2000).

This brief account, which is elaborated on by other contributors to this volume (see Setchell & Lee, this volume), makes it clear that the relative number of males and females profoundly influence the evolution of male and female life histories, morphologies and reproductive strategies in primate groups. Thus, it is important to understand the selective forces that shape the number of males and females in natural populations. Here, we consider how natural selection may have shaped the evolution of birth sex ratios and investment strategies of parents in male and female progeny.

THE EVOLUTION OF BIRTH SEX RATIOS

When individual male and female offspring cost the same amount to produce, natural selection is expected to maintain equal numbers of males and females at birth because the mean fitness of males and females must be the same at equilibrium. This idea was first introduced by R. A. Fisher in 1930. He pointed out that when one sex is less common than the other, the rare sex will have higher mean fitness than the more common sex. This will favour genes that cause parents to bias their birth sex ratio toward the rare sex, and consequently the rare sex will become more common. Any tendency to deviate from an even sex ratio will be countered by selection to balance the sex ratio, as long as the costs of rearing male and female offspring are the same. Subsequent theoretical analyses of this problem confirmed Fisher's insight (Maynard Smith, 1978; Charnov, 1979). Where males and females cost different amounts to rear, the total investment provided to male offspring will equal that provided to female offspring at the end of the period of parental investment. Thus, if one sex is more expensive to rear than the other, the birth sex ratio will be biased toward the cheaper sex, but the average member of the more expensive sex will receive more investment than the average member of the cheaper sex by the end of the period of parental investment.

Extensive analyses of birth sex ratios in outcrossed vertebrates indicate that nature typically conforms to Fisher's rule (Williams, 1979; Clutton-Brock, 1986). For example, despite strong economic incentives to manipulate sex ratios of domestic livestock, little success has been achieved (Rorie, 1999). This is apparently not because of constraints imposed by chromosomal sex determination, so the sex ratios that we see in nature are expected to reflect adaptive processes (West & Sheldon, 2002).

POPULATION-LEVEL AND SPECIES-LEVEL BIRTH SEX RATIOS

Not every species conforms to Fisher's principle; in some species birth sex ratios persistently deviate from unity. For example, fig wasps and mites routinely produce more females than males, and mothers invest more in daughters than sons (Hamilton, 1979; Wilson & Colwell, 1981). This occurs because the evolution of sex ratios can be influenced by the population structure (Hamilton, 1967, 1979; Bulmer & Taylor, 1980; Wilson & Colwell, 1981), a factor that Fisher did not consider. In structured populations, individuals are organised into groups, such as kin groups or 'trait groups' (Wilson, 1977, 1980), and these groupings alter the outcome of selective forces in important ways.

There are two ways of incorporating population structure into evolutionary models: the kin-selection (or inclusive-fitness) approach and the group-selection approach (Hamilton, 1975; Wade, 1978; Uyenoyama & Feldman, 1980). These methods are mathematically equivalent, and produce identical results. Both approaches have been used to model the evolution of skewed sex ratios in animal populations.

It is important to understand how intrademic group selection works, in order to see how selection may favour population-level birth sex-ratio biasing. Intrademic group selection can occur if groups vary in their ability to survive and to reproduce, and if variation is heritable. Then, group selection can increase the frequency of genes that increase group survival and reproductive success. The strength of selection among groups depends on the amount of genetic variation among groups, just as the strength of selection among individuals depends on the amount of genetic variation among individuals. In general, the amount of variation among groups is much smaller than the amount of variation

among individuals, unless groups are very small and there is very little mixing between them. One factor that increases the relative strength of group selection is kinship; when groups are composed of kin, this reduces the amount of genetic variation within groups (because kin are genetically similar) and increases the amount of genetic variation between groups.

Selection acting within groups always favours balanced sex ratios for the reasons that Fisher articulated. But selection acting among groups can favour skewed sex ratios under certain conditions. Biologists have identified several processes that could generate skewed sex ratios. Here we consider two of those: local mate competition and local resource competition.

Local mate competition

Using the inclusive-fitness approach, Hamilton (1967) demonstrated that female-biased sex ratios can evolve when siblings compete for matings among themselves. In such cases, local mate competition exists, and the competitors are all closely related. Hamilton showed that when local mate competition occurs, females benefit by producing just enough male offspring to fertilise each of their female offspring.

To see how this works, consider the case of fig wasps. A single female fig wasp lays all her eggs in a single fig. Her progeny mature and mate with their siblings before they emerge from the fig. After mating, males die and their sisters disperse to lay their own eggs in other figs. Since one male can inseminate all of his sisters, mothers who produce more daughters than sons produce more grandchildren than mothers who produce equal numbers of males and females.

The tension between selection acting within groups and selection acting among groups becomes evident when we consider what happens when multiple females lay their eggs in the same fig. In these circumstances, males also compete with unrelated males for mating opportunities, and this favours members of the rare sex. Under these conditions, genes favouring more even sex ratios will be favoured by selection acting within groups. Consider, for example, what happens when two females lay eggs in the same fig. Suppose that one female produces one male and nine females, while the other female produces five males and five females. In this situation, the mean fitness of sons will be higher than the mean fitness of daughters, and this will favour the genes that cause sex ratios to be more balanced. As the number of female fig wasps that lay eggs in a single fig increases, the more even sex ratios are expected to become (Hamilton, 1967). This prediction has been confirmed in studies of fig wasps in nature (West *et al.*, 2000).

Local resource competition

In most mammalian species, males are the dispersing sex (Greenwood, 1980), while females remain in their natal groups or near the place they were born. Clark (1978) pointed out that when resources are limited locally, related females will compete with one another for access to resources, while dispersing males will compete primarily with unrelated individuals in non-natal groups. Under these circumstances, the extent of local competition would be reduced if females uniformly skewed the sex ratio of their offspring in favour of the dispersing sex. Clark called this process *local resource competition*. Local resource competition and local mate competition are clearly very similar processes and rely on very similar logic.

Using a group-selection model, Silk (1984) showed that local resource competition can favour the evolution of male-biased secondary sex ratios when there is local resource competition acting on females. When the size of local groups is limited by density-dependent processes, groups that produce the largest number of dispersing males contribute the largest numbers of alleles to the global population. Selection acting among groups favours male-biased sex ratios, and the strength of this effect depends on the size of the local groups and the extent of the sex-ratio skew. Skewed sex ratios are more likely to be maintained in small groups than in large ones, and weak skews are maintained at higher equilibrium frequencies than strong skews.

Local resource competition is expected to favour male-biased sex ratios in species with female philopatry, when females compete directly with their daughters and close female kin for access to locally limited resources (Clark, 1978). Clark originally tested her hypothesis with data from *Galago crassicaudatus*, which occupy small, overlapping home-ranges. Clark's fieldwork in South Africa suggested that female galagos settle in home-ranges adjacent to their mothers, while males disperse over greater distances. She found a pronounced male bias in a small sample of births in the wild, and in larger samples of births in zoos and specimens in museum collections. Later, Johnson (1988) found that the extent of local resource competition, measured in terms of the number of immatures per adult female in social groups, was positively related to birth sex ratios in seven primate genera that are characterised by female philopatry.

In these genera, birth sex ratios increase as local resource competition becomes more intense.

Clark's model predicts that sex-ratio biases should be related to species-specific dispersal patterns. However, Johnson (1988) argues that this hypothesis cannot be tested by simply comparing dispersal patterns and birth sex ratios. He wrote, 'When mothers and daughters share home-ranges that do not overlap with other females, all direct competition will involve close kin and selection for male-biased birth sex ratios will be strong. But as the number of females using overlapping home-ranges increases . . . the effects of competition between closely related females will be outweighed by competition from females of different matrilines or groups, and selection for male-biased sex ratios will weaken.' When the extent of home-range overlap is held constant, genera with female philopatry are expected to have higher (more male-biased) birth sex ratios than other genera. This prediction is supported in a sample of 16 primate genera (Johnson, 1988).

Johnson's analyses suggest that local resource competition may influence birth sex ratios in primate groups. However, his data on sex ratios are based on aggregate data from a number of different zoo populations, while information about local resource competition and home-range overlap necessarily comes from studies of free-ranging populations. He argues that sex-ratio data from captive and wild populations tend to be very similar in the few cases, so that it is possible to compare them directly. His analysis relies on the assumption that sex ratios are not facultatively adjusted in relation to local conditions, but reflect species-specific (or genus-specific) adaptations to prevailing ecological conditions.

BIRTH SEX RATIOS WITHIN POPULATIONS

Fisher's rule does not necessarily mean that all females must produce the same number of sons and daughters or invest equally in males and females. Trivers and Willard (1973) recognised this fact and suggested that, under some conditions, selection might favour the ability to adjust progeny sex ratios in relation to the parent's ability to invest in their offspring. Their verbal model is based on several assumptions:

(1) Maternal condition during the period of parental investment is correlated with offspring condition at the end of this period.

(2) Litter size is held constant, so mothers in good condition cannot increase their reproductive output by increasing litter size.

(3) Differences in the condition of offspring at the end of the period of parental investment are maintained into adulthood.

(4) Early development has a greater impact on male reproductive success than on female reproductive success.

Thus, they hypothesised that natural selection will favour genes that facultatively adjust progeny sex ratio in relation to maternal condition, and they predicted that 'parents in better condition would be expected to show a bias toward male offspring' (Trivers & Willard, 1973).

Trivers and Willards' formulation emphasised differences in the variance in fitness among males and females, and this has caused a certain amount of confusion in the literature. But it is not really the variance in reproductive success among males and females that matters. What matters is the effect of a given unit of maternal investment on offspring fitness (Clutton-Brock & Albon, 1982). Thus, if females gain a higher return from investment in one sex than the other, then females with additional resources to invest in their offspring are expected to skew their birth sex ratio towards the sex with the highest rate of return, and provide additional resources to individual offspring of this sex. Females who have few resources to invest in their offspring are expected to skew their birth sex ratio toward the sex that gains little from additional investment. Trivers and Willard suggested that the secondary sex ratio of a population might be even, but the total investment in sons and daughters might be unbalanced, as parents in good condition would provide more care to their male offspring.

Trivers and Willard (1973) clearly expected that the process they outlined would lead females in good physical condition to skew their offspring sex ratio toward males, and females in poor condition to skew their offspring sex ratio toward females. This is based on the fact that females are usually a limiting resource for males, and intrasexual competition favours larger, heavier and more powerful males. In some species, such as red deer (*Cervus elaphus*), maternal condition is positively correlated with male weight at weaning, and male weight at weaning is correlated with male weight in adulthood (Clutton-Brock *et al.*, 1986). As predicted by Trivers and Willard's model, high-ranking female red deer skew their birth sex ratio toward male offspring, while low-ranking females bias their birth sex ratio toward female offspring (Clutton-Brock *et al.*, 1984).

Although Trivers and Willard expected females in good condition to skew offspring sex ratios toward males, formal analyses of the evolution of the relationship between maternal condition and birth sex ratios suggest that this prediction may not always hold. Leimar (1996) constructed a life-history model to evaluate the relationship between maternal rank and offspring sex. In this model, mothers in good condition produce healthier and more successful offspring, and offspring condition is correlated with maternal condition. Leimar found that females in good condition sometimes achieve more reproductive benefits through their daughters than through their sons. This surprising result is owing to the fact that daughters inherit their mothers' condition, and daughters of females in good condition therefore reproduce successfully themselves. But sons of mothers in good condition mate randomly with available females. Thus, when maternal condition is transmitted to offspring, females in good condition will often benefit more from investing in daughters than investing in sons, even if their sons reproduce more successfully than their daughters.

Testing for maternal-condition effects

Trivers and Willard's model was first applied to non-human primates to explain the observation that high-ranking female baboons and macaques produced more females than males, while low-ranking females produced more males than females (Silk *et al.*, 1981; Simpson & Simpson, 1982; Silk, 1983; Altmann *et al.*, 1988). Although this was the inverse of the pattern predicted by Trivers and Willard, some researchers thought that the basic logic of the model fitted the primate data. Their reasoning was based on the following logic.

In many Old World primate species, females acquire their mothers' dominance ranks, and matrilineal dominance hierarchies are formed (reviewed by Chapais, 1992). Matrilineal dominance hierarchies are often stable across decades, and across generations (e.g. Bramblett *et al.*, 1982; Hausfater *et al.*, 1982; Samuels *et al.*, 1987; Silk, 1988; Chapais, 1992; Packer *et al.*, 2000). Moreover, in a number of the same species, dominance is positively correlated with access to resources and with female reproductive success (Harcourt, 1987; Silk, 1987; Silk *et al.*, 1993; van Noordwijk & van Schaik, 1999). This means that high-ranking females are likely to have high-ranking, reproductively successful daughters, while low-ranking females are likely to have low-ranking, reproductively unsuccessful daughters. The prospects for sons, on the other hand, may not be strongly related to maternal rank because males typically leave their natal groups at puberty and attempt to join new groups. According to this reasoning, a given unit of investment by high-ranking females would have a greater effect on daughters' fitness than sons' fitness, while the opposite would be true for low-ranking females. This would explain why high-ranking females skewed their sex ratio toward females, and low-ranking females skewed their sex ratio toward males. This logic is clearly quite similar to the explanation underlying the logic of Leimar's (1996) model published a decade later, and Leimar cited these data as examples of sex-ratio patterns that fit the predictions of his model.

However, other researchers observed different patterns of birth sex ratios and offered different adaptive explanations for their findings. Thus, Meikle and his colleagues (1984) reported that high-ranking female rhesus macaques produced more sons than daughters, while low-ranking females produced more daughters than sons. Meikle et al. (1984) argued that the advantages of high rank had a greater impact on sons' fitness than on daughters' fitness, because the sons of high-ranking females tended to acquire high-ranking positions in non-natal groups. Thus, they predicted that primates would conform to the same patterns that were expected in red deer and other polygynous mammals.

It was possible to come to very different conclusions about the expected effect of maternal rank on offspring sex because very little was known about male life histories and reproductive careers. In most cases, primatologists did not know what became of males after they left their natal groups, and there was considerable dispute about the relationship between male dominance rank and reproductive success (Cowlishaw & Dunbar, 1991). Therefore, the effects of maternal investment on the future fitness of sons and daughters have been difficult to compare. Although we now know much more about males' life histories and reproductive success than we did in the mid 1980s, there is still no consensus about the effect of maternal rank on sons' dominance rank and reproductive success (e.g. Paul et al., 1992; Bercovitch et al., 2000; Packer et al., 2000; van Noordwijk & van Schaik, 2001). In the wake of these provocative empirical findings, a number of other researchers published analyses of the relationship between maternal rank and offspring sex ratio (reviewed by Brown, 2001; Bercovitch, 2002; Brown & Silk, 2002). However, no clear pattern emerged. For example, Berman (1988) found that middle-ranking rhesus macaque females produced a surplus of females, while high- and low-ranking females produced equal numbers of sons and daughters; and Small and Hrdy (1986) found no effect of maternal

rank on offspring sex among rhesus macaques. The situation became progressively more confusing as more studies were published.

Some investigators suspected that the inconsistencies among these studies were the result of stochastic variation in small samples (Rawlins & Kessler, 1986; Altmann et al., 1988; Rhine et al., 1992; Watanabe et al., 1992; Rhine, 1994). However, Clutton-Brock (1991) pointed out that we might expect to see variation in the patterning of sex ratios because the effects of parental investment on the fitness of sons and daughters vary across sites and species. Many researchers continued to offer functional explanations of observed patterns of sex-ratio variation (e.g. Paul & Kuester, 1990; Wasser & Norton, 1993; Dittus, 1998).

Meta-analysis of maternal condition effects on birth sex ratios

Trivers and Willard's model, and Leimar's reformulation, both predict that there will be a uniform effect of maternal condition on birth sex ratios, but the literature contains conflicting claims about the nature and magnitude of such patterns. Meta-analysis provides a means to evaluate the full spectrum of published data. We have conducted a meta-analysis of published studies of the relationship between maternal rank and offspring sex (Brown & Silk, 2002).

In our analysis, we followed procedures outlined by Palmer (2000), who advocates the use of funnel graphs to assess the relationship between effect size and sample size in meta-analyses. Funnel graphs are scatterplots in which effect size is plotted against sample size. At small sample sizes, effect sizes are expected to be quite variable but, as sample sizes increase, effect sizes are expected to converge towards a specific value. If strictly random processes are operating, then as sample size increases, effect sizes will converge toward zero.

By searching the literature, we were able to locate 35 datasets representing 15 species, and 8 genera that contained information about maternal rank, birth sex ratios and sample sizes. The majority of these studies were conducted on species of the genus *Macaca* ($n = 24$) and *Papio* ($n = 5$). Fifteen datasets were drawn from studies of wild populations, and 20 were based on captive and semi-free-ranging populations (further details in Brown & Silk, 2002).

We computed the difference between the proportion of males produced by high- and low-ranking females and plotted this against the total number of births recorded (Fig. 14.1). It is clear from Figure 14.1 that effect sizes

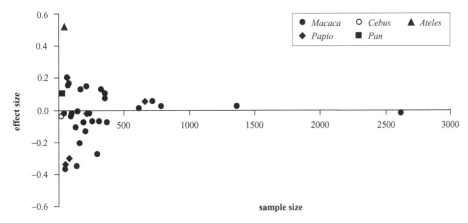

Fig. 14.1   Difference in sex ratio among progeny of high- and low-ranking females. The effect size is equal to the proportion of male offspring produced by high-ranking females minus the proportion of male offspring produced by low-ranking females. When the effect size is 0, high- and low-ranking females produce equal proportions of male infants. Sample size is the total number of offspring produced by high- and low-ranking females. As sample size increases, effect sizes decline toward 0 (for sources of these data, see Brown & Silk, 2002).

depend heavily on sample size; the most extreme results come from the studies with the smallest samples. The weighted mean-effect size is approximately zero. This means that on average there is no difference between the proportion of males produced by high- and low-ranking females. No differences in these results emerged when we limited the analysis to wild populations, captive and semi-free-ranging populations, or to members of the genus *Macaca*.

In some cases, we can assess the relationship between effect size and sample size for specific populations because multiple reports of the relationship between maternal rank and birth sex ratios have been published (Brown & Silk, 2002). There is some consistency in the effect sizes across studies within populations because initial samples are included in subsequent studies. None the less, in seven of eight cases, the magnitude of effect sizes declines as sample sizes increase.

These results clearly demonstrate that there is not a clear and consistent relationship between maternal rank and birth sex ratios in this data set. However, this does not necessarily mean that we can dismiss the idea that primates adjust birth sex ratios in an adaptive fashion. As discussed above, it remains possible that local resource competition biases global sex ratios within populations or that maternal rank

influences birth sex ratios in a more complicated way than initially imagined.

Interaction between local resource competition and maternal rank

Combining Trivers and Willards' ideas about the relationship between maternal rank and offspring sex ratios, and Clark's ideas about the effects of local resource competition, van Schaik and Hrdy (1991) hypothesised that there may be an interaction between the effects of maternal rank and local conditions on birth sex ratios in primates. In populations with female philopatry, the extent of local resource competition may influence the benefits derived from producing sons and daughters. When local resource competition is intense, high-ranking females may gain benefits from having daughters who support them in conflicts and help to maintain their status. At the same time, low-ranking females may have difficulty protecting their daughters from harassment when local resource competition is strong. Thus, when local resource competition is intense, high-ranking females are predicted to produce a smaller proportion of sons than are lower-ranking females.

Van Schaik and Hrdy's ideas differ in two important ways from Clark's model and the assumptions underlying Johnson's analysis. First, van Schaik and Hrdy assume that females facultatively adjust birth sex ratios in relation to current environmental conditions. Second, van Schaik and Hrdy hypothesise that birth sex ratios will vary within groups, as birth sex ratios correspond to maternal rank.

To test their hypothesis, van Schaik and Hrdy computed the annual rate of population growth, a proxy for the intensity of local resource competition, for 11 primate populations in which the relationship between maternal rank and offspring sex ratios was known. Their results suggest that differences in the proportion of males produced by high- and low-ranking females are tightly linked to population growth rates. When population growth rates are low (and the intensity of local resource competition is presumably high), high-ranking females produce relatively fewer sons than low-ranking females, and when population growth rates are higher, the pattern is reversed. In contrast to Johnson's results, van Schaik and Hrdy find no consistent relationship between the overall birth sex ratios of these groups and the extent of local resource competition.

This analysis is appealing because it provides a plausible adaptive explanation for much of the observed variation in the relationship between maternal rank and offspring sex that we described above. However, certain methodological and empirical problems have emerged since the analysis was first published. First, van Schaik and Hrdy's statistical analysis of the relationship between maternal rank and birth sex ratios does not take sample size into account. This means that studies based on small samples are given the same weight in the analysis as studies based on large samples. This is problematic because stochastic processes are expected to generate more variation in small samples. Sample sizes of the 11 studies included in their analysis range from 73 to 719.

Second, Packer et al. (2000) pointed out that most of the effect is based on captive ($n = 5$) or semi-free-ranging ($n = 3$) populations, which are provisioned liberally and sustain unusually high growth rates. The captive and semi-free-ranging populations included in the analysis had annual growth rates of 8–18%. The three wild populations included in their sample have lower growth rates (−9 %, 0 %, and 6 % per year) and do not follow a consistent pattern. While baboons in Amboseli and long-tailed macaques in Ketambe fit the predicted relationship between maternal rank and birth sex ratios closely, vervets in Amboseli deviate markedly from the predicted pattern. Third, subsequent analyses of sex ratios in five wild baboon groups at Gombe (Packer et al., 2000), three free-ranging groups of macaques in Ketambe (van Noordwijk & van Schaik, 1999), and a captive population of ring-tailed lemurs (Nunn & Pereira, 2000) do not fit the pattern van Schaik and Hrdy documented in other populations. At Gombe, for example, there was no consistent relationship between population growth rates, or other measures of local resource competition, and offspring sex (Packer et al., 2001).

How could we integrate the results of our meta-analysis of the relationship between maternal rank and birth sex ratio with the results derived from van Schaik and Hrdy's analysis? Figure 14.1 shows that differences in the birth sex ratios

of high- and low-ranking females converge to zero in large samples. According to van Schaik and Hrdy's analysis this would imply that (1) these populations all have annual growth rates of about 12 per cent, the population growth rate that corresponds to an effect size of zero in their analysis; or (2) population growth rates vary over time and across groups within these samples, and produce no aggregate difference in the birth sex ratios of high- and low-ranking females. At this point it is not clear whether either of these explanations is correct.

It is also unclear how to reconcile van Schaik and Hrdy's results with Johnson's results as to the effects of local resource competition on population sex ratios. Johnson found that there was a strong effect of local resource competition on population sex ratios, but van Schaik and Hrdy's analysis suggests that population sex ratios are not consistently related to the extent of local resource competition. The difference may arise from the fact that these analyses rely on different measures of local resource competition. Van Schaik and Hrdy used annual population growth rates, while Johnson relied on the ratio of immature to adult females. Both seem to be plausible indices of local resource competition, but it is not clear which measure is most biologically meaningful. Differences in their results might also be a function of the taxa included in the analyses. Van Schaik and Hrdy's analysis is limited to three primate genera (*Macaca*, *Papio* and *Cercopithecus*), while Johnson's analysis includes four additional genera (*Presbytis*, *Colobus*, *Erythrocebus* and *Lemur*).

MATERNAL INVESTMENT IN SONS AND DAUGHTERS

Selection is expected to influence patterns of investment in male and female progeny, not just the number of males and females produced. Birth sex ratios are expected to reflect the amount of investment in male and female offspring. If males are more expensive to produce than females, birth sex ratios are expected to be biased toward females, and vice versa. However, it is more difficult to quantify maternal investment than to tabulate sex ratios, so it is much more difficult to construct a comprehensive analysis of sex-biased maternal investment. Researchers have attempted to assess sex differences in maternal investment by assessing gestation length, infant birth weight, growth rates, suckling rates, length of interbirth intervals, and maternal styles (reviewed by Clutton-Brock, 1991; Brown, 2001; Setchell & Lee, this volume).

It is important to point out that interpretations of these kinds of data are complicated by the need to distinguish between selective forces that influence maternal investment strategies and selective forces that shape offspring's strategies for extracting resources from their mothers. Thus, Clutton-Brock (1991) concluded that differences in the costs of rearing sons and daughters in vertebrate species 'appear to be a consequence of sex differences in the behaviour of offspring rather than of differences in parental behaviour. This could suggest that they are a by-product of sexual selection favouring sex differences in juvenile growth rates rather than a consequence of evolved parental strategies . . .' (p. 227).

Johnson (1988) examined the relationship between sex ratios and the extent of sexual dimorphism in adult body weights. He pointed out that high rates of male mortality in sexually dimorphic species might produce male-biased sex ratios, while high growth rates among males in sexually dimorphic species might make males more expensive to produce and skew birth sex ratios in favour of females. However, neither prediction seems to hold consistently. Johnson found no relationship between the extent of adult dimorphism and birth sex ratios in seven primate genera.

Brown's (2001) review indicates that there are few consistent biases in investment in males and females among primates. In macaques, males weigh more at birth than females (reviewed by Bercovitch et al., 2000), and sex differences in body weight are maintained through infancy and into adulthood. In some cases, sons of high-ranking mothers grow faster and mature earlier than sons of low-ranking mothers (Bercovitch, 1993; Alberts & Altmann, 1995; Dixson & Nevison, 1997). As gestation lengths are equal for male and female macaques (Silk et al., 1993), sons may require more investment than daughters in these species. However, 'investment' as defined by Trivers (1972) involves a cost to the mother in terms of her ability to invest in future offspring. If these sex differences in weight were the product of differential maternal investment in sons and daughters, then we would expect to find a consistent relationship between infant sex and milk transfer during suckling bouts, or between infant sex and interbirth intervals. It is very difficult to assess the nutritive content of suckling bouts (Brown, 2001), but easier to measure interbirth intervals. Brown's review of data on the length of interbirth intervals suggests that the relationship between infant sex, maternal rank, and interbirth intervals is quite variable.

Thus, the limited available evidence suggests that mothers do not consistently skew investment toward males or females, and that investment strategies are not systematically influenced by the interaction between maternal rank and infant sex. However, at this point the data are too limited and the sample sizes are too small to be able to draw firm conclusions about adaptive variation in investment patterns.

SUMMARY AND CONCLUSIONS

Adaptive models of birth sex ratios predict that females will adjust the sex ratio of their progeny in relation to their own dominance rank or condition and the extent of local resource competition. While Trivers and Willard originally predicted that females in polygynous, sexually dimorphic species will bias investment in males when they are in good condition and bias investment toward females when they are in poor condition, formal mathematical analysis suggests that the opposite pattern may be favoured by natural selection. Leimar showed that under certain circumstances females in good condition will profit more from investment in daughters than sons, and predicted that when maternal condition is transmitted with high fidelity to offspring, females in good condition will bias investment toward daughters.

Since the 1980s, primatologists have gathered a considerable amount of information about the relationship between maternal rank and birth sex ratios, particularly in Cercopithecine species. These data have been used to support a number of different adaptive hypotheses. There is little dispute about the facts at hand, but considerable room for disagreement about how these facts should be interpreted. We see no simple way to resolve these disputes. However, we caution against assuming that observed variation in primate sex ratios necessarily reflects adaptive strategies. In each case, we must test the null hypothesis before we can safely reject it.

We draw five main conclusions from our review of the primate sex-ratio literature:

(1) Primate females do not adjust the sex ratio of their progeny in relation to their own rank in a uniform way.

(2) Meta-analysis of the relationship between birth sex ratios and maternal rank suggests that the observed variation among populations in the proportion of males produced by high- and low-ranking females may simply be the product of stochastic variation in small samples.

(3) There is some evidence that local resource competition may contribute to variation in birth sex ratios within and across populations, but that claims for adaptive variation in birth sex ratios need to be evaluated with some caution.

(4) To test adaptive hypotheses about variation in primate sex ratios, it is important to use appropriate methods that take sample sizes and effect sizes into account. The null

hypothesis must be seriously considered before adaptive hypotheses are accepted.

(5) It is also important to subject verbal models to formal theoretical analysis because intuition is not always sufficient to generate cogent adaptive predictions. Leimar's formulation of Trivers and Willard's verbal model generates radically different predictions about the effects of maternal condition on offspring sex from the original model.

ACKNOWLEDGEMENTS

We thank Peter Kappeler for his invitation to write about sex-ratio evolution in primates; Dario Maestripieri, Douglas Meikle and Craig Packer for providing us with unpublished data, and Colin Chapman, Olaf Leimar, Andreas Paul and Karen Strier for copies of published papers that were unavailable in our libraries. We have benefited from discussion of these data with Tim Clutton-Brock, Carel van Schaik, Susan Alberts, Jeanne Altmann, Sarah Blaffer Hrdy, Robert Boyd, Peter Hammerstein, Kevin Laland and several anonymous reviewers of the first draft of this manuscript. This chapter was prepared while Joan Silk was on a sabbatical leave in Berlin; she thanks the Wissenschaftskolleg zu Berlin for hospitality and research support. Gillian Brown was supported by the Medical Research Council.

REFERENCES

Alberts, S. A. & Altmann, J. 1995. Preparation and activation: determinants of age at reproductive maturity in male baboons. *Behavioral Ecology and Sociobiology*, **36**, 397–406.

Altmann, J., Hausfater, G. & Altmann, S. A. 1988. Determinants of reproductive success in savannah baboons. In *Reproductive Success*, ed. T. H. Clutton-Brock. Cambridge: Cambridge University Press, pp. 403–18.

Altmann, J., Alberts, S. C., Haines, S. A. *et al.* 1996. Behavior predicts genetic structure in a wild primate group. *Proceedings of the National Academy of Sciences, USA*, **93**, 5797–801.

Bercovitch, F. B. 1993. Dominance rank and reproductive maturation in male rhesus macaques (*Macaca mulatta*). *Journal of Reproduction and Fertility*, **99**, 113–20.

2002. Sex-biased parental investment in primates. *International Journal of Primatology*, **23**, 905–21.

Bercovitch, F. B., Widdig, A. & Nürnberg, P. 2000. Maternal investment in rhesus macaques (*Macaca mulatta*):

reproductive costs and consequences of raising sons. *Behavioral Ecology and Sociobiology*, **48**, 1–11.

Berman, C. M. 1988. Maternal condition and offspring sex ratio in a group of free-ranging rhesus monkeys: an 11-year study. *American Naturalist*, **131**, 307–28.

Bramblett, C. A., Bramblett, S. S., Bishop, D. & Coelho, A. M. Jr. 1982. Longitudinal stability in adult hierarchies among vervet monkeys (*Cercopithecus aethiops*). *American Journal of Primatology*, **2**, 10–19.

Brown, G. R. 2001. Sex-biased investment in nonhuman primates: can Trivers and Willard's theory be tested? *Animal Behaviour*, **61**, 683–94.

Brown, G. R. & Silk, J. B. 2002. Reconsidering the null hypothesis: is maternal rank associated with birth sex ratios in primate groups? *Proceedings of the National Academy of Sciences, USA*, **99**, 11252–5.

Bulmer, M. G. & Taylor, P. D. 1980. Sex ratio under the haystack model. *Journal of Theoretical Biology*, **86**, 83–9.

Chapais, B. 1992. The role of alliances in social inheritance of rank among female primates. In *Coalitions and Alliances in Humans and Other Animals*, ed. A. H. Harcourt & F. B. M. de Waal. Oxford: Oxford Science Publications, pp. 29–59.

Charnov, E. L. 1979. Simultaneous hermaphroditism and sexual selection. *Proceedings of the National Academy of Sciences, USA*, **76**, 2480–4.

Clark, A. B. 1978. Sex ratio and local resource competition in a prosimian primate. *Science*, **201**, 163–5.

Clutton-Brock, T. H. 1986. Sex ratio variation in birds. *Ibis*, **128**, 317–29.

1991. *The Evolution of Parental Care*. Princeton, NJ: Princeton University Press.

Clutton-Brock, T. H. & Albon, S. D. 1982. Parental investment in male and female offspring in mammals. In *Current Problems in Sociobiology*, ed. King's College Sociobiology Group. Cambridge: Cambridge University Press, pp. 223–47.

Clutton-Brock, T. H., Harvey, P. H. & Rudder, B. 1977. Sexual dimorphism, socionomic sex ratio, and body weight in primates. *Nature*, **269**, 797–800.

Clutton-Brock, T. H., Albon, S. D. & Guinness, F. E. 1984. Maternal dominance, breeding success, and birth sex ratios in red deer. *Nature*, **308**, 358–60.

1986. Great expectations: maternal dominance, sex ratios, and offspring reproductive success in red deer. *Animal Behaviour*, **34**, 460–71.

Cowlishaw, G. & Dunbar, R. I. M. 1991. Dominance rank and mating success in male primates. *Animal Behaviour*, **41**, 1045–56.

Darwin, C. 1871. *The Descent of Man and Selection in Relation to Sex*. London: John Murray.

Debyser, I. W. J. 1995a. Prosimian juvenile mortality in zoos and primate centers. *International Journal of Primatology*, 16, 889–907.

1995b. Platyrrhine juvenile mortality in captivity and in the wild. *International Journal of Primatology*, 16, 909–33.

1995c. Catarrhine juvenile mortality in captivity, under seminatural conditions, and in the wild. *International Journal of Primatology*, 16, 935–69.

de Ruiter, J. R., Scheffrahn, W., Trommelen, G. J. J. M. *et al.* 1992. Male social rank and reproductive success in wild long-tailed macaques. In *Paternity in Primates: Genetic Tests and Theories*, ed. R. D. Martin, A. F. Dixson & E. J. Wickings. Basel: Karger, pp. 175–91.

Dittus, W. P. J. 1998. Birth sex ratios in toque macaques and other mammals: integrating the effects of maternal condition and competition. *Behavioral Ecology and Sociobiology*, 44, 149–60.

Dixson, A. F. & Nevison, C. M. 1997. The socioendocrinology of adolescent development in male rhesus monkeys (*Macaca mulatta*). *Hormones and Behavior*, 31, 126–35.

Drews, C. 1996. Contexts and patterns of injuries in free-ranging male baboons (*Papio cynocephalus*). *Behaviour*, 133, 443–74.

Emlen, S. T. & Oring, L. W. 1977. Ecology, sexual selection, and the evolution of mating systems. *Science*, 197, 215–23.

Fedigan, L. M. & Zohar, S. 1997. Sex differences in mortality of Japanese macaques: twenty-one years of data from the Arashiyama West population. *American Journal of Physical Anthropology*, 102, 161–75.

Fisher, R.A. 1930. *The Genetical Theory of Natural Selection*. Oxford: Oxford University Press.

Greenwood, P. J. 1980. Mating systems, philopatry, and dispersal in birds and mammals. *Animal Behaviour*, 28, 1140–62.

Hamilton, W. D. 1967. Extraordinary sex ratios. *Science*, 156, 477–88.

1975. Innate social aptitudes of man: an approach from evolutionary genetics. In *Biosocial Anthropology*, ed. R. Fox. New York, NY: John Wiley, pp. 133–55.

1979. Wingless and fighting males in fig wasps and other insects. In *Sexual Selection and Reproductive Competition in Insects*, ed. M. S. Blum & N. A. Blum. New York, NY: Academic Press, pp. 167–220.

Harcourt, A. H. 1987. Dominance and fertility among female primates. *Journal of Zoology, London*, 213, 471–87.

Harcourt, A. H., Harvey, P. H., Larson, S. G. & Short, R. V. 1981. Testis weight, body weight and breeding system in primates. *Nature*, 293, 55–7.

Harvey, P. H. & Harcourt, A. H. 1984. Sperm competition, testis size and breeding systems in primates. In *Sperm Competition and the Evolution of Animal Mating Systems*, ed. R. L. Smith. New York, NY: Academic Press, pp. 589–600.

Hausfater, G., Altmann, J. & Altmann, S. A. 1982. Long-term consistency of dominance relations among female baboons (*Papio cynocephalus*). *Science*, 217, 752–5.

Hrdy, S. B. & Whitten, P. L. 1987. Patterning of sexual activity. In *Primate Societies*, ed. B. B. Smuts, D. L. Cheney, R. M. Seyfarth, R. W. Wrangham & T. T. Struhsaker. Chicago, IL: University of Chicago Press, pp. 370–84.

Johnson, C. N. 1988. Dispersal and the sex ratio at birth in primates. *Nature*, 332, 726–8.

Launhardt, K., Borries, C., Hardt, C., Epplen, J. T. & Winkler, P. 2001. Paternity analysis of alternative male reproductive routes among the langurs (*Semnopithecus entellus*) of Ramnagar. *Animal Behaviour*, 61, 53–64.

Leimar, O. 1996. Life history analysis of the Trivers and Willard sex-ratio problem. *Behavioral Ecology*, 7, 316–25.

Maynard Smith, J. 1978. *The Evolution of Sex*. Cambridge: Cambridge University Press.

Meikle, D. B., Tilford, B. L. & Vessey, S. H. 1984. Dominance rank, secondary sex ratio, and reproduction of offspring in polygynous primates. *American Naturalist*, 124, 173–88.

Mitani, J. C., Gros-Louis, J. & Richards, A. 1996b. Sexual dimorphism, the operational sex ratio, and the intensity of male competition in polygynous primates. *American Naturalist*, 147, 966–80.

Nunn, C. L. & Pereira, M. E. 2000. Group histories and offspring sex ratios in ringtailed lemurs (*Lemur catta*). *Behavioral Ecology and Sociobiology*, 48, 18–28.

Packer, C., Collins, D. A. & Eberly, L. E. 2000. Problems with primate sex ratios. *Proceedings of the Royal Society of London, Series B*, 355, 1627–35.

Palmer, A. R. 2000. Quasireplication and the contract of error: lessons from sex ratios, heritabilities and fluctuating asymmetry. *Annual Review of Ecology and Systematics*, 31, 441–80.

Palombit, R. A. 1994. Extra-pair copulations in a monogamous ape. *Animal Behaviour*, 47, 721–3.

2000. Infanticide and the evolution of male–female bonds in animals. In *Infanticide by Males and Its Implications*, ed. C. P. van Schaik & C. H. Janson. Cambridge: Cambridge University Press, pp. 239–68.

Palombit, R. A., Seyfarth, R. M. & Cheney, D. L. 1997. The adaptive value of 'friendships' to female baboons: experimental and observational evidence. *Animal Behaviour*, **54**, 599–614.

Paul, A. & Kuester, J. 1990. Adaptive significance of sex ratio adjustment in semifree-ranging Barbary macaques (*Macaca sylvanus*) at Salem. *Behavioral Ecology and Sociobiology*, **27**, 287–93.

Paul, A., Kuester, J. & Arnemann, J. 1992. Maternal rank affects reproductive success of male barbary macaques (*Macaca sylvanus*) – evidence from DNA fingerprinting. *Behavioral Ecology and Sociobiology*, **30**, 337–41.

Pusey, A. E. & Packer, C. 1987. Dispersal and philopatry. In *Primate Societies*, ed. B. B. Smuts, D. L. Cheney, R. M. Seyfarth, R. W. Wrangham & T. T. Struhsaker. Chicago, IL: University of Chicago Press, pp. 250–66.

Rawlins, R. G. & Kessler, M. J. 1986. Secondary sex ratio variation in the Cayo Santiago macaque population. *American Journal of Primatology*, **10**, 9–23.

Rhine, R. J. 1994. A twenty-one-year study of maternal dominance and secondary sex ratio in a colony group of stumptail macaques (*Macaca arctoides*). *American Journal of Primatology*, **32**, 145–8.

Rhine, R. J., Norton, G. W., Rogers, J. & Wasser, S. K. 1992. Secondary sex ratio and maternal dominance rank among wild yellow baboons (*Papio cynocephalus*) of Mikumi National Park, Tanzania. *American Journal of Primatology*, **27**, 261–73.

Rorie, R. W. 1999. Effect of timing on artificial insemination on sex ratio. *Theriogenology*, **52**, 1273–80.

Samuels, A., Silk, J. B. & Altmann, J. 1987. Continuity and change in dominance relations among female baboons. *Animal Behaviour*, **35**, 785–93.

Silk, J. B. 1983. Local resource competition and facultative adjustment of sex ratio in relation to competitive abilities. *American Naturalist*, **121**, 56–66.

1984. Local resource competition and the evolution of male-biased sex ratios. *Journal of Theoretical Biology*, **108**, 203–13.

1987. Social behavior in evolutionary perspective. In *Primate Societies*, ed. B. B. Smuts, D. L. Cheney, R. M. Seyfarth, R. W. Wrangham & T. T. Struhsaker. Chicago, IL: University of Chicago Press, pp. 318–29.

1988. Maternal investment in captive bonnet macaques (*Macaca radiata*). *American Naturalist*, **132**, 1–19.

Silk, J. B., Clark-Wheatley, C. B., Rodman, P. S. & Samuels, A. 1981. Differential reproductive success and facultative adjustment of sex ratios among captive female bonnet macaques (*Macaca radiata*). *Animal Behaviour*, **29**, 1106–20.

Silk, J. B., Short, J., Roberts, J. & Kemnitz, J. 1993. Gestation length in rhesus macaques (*Macaca mulatta*). *International Journal of Primatology*, **14**, 95–104.

Simpson, M. J. A. & Simpson, A. E. 1982. Birth sex ratios and social rank in rhesus monkey mothers. *Nature*, **300**, 440–1.

Small, M. F. & Hrdy, S. B. 1986. Secondary sex ratios by maternal rank, parity, and age in captive rhesus macaques (*Macaca mulatta*). *International Journal of Primatology*, **7**, 289–304.

Smuts, B. B. 1987. Gender, aggression, and influence. In *Primate Societies*, ed. B. B. Smuts, D. L. Cheney, R. M. Seyfarth, R. W. Wrangham & T. T. Struhsaker. Chicago, IL: University of Chicago Press, pp. 400–12.

Soltis, J. & McElreath, R. 2001. Can females gain extra paternal investment by mating with multiple males? A game theoretic approach. *American Naturalist*, **58**, 519–29.

Sommer, V. 2000. The holy wars about infanticide: which side are you on? And why? In *Infanticide by Males and Its Implications*, ed. C. P. van Schaik & C. H. Janson. Cambridge: Cambridge University Press, pp. 9–26.

Sommer, V. & Reichard, U. 2000. Rethinking monogamy: the gibbon case. In *Primate Males: Causes and Consequences of Variation in Group Composition*, ed. P. M. Kappeler. Cambridge: Cambridge University Press, pp. 159–68.

Trivers, R. L. 1972. Parental investment and sexual selection. In *Sexual Selection and the Descent of Man*, ed. B. Campbell. Chicago, IL: Aldine, pp. 136–79.

1985. *Social Evolution*. Menlo Park, CA: Benjamin Cummings.

Trivers, R. L. & Willard, D. 1973. Natural selection of parental ability to vary the sex ratio of offspring. *Science*, **179**, 90–2.

Uyenoyama, M. & Feldman, M. 1980. Theories of kin and group selection: a population genetics perspective. *Theoretical Population Biology*, **17**, 380–414.

van Noordwijk, M. A. & van Schaik, C. P. 1999. The effects of dominance rank and group size on female lifetime reproductive success in wild long-tailed macaques, *Macaca fasicularis*. *Primates*, **40**, 105–30.

2001. Career moves: transfer and rank challenge decision by male long-tailed macaques. *Behaviour*, **138**, 359–95.

van Schaik, C. P. & de Visser, J. A. G. M. 1990. Fragile sons or harassed daughters? Sex differences in mortality among juvenile primates. *Folia Primatologica*, **55**, 10–23.

van Schaik, C. P. & Hrdy, S. B. 1991. Intensity of local resource competition shapes the relationship between maternal rank and sex ratios at birth in cercopithecine primates. *American Naturalist*, **138**, 1555–62.

van Schaik, C. P. & Janson, C. H. (eds.) 2000. *Infanticide by Males and Its Implications*. Cambridge: Cambridge University Press.

Wade, M. J. 1978. A critical review of the models of group selection. *Quarterly Review of Biology*, **53**, 101–14.

Wasser, S. K. & Norton, G. 1993. Baboons adjust secondary sex ratio in response to predictors of sex-specific offspring survival. *Behavioral Ecology and Sociobiology*, **32**, 273–81.

Watanabe, K., Mori, A. & Kawai, M. 1992. Characteristic features of the reproduction of Koshima monkeys, *Macaca fuscata fuscata*: a summary of thirty-four years of observation. *Primates*, **33**, 1–32.

West, S. A. & Sheldon, B. C. 2002. Constraints in the evolution of sex ratio adjustment. *Science*, **295**, 1685–8.

West, S. A., Herre, E. A. & Sheldon, B. C. 2000. The benefits of allocating sex. *Science*, **290**, 288–90.

Williams, G. C. 1979. On the question of adaptive sex ratio in outcrossed vertebrates. *Proceedings of the Royal Society of London, Series B*, **205**, 567–80.

Wilson, D. S. 1977. Structured demes and the evolution of group advantageous traits. *American Naturalist*, **111**, 157–85.

  1980. *The Natural Selection of Populations and Communities*. Menlo Park, CA: Benjamin Cummings.

Wilson, D. S. & Colwell, R. K. 1981. Evolution of sex ratio in structured demes. *Evolution*, **35**, 882–97.

15 • Natural and sexual selection and the evolution of multi-level societies: insights from zebras with comparisons to primates

DANIEL I. RUBENSTEIN
Department of Ecology and Evolutionary Biology
Princeton University
Princeton, NJ, USA

MACE HACK
Nebraska Game and Parks Commission
Lincoln, NE, USA

INTRODUCTION

Animal societies derive from the social relationships that exist among its members (Hinde, 1983). Behavioural ecologists have traditionally focused on the core relationships defining a mating system as a means toward understanding the role of ecology in the evolution of sociality (e.g. Jarman, 1974; Bradbury & Vehrencamp, 1977). Typically, these core relationships emerge from the operation of both natural and sexual selection and how they differently affect the behaviour of females and males (Rubenstein, 1986; van Schaik, 1989). However, emphasis on mating systems has tended to marginalise the importance to social evolution of interactions and relationships that extend beyond the basic breeding unit. This is even true for the small subset of species with multi-level societies, where breeding units and other social subgroups are themselves organised into more complex social groups within a population. By examining how natural and sexual selection operate within multi-level societies, however, a more complete understanding of the function and evolution of sociality emerges than would by investigating the dynamics of mating systems alone (e.g. Dunbar, 1988).

Although the societies of many primate species are multi-levelled, the relative simplicity of societies of plains zebras (*Equus burchelli*) where only two tiers exist – the core breeding units and the herds they often comprise – can provide insights into the rules that give form to multi-level societies. In this chapter we begin by highlighting the environmental and sociosexual factors that shape zebra mating systems and herd dynamics. Then we show how characteristics of zebra herds emerge from individual cost–benefit decisions. Typically, forces of natural selection, because they operate on traits that promote survival via enhanced resource acquisition or reductions in predation risk, are thought to be most

important in shaping patterns of sociality (Jarman, 1974). Yet we show that while these forces play important roles in determining both core mating associations and the dynamics and structure of zebra herds, the forces of sexual selection, with their pressures to maximise mating success, also influence zebra herding behaviour by affecting the social decision-making processes of adult and subadult males. In particular, we show that the need for band, or 'harem', stallions to reduce the risks of cuckoldry by subadult bachelors plays the greatest role in determining when zebra core-mating units should coalesce to form herds. Overall, we show that higher levels of social organisation provide individuals with a source of potential social options for solving ecological and social problems that cannot be solved by adjusting relationships within core social groups. Higher-level connections between groups, however, only appear to remain viable when the relationships holding together the core social units themselves are not disrupted.

ZEBRA MATING SYSTEMS AND HERD DYNAMICS

EQUID MATING SYSTEMS

Our current understanding of plains zebra societies results from studying equid mating systems, of which the seven extant species display two general types (Klingel, 1974). In the African and Asiatic asses (*Equus africanus* and *E. hemionus*, respectively) and Grevy's zebra (*E. grevyi*), adult females and their most recent offspring range over large areas and form only temporary groups (Klingel, 1977; Woodward, 1979; Ginsberg, 1989). Breeding is not highly synchronised and breeding males defend mating territories for most of the year near resources needed by females. Plains zebra, mountain zebra (*Equus zebra*) and horse stallions

Sexual Selection in Primates: New and Comparative Perspectives, ed. Peter M. Kappeler and Carel P. van Schaik. Published by Cambridge University Press. © Cambridge University Press 2004.

(*Equus caballus*) are also polygynous, but defend exclusive mating rights to small year-round harems of one-to-several females (Klingel, 1969a; Joubert, 1972; Groves, 1974; Feist & McCullough, 1976; Miller, 1979; Rubenstein, 1981; Penzhorn, 1984; Berger, 1986; Lloyd & Rasa, 1989). Females form social bonds with each other (Rubenstein, 1986; Rutberg, 1990), often living in the same harem for the majority of their reproductive lives (but see Berger, 1986), and they remain together even after replacement of the stallion. Interestingly, these females are not close relatives, as in most other mammals with stable female groups (Clutton-Brock, 1989), since both sexes disperse from their natal harems at 2 to 4 years of age (Berger, 1987).

Differences in the ecological environments inhabited by equids provide a first step towards explaining the basic dichotomy in their mating systems (Rubenstein, 1986). A fundamental tenet of behavioural ecology is that female movements and associations are driven primarily by the distribution of resources and the risk of predation – both forces of natural selection that shape individual survival and reproductive abilities; while male movements and associations result from the consequent distribution of females – both forces of sexual selection that affect the ability of males to secure matings with multiple females (Bradbury & Vehrencamp, 1977; Emlen & Oring, 1977). All equids are relatively large for ungulates, feed by grazing, and require water in addition to their food. Accordingly, individuals need to consume large amounts of grass and drink daily, for all individuals of the smaller-bodied species and the lactating females of the larger-bodied species (Ginsberg, 1989). The relatively open environments they inhabit and their lack of overt morphological defences can expose them to high rates of predation (Turner, 1992; Scheel, 1993). The asses and Grevy's zebra live in xeric environments where grass occurs in small patches distributed sparsely over a large area. Competition among females for grass presumably prevents their long-term association, and stable groups never form. In addition, the best grazing sites are often far from water (Rubenstein, 1989), weakening bonds between females with different water needs and abilities to travel (cf. lactating and non-lactating). Given the low density of females and their large home-ranges, males maximising reproductive success by defending either access to water or territories encompassing a variety of grassland types, will have at least some productive areas attractive to hungry females (Ginsberg, 1989; Rubenstein, 1994).

The other equid species, including the plains zebra, inhabit more mesic habitats where grass occurs in much larger patches, is distributed more uniformly throughout a patch, and lies in closer proximity to water (Rubenstein, 1994). Consequently, the costs of competition among females are reduced, allowing groups to persist year-round. In addition, higher predator densities in mesic habitats result in greater anti-predator benefits from grouping. However, since female home-ranges are still large, males cannot economically defend the area needed by a group but instead defend the group itself (Klingel, 1969b; Joubert, 1972; Penzhorn, 1984). At least in the population inhabiting the Samburu–Buffalo Springs Game Reserve, a male's defence of his females from harassment by other males or the better access he can provide to contested resources produce reproductive benefits for females that may be the final elements favouring small, highly cohesive groups (Rubenstein, 1986; Linklater *et al.*, 1999). Studies of intraspecific variation confirm the importance of these ecological factors in shaping equid mating systems (Moehlman, 1979; Rubenstein, 1981, 1986; Ginsberg, 1989). A recent review by Linklater (2000), however, questions the tightness of the linkage between these ecological features and the nature of equid mating systems, but it does so by discounting the utility of precisely those studies that most strikingly highlight variations on the two major equid organisational themes. As new details emerge from more standardised comparative studies, alternative explanations determining mating systems based on additional factors, both past and present, may emerge. Yet until they do, habitat variation can be viewed as playing an important role in shaping the mating and social relationships that develop among equids.

ZEBRA HERD DYNAMICS

In contrast to most ungulates, plains zebra live in multi-level societies. A year-round 'harem' or 'band' of several adult females and their recent offspring, defended usually by only a single male, constitutes the basic breeding unit. Non-breeding, but reproductively mature, males also associate, forming 'bachelor' groups that range in size from two to more than 40 individuals. Both types of core social groups are common to equids (Klingel, 1974; Rubenstein, 1986) and other mammals (McCracken & Bradbury, 1981; Clutton-Brock, 1989), but a level of social organisation that is rarely observed in harem-forming species other than primates (reviewed by Stammbach, 1987) is the formation of herds. Harems regularly associate in spatially cohesive herds that vary in both size, from two to over 100 harems (Fig. 15.1a), and in composition, containing harems and

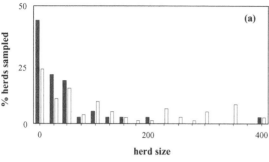

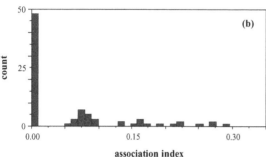

Fig. 15.1   (a) The frequency distribution of plains zebra herds by
size in Ngorongoro Crater, Tanzania. Solid bars represent herds in
the dry season; open bars represent herds in the wet season.
(b) The frequency distribution of index of association among
harems in Ngoronogoro Crater, Tanzania. IA = 2C/(A + B)
where A is the number of times harem A is seen, B is the number
of times harem B is seen and C is the number of times harems A
and B are seen together.

bachelor groups or just harems. Preferential associations
between pairs of harems – observed in different populations ranging from the species' northern limit in the arid
Samburu ecosystem (Rubenstein, 1986) to the relatively lush
grasslands of the Ngorongoro Crater (Fig. 15.1b) – create
temporally stable subgroups within a larger herd and indicate a type of non-randomness to herd formation and structure. These observations, when considered together with
the plains zebra's long life and the potential for close kin of
either sex to reside in neighbouring core groups, suggest that
interactions with individuals outside one's core group reflect
a second tier of social relationships that critically shape the
overall organisation of plains zebra societies.

Social relationships within a breeding group change over
a lifetime. Individuals mature, gain experience and alter their
competitive prowess and abilities to build alliances in a social
context that also change with the comings and goings of
competitors and allies alike, consequently affecting individ-

ual routes to breeding success (Pereira & Fairbanks, 1993;
Feh, 1999). Similarly, the nature and types of relationships
formed beyond the breeding group may critically affect the
particular route an individual takes, as studies on the evolution of helping-behaviour have demonstrated (Brown, 1987).
With respect to true multi-level societies, such as those of
gelada (*Theropithecus gelada*) and hamadryas baboons (*Papio
hamadryas*) (Stammbach, 1987), the association of breeding units into higher-level social groups arises because these
groups fulfil specific functions that breeding units alone do
not (e.g. reduce predation risk in geladas (Dunbar, 1986), or
facilitate mate or resource defence in hamadryas (Kummer,
1968; Sigg *et al.*, 1982)). These different functions could
emerge from the fact that two societies, built from out-
wardly similar core mating groups (harems), and structured
in both cases by kinship, are very different overall organisations – female-centric and matrilineal society in geladas vs.
male-centric and patrilineal society in hamadryas. Thus, an
understanding of the dynamics of social relationships within
breeding groups may provide some understanding of a society's overall dynamics. Feedbacks between levels are likely
to be important and yet remain largely unexplored. Studies
on primate species with multi-level societies clearly show
that two ecological forces – the 'top-down' force of predation and 'bottom-up' forces associated with vegetation – help
determine the nature of higher-level societies that emerge.
Yet a third factor – sociosexual forces – should also come
into play. At least in a plains zebra population inhabiting the
Samburu–Buffalo Springs Game Reserve at the northern
edge of the species' range, the preferential associations that
form among harems are long-lasting and appear to enable
stallions more effectively to keep their females away from the
advances of bachelor males (Rubenstein, 1986). By recording, in the Laikipia region of central Kenya, the size of herds
(total number of individuals or number of harems), their
composition (number of bachelor males relative to breeding males) and the ecological and social context in which
they form, we have now been able to measure the relative
importance of each of these three forces in determining the
nature of zebra herds. By working on private lands, we sample zebra herds living in habitats that vary in overall structure
as measured by the density of trees, as well as by the quantity, quality and diversity of vegetation. Moreover, predators
range freely across ranches. Thus rapid changes in the numbers and types of predators threatening zebras are common.
And lastly, Laikipia ranches vary in the way they legally crop
zebras; some do not crop at all whereas others focus either
on bachelor males or family groups, thus changing the local

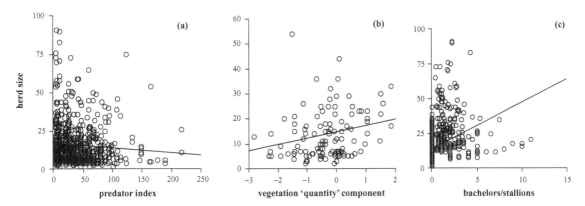

Fig. 15.2   Factors affecting herd sizes of plains zebra: (a) the impact of predator intensity: $y = -0.03x + 17.0$; $F(1, 684) = 4.3$; $p < 0.05$; (b) the impact of vegetation 'quantity': $y = 2.5x + 14.4$; $F(1, 105) = 7.8$; $p < 0.005$; (c) the impact of bachelor males nearby: $y = 2.6x + 10.4$; $F(1, 684) = 1024.5$; $p < 0.0001$.

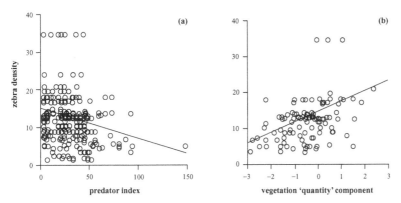

Fig. 15.3   Factors affecting the distribution of plains zebras across adjacent ranches in central Laikipia, Kenya: (a) zebra density on a ranch as a function of the number, type and impact of predators on the ranch (see text for details); (b) zebra density on a ranch as a function of the magnitude of the first-principle component of vegetation that was weighted by 'leaf density' and '% cover', both strong measures of vegetation 'quantity'.

ratio of bachelor to stallion males. Overall, variation is large with respect to all the dependent variables used in the General Linear Model we use to measure the relative importance of 'top-down', 'bottom-up' or sociosexual factors in determining herd size – as measured in terms of number of individuals or core social units; and composition – as measured in terms of bachelor-to-stallion ratios.

SOCIOECOLOGICAL PATTERNS IN LAIKIPIA

The overall analysis shows that all three factors play important roles in determining the dynamics of zebra herds. Although predation is often thought to be the most important factor in bringing individuals together to form groups (Alexander, 1974), our study shows the role of predation is only weakly implicated in determining the actual size of zebra herds (Fig. 15.2a). Yet this does not mean that predation is unimportant. As Fig. 15.3a shows, the risk of predation is a major determinant of where zebras tend to aggregate. Predators typically change location and vary activity throughout the day. Our predator index PI $= \sum_i$ [Abundance ith predator species $\times$ Impact of ith predator species] $\times$ [Habitat Visibility $\times$ Diel Period], where the first part of the expression represents Predator Incidence and the second part Context-Specific Risk, adjusts for these temporal and spatial changes and shows that as the risk of predation increases for a particular ranch, zebra numbers decrease. Thus as predators move into an area, zebras move out.

Once zebras assemble in a particular region – and the quantity of vegetation is a strong draw (Fig. 15.3b) – vegetation characteristics and the number of bachelor males in the vicinity of a harem all influence herd size (Fig. 15.2b, c). Of the three vegetation measures we computed using principle component analysis – 'quantity' 'quality' and 'diversity' – only 'quantity' matters. As the abundance of vegetation increases, so does herd size (Fig. 15.2b). Clearly, high levels of food regardless of its quality or diversity facilitate the coming together of core breeding groups. Apparently, the local abundance of leaves lowers competitive interactions as individuals increase spacing to reduce interference while grazing. But as Fig. 15.2c reveals, even more important in shaping herd size appear to be the effects of sociosexual pressures. As the number of bachelor males to breeding stallions in an area increases, so does herd size (Fig. 15.2c).

In order to determine the relative strength of these factors, partial correlation coefficients were computed and, as Table 15.1 illustrates, the strongest determinant of herd size is the ratio of the number of potentially cuckolding males to stallions in the vicinity of a herd (complete General Linear Model results appear in D. I. Rubenstein, M. Hack & D. Mazo, in preparation). Interestingly, while the 'bottom-up' factor associated with biomass still plays a role, a sociosexual variable plays the strongest role in determining the size of zebra herds. When the same analysis is performed for herd composition, as measured by the proportion of a herd that is composed of harem vs. bachelor groups, it is again a sociosexual variable that has the strongest effect (Table 15.1). In this case, however, it is the number of females in the herd that determines how many bachelor males will also be there. Thus overall it appears that the twin forces of sexual selection – male–male competition and female attraction – play the dominant roles in determining both the size and composition of plains zebra herds.

HERD CHARACTERISTICS AND DECISION-MAKING

Context-specific correlations are useful in suggesting how selection operates to generate particular patterns. Detailed behavioural measurements, however, often reveal how trade-

Table 15.1 *Partial correlation coefficients for herd size and composition versus the six dependent variables in each General Linear Model.*

Herd size vs:	Herd composition vs:
Zebra Density $= 0.09$	Zebra Density $= -0.14$
PC1: 'Quantity' $= 0.18$	PC1: 'Quantity' $= 0.18$
PC2: 'Quality' $= -0.09$	PC2: 'Quality' $= 0.01$
PC3: 'Diversity' $= -0.07$	PC3: 'Diversity' $= 0.06$
Predator Index $= -0.09$	Predator Index $= 0.01$
Bachelors/Stallions $= \underline{0.53}$	No. Females $= \underline{0.26}$

offs are handled and decisions are actually made. By recording the proportion of time that individuals spend foraging and then recording bite-rate during feeding intervals, we compute hourly intake rates. When these are compared across social contexts we can gauge how well individuals are performing in each. Figure 15.4 reveals how intake rate for individual males and females changes as both herd size and composition are altered. Even though males graze about 10 per cent less then females, both males and females feed most efficiently when the harem with which they are associated is alone on the landscape. As harems merge to form larger herds, both males and females experience slight reductions in foraging success but these declines are not statistically significant. In general, the presence of neighbours diminishes bite-rate slightly (Rubenstein, Hack & Mazo, in preparation), and although females spend the same proportion of time grazing as they do when a harem is alone, males end up re-deploying much of the time they normally spend vigilant, and some of the time they spend grazing, to socialising – displaying, fighting, but also mutually grooming – with other breeding stallions. Thus from the perspective of the females, there appears to be little difference in overall intake rate (bite-rate × hourly proportion of time grazing) between being in a herd of one or more harems. The same appears to be the case for males.

Herds, however, also vary in composition, and this is where sexual differences in foraging success appear. When a solitary harem is joined by a group of bachelors, the defending stallion spends virtually all of his time interacting with the intruders (Rubenstein, Hack & Mazo, in preparation) and this significantly reduces intake rate (Fig. 15.4). Not only does he leave his females and join the all-male group, he then proceeds to move among the males performing ritualised displays of dominance until each shows submission.

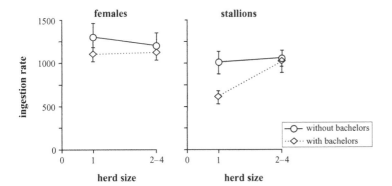

Fig. 15.4   Foraging intake rate for males and females in different social settings. Means and standard errors are shown. Solid line denotes social conditions without bachelor males present, whereas dotted lines depict conditions when bachelor males are present. Neither females nor males show differences in intake rate (bites per hour) when in a solitary harem or in small herds comprising between two and four harems. Males, however, do forage less than females in solitary harems, but the difference disappears when in small herds. When bachelor males join herds, only males show diminished intake rates when herds consist of just one harem, but this reduction disappears when mixed herds consist of more than two harem groups (2-way ANOVA interactions: males $F(3,89) = 3.1$; $p < 0.05$; females $F(3,142) = 2.3$; NS).

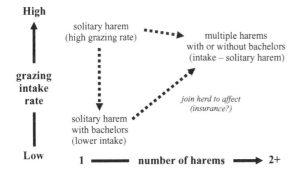

Fig. 15.5   Decision-making by breeding stallions. Stallion decision-making summarised by the impact of different social contexts on foraging success of females and their stallion, as illustrated in Fig. 15.4.

Occasionally, the males cease interacting in order to graze but they quickly return to low levels of jousting. Females, however, largely remain oblivious to these male escapades and continue grazing. As Fig. 15.4 illustrates their intake rate diminishes only slightly from that achieved when bachelor males are absent. Thus it is the stallion that bears the brunt of preventing bachelors from interacting with his females. Given the high foraging cost of protecting against cuckoldry when alone, and the insignificant feeding costs of joining a herd and potentially acquiring the support of other male allies, it is not surprising that herds form.

In fact, patterns associated with intake rate (Fig. 15.5) suggest that herds should form even without the direct instigation of bachelor male incursions, since the costs in terms of reduced foraging efficiency for both males and females are so small whenever harems coalesce into moderate-sized herds. As our study takes place on many ranches where zebra cropping varies in intensity, as well as in terms of the age and sex-class of animals taken, we can compare the average size of herds that form on ranches where bachelor males are rare with those on ranches where they are abundant. On ranches where the bachelor-to-stallion ratio is high (~ 1.5), the size of herds consisting of only harems averages 17.1 ± 1.3 individuals (mean $\pm$ S.E), approximately three harems, which is significantly larger than the herd size of 12.9 ± 0.9 individuals ($\sim$two harems) on ranches where the bachelor-to-stallion ratio is lower (~ 0.7) ($F_{1,256} = 6.7$; $p < 0.01$). It appears that as the probability of encountering bachelor males increases, stallions are more likely to form associations with other stallions to limit the threat bachelors pose. Herds thus appear to be a low-cost form of 'insurance' that serves the sexual interests of breeding stallions.

DECISION-MAKING BY BACHELOR MALES

ONTOGENETIC TRAJECTORIES

While harem stallions appear to adjust their behaviour to the risks imposed by bachelor males, bachelor males themselves are under evolutionary pressure to pass quickly, but efficiently, through this subadult stage of their lives and surround themselves more permanently with females. To become successful breeders, what should 'clever' bachelor males do? Clearly, they must survive; staying in the game is essential. But growing quickly is also important. Without attaining sufficient adult physical stature it will be difficult for a stallion to hold on to females when sexual intrusions occur. But maximising survival and foraging success is likely to be mutually exclusive. Data from our Laikipia zebra populations clearly show that just as for herds composed predominantly of breeders, the size of bachelor groups increases as predation risk increases ($y = 0.91x + 0.9$; $F = 7.1$; $p < 0.01$; $r^2 = 0.10$). Yet where predator protection is likely to be greatest, foraging appears to be most limited (Rubenstein, Hack & Mazo, in preparation). Of seven solitary bachelors seen during our surveys, all were grazing when detected. For those in all-male groups, however, the percentage of grazing individuals drops to near 45 per cent, while for those in herds comprised of both bachelors and harem groups the percentage of bachelors seen grazing falls to just under 20 per cent ($X^2 = 39.6$; $p < 0.0001$; df $= 2$). Apparently, the need for safety and food are largely incompatible.

But to balance staying alive with gaining strength is not likely to be enough. Making the transition to breeding status requires acquiring a variety of social skills. Fighting requires strength but it also depends on agility and mastery of specific tactics designed to outmanoeuvre an opponent (Berger, 1986). Nipping, neck rolling, mock-biting, rump pushing, cutting and turning while running are activities practised repeatedly by bachelors when they are in groups. But mastering the tactics of fighting with other males is itself only part of the social 'tool kit' needed by a maturing bachelor. In equids, successfully breeding males establish strong, long-lasting social bonds with their mates (Rubenstein, 1994) and therefore must learn how to behave toward females to initiate and maintain these bonds. In wild horses, males without extensive experience of interacting with reproductive females frequently lose them to other males, or are less successful in their mating attempts (Stroeh, 2001). Therefore, the successful transition from non-breeder to breeder in plains zebra will require not only physical maturation and growth but the acquisition of critical social skills with both females and other males.

As bachelor males mature they often change social settings. As indicated above, they can be found alone, together with other males or in herds where both males and females are present. Presumably young male zebra vary their social setting because each provides different opportunities to grow,

acquire necessary social skills and survive. Moreover, as bachelor males mature, their requirements may change as their size or condition changes. Assuming that a bachelor is able to choose its preferred social setting independently of other males, which setting or combination of settings should it choose in order to maximise its future breeding success? How should this choice change with age, condition and experience? And does ecological context, in terms of food abundance and predation risk, affect a young male's choice?

MODELLING BACHELOR-MALE DECISION-MAKING

To help gain insights into how maturing bachelor males can best answer these questions, we apply a state-dependent dynamic modelling approach to the non-breeding period of plains zebra bachelor males. Observations on a variety of equid populations reveal that juvenile males disperse from their natal groups when 2 to 3 years old and then usually spend 4 to 6 years in bachelor associations, until physically and socially mature enough to sustain long-lasting relationships with females (see Feh, 1999). During this transition period young zebra males can be found alone, in groups of up to 25 other bachelor males or in mixed herds composed of harem and bachelor groups.

Each social context potentially offers different rewards and potentially extracts different costs. Being solitary should provide the highest foraging gain but also the highest risk of predation and the least opportunity to acquire social skills. Whereas all male groups offer predator protection, feeding rate declines as males engage in activities associated with acquiring male-related social skills (see Fig. 15.4). In mixed herds, predator protection is also great, although the extra 'anti-predator dilution' benefit of having more vulnerable youngsters around may be offset by an increased risk associated with peripheralisation induced by harem stallions. Foraging is probably at its lowest level, again because of peripheralisation to marginal feeding patches, as well as the result of frequent male contests; but in mixed herds the ability to acquire social skills is likely maximised. In such associations not only will bachelors interact with each other but also they will expand their contacts to include breeding stallions as well as their females. If these trade-offs are realistic – and preliminary observations suggest they are – then we can use the model to determine which herd-related developmental strategies are best under particular ecological conditions.

In our model (complete development in Rubenstein & Hack, in preparation) we assume that a young male, after

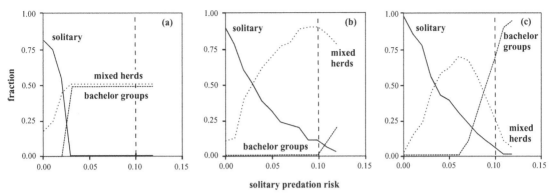

solitary predation risk

Fig. 15.6   Simulations of the stochastic dynamic-programming model on the optimal behaviour of bachelor males. In (a) and (b) predation risk is equal for all-male groups and mixed herds, and equals 0.04; and social experience gained is scaled at 0 for solitary bachelor males, 0.5 for those in all-male groups and 1.0 for males in mixed herds. Where (a) and (b) differ is in terms of weight gain: in (a) solitary males and males in all-male groups graze equally well and gain is scaled at 1, whereas bachelors in mixed herds do less well and gain weight at a rate of 0.5; in (b) weight gain remains the same for bachelors in all-male groups and mixed herds, but solitary males do much better and are scaled at 5. Where (c) differs from (a) and (b) is found in the risk of predation. In (c) bachelors in mixed herds suffer slightly greater risks of being preyed upon, and thus per-capita mortality increases to 0.045.

leaving his natal group, must survive four years as a non-breeder before being able to compete for his own females. During this transition period a bachelor must also grow in size and gain social experience with males and females in order to be competitive. To maximise his chances of success a bachelor may freely switch among social environments, and a male's state at the end of the non-breeding period – as measured by his body size and accrued social experience – determines his eventual lifetime reproductive success relative to the other members of his age cohort. In the model, we have divided the bachelor male's year into four time-steps of equal length corresponding to ecological seasons. Thus the maximum lifetime of a bachelor male is 16 time-steps. Each social environment it chooses entails a risk of predation, so survival to the next time-step is not certain. Also, the rate at which body mass and social experience increases varies with social environment. Because of stochastic events, individuals may differ in state upon arriving at the same decision point. As the iterations depicted in Fig. 15.6 illustrate, the values of the variables associated with each social context used in the model are easily scaled to reflect these trade-offs and adjusted to overall ecological conditions.

The dynamic-programming technique allows us to define the optimal choice of social environment at each time-step, depending upon each male's state values at that time-step. This is achieved by defining fitness in the present time-step (t) as a function of fitness in the next time-step ($t + 1$), contingent on the environmental choice made and the consequent changes in the state in which an individual enters time-step ($t + 1$). Since the fitness in time-step ($t + 1$) is already known for each possible state-value combination, the choice that maximised fitness at ($t + 1$) can easily be determined. How is the fitness at time-step ($t + 1$) already known? We start with the last, or terminal, time-step (T) and assign a terminal fitness value to each possible combination of state values considered. Some terminal fitness values are clearly unlikely since most males that survive to the terminal time-step will have accrued some social experience and gained mass. However, a large enough range of terminal states and their corresponding fitnesses are modelled so as to encompass all possible mass and social experience levels that a male might have accrued over the preceding 15 time-steps. To iterate the dynamic-programming algorithm, we work backwards through time starting at T, with known terminal fitnesses for each state, and determine the optimal action at the immediately previous time-step ($T - 1$) from each possible state at that time-step. Once the optimal actions are specified for each state at T, the fitnesses for each state at ($T - 1$) are also specified and they become the new values from which to determine the optimal action at each state in ($T - 2$). The process is iterated until the first time-step is reached. As such, dynamic programming results in a 'road-map' specifying the optimal action to take at each time-step and for each state in which an individual may be.

Figure 15.6 illustrates three simulations of bachelor-male tendencies under different ecological conditions. In the first two – (a) and (b) – predation risks associated with being in either all-male groups or mixed herds are low and equal (per-capita mortality = 0.04). Again, in both iterations the ability to acquire social experience varies in the same way among social states, and is maximised at 1 for mixed herds, is halved for all-male groups and set to zero for solitary bachelors. In Fig. 15.5a relative mass gain is equalised and set at a maximum of 1 for both solitary bachelors and those in all-male groups; for bachelors in mixed herds, however, relative mass gain is reduced by half. In this iteration it becomes clear that when predation levels are low and, in particular, the risk for solitary individuals is less than 0.025, then bachelors should only be found alone or in mixed herds. And when the risk of being eaten gets very small for solitary individuals, the frequency of solitary bachelors found on a landscape should increase and they should dominate the landscape. What is most striking about the iteration depicted in Fig. 15.6a is that as long as the risk of predation on solitary bachelors is small and lower than when in any group (which would be the case if groups were easier to detect than solitary individuals), then the strategy for bachelors that gives them the highest terminal fitness is one in which they alternate between states. They should go alone to maximise growth rate for a time and then they should join mixed herds to maximise social experience. What they should *not* do is become a 'generalist' and form all-male groups that provide moderate pay-offs in both dimensions. Only when the solitary risk of being eaten reaches 4 per cent should bachelors become indifferent with respect to joining all-male or mixed-sex herds.

The importance of not seeking the middle ground is underscored in the iteration illustrated in Figure 15.6b, where weight gain derived while being solitary exceeds that which can be gained when in all-male groups. When being alone eliminates foraging competition and provides huge relative foraging gains, then the strategy of oscillating between two social environments – one where mass gain, and the other where social skills, can be maximised – is reinforced and extends into regions where the per-capita risk of being eaten when alone soars to 10 per cent.

The importance of predation risk cannot be underestimated. Under conditions when solitary bachelors maintain a large foraging advantage over those joining either type of herd, even a slight decrease of the survival rates to bachelors of joining a mixed herd tilts the balance toward all-male groups (Fig. 15.6c). If increased conspicuousness or less favourable positioning on the edge outweighs the dilution benefit of associating with more vulnerable youngsters, then under moderate levels of risk when solitary, the opti-

mal strategy is still to join mixed herds. But joining all-male groups is clearly becoming common, while wandering alone is becoming rare. By the time the risk of predation on solitary individuals reaches 10 per cent, however, virtually all bachelors do best by joining all-male groups.

Clearly, the simulations show that bachelor males are sensitive to maximising their survival prospects, weight gain and social experience. That they do so by changing social environments is not surprising. That they do so by specialising on one factor at a time, is. The switch to choosing a social environment that provides moderate gains in each dimension *simultaneously* becomes favoured only when the predation risk of being alone, or of being in mixed groups, becomes relatively high.

In our repeated censuses of ranches, we record the frequency in which bachelor males are distributed among these three social states. Overall, only 3 per cent of bachelors are seen alone. Most (67 per cent) are seen in mixed herds, while 30 per cent are seen in all-male groups (Rubenstein, Hack & Mazo, in preparation). Given that all but one ranch hosts at least one pride of lions and one clan of hyaenas, predation risk must be moderately high – at least 4 per cent for those in groups, and higher for those living alone. If the risk to solitary bachelors were to rise as high as 12 per cent – three times as high as being in a group – then the predicted distribution of bachelor males across social states would be roughly in accord with that of the iteration illustrated in Fig. 15.5b in which solitary bachelors experience a large foraging advantage by shunning competitors. Upon inspection, the actual distributions of plains zebra bachelor males are closer to the predictions of this iteration than one in which predation risk is lower for solitary males but higher for bachelors joining mixed herds than for joining all-male groups (Fig. 15.6c).

Risk of predation seems to play a major role in bachelor decision-making. Since predation risk varies with time of day and habitat-openness our censuses also reveal that bachelors are typically found alone when predation intensity scores were low (3.7), but in mixed herds when scores were significantly higher (5.1) ($F_{2,304} = 19.0$; $p < 0.0001$; $r = 0.12$). Moreover, in the one conservancy where predators are regulated and lions are not present, the presence of all-male bachelor groups is low (25 per cent) as predicted, but not significantly lower than on a neighbouring ranch (32 per cent) where predators are more abundant. Overall, predation appears to matter and adjusts the frequencies by which bachelors adopt particular social states, mostly in accord with the predictions of the model.

SYNTHESIS AND COMPARISONS

LESSONS FROM ZEBRAS

The simple multi-level societies of zebras demonstrate that higher levels of sociality can evolve to solve social problems that core mating groups cannot. Core social groups, whether they are uni-male–multi-female 'harems' or closed-membership groups of bachelor males, lower the risks of predation. Moreover, harem groups, by virtue of the protective role of males, also provide enhanced feeding opportunities for females. But in certain environments, males in solitary harem groups cannot easily reduce incursions by bachelor males. Without attaining assistance of other breeding stallions, stallions would otherwise experience higher risks of cuckoldry.

Ecological factors, however, are not unimportant since they also influence, either directly or at times indirectly, the size and composition of herds. Zebras are sensitive to predation risk and seem to respond numerically by avoiding areas where predators are abundant. Yet at the same time they are drawn to areas where leaf biomass is high – just the places where predators hide. Apparently, plains zebra are in a dynamic 'shell game', since simply moving to maximise safety cannot be accomplished without sacrificing access to abundant supplies of food. Thus although predation risk does not directly influence herd size, once all other factors are held constant, predation pressure affects zebra densities, and density does directly affect herd size. Once zebras settle in a particular area, then vegetation abundance, more than any other feature of the landscape after controlling for all other variables, has a positive effect on herd size and composition. As leaf density increases, so does the size of herds, and the fraction of bachelor males that those herds comprise increases as well. But overall, the factor that has the greatest direct influence on herd size is the pressure of intruding, and potentially cuckolding, bachelor males. When their numbers are high, herds tend to be large. Apparently, a sexual problem is best solved socially, but at a level above the core social unit.

The actual decision-making process appears to be one of adjusting both proximate and ultimate costs and benefits among alternative social states. Breeding stallions appear to band together pre-emptively with their females to reduce the chances of successful incursions by bachelors and their ultimate, and potentially debilitating, reproductive consequences (Rubenstein, 1986); moreover, banding together with other stallions also reduces the proximate costs associated with foraging reductions that males alone incur when trying to reduce the chances of being cuckolded. Fortunately, females in most herds with 60 or fewer animals suffer few reductions in foraging performance. Thus they are seemingly indifferent to being in a solitary harem or in a herd of moderate size. When such sexual conflicts of interest are eliminated, the pressure for stallions to aggregate is unopposed by the interests, or actions, of their females. In this particular context the forces of sexual selection operating on males do not come into conflict with the forces of natural selection acting on females, and the new social state is stable.

Stochastic dynamic modelling predicts that bachelor males should also alter their social environment, based on relative costs and benefits associated with alternative social states, but on a longer time scale than that of stallions. The eco-correlates analysis shows that bachelors are drawn to herds with many females (Table 15.1), but the model demonstrates that the attractiveness of herds is not likely to be universal. Over a time-horizon longer than that over which breeding stallions make decisions about social states, the model predicts the adoption of a diversified strategy of alternating between specialist tactics, each of which maximises gains in one dimension – physical or social growth – for short periods of time. This model predicts oscillations among social states, unless predation levels for solitary bachelors are very high or predation levels for bachelors in mixed herds are somewhat higher than those for bachelors in all-male groups. Actual distributions of bachelors among these alternatives match reasonably well the predictions of a stochastic dynamic-programming model the parameters of which assume that solitary males gain a strong, fitness-enhancing foraging advantage over social males, but that such solitary males also face predation levels three times higher than those in groups. Even with respect to developmental strategies that are under the influence of sexual selection, the model shows that factors of natural selection are likely to constrain options. Nevertheless, the modelling of social relationships within multi-level societies of zebras suggests that during ontogeny the acquisition of social skills should be as important as increasing body size. Since maximising the two-in-one social state appears difficult, changing social states, and in turn social relationships, are likely to be important features of male life histories.

PRIMATE COMPARISONS

The dynamics of higher-order primate social systems, especially those of baboons, correspond to those of zebras in many ways. Savannah (*Papio cynocephalus*), hamadryas and gelada baboons appear to exhibit strikingly different patterns of social organisation. Savannah baboons typically live in multi-male–multi-female groups, whereas hamadryas and gelada baboons, both close evolutionary relations, live in harem groups comprising one male and many females. Yet as in plains zebras, both of these harem-based systems often show higher levels of social organisation since the harem groups often coalesce, forming herds.

Studies by Kummer (1968), Dunbar (1986, 1988), Stambach (1987) and Barton (2000) illustrate how these apparently different societies appear to be variations on a common theme; the entities that emerge are novel responses to different ecological conditions. Barton (2000) has proposed a model based on the interplay of five factors that can account for the differences among the species. He argues that two factors – *sexual dimorphism* (via the large size of males, making them socially important to females) and *male polygyny* (the tendency for males to bond to as many females as possible and exclude other males from joining) – are universal and influence the social structure of all baboon societies. Together, they provide a force that generates strong male–female bonds (e.g. 'cross-sex bonding'; Byrne *et al.*, 1990) and encourage males to bond with as many females as possible, while segregating themselves from other males. But whether such a segregating tendency actually leads to isolated uni-male–multi-female core groups and fission–fusion herds typified by hamadryas and gelada baboons, depends on three additional factors that vary in strength with ecological circumstances. First is the *risk of predation*. This sets the lower limit to group size, which should increase as predation intensity increases. And as group sizes increase so should the number of males associating continually with females (Andleman, 1986; Altmann, 1990). Second is the *availability of food*. This sets the upper limit to group size, and when food becomes scarce the maximal size of groups should shrink. And third is the strength of *female–female alliances*. As the magnitude of intragroup contest competition increases, so should the strength of female–female alliances since such associations will determine the outcome of within-group competition. Barton's model suggests that it is the balance between the strength of female–female and male–female bonds that ultimately determines whether or not baboons are organised into multi-male or uni-male core social units.

Whenever predation risk is high, groups should be large, but 'bigness' can come in two varieties. In one, large groups can be cohesive with strong female–female and female–male links, as typified by olive (*Papio anubis*) and annubis baboons. In the other, normally separate uni-male groups can create large groupings by forming bands in loose and opportunistic associations. According to Barton's model, what determines one from the other is the abundance and dispersion of resources. As the patchiness of food increases, alliances among females come to determine the outcome of contests for monopolisable resources. As a result, strong female–female bonds prevent males, who themselves are tied strongly to particular females, from removing those females from the cohesive group. Alternatively, when food is more sparsely and evenly distributed, contest competition is reduced and female alliances, along with the corresponding bonds that develop among females, tend to be weak. As a result, the segregating tendencies of males predominate. Bands or herd-like structures will still form when predation risk becomes high, but the solution to the predation problem is solved at a higher level than the core group.

This model accurately accounts for the fission–fusion nature of hamadryas society and it even explains the breakdown of the typical multi-male–multi-female groups of chacma baboons (*Papio ursinus*) when they inhabit subalpine habitats in South Africa. With feeding competition at extremely low levels in such habitats, many chacma baboons live in uni-male groups (Barton, 2000). Just like hamadryas baboons in the absence of predation, the chacma core units remain apart and are only reported to fuse into bands in regions where leopards are present (Barton, 2000).

Applying Barton's model to account for the fission–fusion dynamics of gelada baboons is not as straightforward. Dunbar (1986) characterises the core harem groups of geladas as having both strong male–female and female–female ties. As such they would be predicted to live in large multi-male–multi-female groups. These female–female associations, however, are among close kin and they do not extend outside the core group. Hence, when harems coalesce into bands the extended female–female networks exhibited by savannah baboons are absent and, unlike savannah baboons, large aggregations of gelada baboons readily break up. Dunbar (1986) argues that coalitional support, even if limited in extent, is necessary when feeding competition does occur. Since gelada forage on grass-like lawns, he notes that such competition only occurs in large groups that routinely form only on open grasslands where predation risk is high. When foraging in less risky habitats, such as on grassy slopes, bands break up into segregated harem groups, and feeding competition is reduced. Thus the existence of even modest female alliances provides sufficient social flexibility on the part of geladas to either maximise foraging efficiency in relatively safe habitats by segregating to reduce competition, and by relying on coalitional support to do the same in riskier habitats where safety in numbers also intensifies competition.

What is striking about Barton's model and its ability to explain how ecological factors interact to favour flexible social patterns and the evolution of higher levels of social organisation in baboons, is that with some minor modifications it can also account for the social dynamics of plains zebras. First, Barton's two universal factors apply to zebras as well. Males are important since they play a vital role in protecting females from harassment and give them extra time to forage; and males also strive to acquire as many females as possible. Thus strong male–female bonds are common and males in most harem-living equids avoid each other. Second, Barton's variable factors also come into play, but with some interesting twists. Unlike the strong bonds that exist between males and females, the bonds among females in most harem-living equids are weak. This pattern most likely emerges from the fact that equids forage by clipping vegetation and, unlike most baboons, rarely invest in digging for rhizomes or spending much time or energy in acquiring individual food items. Contest competition is thus rare, dominance hierarchies are weak and the need for female–female alliances is virtually non-existent. Moreover, food availability is high. Not only are grasses distributed relatively evenly on landscapes, but also the hindgut fermentation system of equids enables them to process food continuously. And, lastly, predation risk is high.

But it is on this point that the biology of equids and baboons diverges. Although a large array of group-hunting predators prey on zebras, the relatively large body size of zebras, their fleetness and the fact that males are highly vigilant mitigates the per-capita risk of dying for those zebras inhabiting closed-membership harem groups consisting of, on average, 10 to 12 individuals. And as the eco-correlate analysis has shown, the ability of zebras to range widely enables them to move away from predators fairly rapidly. Overall, the many counter-strategies of zebras living in closed-membership harem groups help keep per-capita predator risk relatively low. Yet predation risk would be high if core groups were smaller and less cohesive. For horses, the other harem-forming equid, bachelor males live in small open-membership groups (Rubenstein, 1981). This is quite different from zebra bachelor males. In both the Laikipia and Samburu populations, associations among bachelor males are strong, and groups typically range in size from six to nine individuals (personal observation). Therefore, it appears as if this species-specific transformation in the nature of bachelor-male relationships is a direct effect of current

predation pressures being greater for zebras than for horses. But what is most intriguing about this change is that although the eco-correlate analysis does not show a strong direct effect of predation risk on zebra herd size, predation apparently does exert an effect, only it appears to do so indirectly. Predation risk, by inducing bachelor males to live in large groups where long-term associations enable the development of coordinated action, appears to increase the risk to stallions of being cuckolded and of wasting valuable time and energy in trying to reduce this risk. As a result, stallions are driven to aggregate to lower these actual, or potential, costs. Because of the indirect way predation pressure acts on zebra herd dynamics through the sexual behaviour of bachelor males, the forces of sexual selection appear to play a greater role in shaping higher-level features of zebra societies than they do in shaping similar patterns of primate sociality. In primates, ecological factors shaped by the forces of natural selection appear to dominate.

CONCLUSION

Overall, zebra herds appear to form in order to solve social and ecological problems that emerge somewhat unpredictably. That sociosexual problems underlie the formation of herds should not be too surprising. Mediating complex sets of social tensions is not easily done and appears to require forming and dissolving social bonds involving conspecifics outside an individual's immediate social sphere. Higher levels of sociality – herds for zebras – clearly provide a diverse array of potential options to call upon when solving novel social challenges created by changing environmental circumstances. As long as bonds central to holding together the underlying core social groups are not jeopardised, multiple levels of social organisation will evolve. While the structure of some, such as those of hamadryas and even gelada baboons, may be relatively long-lasting, others, like the herds of plains zebra, may be more short-lived enabling rapid adjustments to short-term problems.

ACKNOWLEDGEMENTS

We thank the National Science Foundation (IBN-9874523) and the National Parks Service, St Louis Zoo, EarthWatch Institute and the Laikipia Research Project for financially supporting our research. Cassandra Nunez, David Saltz, Jessica Rogers, Mark Cornwall, Dana Mazo and Geoffrey Chege helped in gathering valuable data, and the Mpala Research Center, The National Parks Service, The Nature

Reserves Authority, Ol Jogi Conservancy, Lewa Wildlife Conservancy, Segera Ranch and El Karama Ranch enabled us to study equid populations on their lands. But most importantly, we thank the government and people of Kenya for enabling us to study their wonderful wildlife. Comments from Peter Kappeler, Carel van Schaik and anonymous reviewers helped improve the manuscript.

REFERENCES

Alexander, R. D. 1974. The evolution of social behavior. *Annual Review of Ecological Systems*, 5, 325–83.

Altmann, J. 1990. Primate males go where females are. *Animal Behaviour*, 39, 193–4.

Andleman, S. J. 1986. Ecological and social determinants of Cercopithecine mating patterns. In *Ecological Aspects of Social Evolution: Birds and Mammals*, ed. D. I. Rubenstein & R. W. Wrangham. Princeton, NJ: Princeton University Press, pp. 201–16.

Barton, R. A. 2000. Socioecology of baboons: the interaction of male and female strategies. In *Primate Males: Causes and Consequences of Variation in Group Composition*, ed. P. M. Kappeler. Cambridge: Cambridge University Press, pp. 97–107.

Berger, J. 1986. *Wild Horses of the Great Basin: Social Competition and Population Size*. Chicago, IL: University of Chicago Press.

1987. Reproductive fates of dispersers in a harem-dwelling ungulate: the wild horse. In *Mammalian Dispersal Patterns: The Effects of Social Structure on Population Genetics*, ed. B. D. Chepko-Sade & Z. T. Halpin. Chicago, IL: University of Chicago, pp. 41–54.

Bradbury, J. W. & Vehrencamp, S. L. 1977. Social organization and foraging in emballonurid bats. III. Mating systems. *Behavioral Ecology and Sociobiology*, 2, 1–17.

Brown, J. L. 1987. *Helping and Communal Breeding in Birds*. Princeton, NJ: Princeton University.

Byrne, R., Whiten, A. & Henzi, S. 1990. Social relationships in mountain baboons: leadership and affiliation in a non-female-bonded monkey. *American Journal of Primatology*, 20, 313–329.

Clutton-Brock, T. H. 1989. Mammalian mating systems. *Proceedings of the Royal Society of London, Series B*, 236, 339–72.

Dunbar, R. I. M. 1986. The social ecology of gelada baboons. In *Ecological Aspects of Social Evolution*, ed. D. I. Rubenstein & R. W. Wrangham. Princeton, NJ: Princeton University, pp. 332–51.

1988. *Primate Social Systems*. Ithaca: Cornell University.

Emlen, S. T. & Oring, L. W. 1977. Ecology, sexual selection, and the evolution of mating systems. *Science*, 197, 215–23.

Feist, J. D. & McCullough, D. R. 1976. Behavior patterns and communication in feral horses. *Zeitschrift für Tierpsychologie*, 41, 337–71.

Feh, C. 1999. Alliances and reproductive success in Camargue horses. *Animal Behaviour*, 57, 705–13.

Ginsberg, J. R. 1989. The ecology of female behaviour and male mating success in the Grevy's zebra. *Symposia of the Zoological Society of London*, 61, 89–110.

Groves, C. P. 1974. *Horses, Asses, and Zebras in the Wild*. Hollywood, FL: Curtis Books.

Hinde, R. A. 1983. A conceptual framework. In *Primate Social Relationships*, ed. R. A. Hinde. Oxford: Blackwell, pp. 1–7.

Jarman, P. J. 1974. The social organisation of antelope in relation to their ecology. *Behaviour*, 48, 215–67.

Joubert, E. 1972. The social organisation and associated behaviour in the Hartmann zebra *Equus zebra hartmannae*. *Madoqua*, 1, 17–56.

Klingel, H. 1969a. Reproduction in the plains zebra, *Equus burchelli boehmi*: behaviour and ecological factors. *Journal of Reproduction and Fertility*, Supplement, 6, 339–45.

1969b. The social organisation and population ecology of the plains zebra (*Equus quagga*). *Zoologica Africana*, 4, 249–63.

1974. A comparison of the social behaviour of the Equidae. In *The Behaviour of Ungulates and Its Relation to Management*, ed. V. Geist & F. Walther. Morges, Switzerland: IUCN Publications, pp. 124–32.

1977. Observations on social organization and behaviour of African and Asiatic wild asses. *Zeitschrift für Tierpsychologie*, 44, 323–31.

Kummer, H. 1968. *Social Organization of Hamadrayas Baboons. A Field Study*. Chicago, IL: University of Chicago.

Linklater, W. L. 2000. Adaptive explanation in socio-ecology: lessons from the Equidae. *Biological Reviews of the Cambridge Philosophical Society*, 75, 1–20.

Linklater, W. L., Cameron, E. Z., Minot, E. O. & Stafford, K. J. 1999. Stallion harassment and the mating system of horses. *Animal Behaviour*, 58, 295–306.

Lloyd, P. H. & Rasa, O. A. E. 1989. Status, reproductive success and fitness in Cape mountain zebra (*Equus zebra zebra*). *Behavioral Ecology and Sociobiology*, 25, 411–20.

McCracken, G. F. & Bradbury, J. W. 1981. Social organization and kinship in the polygynous bat *Phyllostomus hastatus*. *Behavioral Ecology and Sociobiology*, 8, 11–34.

Miller, R. 1979. Band organization and stability in Red Desert feral horses. In *Proceedings of a Conference on the Ecology and Behavior of Feral Equids*, ed. R. H. Denniston. Laramie, WY: University of Wyoming, pp. 113–23.

Moehlman, P. D. 1979. Behavior and ecology of feral asses (*Equus asinus*). *National Geographic Society Research Reports*, 405–11.

Penzhorn, B. L. 1984. A long-term study of social organisation and behaviour of Cape mountain zebras *Equus zebra zebra*. *Zeitschrift für Tierpsychologie*, **64**, 97–146.

Pereira, M. E. & Fairbanks, L. A. 1993. *Juvenile Primates*. New York, NY: Oxford University Press.

Rubenstein, D. I. 1981. Behavioural ecology of island feral horses. *Equine Veterinary Journal*, **13**, 27–34.

1986. Ecology and sociality in horses and zebras. In *Ecological Aspects of Social Evolution*, ed. D. I. Rubenstein & R. W. Wrangham. Princeton, NJ: Princeton University, pp. 282–302.

1989. Life history and social organization in arid adapted ungulates. *Journal of Arid Environments*, **17**, 145–56.

1994. The ecology of female social behaviour in horses, zebras and asses. In *Animal Societies: Individuals, Interactions and Organisation*, ed. P. J. Jarman & A. Rossiter. Kyoto: Kyoto University, pp. 13–28.

Rutberg, A. T. 1990. Inter-group transfer in Assateague pony mares. *Animal Behaviour*, **40**, 945–52.

Scheel, D. 1993. Profitability, encounter rates, and prey choice of African lions. *Behavioral Ecology*, **4**, 90–7.

Sigg, H., Stolba, A., Abegglen, J.-J. & Dasser, V. 1982. Life history of hamadryas baboons: physical development, infant mortality, reproductive parameters and family relationships. *Primates*, **23**, 473–87.

Stammbach, E. 1987. Desert, forest, and montane baboons: multilevel societies. In *Primate Societies*, ed. B. B. Smuts, D. L. Cheney, R. M. Seyfarth, R. W. Wrangham & T. T. Struhsaker. Chicago, IL: University of Chicago, pp. 112–20.

Stroeh, O. 2001. The Effects of Management Strategies on the Behavioural Ecology of the Shackleford Banks Male Horses. B. Sc. thesis.

Turner, J. W. J. 1992. Seasonal mountain lion predation on a feral horse population. *Canadian Journal of Zoology*, **70**, 929–34.

van Schaik, C. P. 1989. The ecology of social relationships among female primates. In *Comparative Socioecology*, ed. V. Standen & R. A. Foley. Oxford: Blackwell, pp. 195–218.

Woodward, S. L. 1979. The social system of feral asses (*Equus asinus*). *Zeitschrift für Tierpsychologie*, **49**, 304–16.

Index